T0145253

Natural Computing Series

Series Editors: Thomas Bäck Lila Kari

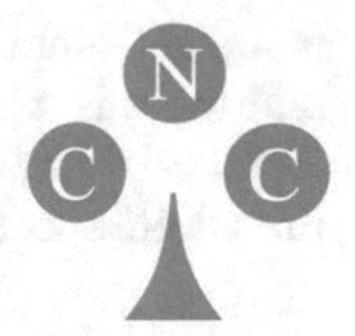

Natural Computing is one of the most exciting developments in computer science, and there is a growing consensus that it will become a major field in this century. This series includes monographs, textbooks, and state-of-the-art collections covering the whole spectrum of Natural Computing and ranging from theory to applications.

More information about this series at http://www.springer.com/series/4190

Benjamin Doerr • Frank Neumann
Editors

Theory of Evolutionary Computation

Recent Developments in Discrete Optimization

Editors
Benjamin Doerr
Laboratoire d'Informatique (LIX) - UMR 7161
École Polytechnique
Palaiseau, France

Frank Neumann
School of Computer Science
The University of Adelaide
Adelaide, SA, Australia

ISSN 1619-7127
Natural Computing Series
ISBN 978-3-030-29416-8 ISBN 978-3-030-29414-4 (eBook)
https://doi.org/10.1007/978-3-030-29414-4

This Springer imprint is published by the registered company Springer Nature Switzerland AG
The registered company address is: Gewerbestrasse 11, 6330 Cham, Switzerland

Preface

The theory of evolutionary computation, or, more generally, randomized search heuristics, is aimed at understanding how these methods work and why they are so successful in many applications. While there has always been theoretical work in this field, and even more since Ingo Wegener (1950–2008) pushed for a mathematical approach inspired by the classical field of randomized algorithms, this research area remains young and many astonishing advances have only been made in the last five to ten years. These include new and more powerful methods, the solution of long-standing open problems, and the analysis of heuristics that could not be analyzed before. Not only have the topics changed and become closer to what is the state of the art in applications, but also the field has progressed from only analyzing existing methods to finding unexpected and more powerful parameter choices, designing new building blocks such as mutation operators, selection operators, and mechanisms that adjust parameters on the fly, and even proposing completely new heuristics.

In this edited book, we report on some of these recent developments. Our aim is to give a concise summary of the state of the art to experts in the field and to make this exciting area more accessible to students and researchers in related fields.

The book starts with two chapters on mathematical methods that are often used in the analysis of randomized search heuristic. These are followed by three chapters on how to measure the complexity of a search heuristic: we discuss black-box complexity, a counterpart of classical complexity theory in black-box optimization, parameterized complexity, aimed at a more fine-grained view of the difficulty of problems, and the fixed-budget perspective, which answers the question of how good a solution will be after investing a certain computational budget. We then describe theoretical results on three important questions in evolutionary computation, namely how to profit from changing the parameters during the run of an algorithm, how evolutionary algorithms are able to cope with dynamically changing or stochastic environments, and how population diversity influences performance. Finally, we look at three algorithm classes that have only recently become the focus of theoretical work, namely estimation-of-distribution algorithms, artificial immune systems, and genetic programming.

We hope that this book will help students and researchers in the field and around it to access these topics, to deepen their understanding, and possibly to join this young and exciting area, in which many very fundamental questions are still wide open.

We thank all authors for accepting the time-consuming task of writing a book chapter and for completing this task to perfection. We are very grateful to the reviewers of each chapter, whose careful reading is a guarantee of the high quality we aim at. Our final thanks go to the publisher, and, in particular, Ronan Nugent, for all their help and responsiveness, and to the copyeditor, Douglas Meekison, for his work on the book manuscript.

Palaiseau, Adelaide, *Benjamin Doerr*
May 2019 *Frank Neumann*

Contents

List of Contributors

Benjamin Doerr
LIX – UMR 7161, 1 rue Honoré d'Estienne d'Orves, Bâtiment Alan Turing, Campus de l'École Polytechnique, 91120 Palaiseau, France, e-mail: doerr@lix.polytechnique.fr

Carola Doerr
Sorbonne Université, CNRS, Laboratoire d'informatique de Paris 6 (LIP6), 75252 Paris, France, e-mail: Carola.Doerr@lip6.fr

Thomas Jansen
Department of Computer Science, Llandinam Building, Aberystwyth University, Aberystwyth SY23 3DB, Ceredigion, United Kingdom, e-mail: t.jansen@aber.ac.uk

Martin S. Krejca
Hasso-Plattner-Institut, Prof.-Dr.-Helmert-Str. 2–3, 14482 Potsdam, Germany, e-mail: martin.krejca@hpi.de

Johannes Lengler
ETH Zürich, Department of Computer Science, Universitätstrasse 6, 8092 Zürich, Switzerland, e-mail: johannes.lengler@inf.ethz.ch

Andrei Lissovoi
Rigorous Research, Department of Computer Science, University of Sheffield, 211 Portobello, Sheffield S1 4DP, United Kingdom, e-mail: a.lissovoi@sheffield.ac.uk

Frank Neumann
Optimisation and Logistics, School of Computer Science, The University of Adelaide, SA 5005, Australia, e-mail: frank.neumann@adelaide.edu.au

Pietro S. Oliveto
Rigorous Research, Department of Computer Science, University of Sheffield, 211 Portobello, Sheffield S1 4DP, United Kingdom, e-mail: p.oliveto@sheffield.ac.uk

Mojgan Pourhassan
Optimisation and Logistics, School of Computer Science, The University of Adelaide, SA 5005, Australia, e-mail: mojgan.pourhassan@adelaide.edu.au

Vahid Roostapour
Optimisation and Logistics, School of Computer Science, The University of Adelaide, SA 5005, Australia, e-mail: vahid.roostapour@adelaide.edu.au

Dirk Sudholt
Department of Computer Science, University of Sheffield, 211 Portobello, Sheffield S1 4DP, United Kingdom, e-mail: d.sudholt@sheffield.ac.uk

Andrew M. Sutton
Department of Computer Science, University of Minnesota Duluth, 1049 University Drive, Duluth, MN 55812, USA, e-mail: amsutton@umn.edu

Carsten Witt
DTU Compute, Technical University of Denmark, Richard Petersens Plads, Building 322, 2800 Kgs. Lyngby, Denmark, e-mail: cawi@dtu.dk

Christine Zarges
Department of Computer Science, Llandinam Building, Aberystwyth University, Aberystwyth SY23 3DB, Ceredigion, United Kingdom, e-mail: c.zarges@aber.ac.uk

Chapter 1
Probabilistic Tools for the Analysis of Randomized Optimization Heuristics

Benjamin Doerr

Abstract This chapter collects several probabilistic tools that have proven to be useful in the analysis of randomized search heuristics. This includes classic material such as the Markov, Chebyshev, and Chernoff inequalities, but also lesser-known topics such as stochastic domination and coupling, and Chernoff bounds for geometrically distributed random variables and for negatively correlated random variables. Most of the results presented here have appeared previously, but some only in recent conference publications. While the focus is on presenting tools for the analysis of randomized search heuristics, many of these may be useful as well for the analysis of classic randomized algorithms or discrete random structures.

1.1 Introduction

Unlike in the field of classic randomized algorithms for discrete optimization problems, where theory has always supported (and, in fact, often led) the development and understanding of new algorithms, the theoretical analysis of nature-inspired search heuristics is much younger than the use of these heuristics. The use of nature-inspired heuristics can easily be traced back to the 1960s; their rigorous analysis with proven performance guarantees only started in the late 1990s. Propelled by impressive results, most notably from the German computer scientist Ingo Wegener (1950–2008) and his students, theoretical studies became quickly accepted in the field of nature-inspired algorithms and now form an integral part of it. They help to understand these algorithms, guide the choice of their parameters, and even (as in the field of classic algorithms) suggest new promising algorithms. It is safe to say that Wegener's vision that nature-inspired heuristics are nothing more

Benjamin Doerr
École Polytechnique, CNRS, Laboratoire d'Informatique (LIX), Palaiseau, France

B. Doerr, F. Neumann (eds.), *Theory of Evolutionary Computation*,
Natural Computing Series, https://doi.org/10.1007/978-3-030-29414-4_1

than a particular class of randomized algorithms, which therefore should be analyzed with the same rigor as other randomized algorithms, has come true.

After around 20 years of theoretical analysis of nature-inspired algorithms, however, we have to note that the methods used here are different from those used in the analysis of classic randomized algorithms. This is most visible for particular methods such as the fitness level method or drift analysis, but applies even to the elementary probabilistic tools employed throughout the field.

The aim of this chapter is to collect those elementary tools which have often been used over the past 20 years. This includes classic material such as expectations, variances, the coupon collector process, Markov's inequality, Chebyshev's inequality and Chernoff–Hoeffding bounds for sums of independent random variables, but also topics that are used rarely outside the analysis of nature-inspired heuristics such as stochastic domination, Chernoff–Hoeffding bounds for sums of independent geometrically distributed random variables, and Chernoff–Hoeffding bounds for sums of random variables which are not fully independent. For many results, we also sketch a typical application or refer to applications in the literature.

The large majority of the results and applications presented in this chapter have appeared previously, some in textbooks, some in recent conference publications. The following results, while not necessarily very deep, are original to the best of our knowledge.

- The result that all known Chernoff bounds, when applied to binary random variables, hold as well for negatively correlated random variables. More precisely, for bounds on the upper tail, we need only 1-negative correlation, and for bounds on the lower tail, we need only 0-negative correlation (Section 1.10.2.2).
- The insight that all commonly known Chernoff bounds can be deduced from only two bounds (Section 1.10.1.5).
- A version of the method of bounded differences which requires only that the t-th random variable has a bounded influence on the expected result stemming from variables $t+1$ to n. This appears to be an interesting compromise between the classic method of bounded differences, which is hard to use for iterative algorithms, and martingale methods, which require familiarity with martingales (Theorem 1.10.28).
- Via an elementary two-stage rounding trick, we give simple proofs of the facts that (i) a sum X of independent binary random variables with $\mathrm{Var}[X] \geq 1$ exceeds its expectation with constant probability by at least $\Omega(\sqrt{\mathrm{Var}[X]})$ and (ii) it attains a particular value at most with probability $2/\sqrt{\mathrm{Var}[X]}$ (Lemmas 1.10.16 and 1.10.17). Both results were proven earlier by deeper methods, for example, an approximation via the normal distribution.

This chapter is intended to serve both as an introduction for newcomers to the field and as a reference book for regular users of these methods. With both addressees in mind, we have not shied away from also stating elementary reformulations of results or explicitly formulating statements that rely only on elementary mathematics, such as:

- how to choose the deviation parameter δ in the strong multiplicative Chernoff bound so that the tail probability $(e/\delta)^\delta$ is below a desired value (Lemma 1.10.2), and
- how to translate a tail bound into an expectation (Corollary 1.6.2).

We hope that this will save all users of this chapter some time, which can be better spent on understanding the challenging random processes that arise in the analysis of nature-inspired heuristics.

1.2 Notation

All notation in this chapter is standard and should not need much additional explanation. We use $\mathbb{N} := \{1,2,\dots\}$ to denote the positive integers. We write $\mathbb{N}_0 := \mathbb{N} \cup \{0\}$. For intervals of integers, we write $[a..b] := \{x \in \mathbb{Z} \mid a \le x \le b\}$. We use the standard definition $0^0 := 1$ (and not $0^0 = 0$).

1.3 Elementary Probability Theory

We shall assume that the reader has some basic understanding of the concepts of *probability spaces*, *events*, and *random variables*. As is usual in probability theory and is very convenient in the analysis of algorithms, we shall almost never explicitly state the probability space we are working in. Hence an intuitive understanding of the notion of a random variable should be enough to follow this exposition.

While many results presented in the following naturally extend to continuous probability spaces, in the interests of simplicity and accessibility to a discrete-optimization audience, we shall assume that all random variables in this book are *discrete*, that is, they take at most a countable number of values. As a simple example, consider the random experiment of independently rolling two distinguishable dice. Let X_1 denote the outcome of the first roll, that is, the number between 1 and 6 which the first die displays. Likewise, let X_2 denote the outcome of the second roll. These are already two random variables. We formalize the statement that with probability $\frac{1}{6}$ the first die shows a one by saying $\Pr[X_1 = 1] = \frac{1}{6}$. Also, the probability that both dice show the same number is $\Pr[X_1 = X_2] = \frac{1}{6}$. The *complementary event* that they show different numbers naturally has a probability of $\Pr[X_1 \neq X_2] = 1 - \Pr[X_1 = X_2] = \frac{5}{6}$.

We can add random variables (defined over the same probability space), e.g., $X := X_1 + X_2$ is the sum of the numbers shown by the two dice, and we can multiply a random variable by a number, e.g., $X := 2X_1$ is twice the number shown by the first die.

The most common type of random variable we shall encounter in this book is an extremely simple one called a *binary random variable* or *Bernoulli random variable*. It takes the values 0 and 1 only. In consequence, the probability distribution of a binary random variable X is fully described by its probability $\Pr[X=1]$ of being one, since $\Pr[X=0] = 1 - \Pr[X=1]$.

Binary random variables often show up as *indicator random variables* for random events. For example, if the random experiment is a simple roll of a die, we may define a random variable X by setting $X=1$ if the die shows a six, and $X=0$ otherwise. We say that X is the indicator random variable for the event "die shows a six."

Indicator random variables are useful for counting. If we roll a die n times and $X_1,\dots,X_n$ are the indicator random variables for the events that the corresponding roll showed a 6 (considered as a *success*), then $\sum_{i=1}^n X_i$ is a random variable describing the number of times we saw a 6 in these n rolls. In general, a random variable X that is the sum of n independent binary random variables that all are one with equal probabilities p is called a *binomial random variable* (with success probability p). We denote this distribution by $\mathrm{Bin}(n,p)$ and write $X \sim \mathrm{Bin}(n,p)$ to denote that X has this distribution. We have

$$\Pr[X=k] = \binom{n}{k} p^k (1-p)^{n-k}$$

for all $k \in [0..n]$. See Section 1.4.3 for the definition of the binomial coefficient.

A different question is how long we have to wait until we roll a 6. Assume that we have an infinite sequence of die rolls and $X_1, X_2, \dots$ are the indicator random variables for the event that the corresponding roll shows a six (*success*). Then we are interested in the random variable $Y = \min\{k \in \mathbb{N} \mid X_k = 1\}$. Again for the general case of all X_i being one independently with probability $p > 0$, this random variable Y is called a *geometric random variable* (with success probability p). We denote this distribution by $\mathrm{Geom}(p)$ and write $Y \sim \mathrm{Geom}(p)$ to indicate that Y is geometrically distributed (with parameter p). We have

$$\Pr[Y=k] = (1-p)^{k-1} p$$

for all $k \in \mathbb{N}$. We note that an equally established definition is to count only the failures, that is, to consider the random variable $Y-1$. So, some care is necessary when comparing results from different sources.

1.4 Useful Inequalities

Before starting our presentation of probabilistic tools useful in the analysis of randomized search heuristics, let us briefly mention a few inequalities that are often needed to estimate probabilities arising naturally in this area.

1.4.1 Switching Between Exponential and Polynomial Terms

When dealing with events occurring with a small probability $\varepsilon > 0$, we often encounter expressions such as $(1-\varepsilon)^n$. Such a mix of a polynomial term $(1-\varepsilon)$ with an exponentiation is often hard to work with. It is therefore very convenient that $1-\varepsilon \approx e^{-\varepsilon}$, so that the above expression becomes approximately the purely exponential term $e^{-\varepsilon n}$. In this section, we collect a few estimates of this flavor. With the exception of the second inequality in (1.4.9), a sharper version of a Weierstrass product inequality, all are well known and can be derived via elementary arguments.

Lemma 1.4.1. *For all $x \in \mathbb{R}$,*

$$1+x \leq e^x.$$

We give a canonical proof as an example of a proof method that is often useful for such estimates.

Proof. Define a function $f : \mathbb{R} \to \mathbb{R}$ by $f(x) = e^x - 1 - x$ for all $x \in \mathbb{R}$. Since $f'(x) = e^x - 1$, we have $f'(x) = 0$ if and only if $x = 0$. Since $f''(x) = e^x > 0$ for all x, we see that $x = 0$ is the unique minimum of f. Since $f(0) = 0$, we have $f(x) \geq 0$ for all x, which is equivalent to the claim of the lemma. □

Applying Lemma 1.4.1 to $-x$ and taking reciprocals, we immediately derive the first of the following two upper bounds on the exponential function. The second bound again follows from elementary calculus. Obviously, the first estimate is better for $x < 0$, and the second is better for $x > 0$.

Lemma 1.4.2. *(a) For all $x < 1$,*

$$e^x \leq \frac{1}{1-x} = 1 + \frac{x}{1-x} = 1 + x + \frac{x^2}{1-x}. \tag{1.4.1}$$

In particular, for $0 \leq x \leq 1$, we have $e^{-x} \leq 1 - \frac{x}{2}$.
(b) For all $x < 1.79$,

$$e^x \leq 1 + x + x^2. \tag{1.4.2}$$

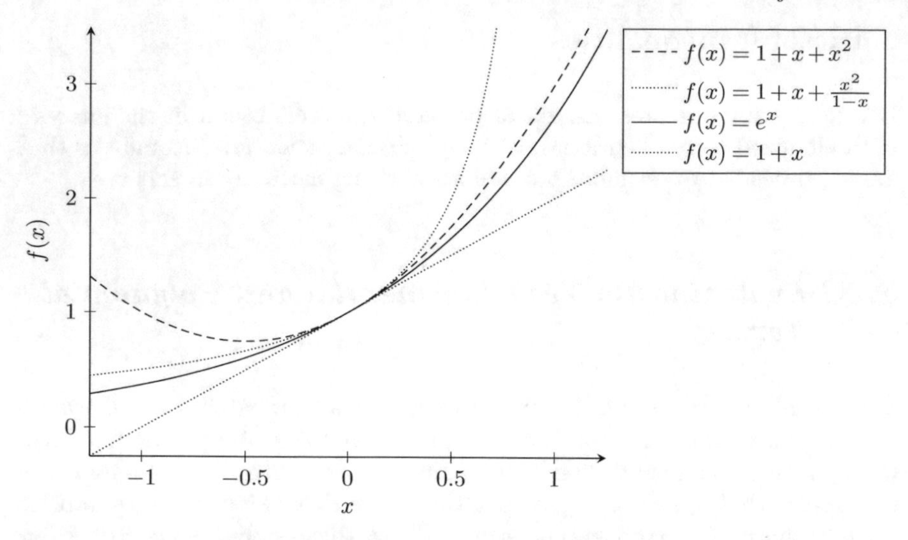

Fig. 1.1 Plot of the estimates of Lemmas 1.4.1 and 1.4.2.

As is visible also from Figure 1.1, these estimates are strongest for x close to zero. By combining Lemmas 1.4.1 and 1.4.2, the following useful estimate was obtained in [87, Lemma 8].

Corollary 1.4.3. *For all* $x \in [0,1]$ *and* $y > 0$, $(1-x)^y \leq \frac{1}{1+xy}$.

Replacing x with $\frac{x}{1+x}$ in the first inequality of Lemma 1.4.2 gives the following bounds.

Corollary 1.4.4. *For all* $x > -1$, *we have*

$$e^{\frac{x}{1+x}} \leq 1+x \leq e^x. \tag{1.4.3}$$

For all $x, y > 0$,

$$e^{\frac{xy}{x+y}} \leq (1+\tfrac{x}{y})^y \leq e^x. \tag{1.4.4}$$

The first bound in (1.4.3) can, with different arguments and for a smaller range of x, be sharpened to the following estimate [28, Lemma 8(c)].

Lemma 1.4.5. *For all* $x \in [0, \frac{2}{3}]$, $e^{-x-x^2} \leq 1-x$.

A reformulation of (1.4.3) often useful in the context of standard bit mutation (mutating a bit string by flipping each bit independently with a small probability such as $\frac{1}{n}$) is the following (see Figure 1.2 for some related plots). Note that the first bound holds for all $r \geq 1$, while it is often only stated for $r \in \mathbb{N}$. For the (not so interesting) boundary case $r = 1$, recall that we use the common convention $0^0 := 1$.

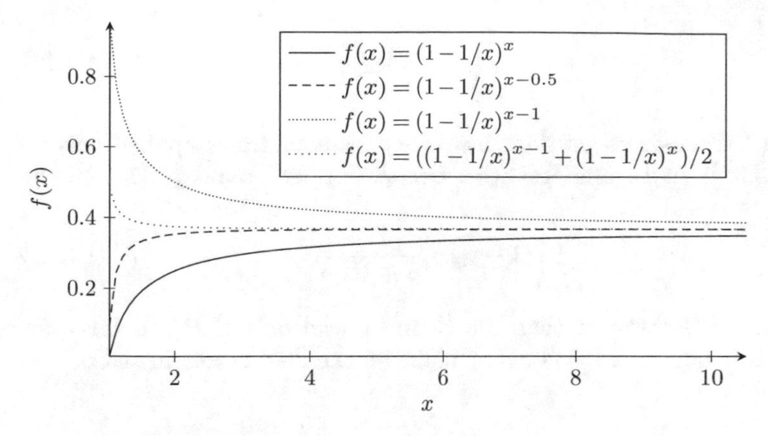

Fig. 1.2 Plots related to Corollary 1.4.6.

Corollary 1.4.6. *For all $r \geq 1$ and $0 \leq s \leq r$,*

$$(1-\tfrac{1}{r})^r \leq \tfrac{1}{e} \leq (1-\tfrac{1}{r})^{r-1}, \tag{1.4.5}$$
$$(1-\tfrac{s}{r})^r \leq e^{-s} \leq (1-\tfrac{s}{r})^{r-s}. \tag{1.4.6}$$

Occasionally, it is useful to know that $(1-\frac{1}{r})^r$ is monotonically increasing and that $(1-\frac{1}{r})^{r-1}$ is monotonically decreasing in r (and thus both converge to $\frac{1}{e}$).

Lemma 1.4.7. *For all $1 \leq s \leq r$, we have*

$$(1-\tfrac{1}{s})^s \leq (1-\tfrac{1}{r})^r, \tag{1.4.7}$$
$$(1-\tfrac{1}{s})^{s-1} \geq (1-\tfrac{1}{r})^{r-1}. \tag{1.4.8}$$

Finally, we mention Bernoulli's inequality and a related result. Lemma 1.4.8(b) below will be proven at the end of Section 1.5.2, both to show how probabilistic arguments can be used to prove non-probabilistic results and because we have not found a proof for the upper bound in the literature.

Lemma 1.4.8. *(a) Bernoulli's inequality. Let $x \geq -1$ and $r \in \{0\} \cup [1,\infty)$. Then $(1+x)^r \geq 1+rx$.*
(b) Weierstrass product inequalities. Let $p_1,\dots,p_n \in [0,1]$. Let $P := \sum_{i=1}^n p_i$. Then

$$1-P \leq \prod_{i=1}^{n}(1-p_i) \leq 1-P+\sum_{i<j} p_i p_j \leq 1-P+\tfrac{1}{2}P^2. \tag{1.4.9}$$

If in addition $P<1$, then

$$1+P \le \prod_{i=1}^{n}(1+p_i) \le \frac{1}{1-P}. \tag{1.4.10}$$

The term "Weierstrass product inequality" is sometimes applied only to the lower bounds in Lemma 1.4.8(b). For the upper bound in (1.4.9), the estimate

$$\prod_{i=1}^{n}(1-p_i) \le \frac{1}{1+P} \tag{1.4.11}$$

is well known. It is stronger than our bound if and only if $P>1$. Since for $P \ge 1$ the lower bound is trivial, this might be the less interesting case.

1.4.2 Harmonic Number

Quite frequently in the analysis of randomized search heuristics, we will encounter the *harmonic number* H_n. For all $n \in \mathbb{N}$, it is defined by $H_n = \sum_{k=1}^{n} \frac{1}{k}$. Approximating this sum via integrals, namely by

$$\int_1^{n+1} \frac{1}{x} dx \le H_n \le 1 + \int_1^n \frac{1}{x} dx,$$

we obtain the estimate

$$\ln n < H_n \le 1 + \ln n, \tag{1.4.12}$$

valid for all $n \ge 1$. Sharper estimates involving the Euler–Mascheroni constant $\gamma \approx 0.5772156649$ are known, e.g.,

$$H_n = \ln n + \gamma \pm O(\tfrac{1}{n}),$$
$$H_n = \ln n + \gamma + \tfrac{1}{2n} \pm O(\tfrac{1}{n^2}).$$

For non-asymptotic statements, it is helpful to know that $H_n - \ln n$ is monotonically decreasing (with limit γ, obviously). In most cases, however, the simple estimate (1.4.12) will be sufficient.

1.4.3 Binomial Coefficients and Stirling's Formula

Since discrete probability is strongly related to counting, we often encounter the binomial coefficients, defined by

$$\binom{n}{k} := \frac{n!}{k!\,(n-k)!}$$

for all $n \in \mathbb{N}_0$, $k \in [0..n]$. The binomial coefficient $\binom{n}{k}$ equals the number of k-element subsets of a given n-element set. For this reason, the above definition is often extended to $\binom{n}{k} := 0$ for $k > n$.

In this section, we give several useful estimates for binomial coefficients. We start by remarking that, while very precise estimates are available, in the analysis of randomized search heuristics crude estimates are often sufficient.

The following lemma lists some estimates which all can be proven by elementary means. To prove the second inequality of (1.4.16), note that $e^k = \sum_{i=0}^{\infty} \frac{k^i}{i!} \geq \frac{k^k}{k!}$ gives the elementary estimate

$$\left(\frac{k}{e}\right)^k \leq k! \leq k^k. \tag{1.4.13}$$

To prove (1.4.17), note that for even n we have $\binom{n}{n/2} = \frac{n!}{(n/2)!(n/2)!} = \prod_{i=1}^{n/2} \frac{2i(2i-1)}{i^2} = 2^n \prod_{i=1}^{n/2}(1 - \frac{1}{2i}) \leq 2^n \exp(-\frac{1}{2}\sum_{i=1}^{n/2} \frac{1}{i}) \leq 2^n \exp(-\frac{1}{2} \ln \frac{n}{2}) = 2^n \sqrt{\frac{2}{n}}$ (see Lemma 1.4.1 and (1.4.12)), while for odd n we have $\binom{n}{\lfloor n/2 \rfloor} = \frac{1}{2}\binom{n+1}{(n+1)/2} \leq 2^n \sqrt{\frac{2}{n+1}}$.

Lemma 1.4.9. *For all $n \in \mathbb{N}$ and $k \in [1..n]$, we have*

$$\binom{n}{k} \leq 2^n, \tag{1.4.14}$$

$$\left(\frac{n}{k}\right)^k \leq \binom{n}{k} \leq n^k, \tag{1.4.15}$$

$$\binom{n}{k} \leq \frac{n^k}{k!} \leq \left(\frac{ne}{k}\right)^k, \tag{1.4.16}$$

$$\binom{n}{k} \leq \binom{n}{\lfloor n/2 \rfloor} \leq 2^n \sqrt{\frac{2}{n}}. \tag{1.4.17}$$

Stronger estimates, giving also the well-known version

$$\binom{n}{k} \leq \binom{n}{\lfloor n/2 \rfloor} \leq 2^n \sqrt{\frac{2}{\pi n}} \tag{1.4.18}$$

of (1.4.17), can be obtained from the following estimate, known as Stirling's formula.

Theorem 1.4.10 (Robbins [86]). *For all $n \in \mathbb{N}$,*

$$n! = \sqrt{2\pi n}\left(\tfrac{n}{e}\right)^n R_n,$$

where $1 < \exp(\frac{1}{12n+1}) < R_n < \exp(\frac{1}{12n}) < 1.08690405$.

Corollary 1.4.11. *For all $n \in \mathbb{N}$ and $k \in [1..n-1]$,*

$$\binom{n}{k} = \frac{1}{\sqrt{2\pi}} \sqrt{\frac{n}{k(n-k)}} \left(\frac{n}{k}\right)^k \left(\frac{n}{n-k}\right)^{n-k} R_{nk},$$

where $0.88102729\ldots = \exp(-\frac{1}{6} + \frac{1}{25}) \le \exp(-\frac{1}{12k} - \frac{1}{12(n-k)} + \frac{1}{12n+1}) < R_{nk} < \exp(-\frac{1}{12k+1} - \frac{1}{12(n-k)+1} + \frac{1}{12n}) < 1$.

We refer to [55] for an analysis of randomized search heuristics which clearly requires Stirling's formula. Stirling's formula was also used in [43, proof of Lemma 8] to compute another useful fact, namely that all binomial coefficients that are $O(\sqrt{n})$ away from the middle one have the same asymptotic order of magnitude of $\Theta(2^n n^{-1/2})$. Here the upper bound is simply (1.4.17).

Corollary 1.4.12. *Let $\gamma \ge 0$. Let $n \in \mathbb{N}$ and $\ell = \frac{n}{2} \pm \gamma\sqrt{n}$. Then $\binom{n}{\ell} \ge (1 - o(1)) \frac{2^n}{2\sqrt{\pi n}} e^{-4\gamma^2}$.*

When working with mutation rates different from the classical choice of $\frac{1}{n}$, the following estimates can be useful.

Lemma 1.4.13. *Let $n \in \mathbb{N}$, $k \in [0..n]$, and $p \in [0,1]$. Let $X \sim \mathrm{Bin}(n,p)$.*

(a) Let $Y \sim \mathrm{Bin}(n, \frac{k}{n})$. Then $\Pr[X = k] \le \Pr[Y = k]$. This inequality is strict except for the trivial case $p = \frac{k}{n}$.
(b) For $k \in [1..n-1]$, $\Pr[X = k] \le \frac{1}{\sqrt{2\pi}} \sqrt{\frac{n}{k(n-k)}}$.

Proof. The first part follows from $\Pr[X = k] = \binom{n}{k} p^k (1-p)^{n-k}$ and noting that $p \mapsto p^k(1-p)^{n-k}$ has a unique maximum in the interval $[0,1]$, namely at $p = \frac{k}{n}$. The second part follows from the first and from using Corollary 1.4.11 to estimate the binomial coefficient in the expression $\Pr[Y = k] = \binom{n}{k} (\frac{1}{k})^k (1 - \frac{n}{k})^{n-k}$. □

For the special case where $np = k$, the second part of the lemma above was shown in [94, Lemma 10 of the arXiv version]. For $k \in \{\lfloor np \rfloor, \lceil np \rceil\}$ but $np \ne k$, a bound larger than ours by a factor of e was shown there as well.

Finally, we note that to estimate sums of binomial coefficients, large-deviation bounds (to be discussed in Section 1.10) can be an elegant tool. Imagine we need an upper bound on $S = \sum_{k=a}^{n} \binom{n}{k}$, where $a > \frac{n}{2}$. Let X be a random variable with distribution $\mathrm{Bin}(n, \frac{1}{2})$. Then $\Pr[X \ge a] = 2^{-n} S$. Using the additive Chernoff bound of Theorem 1.10.7, we also see that $\Pr[X \ge a] = \Pr[X \ge E[X] + (a - \frac{n}{2})] \le \exp(-\frac{2(a-\frac{n}{2})^2}{n})$. Consequently, $S \le 2^n \exp(-\frac{2(a-\frac{n}{2})^2}{n})$.

The same argument can even be used to estimate single binomial coefficients, in particular, those not to close to the middle one. Note that by

Lemma 1.10.38, $S = \sum_{k=a}^{n} \binom{n}{k}$ and $\binom{n}{a}$ are quite close when a is not too close to $\frac{n}{2}$. Hence

$$\binom{n}{a} \leq 2^n \exp\left(-\frac{2(a-\frac{n}{2})^2}{n}\right) \tag{1.4.19}$$

is a good estimate in this case.

1.5 Union Bound

The *union bound*, sometimes called Boole's inequality, is a very elementary consequence of the axioms of a probability space, in particular, the σ-additivity of the probability measure.

Lemma 1.5.1 (union bound). *Let $E_1, \ldots, E_n$ be arbitrary events in some probability space. Then*

$$\Pr\left[\bigcup_{i=1}^{n} E_i\right] \leq \sum_{i=1}^{n} \Pr[E_i].$$

Despite its simplicity, the union bound is a surprisingly powerful tool in the analysis of randomized algorithms. It draws its strength from the fact that it does not need any additional assumptions. In particular, the events E_i are not required to be independent. Here is an example of such an application of the union bound.

1.5.1 *Example: The (1+1) EA Solving the Needle Problem*

The *needle function* is the fitness function $f : \{0,1\}^n \to \mathbb{Z}$ defined by $f(x) = 0$ for all $x \in \{0,1\}^n \setminus \{(1,\ldots,1)\}$ and $f((1,\ldots,1)) = 1$. It is neither surprising nor difficult to prove that all reasonable randomized search heuristics need time exponential in n to find the maximum of the needle function. To give a simple example of the use of the simplified drift theorem, it was shown in [79] that the classic $(1+1)$ EA, within a sufficiently small exponential time, does not even get close to the optimum of the needle function (see Theorem 1.5.2 below). We now show that the same result (and in fact a stronger one) can be shown via the union bound.

The $(1+1)$ EA, described in Algorithm 1.1, is a simple randomized search heuristic that starts with a random search point $x \in \{0,1\}^n$. Then, in each iteration, it generates from x a new search point y by copying x into y and flipping each bit independently with probability $\frac{1}{n}$. If the new search point ("offspring") y is at least as good as the parent x, that is, if $f(y) \geq f(x)$ for an objective function to be maximized, then x is replaced by y; that is, we set $x := y$. Otherwise, y is discarded.

Algorithm 1.1: The (1+1) EA for maximizing $f\colon \{0,1\}^n \to \mathbb{R}$

```
1 Choose x ∈ {0,1}^n uniformly at random;
2 for t = 1,2,3,... do
3     y ← x;
4     for i ∈ [1..n] do
5         with probability 1/n do y_i ← 1 − y_i;
6     if f(y) ≥ f(x) then x ← y;
```

The precise result of [79, Theorem 5] is the following.

Theorem 1.5.2. *For all $\eta > 0$ there are $c_1, c_2 > 0$ such that with probability $1 - 2^{c_1 n}$ the first $2^{c_2 n}$ search points generated in a run of the (1+1) EA on the needle function all have a Hamming distance of more than $(\frac{1}{2} - \eta)n$ from the optimum.*

The proof of this theorem in [79] argues as follows. Denote by $x^{(0)}, x^{(1)}, \ldots$ the search points generated in a run of the $(1+1)$ EA. Denote by x^* the optimum of the needle function. For all $i \geq 0$, let $X_i := H(x^{(i)}, x^*) := |\{j \in [1..n] \mid x_j^{(i)} \neq x_j^*\}|$ be the Hamming distance of $x^{(i)}$ from the optimum. The random initial search point $x^{(0)}$ has an expected Hamming distance of $\frac{n}{2}$ from the optimum. By a simple Chernoff bound argument (Theorem 1.10.7), we see that, with probability $1 - \exp(-2\eta^2 n)$, we have $X_0 = H(x^{(0)}, x^*) > (\frac{1}{2} - \eta)n$. Now a careful analysis of the random process $(X_i)_{i \geq 0}$ via a new "simplified drift theorem" gives the claim.

We now show that the Chernoff bound argument plus a simple union bound is sufficient to prove the theorem. We show the following more explicit bound, which also applies to all other unbiased algorithms in the sense of Lehre and Witt [66] (roughly speaking, all algorithms which treat the bit positions $[1..n]$ and the bit values $\{0,1\}$ in a symmetric fashion).

Theorem 1.5.3. *For all $\eta > 0$ and $c > 0$ we have that with probability at least $1 - 2^{(c - 2\ln(2)\eta^2)n}$ the first $L := 2^{cn}$ search points generated in a run of the (1+1) EA (or any other unbiased black-box optimization algorithm) on the needle function all have a Hamming distance of more than $(\frac{1}{2} - \eta)n$ from the optimum.*

Proof. The key observation is that as long as the $(1+1)$ EA has not found the optimum, any search point x generated by the $(1+1)$ EA is uniformly distributed in $\{0,1\}^n$. Hence $\Pr[H(x, x^*) \leq (\frac{1}{2} - \eta)n] \leq \exp(-2\eta^2 n)$ by Theorem 1.10.7. By the union bound, the probability that one of the first $L := 2^{cn}$

search points generated by the (1+1) EA has a distance $H(x,x^*)$ of at most $(\frac{1}{2}-\eta)n$ is at most $L\exp(-2\eta^2 n) = 2^{(c-2\ln(2)\eta^2)n}$.

To be more formal, let $x^{(0)}, x^{(1)}, \ldots$ be the search points generated in a run of the (1+1) EA. Let $T = \min\{t \in \mathbb{N}_0 \mid x^{(t)} = x^*\}$. Define a sequence $y^{(0)}, y^{(1)}, \ldots$ of search points by setting $y^{(t)} := x^{(t)}$ for all $t \le T$. For all $t > T$, let $y^{(t)}$ be obtained from $y^{(t-1)}$ by flipping each bit independently with probability $\frac{1}{n}$. With this definition, and since $x^{(t)} = x^*$ for all $t \ge T$, we have

$$\begin{aligned}\{x^{(t)} \mid t \in [0..L-1]\} &= \{x^{(t)} \mid t \in [0..\min\{T, L-1\}]\}\\ &= \{y^{(t)} \mid t \in [0..\min\{T, L-1\}]\} \subseteq \{y^{(t)} \mid t \in [0..L-1]\}.\end{aligned}$$

Consequently,

$$\begin{aligned}&\Pr[\exists t \in [0..L-1] : H(x^{(t)}, x^*) \le (\tfrac{1}{2}-\eta)n]\\ &\quad\le \Pr[\exists t \in [0..L-1] : H(y^{(t)}, x^*) \le (\tfrac{1}{2}-\eta)n].\end{aligned}$$

By the union bound,

$$\Pr[\exists t \in [0..L-1] : H(y^{(t)}, x^*) \le (\tfrac{1}{2}-\eta)n] \le \sum_{t=0}^{L-1} \Pr[H(y^{(t)}, x^*) \le (\tfrac{1}{2}-\eta)n].$$

Note that when $y^{(t)}$ is a search point uniformly distributed in $\{0,1\}^n$, then so is $y^{(t+1)}$. Since $y^{(0)}$ is uniformly distributed, all $y^{(t)}$ are. Hence, by Theorem 1.10.7, we have $\Pr[H(y^{(t)}, x^*) \le (\frac{1}{2}-\eta)n] \le \exp(-2\eta^2 n)$ for all t and thus

$$\sum_{t=0}^{L-1} \Pr[H(y^{(t)}, x^*) \le (\tfrac{1}{2}-\eta)n] \le L\exp(-2\eta^2 n) = 2^{(c-2\ln(2)\eta^2)n}.$$

This proof immediately extends to all algorithms which, when optimizing the needle function, generate uniformly distributed search points until the optimum is found. These are, in particular, all unbiased algorithms in the sense of Lehre and Witt [66]. □

Note that the y_t in the proof above are heavily correlated. For all t, the search points y_t and y_{t+1} have an expected Hamming distance of exactly one. Nevertheless, we could apply the union bound to the events "$H(y_t, x^*) < (\frac{1}{2}-\eta)n$" and from this obtain a very elementary proof of Theorem 1.5.3.

1.5.2 Lower Bounds, Bonferroni Inequalities

The union bound is tight, that is, it holds with equality, when the events E_i are disjoint. In this case, the union bound simply reverts to the σ-additivity of the probability measure. The *second Bonferroni inequality* gives a lower bound on the probability of a union of events also when they are not disjoint.

Lemma 1.5.4. *Let $E_1,\dots,E_n$ be arbitrary events in some probability space. Then*

$$\Pr\left[\bigcup_{i=1}^{n} E_i\right] \ge \sum_{i=1}^{n} \Pr[E_i] - \sum_{i=1}^{n-1} \sum_{j=i+1}^{n} \Pr[E_i \cap E_j].$$

As an illustration, let us consider the performance of *blind random search* on the needle function; that is, we let $x^{(1)}, x^{(2)}, \dots$ be independent random search points from $\{0,1\}^n$ and ask ourselves what is the first hitting time $T = \min\{t \in \mathbb{N} \mid x^{(t)} = (1,\dots,1)\}$ of the maximum $x^* = (1,\dots,1)$ of the needle function (any other function $f : \{0,1\}^n \to \mathbb{R}$ with a unique global optimum would do as well). This is easy to compute directly. We see that T has a geometric distribution with success probability 2^{-n}, so the probability that L iterations do not suffice to find the optimum is $\Pr[T > L] = (1-2^{-n})^L$.

Let us nevertheless see what we can derive from the union bound and the second Bonferroni inequality. Let E_t be the event $x^{(t)} = x^*$. Then the union bound gives

$$\Pr[T \le L] = \Pr\left[\bigcup_{t=1}^{L} E_t\right] \le L2^{-n},$$

and the second Bonferroni inequality yields

$$\Pr[T \le L] = \Pr\left[\bigcup_{t=1}^{L} E_t\right] \ge L2^{-n} - \frac{L(L-1)}{2} 2^{-2n}.$$

Hence, if $L = o(2^n)$, that is, L is of smaller asymptotic order than 2^n, then $\Pr[T \le L] = (1-o(1))L2^{-n}$; that is, the union bound estimate is asymptotically tight.

For the sake of completeness, we now state the full set of Bonferroni inequalities. Note that the case $k=1$ is the union bound and the case $k=2$ is the lemma above.

Lemma 1.5.5. *Let $E_1,\dots,E_n$ be arbitrary events in some probability space. For all $k \in [1..n]$, let*

$$S_k := \sum_{1 \le i_1 < \dots < i_k \le n} \Pr[A_{i_1} \cap \dots \cap A_{i_k}].$$

Then, for all $k \in [1..n]$, we have

- $\Pr\left[\bigcup_{i=1}^{n} E_i\right] \le \sum_{j=1}^{k}(-1)^{j-1}S_j$ *for* $k \in [1..n]$ *odd,*
- $\Pr\left[\bigcup_{i=1}^{n} E_i\right] \ge \sum_{j=1}^{k}(-1)^{j-1}S_j$ *for* $k \in [1..n]$ *even.*

In simple terms, the Bonferroni inequalities state that when we omit the terms for $j > k$ in the inclusion–exclusion formula

$$\Pr\left[\bigcup_{i=1}^{n} E_i\right] = \sum_{j=1}^{n}(-1)^{j-1}S_j,$$

then the first of the omitted terms (that is, the one for $j = k+1$) dominates the error. So, if k is odd and thus the first omitted term is negative, then we obtain a "$\le$" inequality, and the reverse for k even.

We now use the Bonferroni inequalities to prove two of the inequalities given in Lemma 1.4.8(b).

Proof (of (1.4.9)*).* Consider some probability space with independent events $E_1,\ldots,E_n$ having $\Pr[E_i] = p_i$. Due to the independence,

$$\prod_{i=1}^{n}(1-p_i) = \Pr[\forall i \in [1..n] : \neg E_i] = 1 - \Pr[\exists i \in [1..n] : E_i]. \tag{1.5.1}$$

By the union bound, the right-hand side of (1.5.1) is at least $1 - \sum_{i=1}^{n} p_i = 1 - P$. By the Bonferroni inequality for $k = 2$ and again the independence, the right-hand side of (1.5.1) is at most

$$\begin{aligned} 1 - P + \sum_{i<j} \Pr[E_i \cap E_j] &= 1 - P + \sum_{i<j} p_i p_j \le 1 - P + \frac{1}{2}\sum_{i=1}^{n}\sum_{j=1}^{n} p_i p_j \\ &= 1 - P + \tfrac{1}{2}P^2. \end{aligned}$$

□

Note that the slack in the last inequality is only the term $\frac{1}{2}\sum_{i=1}^{n} p_i^2$, so there is not much reason to prefer the stronger upper bound $1 - P + \sum_{i<j} p_i p_j$ over the bound $1 - P + \frac{1}{2}P^2$.

1.6 Expectation and Variance

The expectation and variance are two key characteristic numbers of a random variable.

1.6.1 Expectation

The *expectation* (or mean) of a random variable X taking values in some set $\Omega \subseteq \mathbb{R}$ is defined by $E[X] = \sum_{\omega \in \Omega} \omega \Pr[X = \omega]$, where we shall always assume that the sum exists and is finite. As a trivial example, we immediately see that if X is a binary random variable, then $E[X] = \Pr[X = 1]$.

For non-negative integral random variables, the expectation can also be computed by the following formula (which is valid also when $E[X]$ is not finite).

Lemma 1.6.1. *Let X be a random variable taking values in the non-negative integers. Then*

$$E[X] = \sum_{i=1}^{\infty} \Pr[X \geq i].$$

If X takes values in $(-\infty, 0] \cup \mathbb{N}$, then $E[X] \leq \sum_{i=1}^{\infty} \Pr[X \geq i]$ still holds.

This lemma, among others, allows one to conveniently transform information about the tail bound of a distribution into a bound on its expectation. This was done, for example, in [47, proof of Lemma 10] for lower bounds, in [42, proof of Theorem 2] in a classic runtime analysis, and in [29, proof of Theorem 5] in the simplified proof of the multiplicative drift theorem.

Lemma 1.6.1 can also be employed to conveniently derive from information about the upper tail of a random variable an estimate of its expectation, as is done in the following elementary result.

Corollary 1.6.2 (expectations from exponential tail bounds). *Let $\alpha, \beta > 0$ and $T \geq 0$. Let X be an integer random variable and Y be a non-negative integer random variable.*

(a) If $\Pr[X \geq T + \lambda] \leq \alpha \exp(-\frac{\lambda}{\beta})$ for all $\lambda \in \mathbb{N}$, then $E[X] \leq T + \alpha\beta$.
(b) If $\Pr[Y \leq T - \lambda] \leq \alpha \exp(-\frac{\lambda}{\beta})$ for all $\lambda \in [1..T]$, then $E[Y] \geq T - \alpha\beta$.
(c) If $\Pr[X \geq (1+\varepsilon)T] \leq \alpha \exp(-\frac{\varepsilon}{\beta})$ for all $\varepsilon > 0$, then $E[X] \leq (1+\alpha\beta)T$.
(d) If $\Pr[X \leq (1-\varepsilon)T] \leq \alpha \exp(-\frac{\varepsilon}{\beta})$ for all $\varepsilon \in (0,1]$, then $E[X] \geq (1-\alpha\beta)T$.

Proof. By Lemma 1.6.1, we compute

$$\begin{aligned} E[X] &\leq \sum_{i=1}^{\infty} \Pr[X \geq i] \leq T + \sum_{i=T+1}^{\infty} \alpha \exp\left(-\frac{i-T}{\beta}\right) \\ &= T - \alpha + \alpha \frac{1}{1 - \exp(-1/\beta)} \leq T + \alpha\beta, \end{aligned}$$

where the last estimate uses (1.4.1).

Similarly, we compute

$$E[Y] = \sum_{i=1}^{T} \Pr[Y \geq i] \geq \sum_{i=1}^{T} (1 - \Pr[Y \leq i-1])$$
$$\geq \sum_{\lambda=1}^{T} (1 - \alpha \exp(-\tfrac{\lambda}{\beta})) \geq T - \sum_{\lambda=1}^{\infty} \alpha \exp(-\tfrac{\lambda}{\beta}) \geq T - \alpha\beta.$$

The last two claims are simple reformulations of the first two. □

In a similar vein, Lemma 1.6.1 yields an elegant analysis of the expectation of a geometric random variable. Let X be a geometric random variable with success probability p. Intuitively, we feel that the expected waiting time for a success should be $\frac{1}{p}$. This intuition is guided by the fact that after $\frac{1}{p}$ repetitions of the underlying binary random experiment, the expected number of successes is exactly one. This intuition leads to the right result; the "proof", however, is not correct. The correct proof uses either standard results in Markov chain theory, elementary but non-trivial calculations, or (as done below) the same reasoning as in the lemma above.

Lemma 1.6.3 (waiting-time argument). *Let X be a geometric random variable with success probability $p > 0$. Then $E[X] = \frac{1}{p}$.*

Proof. We have $\Pr[X \geq i] = (1-p)^{i-1}$, since $X \geq i$ is the event of having no success in the first $i-1$ rounds of the random experiment. Now Lemma 1.6.1 gives

$$E[X] = \sum_{i=1}^{\infty} \Pr[X \geq i] = \sum_{i=1}^{\infty} (1-p)^{i-1} = \frac{1}{1-(1-p)} = \frac{1}{p}.$$

□

An elementary, but very useful property is that the expectation is linear.

Lemma 1.6.4 (linearity of expectation). *Let $X_1, \ldots, X_n$ be arbitrary random variables and $a_1, \ldots, a_n \in \mathbb{R}$. Then*

$$E\left[\sum_{i=1}^{n} a_i X_i\right] = \sum_{i=1}^{n} a_i E[X_i].$$

This fact is very convenient when we can write a complicated random variable as sum of simpler ones. For example, let X be a binomial random variable with parameters n and p, that is, we have $\Pr[X = k] = \binom{n}{k} p^k (1-p)^{n-k}$. Since X counts the number of successes in n (independent) trials, we can write $X = \sum_{i=1}^{n} X_i$ as the sum of (independent) binary random variables $X_1, \ldots, X_n$, each with $\Pr[X_i = 1] = p$. Here X_i is the indicator random variable for the event that the i-th trial is a success. Using linearity of expectation, we compute

$$E[X] = E\left[\sum_{i=1}^{n} X_i\right] = \sum_{i=1}^{n} E[X_i] = np.$$

We have just proved the following.

Lemma 1.6.5 (expectation of binomial random variables). *Let X be a binomial random variable with parameters n and p. Then $E[X] = pn$.*

In the same fashion, we can compute the following elementary facts.

Lemma 1.6.6. *Let $x, y, x^* \in \{0,1\}^n$. Denote by $H(x,y) := |\{i \in [1..n] \mid x_i \neq y_i\}|$ the Hamming distance between x and y.*

(a) Let z be obtained from x via standard bit mutation with rate $p \in [0,1]$, that is, by flipping each bit of x independently with probability p. Then $E[H(x,z)] = pn$ and $E[H(z,x^*)] = H(x,x^*) + p(n - 2H(x,x^*))$.
(b) Let z be obtained from x and y via uniform crossover, that is, for each $i \in [1..n]$ independently, we set $z_i = x_i$ or $z_i = y_i$ each with probability $\frac{1}{2}$. Then $E[H(x,z)] = \frac{1}{2}H(x,y)$ and $E[H(z,x^*)] = \frac{1}{2}(H(x,x^*) + H(y,x^*))$.
(c) Let z be obtained from the unordered pair $\{x,y\}$ via 1-point crossover; that is, we choose r uniformly at random from $[0..n]$ and then, with probability $\frac{1}{2}$ each,

- *define z by $z_i = x_i$ for $i \leq r$ and $z_i = y_i$ for $i > r$, or*
- *define z by $z_i = y_i$ for $i \leq r$ and $z_i = x_i$ for $i > r$.*

Then $E[H(x,z)] = \frac{1}{2}H(x,y)$ and $E[H(z,x^*)] = \frac{1}{2}(H(x,x^*) + H(y,x^*))$.

The fact that the results for the two crossover operators are identical shows again that linearity of expectation does not care about possible dependencies. We have $\Pr[z_i = x_i] = \frac{1}{2}$ in both cases, and this is what is important for the result, whereas the fact that the events "$z_i = x_i$" are independent for uniform crossover and strongly dependent for 1-point crossover has no influence on the result.

1.6.2 Markov's Inequality

Markov's inequality is an elementary large-deviation bound valid for *all* non-negative random variables.

Lemma 1.6.7 (Markov's inequality). *Let X be a non-negative random variable with $E[X] > 0$. Then, for all $\lambda > 0$,*

$$\Pr[X \geq \lambda E[X]] \leq \frac{1}{\lambda}, \tag{1.6.1}$$

$$\Pr[X \geq \lambda] \leq \frac{E[X]}{\lambda}. \tag{1.6.2}$$

Proof. We have

$$E[X] = \sum_{\omega} \omega \Pr[X = \omega] \geq \sum_{\omega \geq \lambda} \lambda \Pr[X = \omega] = \lambda \Pr[X \geq \lambda],$$

proving (1.6.2). □

We note that (1.6.2) also holds without the assumption $E[X] > 0$. More interestingly, the proof above shows that Markov's inequality is always strict (that is, it holds with "$<$" instead of "$\leq$") when X takes at least three different values with positive probability.

It is important to note that Markov's inequality, without further assumptions, only gives information about deviations above the expectation. If X is a (not necessarily non-negative) random variable taking only values not larger than some $u \in \mathbb{R}$, then the random variable $u - X$ is non-negative and Markov's inequality gives the bound

$$\Pr[X \leq \lambda] \leq \frac{u - E[X]}{u - \lambda}, \tag{1.6.3}$$

which is sometimes called the *reverse Markov's inequality*. An equivalent formulation of this bound is

$$\Pr[X > \lambda] \geq \frac{E[X] - \lambda}{u - \lambda}. \tag{1.6.4}$$

Markov's inequality is useful if not much information is available about the random variable under consideration. Also, when the expectation of X is very small, the following elementary corollary is convenient and, in fact, often quite tight.

Corollary 1.6.8 (first moment method). *If X is a non-negative random variable, then* $\Pr[X \geq 1] \leq E[X]$.

Corollary 1.6.8 together with linearity of expectation often gives the same results as the union bound. For an example, recall that in Section 1.5.2 we observed that in a run of the blind random search heuristic, the probability that the t-th search point x_t is the unique optimum of a given function $f : \{0,1\}^n \to \mathbb{R}$ is 2^{-n}. Denote this event by E_t and let X_t be the indicator random variable for this event. Then the probability that one of the first L search points is the optimum can be estimated equally well via the union bound or via the above corollary and linearity of expectation:

$$\Pr\left[\bigcup_{t=1}^{L} E_t\right] \leq \sum_{t=1}^{L} \Pr[E_t] = L2^{-n},$$

$$\Pr\left[\sum_{t=1}^{L} X_t \geq 1\right] \leq E\left[\sum_{t=1}^{L} X_t\right] = \sum_{t=1}^{L} E[X_t] = L2^{-n}.$$

1.6.3 Chebyshev's Inequality

The second elementary large-deviation bound is Chebyshev's inequality, sometimes called the Bienaymé–Chebyshev inequality as it was first stated by Bienaymé [7] and later proven by Chebyshev [95]. It seems less often used in the theory of randomized search heuristics (exceptions being [35, 74]).

Recall that the *variance* of a discrete random variable X is

$$\mathrm{Var}[X] = E[(X - E[X])^2] = E[X^2] - E[X]^2. \tag{1.6.5}$$

Just by definition, the variance is a measure of how well X is concentrated around its mean.

From the variance, we also obtain a bound on the expected (absolute) deviation from the mean. Applying the well-known estimate $E[X]^2 \le E[X^2]$, which follows from the second equality in (1.6.5), to the random variable $|X - E[X]|$, we obtain

$$E[|X - E[X]|] \le \sqrt{E[(X - E[X])^2]} = \sqrt{\mathrm{Var}[X]}. \tag{1.6.6}$$

More often, we use the variance to bound the probability of deviating from the expectation by a certain amount. Applying Markov's inequality to the random variable $(X - E[X])^2$ easily yields the following very useful inequality.

Lemma 1.6.9 (Chebyshev's inequality). *Let X be a random variable with* $\mathrm{Var}[X] > 0$. *Then, for all $\lambda > 0$,*

$$\Pr\left[|X - E[X]| \ge \lambda\sqrt{\mathrm{Var}[X]}\right] \le \tfrac{1}{\lambda^2}, \tag{1.6.7}$$

$$\Pr\left[|X - E[X]| \ge \lambda\right] \le \tfrac{\mathrm{Var}[X]}{\lambda^2}. \tag{1.6.8}$$

Similarly to Markov's inequality, the second estimate is valid also without the assumption $\mathrm{Var}[X] > 0$. Note that Chebyshev's inequality automatically yields a two-sided tail bound (that is, a bound for $|X - E[X]|$), as opposed to Markov's inequality (which just gives a bound for exceeding the expectation). There is a one-sided version of Chebyshev's inequality that is often attributed to Cantelli, though Hoeffding [54] sees Chebyshev [96] as its inventor.

Lemma 1.6.10 (Cantelli's inequality). *Let X be a random variable with* $\mathrm{Var}[X] > 0$. *Then for all $\lambda > 0$,*

$$\Pr\left[X \ge E[X] + \lambda\sqrt{\mathrm{Var}[X]}\right] \le \tfrac{1}{\lambda^2+1}, \tag{1.6.9}$$

$$\Pr\left[X \le E[X] - \lambda\sqrt{\mathrm{Var}[X]}\right] \le \tfrac{1}{\lambda^2+1}. \tag{1.6.10}$$

In many applications, the slightly better bound of Cantelli's inequality is not very interesting. Cantelli's inequality has, however, the charm that the right-hand side is always less than one, and hence one can also obtain non-trivial probabilities for deviations smaller than $\sqrt{\mathrm{Var}[X]}$. We shall exploit this in the proof of Lemma 1.10.16.

While Markov's inequality can be used to show that a non-negative random variable X rarely is positive (first moment method), Chebyshev's inequality can serve the opposite purpose, namely showing that X is positive with good probability. By taking $\lambda = E[X]$ in (1.6.8), we obtain the first estimate of the following lemma. Using the Cauchy–Schwarz inequality and computing

$$E[X]^2 = E[X\mathbf{1}_{X\neq 0}]^2 \leq E[X^2]E[\mathbf{1}_{X\neq 0}] = E[X^2]\Pr[X \neq 0],$$

we obtain the second estimate, which has the nice equivalent formulation

$$\Pr[X \neq 0] \geq \frac{E[X]^2}{E[X^2]}. \tag{1.6.11}$$

Since $E[X^2] \geq E[X]^2$, the second estimate gives a stronger bound on $\Pr[X = 0]$ than the first. While the lemma below does not require that X is non-negative, the typical application of showing that X is positive requires that X is non-negative in the second bound, so that $\Pr[X \neq 0] = \Pr[X > 0]$.

Lemma 1.6.11 (second moment method). *For a random variable X with $E[X] \neq 0$,*

$$\Pr[X = 0] \leq \Pr[X \leq 0] \leq \frac{\mathrm{Var}[X]}{E[X]^2}, \tag{1.6.12}$$

$$\Pr[X = 0] \leq \frac{\mathrm{Var}[X]}{E[X^2]}. \tag{1.6.13}$$

In the (purely academic) example of finding a unique global optimum via blind random search (see Section 1.5.2), let X_t be the indicator random variable for the event that the t-th search point is the optimum. Let $X = \sum_{t=1}^{L} X_t$. Then the probability that the optimum is found within the first L iterations is

$$\Pr[X > 0] = 1 - \Pr[X = 0] \geq 1 - \frac{\mathrm{Var}[X]}{E[X]^2}.$$

The variance of a sum of binary random variables is

$$\mathrm{Var}[X] = \sum_{t=1}^{L} \mathrm{Var}[X_t] + \sum_{s<t} \mathrm{Cov}[X_s, X_t] \leq E[X] + \sum_{s<t} \mathrm{Cov}[X_s, X_t],$$

where we recall the definition of the *covariance*,

$$\mathrm{Cov}[U,V] := E[UV] - E[U]E[V],$$

of two arbitrary random variables U and V. Here we have $\mathrm{Cov}[X_s, X_t] = 0$, since the X_t are independent. Consequently,

$$\Pr[X > 0] \geq 1 - \frac{1}{E[X]}.$$

Hence the probability of finding the optimum within L iterations is $\Pr[T \leq L] = \Pr[X > 0] \geq 1 - \frac{1}{L2^{-n}}$. Note that this estimate is, for the interesting case where $E[X]$ is large, much better than the bound $\Pr[T \leq L] \geq L2^{-n} - \frac{L(L-1)}{2}2^{-2n}$ which we obtained from the second Bonferroni inequality.

1.7 Conditioning

In the analysis of randomized heuristics, we often want to argue that a certain desired event C already holds, and then continue arguing under this condition. Formally, this gives rise to a new probability space where each of the original events A now has a probability of

$$\Pr[A \mid C] := \frac{\Pr[A \cap C]}{\Pr[C]}.$$

Obviously, this only makes sense for events C with $\Pr[C] > 0$. In an analogous fashion, we define the expectation of a random variable X conditional on C by $E[X \mid C] = \sum_{\omega \in C} X(\omega) \Pr[\omega \mid C]$. The random variable behind this definition, which takes a value x with probability $\Pr[X = x]/\Pr[C]$, is sometimes denoted by $(X \mid C)$.

While we shall not use this notation, we still feel the need to warn the reader that there is a related notion of the conditional expectation with respect to a random variable, which sometimes creates confusion. If X and Y are two random variables defined on the same probability space, then $E[X \mid Y]$ is a function (that is, a random variable) defined on the range of Y by $E[X \mid Y](y) = E[X \mid Y = y]$.

Conditioning as a proof technique has many faces, among them the following.

1.7.1 Decomposing Events

If we can write some event A as the intersection of two events A_1 and A_2, then it can be useful to compute first the probability of A_1 and then the probability of A_2 conditional on A_1. Directly from the definition, we have $\Pr[A_1 \cap A_2] = \Pr[A_1]\Pr[A_2 \mid A_1]$. Of course, this requires that we have some direct way of computing $\Pr[A_1 \mid A_2]$.

1.7.2 Case Distinctions

Let $C_1,\ldots,C_k$ be a partition of our probability space. If it is easy to analyze our problem conditional on each of these events ("in the that case C_i holds"), then the following *law of total probability* and *law of total expectation* are useful.

Lemma 1.7.1 (laws of total probability and total expectation). *Let $C_1,\ldots,C_k$ be a partition of our probability space. Let A be some event and X be some random variable. Then*

$$\Pr[A] = \sum_{i=1}^{k} \Pr[A \mid C_i] \Pr[C_i],$$

$$E[X] = \sum_{i=1}^{k} E[X \mid C_i] \Pr[C_i].$$

1.7.3 Excluding Rare Events

Quite often, in the analysis of nature-inspired search heuristics, we would like to exclude some rare unwanted event. For example, assume that we are analyzing an evolutionary algorithm using standard bit mutation with mutation rate $\frac{1}{n}$. Then it is very unlikely that in an application of this mutation operator more than $n^{1/4}$ bits are flipped. So it could be convenient to exclude this rare event, say by stating that "with probability $1-2^{-\Omega(n^{1/4})}$, in none of the first n^2 applications of the mutation operator more than $n^{1/4}$ bits are flipped; let us condition on this event in the following." The proofs of Theorems 7 and 8 in [47] are examples of the use of such reasoning.

What could be a problem with this approach is that as soon as we condition on such an event, we change the probability space and thus arguments that are valid in the unconditional setting are not valid anymore. As a simple example, note that once we condition on the event that we flip at most $n^{1/4}$ bits, the events E_i that the i-th bit is flipped are not independent anymore. Fortunately, we can safely ignore this in most cases (and many authors do so without saying a word on this matter). The reason is that when we condition on an almost sure event, then the probabilities of all events change only very little (see the lemma below for this statement made precise). Hence, in our example, we can compute the probability of some event assuming that the bit flips are independent and then correct this probability by a minor amount.

Lemma 1.7.2. *Let C be some event with probability $1-p$. Let A be any event. Then*

$$\frac{\Pr[A]-p}{1-p} \le \Pr[A \mid C] \le \frac{\Pr[A]}{1-p}.$$

In particular, for $p \le \frac{1}{2}$, *we have* $\Pr[A] - p \le \Pr[A \mid C] \le \Pr[A] + 2p$.

The proof of this lemma follows directly from the definition of conditional probabilities and the elementary estimate $\Pr[A] - p \le \Pr[A \setminus \overline{C}] = \Pr[A \cap C] \le \Pr[A]$, where $\overline{C}$ denotes the complement of C. From this, we also observe the natural fact that when $A \subseteq C$, that is, the event A implies C, then conditioning on C does not decrease the probability of A:

$$\Pr[A \mid C] = \frac{\Pr[A \cap C]}{\Pr[C]} = \frac{\Pr[A]}{\Pr[C]} \ge \Pr[A]. \tag{1.7.1}$$

Likewise, when $A \supseteq \overline{C}$, then

$$\Pr[A \mid C] = \frac{\Pr[A \setminus \overline{C}]}{\Pr[C]} = \frac{\Pr[A] - p}{1 - p} \le \Pr[A]. \tag{1.7.2}$$

For example, if X is the number of bits flipped in an application of standard bit mutation, then

$$\Pr[X \le 10 \mid X \le \tfrac{n}{2}] \ge \Pr[X \le 10],$$
$$\Pr[X \ge 10 \mid X \le \tfrac{n}{2}] \le \Pr[X \ge 10].$$

1.7.4 Conditional Binomial Random Variables

We occasionally need to know the expected value of a binomially distributed random variable $X \sim \mathrm{Bin}(n,p)$ conditional on the variable having at least a certain value k. An intuitive (but wrong) argument is that $E[X \mid X \ge k]$ should be around $k + p(n-k)$, because we know already that k of the n independent trials are successes and the remaining $(n-k)$ trials still have their independent success probability of p. While this argument is wrong (as we might need more than k trials to have k successes), the result is correct as an upper bound, as shown in this lemma from [23, Lemma 1].

Lemma 1.7.3. *Let X be a random variable that is binomially distributed with parameters n and $p \in [0,1]$. Let $k \in [0..n]$. Then*

$$E[X \mid X \ge k] \le k + (n-k)p \le k + E[X].$$

Proof. Let $X = \sum_{i=1}^n X_i$ with $X_1, \ldots, X_n$ being independent binary random variables with $\Pr[X_i = 1] = p$ for all $i \in [1..n]$. Conditioning on $X \ge k$, let $\ell := \min\{i \in [1..n] \mid \sum_{j=1}^i X_j = k\}$. Then

$$E[X \mid X \ge k] = \sum_{i=1}^n \Pr[\ell = i \mid X \ge k] E[X \mid \ell = i].$$

Note that $\ell \geq k$ by definition. Note also that $(X \mid \ell = i) = k + \sum_{j=i+1}^{n} X_j$ with unconditioned X_j. In particular, $E[X \mid \ell = i] = k + (n-i)p$. Consequently,

$$E[X \mid X \geq k] = \sum_{i=1}^{n} \Pr[\ell = i \mid X \geq k] E[X \mid \ell = i]$$
$$\leq \sum_{i=k}^{n} \Pr[\ell = i \mid X \geq k](k + (n-k)p) = k + (n-k)p.$$

□

We note that, in the language introduced in the following section, we have actually shown the stronger statement that $(X \mid X \geq k)$ is dominated by $k + \mathrm{Bin}(n-k, p)$. This stronger version can be useful for obtaining tail bounds for $(X \mid X \geq k)$.

1.8 Stochastic Domination and Coupling

In this section, we discuss two concepts that are not too often used explicitly, but where we feel that mastering them can greatly help in the analysis of randomized search heuristics. The first of these is *stochastic domination*, which is a very strong way of saying that one random variable is better than another even when they are not defined on the same probability space. The second concept is *coupling*, which means defining two random variables suitably over the same probability space to facilitate comparing them. These two concepts are strongly related: if a random variable Y dominates X, then X and Y can be coupled in such a way that Y is pointwise not smaller than X, and vice versa. The results of this section and some related ones have appeared, in a more condensed form, in [20].

1.8.1 The Notion of Stochastic Domination

Possibly the first to use the notion of stochastic domination in the rigorous analysis of an evolutionary algorithm was Droste, who employed it in [45, 46] to make precise an argument often used in an informal manner, namely that some artificial random process is not faster than the process describing a run of the algorithm under investigation.

Definition 1.8.1 (stochastic domination). Let X and Y be two random variables not necessarily defined on the same probability space. We say that Y stochastically dominates X, written as $X \preceq Y$, if for all $\lambda \in \mathbb{R}$ we have $\Pr[X \leq \lambda] \geq \Pr[Y \leq \lambda]$.

If Y dominates X, then the cumulative distribution function of Y is pointwise not larger than that of X. The definition of domination is equivalent to

$$\forall \lambda \in \mathbb{R} : \Pr[X \geq \lambda] \leq \Pr[Y \geq \lambda],$$

which is maybe a formulation that makes it more visible why we feel that Y is at least as large as X.

Concerning nomenclature, we remark that some research communities require in addition that the inequality is strict for at least one value of λ. Hence, intuitively speaking, Y is strictly larger than X. From the mathematical perspective, this appears not to be very practical. Consequently, our definition above is more common in computer science. We also note that stochastic domination is sometimes called *first-order stochastic domination.* For an extensive treatment of various forms of stochastic orders, we refer to [72].

The usual way of explaining stochastic domination is via games. Let us consider the following three games.

Game A. With probability $\frac{1}{2}$ in each case, you win 500 or 1500.
Game B. With probability $\frac{1}{3}$, you win 500, with probability $\frac{1}{6}$, you win 800, and with probability $\frac{1}{2}$, you win 1500.
Game C. With probability $\frac{1}{1000}$, you win 2,000,000. Otherwise, you win nothing.

Which of these games is best to play? It is intuitively clear that you would prefer Game B over Game A. However, it is not clear whether you should prefer Game C over Game B. Clearly, the expected win in Game C is 2000, compared with only 1050 in Game B. However, the chance of winning anything at all is really small in Game C. If you do not like to go home empty-handed, you might prefer Game B.

The mathematical take on these games is that the random variable X_{B} describing the win in Game B stochastically dominates the variable X_{A} for Game A. This captures our intuitive feeling that it cannot be wrong to prefer Game B over Game A. For Games B and C, neither of X_{B} and X_{C} dominates the other. Consequently, it depends on the precise utility function of the player which game the player prefers. This statement is made precise in the following lemma.

Lemma 1.8.2. *The following two conditions are equivalent.*

(a) $X \preceq Y$.
(b) For all monotonically non-decreasing functions $f : \mathbb{R} \to \mathbb{R}$, *we have*

$$E[f(X)] \leq E[f(Y)].$$

As a simple corollary, we note the following.

Corollary 1.8.3. *If* $X \preceq Y$, *then* $E[X] \leq E[Y]$.

We note another simple, but useful property.

Lemma 1.8.4. *Let $X_1, \dots, X_n$ be independent random variables defined over some common probability space. Let $Y_1, \dots, Y_n$ be independent random variables defined over a possibly different probability space. If $X_i \preceq Y_i$ for all $i \in [1..n]$, then*

$$\sum_{i=1}^{n} X_i \preceq \sum_{i=1}^{n} Y_i.$$

For discrete random variables, this result is a special case of Lemma 1.8.8 stated further below.

Finally, we note two trivial facts.

Lemma 1.8.5. *Let X and Y be random variables.*

(a) If X and Y are defined on the same probability space and $X \le Y$, then $X \preceq Y$.
(b) If X and Y are identically distributed, then $X \preceq Y$.

1.8.2 Stochastic Domination in Runtime Analysis

From the perspective of algorithm analysis, stochastic domination allows one to state very clearly that one algorithm is better than another. If the runtime distribution X_{A} of algorithm A dominates the distribution X_{B} of algorithm B, then from the runtime perspective algorithm B is always preferable to algorithm A.

In a similar vein, we can also use domination to give more detailed descriptions of the runtime of an algorithm. For almost all algorithms, we will not be able to determine precisely the runtime distribution. However, via stochastic domination, we can give a lot of useful information beyond, say, just the expectation. We demonstrate this via an extension of the classic fitness level method, which is implicit in the work of Zhou, Luo, Lu, and Han [104].

Theorem 1.8.6 (domination version of the fitness level method). *Consider an iterative randomized search heuristic $\mathcal{A}$ maximizing a function $f : \Omega \to \mathbb{R}$. Let $A_1, \dots, A_m$ be a partition of Ω such that for all $i, j \in [1..m]$ with $i < j$ and all $x \in A_i$, $y \in A_j$, we have $f(x) < f(y)$. Set $A_{\ge i} := A_i \cup \dots \cup A_m$. Let $p_1, \dots, p_{m-1}$ be such that for all $i \in [1..m-1]$ we have that if the best-so-far search point is in A_i, then, regardless of the past, $\mathcal{A}$ has a probability of at least p_i of generating a search point in $A_{\ge i+1}$ in the next iteration.*

Denote by T the (random) number of iterations $\mathcal{A}$ takes to generate a search point in A_m. Then

$$T \preceq \sum_{i=1}^{m-1} \mathrm{Geom}(p_i),$$

where this sum is to be understood as a sum of independent geometric distributions.

To prove this theorem, we need a technical lemma, which we defer to the next subsection to ease reading this part.

Proof. Consider a run of the algorithm $\mathcal{A}$. For all $i \in [1..m]$, let T_i be the first time (iteration) when $\mathcal{A}$ has generated a search point in $A_{\geq i}$. Then $T = T_m = \sum_{i=1}^{m-1}(T_{i+1} - T_i)$. By assumption, $T_{i+1} - T_i$ is dominated by a geometric random variable with parameter p_i regardless of what happened before time T_i. Consequently, Lemma 1.8.8 gives the claim. □

Note that a result such as Theorem 1.8.6 implies various statements about the runtime. By Corollary 1.8.3, the expected runtime satisfies $E[T] \leq \sum_{i=1}^{m-1} \frac{1}{p_i}$, which is the common version of the fitness level theorem [97]. By using tail bounds for sums of independent geometric random variables (see Section 1.10.4), we also obtain runtime bounds that hold with high probability. This was first proposed in [104]. We defer a list of examples where previous results can profitably be turned into a domination statement to Section 1.10.4, where we will also have the large-deviation bounds needed to exploit such statements.

1.8.3 Domination by Independent Random Variables

A situation often encountered in the analysis of algorithms is that a sequence of random variables is not independent, but that each member of the sequence has a good chance of having a desired property no matter what the outcome of its predecessors was. In this case, the random variables in some sense can be treated as if they were independent.

Lemma 1.8.7. *Let $X_1, \ldots, X_n$ be arbitrary binary random variables and let $X_1^*, \ldots, X_n^*$ be independent binary random variables.*

(a) If we have

$$\Pr[X_i = 1 \mid X_1 = x_1, \ldots, X_{i-1} = x_{i-1}] \leq \Pr[X_i^* = 1]$$

for all $i \in [1..n]$ and all $x_1, \ldots, x_{i-1} \in \{0,1\}$ with $\Pr[X_1 = x_1, \ldots, X_{i-1} = x_{i-1}] > 0$, then

$$\sum_{i=1}^{n} X_i \preceq \sum_{i=1}^{n} X_i^*.$$

(b) If we have

$$\Pr[X_i = 1 \mid X_1 = x_1, \ldots, X_{i-1} = x_{i-1}] \geq \Pr[X_i^* = 1]$$

for all $i \in [1..n]$ *and all* $x_1,\dots,x_{i-1} \in \{0,1\}$ *with* $\Pr[X_1 = x_1,\dots,X_{i-1} = x_{i-1}] > 0$, *then*

$$\sum_{i=1}^{n} X_i^* \preceq \sum_{i=1}^{n} X_i.$$

Note that here and in the following, we view "$X_1 = x_1,\dots,X_{i-1} = x_{i-1}$" for $i = 1$ as an empty intersection of events, that is, an intersection over an empty index set. As in most textbooks, we define this to be the whole probability space.

Both parts of the lemma are simple corollaries of the following, slightly technical, general result, which might be of independent interest.

For two sequences $(X_1,\dots,X_n)$ and $(X_1^*,\dots,X_n^*)$ of random variables, we say that $(X_1^*,\dots,X_n^*)$ *unconditionally sequentially dominates* $(X_1,\dots,X_n)$ if for all $i \in [1..n]$ and all $x_1,\dots,x_{i-1} \in \mathbb{R}$ with $\Pr[X_1 = x_1,\dots,X_{i-1} = x_{i-1}] > 0$, we have $(X_i \mid X_1 = x_1,\dots,X_{i-1} = x_{i-1}) \preceq X_i^*$. Analogously, we speak of *unconditional sequential subdomination* if the last condition is replaced by $X_i^* \preceq (X_i \mid X_1 = x_1,\dots,X_{i-1} = x_{i-1})$.

The following lemma shows that unconditional sequential (sub)domination and independence of the X_i^* imply (sub)domination for the sums of these random variables. Note that unconditional sequential (sub)domination is inherited by subsequences, so the following lemma immediately extends to sums over arbitrary subsets I of the index set $[1..n]$.

Lemma 1.8.8. *Let* $X_1,\dots,X_n$ *be arbitrary discrete random variables. Let* $X_1^*,\dots,X_n^*$ *be independent discrete random variables.*

(a) If $(X_1^*,\dots,X_n^*)$ *unconditionally sequentially dominates* $(X_1,\dots,X_n)$, *then* $\sum_{i=1}^{n} X_i \preceq \sum_{i=1}^{n} X_i^*$.
(b) If $(X_1^*,\dots,X_n^*)$ *unconditionally sequentially subdominates* $(X_1,\dots,X_n)$, *then* $\sum_{i=1}^{n} X_i^* \preceq \sum_{i=1}^{n} X_i$.

Proof. The two parts of the lemma imply each other (as can be seen by multiplying the random variables by -1), so it suffices to prove the first statement.

Since the statement of the theorem is independent of the correlation between the X_i and the X_i^*, we may assume that they are independent. Let $\lambda \in \mathbb{R}$. Define

$$P_j := \Pr\left[\sum_{i=1}^{j} X_i + \sum_{i=j+1}^{n} X_i^* \geq \lambda\right]$$

for $j \in [0..n]$. We show $P_{j+1} \leq P_j$ for all $j \in [0..n-1]$.

For $m \in \mathbb{R}$, let Ω_m denote the set of all $(x_1,\dots,x_j,x_{j+2},\dots,x_n) \in \mathbb{R}^{n-1}$ such that $\Pr[X_1 = x_1,\dots,X_j = x_j] > 0$ and $\sum_{i \in [1..n] \setminus \{j+1\}} x_i = \lambda - m$. Let $M := \{m \in \mathbb{R} \mid \Omega_m \neq \emptyset\}$. Then

$$\begin{aligned}
P_{j+1} &= \Pr\left[\sum_{i=1}^{j+1} X_i + \sum_{i=j+2}^{n} X_i^* \geq \lambda\right] \\
&= \sum_{m \in M} \Pr\left[\sum_{i=1}^{j} X_i + \sum_{i=j+2}^{n} X_i^* = \lambda - m \wedge X_{j+1} \geq m\right] \\
&= \sum_{m \in M} \sum_{(x_1,\ldots,x_j,x_{j+2},\ldots,x_n) \in \Omega_m} \Pr[X_1 = x_1, \ldots, X_j = x_j] \cdot \\
&\qquad \Pr\left[X_{j+1} \geq m \,\middle|\, X_1 = x_1, \ldots, X_j = x_j\right] \cdot \prod_{i=j+2}^{n} \Pr[X_i^* = x_i] \\
&\leq \sum_{m \in M} \Pr\left[\sum_{i=1}^{j} X_i + \sum_{i=j+2}^{n} X_i^* = \lambda - m\right] \cdot \Pr\left[X_{j+1}^* \geq m\right] \\
&= \Pr\left[\sum_{i=1}^{j} X_i + \sum_{i=j+1}^{n} X_i^* \geq \lambda\right] \\
&= P_j.
\end{aligned}$$

Thus, we have

$$\Pr\left[\sum_{i=1}^{n} X_i \geq \lambda\right] = P_n \leq P_{n-1} \leq \cdots \leq P_1 \leq P_0 = \Pr\left[\sum_{i=1}^{n} X_i^* \geq \lambda\right].$$

□

1.8.4 Coupling

Coupling is an analysis technique that consists of defining two unrelated random variables over the same probability space to ease comparing them. As an example, let us consider standard bit mutation with rate p and with rate q, where $p < q$. Intuitively, it seems obvious that we will flip more bits when using the higher rate q. We could make this precise by looking at the distributions of the random variables X_p and X_q describing the numbers of bits that flip and computing that $X_p \preceq X_q$. For that, we would need to show that for all $k \in [0..n]$, we have

$$\sum_{i=0}^{k} \binom{n}{i} p^i (1-p)^{n-i} \geq \sum_{i=0}^{k} \binom{n}{i} q^i (1-q)^{n-i}.$$

Coupling is a way to get the same result in a more natural manner.

Consider the following random experiment. For each $i \in [1..n]$, let r_i be a random number chosen independently and uniformly distributed in $[0,1]$. Let $\tilde{X}_p$ be the number of the r_i that are less than p and let X_q be the number of the r_i that are less than q. We immediately see that $\tilde{X}_p \sim \text{Bin}(n,p)$ and $\tilde{X}_q \sim \text{Bin}(n,q)$. However, we know more. We have defined $\tilde{X}_p$ and $\tilde{X}_q$ over a common probability space in such a way that we have $\tilde{X}_p \le \tilde{X}_q$ with probability one: X_p and X_q, viewed as functions on the (hidden) probability space $\Omega = \{(r_1,\dots,r_n) \mid r_1,\dots,r_n \in [0,1]\}$, satisfy $\tilde{X}_p(\omega) \le \tilde{X}_q(\omega)$ for all $\omega \in \Omega$. Consequently, by the trivial Lemma 1.8.5, we have $X_p \preceq \tilde{X}_p \preceq \tilde{X}_q \preceq X_q$ and hence $X_p \preceq X_q$.

The same argument works for geometric distributions. We summarize these findings (and two more) in the following lemma. Part (b) follows from the obvious embedding (which is a coupling as well) of the $\text{Bin}(n,p)$ probability space into that of $\text{Bin}(m,p)$. The first inequality of part (c) is easily computed directly from the definition of domination (and holds in fact for all random variables); the second part was proven in [63, Lemma 1].

Lemma 1.8.9. *Let X and Y be two random variables. Let $p,q \in [0,1]$ with $p \le q$.*

(a) If $X \sim \text{Bin}(n,p)$ and $Y \sim \text{Bin}(n,q)$, then $X \preceq Y$.
(b) If $n \le m$, $X \sim \text{Bin}(n,p)$, and $Y \sim \text{Bin}(m,p)$, then $X \preceq Y$.
(c) If $X \sim \text{Bin}(n,p)$ and $x \in [0..n]$, then $X \preceq (X \mid X \ge x) \preceq (X+x)$.
(d) If $p > 0$, $X \sim \text{Geom}(p)$, and $Y \sim \text{Geom}(q)$, then $X \preceq Y$.

Let us now formally define what we mean by coupling. Let X and Y be two random variables, not necessarily defined over the same probability space. We say that $(\tilde{X},\tilde{Y})$ is a *coupling* of (X,Y) if $\tilde{X}$ and $\tilde{Y}$ are defined over a common probability space and if X and X' as well as Y and Y' are identically distributed.

This definition itself is very weak. (X,Y) have many couplings and most of them are not interesting. So, the art of using coupling as a proof and analysis technique is to find a coupling of (X,Y) that allows one to derive some useful information.

It is not a coincidence that we could use coupling to prove stochastic domination. The following theorem is well known.

Theorem 1.8.10. *Let X and Y be two random variables. Then the following two statements are equivalent.*

(a) $X \preceq Y$.
(b) There is a coupling $(\tilde{X},\tilde{Y})$ of (X,Y) such that $\tilde{X} \le \tilde{Y}$.

We remark, without giving much detail, that coupling as a proof technique has found numerous powerful applications beyond its connection to stochastic domination. In the analysis of population-based evolutionary algorithms, a powerful strategy to prove lower bounds is to couple the true population of the algorithm with the population of an artificial process without selection, and by this overcome the difficult dependencies introduced by the variation–selection cycle of the algorithm. This was first done in [99, 100] for the analysis of the $(\mu+1)$ EA and an elitist steady-state genetic algorithm. This technique then found applications for memetic algorithms [91], aging mechanisms [59], non-elitist algorithms [68], multi-objective evolutionary algorithms [39], and the $(\mu+\lambda)$ EA [1].

1.8.5 Domination in Fitness or Distance

So far, we have used stochastic domination to compare runtime distributions. We now show that stochastic domination is a powerful proof tool also when applied to other distributions. To do so, we give a short and elegant proof of a result of Witt [101] that compares the runtimes of mutation-based algorithms. The main reason why our proof is significantly shorter than that of Witt is that we use the notion of stochastic domination for the distance from the optimum also. This will also be an example where we exploit heavily the connection between coupling and stochastic domination (Theorem 1.8.10).

To state this result, we need the notion of a (μ,p) *mutation-based algorithm* introduced in [92]. This class of algorithms is called only *mutation-based* in [92], but since (i) it does not include all adaptive algorithms using mutation only, for example, those considered in [3, 10, 17, 28, 44, 58, 78], (ii) it does not include all algorithms using a different mutation operator than standard bit mutation, for example, those in [24, 25, 41, 69], and (iii) this notion collides with the notion of unary unbiased black-box complexity algorithms (see [66]), which without greater justification could also be called the class of mutation-based algorithms, we feel that a notion making these restrictions precise is more appropriate.

The class of (μ,p) mutation-based algorithms comprises all algorithms which first generate a set of μ search points uniformly and independently at random from $\{0,1\}^n$ and then repeat generating new search points from any of the previous ones via standard bit mutation with probability p. This class includes all $(\mu+\lambda)$ and (μ,λ) EAs which use only standard bit mutation with static mutation rate p.

We denote by $(1+1)$ EA_μ the following algorithm in this class. It first generates μ random search points. From these, it selects uniformly at random one with highest fitness and then continues from this search point like the $(1+1)$ EA, that is, it repeatedly generates a new search point from the current one via standard bit mutation with rate p and replaces the previous search point with the new one if the new one is not worse (in terms of fitness). This algorithm was called "$(1+1)$ EA with BestOf(μ) initialization" in [64].

For any algorithm $\mathcal{A}$ from the class of (μ,p) mutation-based algorithms and any fitness function $f:\{0,1\}^n \to \mathbb{R}$, let us denote by $T(\mathcal{A},f)$ the runtime of the algorithm $\mathcal{A}$ on the fitness function f, that is, the number of the first individual generated that is an optimal solution. Usually, this will be μ plus the number of the iteration in which the optimum was generated. To cover also the case where one of the random initial individuals is optimal, let us assume that these initial individuals are generated sequentially. As a final technicality, for reasons of convenience, let us assume that the $(1+1)$ EA_μ, in iteration $\mu+1$, does not choose as parent a random previous search point with maximal fitness, but the last one with maximal fitness. Since the initial μ individuals are generated independently, this modification does not change the distribution of this parent.

In this language, Witt [101, Theorem 6.2] showed the following remarkable result.

Theorem 1.8.11. *For any* (μ,p) *mutation-based algorithm* $\mathcal{A}$ *and any* $f : \{0,1\}^n \to \mathbb{R}$ *with unique global optimum,*

$$T((1+1)\ EA_\mu, \textsc{OneMax}) \preceq T(\mathcal{A}, f).$$

This result significantly extends results of a similar flavor in [9, 38, 92]. The importance of such types of result is that they allow one to prove lower bounds for the performance of many algorithms on essentially arbitrary fitness functions by considering just the performance of the $(1+1)$ EA_μ on OneMax.

Let us denote by $|x|_1$ the number of ones in the bit string $x \in \{0,1\}^n$. In other words, $|x|_1 = \|x\|_1$, but the former is nicer to read. Witt [101, Lemma 6.1] has shown the following natural domination relation between offspring generated via standard bit mutation.

Lemma 1.8.12. *Let* $x, y \in \{0,1\}^n$. *Let* $p \in [0, \frac{1}{2}]$. *Let* x', y' *be obtained from* x, y *via standard bit mutation with rate* p. *If* $|x|_1 \le |y|_1$, *then* $|x'|_1 \preceq |y'|_1$.

We are now ready to give our alternate proof of Theorem 1.8.11. While it is clearly shorter than the original one in [101], we also feel that it is more natural. In very simple words, it shows that $T(\mathcal{A}, f)$ dominates $T((1+1)\ \mathrm{EA}_\mu, \textsc{OneMax})$ because the search points generated in a run of the $(1+1)$ EA_μ on OneMax are always at least as close to the optimum (in the domination or coupling sense) as those in a run of $\mathcal{A}$ on f.

Proof. Since $\mathcal{A}$ treats bit positions and bit values in a symmetric fashion, we may assume without loss of generality that the unique optimum of f is $(1,\ldots,1)$.

Let $x^{(1)}, x^{(2)}, \ldots$ be the sequence of search points generated in a run of $\mathcal{A}$ on the fitness function f. Hence $x^{(1)}, \ldots, x^{(\mu)}$ are independently and uniformly distributed in $\{0,1\}^n$ and all subsequent search points are generated from suitably chosen previous ones via standard bit mutation with rate p. Let $y^{(1)}, y^{(2)}, \ldots$ be the sequence of search points generated in a run of the $(1+1)$ EA_μ on the fitness function OneMax.

We now show how to couple these random sequences of search points in such a way that $|\tilde{x}^{(t)}|_1 \le |\tilde{y}^{(t)}|_1$ for all $t \in \mathbb{N}$. We take as the common probability space Ω simply the space that $(x^{(t)})_{t \in \mathbb{N}}$ is defined on and let $\tilde{x}^{(t)} = x^{(t)}$ for all $t \in \mathbb{N}$.

We define the $\tilde{y}^{(t)}$ inductively as follows. For $t \in [1..\mu]$, let $\tilde{y}^{(t)} = x^{(t)}$. Note that this trivially implies $|\tilde{x}^{(t)}|_1 \le |\tilde{y}^{(t)}|_1$ for these search points. Let $t > \mu$ and assume that $|\tilde{x}^{(t')}|_1 \le |\tilde{y}^{(t')}|_1$ for all $t' < t$. Let $s \in [1..t-1]$ be maximal such that $\tilde{y}^{(s)}$ has the maximal OneMax-fitness among $\tilde{y}^{(1)}, \ldots, \tilde{y}^{(t-1)}$. Let $r \in [1..t-1]$ be such that $x^{(t)}$ was generated from $x^{(r)}$ in the run of $\mathcal{A}$ on f. By induction, we have $|x^{(r)}|_1 \le |\tilde{y}^{(r)}|_1$. By the choice of s, we have $|\tilde{y}^{(r)}|_1 \le |\tilde{y}^{(s)}|_1$. Consequently, we have $|x^{(r)}|_1 \le |\tilde{y}^{(s)}|_1$. By Lemma 1.8.12 and Theorem 1.8.10, there is a random $\tilde{y}^{(t)}$ (defined on Ω) such that $\tilde{y}^{(t)}$ has the distribution of being obtained from $\tilde{y}^{(s)}$ via standard bit mutation with rate p and such that $|x^{(t)}|_1 \le |\tilde{y}^{(t)}|_1$.

With this construction, the sequence $(\tilde{y}^{(t)})_{t \in \mathbb{N}}$ has the same distribution as $(y^{(t)})_{t \in \mathbb{N}}$. This is because the first μ elements are random and then each subsequent one is generated via standard bit mutation from the current best one, which is just the way the $(1+1)$ EA$_\mu$ is defined. At the same time, we have $|\tilde{x}^{(t)}|_1 \le |\tilde{y}^{(t)}|_1$ for all $t \in \mathbb{N}$. Consequently, we have $\min\{t \in \mathbb{N} \mid |\tilde{y}^{(t)}|_1 = n\} \le \min\{t \in \mathbb{N} \mid |x^{(t)}|_1 = n\}$. Since $T((1+1)\text{ EA}_\mu, \text{OneMax})$ and $\min\{t \in \mathbb{N} \mid |\tilde{y}^{(t)}|_1 = n\}$ are identically distributed and also $T(\mathcal{A}, f)$ and $\min\{t \in \mathbb{N} \mid |x^{(t)}|_1 = n\}$ are identically distributed, we have $T((1+1)\text{ EA}_\mu, \text{OneMax}) \preceq T(\mathcal{A}, f)$. □

While not explicitly using the notion of stochastic domination, the result and proof in [9] bear some similarity to those above. In very simple words and omitting many details, the result [9, Theorem 1] states the following. Assume that you run the $(1+1)$ EA and some other algorithm A (from a relatively large class of algorithms) to maximize a function f. Denote by $x^{(t)}$ and $y^{(t)}$ the best individuals produced by the $(1+1)$ EA and A up to iteration t. Assume that for all t and all possible runs of the algorithms up to iteration t we have that $f(x^{(t)}) \ge f(y^{(t)})$ implies $f(x^{(t+1)}) \succeq f(y^{(t+1)})$. Assume further that the random initial individual of the $(1+1)$ EA is at least as good (in terms of f) as all initial individuals of algorithm A. Then $f(x^{(t)}) \succeq f(y^{(t)})$ for all t.

The proof of this result (like that of the fitness domination statement in our proof of Theorem 1.8.11) uses induction over the time t. Since [9] does not use the notion of stochastic domination explicitly, there the two processes cannot simply be coupled, but instead the two distributions have to be compared using an argument called Abel transform.

1.9 The Coupon Collector Process

The *coupon collector process* is one of the central building blocks in the analysis of randomized algorithms. It is particularly important in the theory of randomized search heuristics, where it often appears as a subprocess.

The coupon collector process is the following simple randomized process. Assume that there are n types of coupons available. Whenever you buy a certain product, you get one coupon of a type chosen uniformly at random from the n types. How long does it take until you have a coupon of each type? In this section, we denote the random variable describing the first round after which we have all types by T_n and call it the *coupon-collecting time*. In simple words, this is the number of rounds it takes to obtain all types.

As an easy example showing how the coupon collector problem arises in the theory of randomized search heuristics, let us regard how the *randomized local search* heuristic (RLS) optimizes strictly monotonically increasing functions. The RLS heuristic, when maximizing a given function $f : \{0,1\}^n \to \mathbb{R}$, starts with a random search point. Then, in each iteration of the process, a single random bit is flipped in the current solution. If this gives a solution worse than the current one (in terms of f), then the new solution is discarded. Otherwise, the process is continued from this new solution.

Assume that f is *strictly monotonically increasing*, that is, flipping any 0-bit to 1 increases the function value. Then the optimization process of RLS on f strongly resembles a coupon collector process. In each round, we flip a random bit. If this bit was 1 in our current solution, then nothing changes (we discard the new solution as it has a smaller f-value). If this bit was 0, then we keep the new solution, which now has one extra 1. Hence, taking the 1-bits as coupons, we obtain a random coupon in each round. This has no effect if we have this coupon already, but is good if we do not.

We observe that the optimization time (the number of solutions evaluated until the optimal solution is found) of RLS on strictly monotonic functions is exactly the coupon-collecting time when we start with an initial stake of coupons that follows a $\mathrm{Bin}(n, \frac{1}{2})$ distribution. This shows that the optimization time is at most the ordinary coupon-collector time (where we start with no coupons). See [22] for a very precise analysis of this process.

The expectation of the coupon-collecting time is easy to determine. Recall from Section 1.4.2 the definition of the harmonic number $H_n := \sum_{k=1}^{n} \frac{1}{k}$.

Theorem 1.9.1 (coupon collector, expectation). *The expected time to collect all n coupons is $E[T_n] = nH_n = (1+o(1))n \ln n$.*

Proof. Given that we already have k different coupons for some $k \in [0..n-1]$, the probability that the next coupon is one that we do not already have is $\frac{n-k}{n}$. By the waiting-time argument (Lemma 1.6.3), we see that the time $T_{n,k}$ needed to obtain a new coupon, given that we have exactly k different ones, satisfies $E[T_{n,k}] = \frac{n}{n-k}$. Clearly, the total time T_n needed to obtain all coupons is $\sum_{k=0}^{n-1} T_{n,k}$. Hence, by linearity of expectation (Lemma 1.6.4), $E[T_n] = \sum_{k=0}^{n-1} E[T_{n,k}] = nH_n$. □

We proceed by trying to gain more information about T_n than just the expectation. The tools discussed so far (and one to come in a later section) lead to the following results.

- Markov's inequality (Lemma 1.6.7) gives $\Pr[T_n \geq \lambda n H_n] \leq \frac{1}{\lambda}$ for all $\lambda \geq 1$.
- Chebyshev's inequality (Lemma 1.6.9) can be used to prove $\Pr[|T_n - nH_n| \geq \varepsilon n] \leq \frac{\pi^2}{6\varepsilon^2}$ for all $\varepsilon \geq \frac{6}{\pi^2} \approx 0.6079$. This builds on the fact (implicit in the proof above) that the coupon-collecting time is the sum of independent geometric random variables $T_n = \sum_{k=0}^{n-1} \text{Geom}(\frac{n-k}{n})$. Hence the variance is $\text{Var}[T_n] = \frac{\pi^2 n^2}{6}$.
- Again exploiting $T_n = \sum_{k=0}^{n-1} \text{Geom}(\frac{n-k}{n})$, Witt's Chernoff bound for geometric random variables (Theorem 1.10.34) gives

$$\Pr[T_n \geq E[T_n] + \varepsilon n] \leq \begin{cases} \exp(-\frac{3\varepsilon^2}{\pi^2}) & \text{if } \varepsilon \leq \frac{\pi^2}{6}, \\ \exp(-\frac{\varepsilon}{4}) & \text{if } \varepsilon > \frac{\pi^2}{6}, \end{cases}$$
$$\Pr[T_n \leq E[T_n] - \varepsilon n] \leq \exp(-\tfrac{3\varepsilon^2}{\pi^2})$$

  for all $\varepsilon \geq 0$. See [102] for details.

Interestingly, asymptotically stronger tail bounds for T_n can be derived by fairly elementary means. The key idea is to consider not how the number of coupons increases over time, but instead the event that we miss a particular coupon for some period of time. Note that the probability that a particular coupon is not obtained in t rounds is $(1-\frac{1}{n})^t$. By a union bound argument (see Lemma 1.5.1), the probability that there is a coupon that is not obtained within t rounds, and, equivalently, that $T_n > t$, satisfies

$$\Pr[T_n > t] \leq n(1-\tfrac{1}{n})^t.$$

Using the simple estimate of Lemma 1.4.1, we obtain the following (equivalent) bounds.

Theorem 1.9.2 (coupon collector, upper tail). *For all* $\varepsilon \geq 0$,

$$\Pr[T_n \geq (1+\varepsilon) n \ln n] \leq n^{-\varepsilon}, \tag{1.9.1}$$
$$\Pr[T_n \geq n \ln n + \varepsilon n] \leq \exp(-\varepsilon). \tag{1.9.2}$$

Surprisingly, prior to the following result from [18], no good lower bound for the coupon-collecting time was published.

Theorem 1.9.3 (coupon collector, lower tail). *For all* $\varepsilon \geq 0$,

$$\Pr[T_n \leq (1-\varepsilon)(n-1) \ln n] \leq \exp(-n^{\varepsilon}), \tag{1.9.3}$$
$$\Pr[T_n \leq (n-1) \ln n - \varepsilon(n-1)] \leq \exp(-e^{\varepsilon}). \tag{1.9.4}$$

Theorem 1.9.3 was proven in [18] by showing that the events of having a coupon after a certain time are 1-negatively correlated. The following proof defers this task to Lemma 1.10.26.

Proof. Let $t=(1-\varepsilon)(n-1)\ln n$. For $i \in [1..n]$, let X_i be the indicator random variable for the event that a coupon of type i is obtained within the first t rounds. Then $\Pr[X_i=1]=1-(1-\frac{1}{n})^t \le 1-\exp(-(1-\varepsilon)\ln n)=1-n^{-1+\varepsilon}$, where the estimate follows from Corollary 1.4.6.

Since, in the coupon collector process, in each round j we choose a random set S_j of cardinality 1, by Lemma 1.10.26 the X_i are 1-negatively correlated. Consequently,

$$\begin{aligned}\Pr[T_n \le (1-\varepsilon)(n-1)\ln n] &= \Pr[\forall i \in [1..n] : X_i = 1]\\ &\le \prod_{i=1}^{n} \Pr[X_i=1]\\ &\le (1-n^{-1+\varepsilon})^n \le \exp(-n^{\varepsilon})\end{aligned}$$

by Lemma 1.4.1. □

We may remark that a good mathematical understanding of the coupon collector process is important not only because such processes directly show up in some randomized algorithms, but also because it might give us the right intuitive understanding of other processes. Consider, for example, a run of the $(1+1)$ EA on some pseudo-Boolean function $f:\{0,1\}^n \to \mathbb{R}$ with a unique global maximum.

The following intuitive consideration leads us to believe that the $(1+1)$ EA, with high probability, needs at least roughly $n\ln\frac{n}{2}$ iterations to find the optimum of f: By the strong concentration of the binomial distribution, the initial search point differs in at least roughly $\frac{n}{2}$ bits from the global optimum. To find the global optimum, it is necessary (but clearly not sufficient) that each of these missing bits is flipped at least once in some mutation step. Now that the $(1+1)$ EA on average flips one bit per iteration, this looks like a coupon collector process started with an initial stake of $\frac{n}{2}$ coupons, so we expect to need at least roughly $n\ln\frac{n}{2}$ iterations to perform the $n\ln\frac{n}{2}$ bit flips necessary to have each missing bit flipped at least once. Clearly, this argument is not rigorous, but it suggests the right answer to us.

Theorem 1.9.4. *The optimization time T of the* $(1+1)$ *EA on any function* $f:\{0,1\}^n \to \mathbb{R}$ *with a unique global maximum satisfies*

$$\Pr[T \le (1-\varepsilon)(n-1)\ln\tfrac{n}{2}] \le \exp(-n^{\varepsilon}).$$

Proof. By symmetry, we may assume that the unique global optimum of f is $(1,\dots,1)$. Let $t=(1-\varepsilon)(n-1)\ln\frac{n}{2}$. For all $i \in [1..n]$, let Y_i denote the event that the i-th bit is zero in the initial search point and is not flipped in any application of the mutation operator in the first t iterations. Let $X_i = 1-Y_i$.

Then $\Pr[X_i = 1] = 1 - \frac{1}{2}(1-\frac{1}{n})^t \le 1 - n^{-1+\varepsilon}$. The events X_i are independent, so we compute

$$\begin{aligned}\Pr[T \le t] &\le \Pr[\forall i \in [1..n] : X_i = 1] \\ &= \prod_{i=1}^{n} \Pr[X_i = 1] \\ &= (1 - n^{-1+\varepsilon})^n = \exp(-n^{\varepsilon}).\end{aligned}$$

□

We have stated the above theorem to give a simple example of how understanding the coupon collector process can help also in understanding randomized search heuristics that do not directly simulate a coupon collecting process. We remark that the theorem above is not the best possible; in particular, it does not rule out an expected optimization time of $n \ln \frac{n}{2}$. In contrast, it is known that the optimization time of the (1+1) EA on the ONEMAX function is $E[T] \ge en \ln n - O(n)$ [27], improving on the minimally weaker bound $E[T] \ge en \ln n - O(n \log \log n)$ from, independently, [26] and [92]. By Theorem 1.8.11, this lower bound holds for the performance of the (1+1) EA on any function $f : \{0,1\}^n \to \mathbb{R}$ with a unique optimum.

1.10 Large-Deviation Bounds

Often, not only we are interested in the expectation of some random variable, but we also need a bound that holds with high probability. We have seen in the proof of Theorem 1.5.3 that such high-probability statements can be very useful: if a certain bad event occurs in each iteration with a very small probability only, then a simple union bound is enough to argue that this event is unlikely to occur even over a large number of iterations. The better the original high-probability statement is, the more iterations we can cover. For this reason, the tools discussed in this chapter are among those most often employed in the theory of randomized search heuristics.

Since computing the expectation is often easy, a very common approach is to first compute the expectation of a random variable and then bound the probability that the random variable deviates from this expectation by too large an amount. The tools for this second step are called *tail inequalities* or *large-deviation inequalities*, and this is the topic of this section. In a sense, Markov's and Chebyshev's inequalities, discussed in Section 1.6, can be seen as large-deviation inequalities as well, but usually the term is reserved for exponential tail bounds.

A large number of large-deviation bounds have been developed in the past. They differ in the situations they are applicable to, and also in their sharpness. Often, the sharpest bounds give expressions for the tail probability that are very difficult to work with. Hence some experience is needed to choose a tail bound that is not overly complicated, but sharp enough to give the desired result.

To give the novice to this topic some orientation, here is a short list of results that are particularly useful and which are sufficient in many situations.

(a) The simple multiplicative Chernoff bounds (1.10.5) and (1.10.12), showing that for sums of independent $[0, 1]$ random variables, a constant-factor deviation from the expectation occurs only with a probability negatively exponential in the expectation.
(b) The additive Chernoff bound of Theorem 1.10.7, showing that a sum of n independent $[0, 1]$ random variables deviates from the expectation by more than an additive term of λ only with probability $\exp(-2\lambda^2/n)$.
(c) The fact that essentially all large-deviation bounds can be used also with a pessimistic estimate for the expectation instead of the precise expectation (Section 1.10.1.8).
(d) The method of bounded differences (Theorem 1.10.27), which states that the additive Chernoff bounds remain valid if X is functionally dependent on independent random variables each having a small influence on X.

For the experienced reader, the following results may be interesting as they go beyond what most introductions to tail bounds cover.

(a) In Section 1.10.2.2, we show that essentially all of the large-deviation bounds usually stated for sums of independent random variables are also valid for negatively correlated random variables. An important application of this result is to distributions arising from sampling without replacement or with partial replacement.
(b) In Section 1.10.4, we present a number of large-deviation bounds for sums of independent geometrically distributed random variables. These seem to be particularly useful in the analysis of randomized search heuristics, whereas they are rarely used with classic randomized algorithms.
(c) In Theorem 1.10.28, we present a version of the bounded-differences method which requires only that the t-th random variable has a bounded influence on the *expected* outcome resulting from variables $t+1$ to n. This is much weaker than the common bounded-differences assumption that each random variable, regardless of how we condition on the remaining variables, has a bounded influence on the result. We feel that this new version (which is an easy consequence of known results) may be very useful in the analysis of iterative improvement heuristics. In particular, it may lead to elementary proofs for results which so far can only be proven via tail bounds for martingales.

1.10.1 Chernoff Bounds for Sums of Independent Bounded Random Variables

In this rather long subsection, we assume that our random variable of interest is the sum of n independent random variables, each taking values in some bounded range, often $[0,1]$. While some textbooks present these bounds for discrete random variables, e.g., taking the values 0 and 1 only, all the results are true without this restriction.

The bounds presented below are all known under names such as *Chernoff* or *Hoeffding* bounds, referring to the seminal papers by Chernoff [13] and Hoeffding [54]. Since the first bounds of this type were proven by Bernstein [6] – via the so-called exponential moments method that is used in essentially all proofs of such results (see Section 1.10.1.7) – the name "Bernstein inequalities" would be more appropriate. We shall not be that precise, and instead use the most common name "Chernoff inequalities" for all such bounds.

For the reader's convenience, as in the remainder of this chapter, we shall not be shy to write out minor reformulations of some results. We believe that it helps a lot to have seen such reformulations and we think that it is convenient, both for using the bounds and for referring to them, if all natural versions are visible in the text.

1.10.1.1 Multiplicative Chernoff Bounds for the Upper Tail

The multiplicative Chernoff bounds presented in this and the next section bound the probability of deviating from the expectation by at least a given factor. Since in many algorithm analyses we are interested only in the asymptotic order of magnitude of some quantity, a constant-factor deviation can be easily tolerated, and knowing that larger deviations are very unlikely is just what we want to know. For this reason, the multiplicative Chernoff bounds are often the right tool.

The following theorem collects a number of bounds for the upper tail, that is, for deviations above the expectation. Some of the bounds are visualized in Figure 1.3.

Theorem 1.10.1. *Let $X_1,\ldots,X_n$ be independent random variables taking values in $[0,1]$. Let $X=\sum_{i=1}^n X_i$. Let $\delta \ge 0$. Then*

$$\Pr[X \ge (1+\delta)E[X]]$$

$$\le \left(\frac{1}{1+\delta}\right)^{(1+\delta)E[X]} \left(\frac{n-E[X]}{n-(1+\delta)E[X]}\right)^{n-(1+\delta)E[X]} \tag{1.10.1}$$

$$\le \left(\frac{e^\delta}{(1+\delta)^{1+\delta}}\right)^{E[X]} = \exp(-((1+\delta)\ln(1+\delta)-\delta)E[X]) \tag{1.10.2}$$

$$\leq \exp\left(-\frac{\delta^2 E[X]}{2+\frac{2}{3}\delta}\right) \tag{1.10.3}$$

$$\leq \exp\left(-\frac{\min\{\delta^2,\delta\}E[X]}{3}\right), \tag{1.10.4}$$

where the bound in (1.10.1) *is read as* 0 *for* $\delta > \frac{n-E[X]}{E[X]}$ *and as* $(\frac{E[X]}{n})^n$ *for* $\delta = \frac{n-E[X]}{E[X]}$. *For* $\delta \leq 1$, (1.10.4) *simplifies to*

$$\Pr[X \geq (1+\delta)E[X]] \leq \exp\left(-\frac{\delta^2 E[X]}{3}\right). \tag{1.10.5}$$

The first and strongest bound (1.10.1) was first stated explicitly by Hoeffding [54]. It improves over Chernoff's [13] tail bounds in particular by not requiring that the X_i are identically distributed. Hoeffding also showed that (1.10.1) is the best bound that can be shown via the exponential moments methods under the assumptions of Theorem 1.10.1.

For $E[X]$ small, say $E[X] = o(n)$ when taking a view asymptotic in $n \to \infty$, the second bound (1.10.2) is easier to use, but essentially as strong as (1.10.1). More precisely, it is larger by only a factor of $(1+o(1))^{E[X]}$, since we have estimated

$$\left(\frac{n-E[X]}{n-(1+\delta)E[X]}\right)^{n-(1+\delta)E[X]} = \left(1+\frac{\delta E[X]}{n-(1+\delta)E[X]}\right)^{n-(1+\delta)E[X]}$$
$$\leq e^{\delta E[X]} \tag{1.10.6}$$

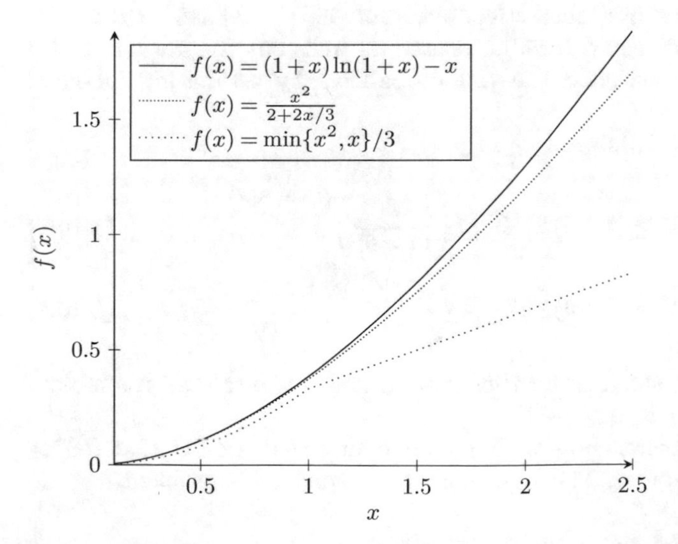

Fig. 1.3 Visual comparison of the bounds (1.10.2), (1.10.3), and (1.10.4). The term $f(x)$ leading to the bound $\Pr[X \geq (1+x)E[X]] \leq \exp(-f(x)E[X])$ is depicted.

using Lemma 1.4.1.

Equation (1.10.3) is derived from (1.10.2) by noting that $(1+\delta)\ln(1+\delta)-\delta \geq \frac{3\delta^2}{6+2\delta}$ holds for all $\delta \geq 0$; see Theorem 2.3 and Lemma 2.4 in McDiarmid [70]. Equations (1.10.4) and (1.10.5) are trivial simplifications of (1.10.3).

In general, to successfully use Chernoff bounds in one's research, it greatly helps to look a little behind the formulas and understand their meaning. Very roughly speaking, we can distinguish three different regimes relative to δ, namely that the tail probability is of order $\exp(-\Theta(\delta\log(\delta)E[X]))$, $\exp(-\Theta(\delta E[X]))$, and $\exp(-\Theta(\delta^2 E[X]))$. Here, in principle, the middle regime, referring to the case of δ constant, could be seen as a subcase of either of the other two regimes. Since this case of constant-factor deviations from the expectation occurs very frequently, however, we discuss it separately.

Superexponential Regime

Equation (1.10.2) shows a tail bound of order $\delta^{-\Theta(\delta E[X])} = \exp(-\Theta(\delta\log(\delta)E[X]))$, where the asymptotics are for $\delta \to \infty$. In this regime, the deviation $\delta E[X]$ from the expectation $E[X]$ is much larger than the expectation itself. It is not very often that we need to analyze such large deviations, so this Chernoff bound is rarely used. It can be useful in the analysis of evolutionary algorithms with larger offspring populations, where the most extreme behavior among the offspring can deviate significantly from the expected behavior. See [23, 40] for examples of how to use Chernoff bounds for the large deviations occurring in the analysis of the $(1+\lambda)$ EA. Note that in [56], the first theoretical study of the $(1+\lambda)$ EA, and in [51] such Chernoff bounds could have been used as well, but the authors found it easier to directly estimate the tail probability by estimating binomial coefficients.

Weaker forms of (1.10.2) are

$$\Pr[X \geq (1+\delta)E[X]] \leq \left(\frac{e}{(1+\delta)}\right)^{(1+\delta)E[X]}, \tag{1.10.7}$$

$$\Pr[X \geq (1+\delta)E[X]] \leq \left(\frac{e}{\delta}\right)^{\delta E[X]}, \tag{1.10.8}$$

where the first one is stronger for those values of δ where the tail probability is less than one (that is, $\delta > e-1$).

It is not totally obvious how to find a value for δ that ensures that $(\frac{e}{\delta})^\delta$ is less than a desired bound. The following lemma solves this problem.

Lemma 1.10.2. *Let* $t \geq e^{e^{1/e}} \approx 4.24044349\ldots$. *Let*

$$\delta = \frac{\ln t}{\ln(\frac{\ln t}{e \ln\ln t})}.$$

Then $(\frac{e}{\delta})^\delta \le \frac{1}{t}$.

Proof. We compute $\delta \ln \frac{\delta}{e} = \delta \ln\big(\frac{\ln t}{e\ln(\frac{\ln t}{e\ln\ln t})}\big) \ge \delta \ln(\frac{\ln t}{e \ln\ln t}) = \ln t$. □

We can use this estimate to bound the number of bits flipped in an application of the standard bit mutation operator defined in Lemma 1.6.6. By linearity of expectation, it is clear that the expected Hamming distance $H(x,y)$ between a parent x and an offspring y is $E[H(x,y)] = \alpha$ when the mutation rate is $\frac{\alpha}{n}$; see Lemma 1.6.6. Using Chernoff bounds, we now give an upper bound on how far we can exceed this value. Such arguments are often useful in the analysis of evolutionary algorithms; see, e.g., Lemma 26 in [40] for an example.

Lemma 1.10.3. *(a) Let* $x \in \{0,1\}^n$ *and* y *be obtained from* x *via standard bit mutation with mutation rate* $\frac{\alpha}{n}$. *Then* $\Pr[H(x,y) \ge k] \le (\frac{e\alpha}{k})^k$.
(b) Let $0 < p \le \exp(-\alpha \exp(\frac{1}{e}))$. *Let*

$$k \ge k_p := \frac{\ln(1/p)}{\ln\left(\frac{\ln(1/p^{1/\alpha})}{e \ln\ln(1/p^{1/\alpha})}\right)}.$$

Then $\Pr[H(x,y) \ge k] \le p$.
(c) Let $T \in \mathbb{N}$ *and* $0 < p \le \frac{1}{T}\exp(-\alpha\exp(\frac{1}{e}))$. *Let* $y_1,\dots,y_T$ *be obtained from* $x_1,\dots,x_T$, *respectively, via standard bit mutation. Let*

$$k \ge \frac{\ln(T/p)}{\ln\left(\frac{\ln((T/p)^{1/\alpha})}{e\ln\ln((T/p)^{1/\alpha})}\right)}.$$

Then $\Pr[\exists i \in [1..T] : H(x_i,y_i) \ge k] \le p$.

Proof. Note that $H(x,y) \sim \mathrm{Bin}(n, \frac{\alpha}{n})$, and hence $H(x,y)$ can be written as a sum of n independent random variables $X_1,\dots,X_n$ with $\Pr[X_i = 1] = \frac{\alpha}{n}$ and $\Pr[X_i = 0] = 1 - \frac{\alpha}{n}$ for all $i \in [1..n]$. Since $E[H(x,y)] = \alpha$, we can apply (1.10.7) with $(\delta + 1) = \frac{k}{\alpha}$. This proves (a).

For part (b), we use part (a) and Lemma 1.10.2 and compute $\Pr[H(x,y) \ge k] \le \Pr[H(x,y) \ge k_p] \le ((\frac{e}{k_p/\alpha})^{k_p/\alpha})^\alpha \le (p^{1/\alpha})^\alpha = p$. Similarly, for (c) we obtain $\Pr[H(x_i,y_i) \ge k] \le \frac{p}{T}$ and use the union bound (Lemma 1.5.1). □

Observe that the bounds in Lemma 1.10.3 are independent of n. Also, the bounds in parts (b) and (c) depend only mildly on α. By applying part (c) with $p = n^{-c_1}$ and $T = n^{c_2}$, we see that the probability that an evolutionary algorithm using standard bit mutation with rate $\frac{\alpha}{n}$, where α is a constant, flips more than $(c_1 + c_2 + o(1))\frac{\ln n}{\ln\ln n}$ bits in any of the first n^{c_2} applications of the mutation operator is at most n^{-c_1}.

We gave the results above to demonstrate the use of Chernoff bounds for sums of independent bounded random variables. Since the number of bits that are flipped in standard bit mutation follows a binomial distribution, similar bounds can also (and by more elementary arguments) be obtained from analyzing the binomial distribution. See Lemma 1.10.37 for an example.

Exponential Regime

When $\delta = \Theta(1)$, all bounds give a tail probability of order $\exp(-\Theta(\delta E[X]))$. Note that the difference between these bounds is often not very large. For $\delta = 1$, the bounds in (1.10.2), (1.10.3), and (1.10.4) become $(0.67957\ldots)^{E[X]}$, $(0.68728\ldots)^{E[X]}$, and $(0.71653\ldots)^{E[X]}$, respectively. So there is often no reason to use the unwieldy equation (1.10.2).

We remark that also for large δ, where the bound (1.10.2) gives the better asymptotics $\exp(-\Theta(\delta \log(\delta) E[X]))$, one can, with the help of Section 1.10.1.8, resort to the easier-to-use bounds (1.10.3) and (1.10.4) when the additional logarithmic term is not needed. For example, when X is again the number of bits that flip in an application of the standard bit mutation operator with mutation rate $p = \frac{\alpha}{n}$, then, for all $c > 0$ and $n \in \mathbb{N}$ with $c \ln n \geq \alpha$, equation (1.10.4) with $E[X] \leq \mu^+ := c \ln n$ and the argument of Section 1.10.1.8 give $\Pr[X \geq 2c \ln n] = \Pr[X \geq (1+1)\mu^+] \leq \exp(-\frac{1}{3}\mu^+) = n^{-c/3}$, which in many applications is fully sufficient.

A different way of stating an $\exp(-\Theta(\delta E[X]))$ tail bound, following directly from applying (1.10.7) for $\delta \geq 2e - 1$, is the following.

Corollary 1.10.4. *Under the assumptions of Theorem 1.10.1, we have*

$$\Pr[X \geq k] \leq 2^{-k} \tag{1.10.9}$$

for all $k \geq 2eE[X]$.

Sub-exponential Regime

Since Chernoff bounds give very low probabilities for the tail events, we can often work with $\delta = o(1)$ and still obtain sufficiently low probabilities for the deviations. Therefore, this regime occurs frequently in the analysis of randomized search heuristics. Since the tail probability is of order $\exp(-\Theta(\delta^2 E[X]))$, we need δ to be at least of order $(E[X])^{-1/2}$ to obtain useful statements. Note that for $E[X]$ close to $\frac{n}{2}$, Theorem 1.10.7 below gives slightly stronger bounds. A typical application in this regime is showing that the random initial search points of an algorithm with high probability all have a Hamming distance of at least $\frac{n}{2}(1 - o(1))$ from the optimum. See Lemma 1.10.8 below for further details.

1.10.1.2 Multiplicative Chernoff Bounds for the Lower Tail

In principle, of course, there is no difference between bounds for the upper and lower tails. If, in the situation of Theorem 1.10.1, we set $Y_i := 1 - X_i$, then the Y_i are independent random variables taking values in $[0,1]$, and any upper tail bound for X turns into a lower tail bound for $Y := \sum_{i=1}^n Y_i$ via $\Pr[Y \le t] = \Pr[X \ge n - t]$. However, since this transformation also changes the expectation, that is, $E[Y] = n - E[X]$, a convenient bound such as (1.10.5) becomes the cumbersome estimate $\Pr[Y \le (1-\delta)E[Y]] \le \exp(-\frac{1}{3}(1+\delta\frac{E[Y]}{n-E[Y]})^2(n-E[Y]))$.

For this reason, usually the tail bounds for the lower tail either are proven completely separately (but using similar ideas) or are derived by significantly simplifying the results stemming from applying the above symmetry argument to (1.10.1). Either approach can be used to show the following bounds. As a visible result of the asymmetry of the situation for upper and lower bounds, note the better constant of $\frac{1}{2}$ in the exponent of (1.10.12) as compared with the $\frac{1}{3}$ in (1.10.5). Two of the terms appearing in this result are visualized in Figure 1.4.

Theorem 1.10.5. *Let $X_1,\dots,X_n$ be independent random variables taking values in $[0,1]$. Let $X = \sum_{i=1}^n X_i$. Let $\delta \in [0,1]$. Then*

$$\Pr[X \le (1-\delta)E[X]]$$

$$\le \left(\frac{1}{1-\delta}\right)^{(1-\delta)E[X]} \left(\frac{n-E[X]}{n-(1-\delta)E[X]}\right)^{n-(1-\delta)E[X]} \tag{1.10.10}$$

$$\le \left(\frac{e^{-\delta}}{(1-\delta)^{1-\delta}}\right)^{E[X]} \tag{1.10.11}$$

$$\le \exp\left(-\frac{\delta^2 E[X]}{2}\right), \tag{1.10.12}$$

where the first bound reads as $(1-\frac{E[X]}{n})^n$ for $\delta = 1$.

For the not-so-interesting boundary cases, recall our definition $0^0 := 1$. The first bound (1.10.10) follows from (1.10.1) by regarding the random variables $Y_i := 1 - X_i$. Allowing the following easy derivation is maybe the main strength of (1.10.1). Setting $Y = \sum_{i=1}^n Y_i$ and $\delta' = \delta\frac{E[X]}{E[Y]}$, we compute

$$\begin{aligned}\Pr[X \le (1-\delta)E[X]] &= \Pr[Y \ge (1+\delta')E[Y]] \\ &\le \left(\frac{1}{1+\delta'}\right)^{(1+\delta')E[Y]} \left(\frac{n-E[Y]}{n-(1+\delta')E[Y]}\right)^{n-(1+\delta')E[Y]} \\ &= \left(\frac{n-E[X]}{n-(1-\delta)E[X]}\right)^{n-(1-\delta)E[X]} \left(\frac{1}{1-\delta}\right)^{(1-\delta)E[X]}.\end{aligned}$$

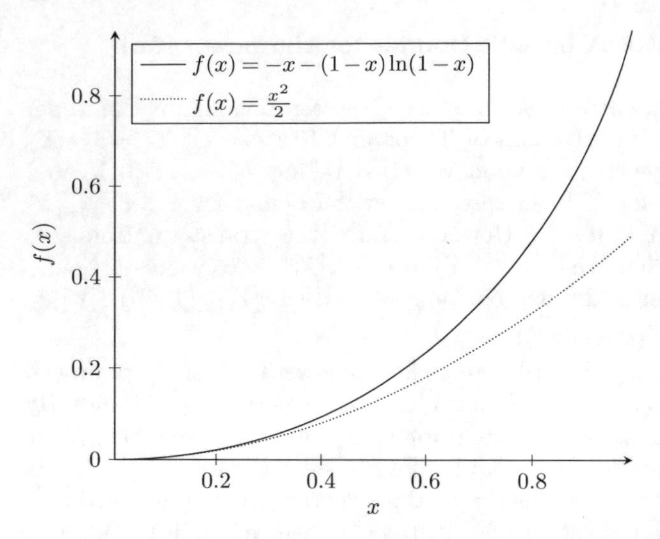

Fig. 1.4 Visual comparison of the bounds (1.10.11) and (1.10.12). The term $f(x)$ leading to the bound $\Pr[X \le (1-x)E[X]] \le \exp(-f(x)E[X])$ is depicted.

Obviously, in an analogous fashion, (1.10.1) can be derived from (1.10.10), so the two bounds are equivalent. Equation (1.10.11) follows from (1.10.10) using an elementary estimate analogous to (1.10.6). Equation (1.10.12) follows from (1.10.11) using elementary calculus; see, for example, the proof of Theorem 4.5 in [71].

Theorems 1.10.1 and 1.10.5 show in particular that constant-factor deviations from the expectation appear only with exponentially small probability.

Corollary 1.10.6. *Let $X_1, \ldots, X_n$ be independent random variables taking values in* $[0,1]$. *Let* $X = \sum_{i=1}^n X_i$. *Let* $\delta \in [0,1]$. *Then*

$$\Pr\left[|X - E[X]| \ge \delta E[X]\right] \le 2\exp\left(-\frac{\delta^2 E[X]}{3}\right).$$

1.10.1.3 Additive Chernoff Bounds

We now present a few bounds for the probability that a random variable deviates from its expectation by an additive term independent of the expectation. The advantage of such bounds is that they are identical for the upper and lower tails and that they are invariant under additive rescalings.

From (1.10.1) in Theorem 1.10.1, by careful estimates (see, e.g., Hoeffding [54]) and exploiting the obvious symmetry, we obtain the following estimates. As mentioned earlier, when $E[X]$ is close to $\frac{n}{2}$, this additive

Chernoff bound gives (slightly) stronger results than the simplified bounds of Theorems 1.10.1 and 1.10.5.

Theorem 1.10.7. *Let $X_1,\dots,X_n$ be independent random variables taking values in* $[0,1]$. *Let* $X=\sum_{i=1}^n X_i$. *Then, for all* $\lambda\ge 0$,

$$\Pr[X\ge E[X]+\lambda]\le \exp\left(-\frac{2\lambda^2}{n}\right), \tag{1.10.13}$$

$$\Pr[X\le E[X]-\lambda]\le \exp\left(-\frac{2\lambda^2}{n}\right). \tag{1.10.14}$$

A second advantage of additive Chernoff bounds is that they are often very easy to apply. As a typical application in evolutionary computation, let us consider the Hamming distance $H(x,x^*)$ of a random search point $x\in\{0,1\}$ from a given search point x^*. This could be, for example, the distance of a random initial solution from the optimum.

Lemma 1.10.8. *Let* $x^*\in\{0,1\}^n$. *Let* $x\in\{0,1\}^n$ *be chosen uniformly at random. Then, for all* $\lambda\ge 0$,

$$\Pr\left[\left|H(x,x^*)-\frac{n}{2}\right|\ge\lambda\right]\le 2\exp\left(-\frac{2\lambda^2}{n}\right).$$

Proof. Note that if $x\in\{0,1\}^n$ is uniformly distributed, then the x_i are independent random variables uniformly distributed in $\{0,1\}$. Hence, regardless of x^*, the indicator random variables X_i for the event that $x_i\ne x_i^*$ are also independent random variables uniformly distributed in $\{0,1\}$. Since $H(x,x^*)=\sum_{i=1}^n X_i$, the claim follows immediately from applying Theorem 1.10.7 to the events "$H(x,x^*)\ge E[H(x,x^*)]+\lambda$" and "$H(x,x^*)\le E[H(x,x^*)]-\lambda$". □

This lemma implies that even among a polynomial number of initial search points there is none which is closer to the optimum than $\frac{n}{2}-O(\sqrt{n\log n})$. This argument has been used numerous times in lower-bound proofs. This argument is also the reason why the best-known black-box algorithm for the optimization of OneMax, namely repeatedly sampling random search points until the fitness values observed determine the optimum, also works well for jump functions [12].

The following theorem, again due to Hoeffding [54], non-trivially extends Theorem 1.10.7 by allowing the X_i to take values in arbitrary intervals $[a_i,b_i]$.

Theorem 1.10.9. *Let $X_1,\dots,X_n$ be independent random variables. Assume that each X_i takes values in a real interval* $[a_i,b_i]$ *of length* $c_i:=b_i-a_i$. *Let* $X=\sum_{i=1}^n X_i$. *Then, for all* $\lambda>0$,

$$\Pr[X\ge E[X]+\lambda]\le \exp\left(-\frac{2\lambda^2}{\sum_{i=1}^n c_i^2}\right), \tag{1.10.15}$$

$$\Pr[X \leq E[X] - \lambda] \leq \exp\left(-\frac{2\lambda^2}{\sum_{i=1}^n c_i^2}\right). \tag{1.10.16}$$

For comparison, we now reformulate Theorems 1.10.1 and 1.10.5 as additive bounds. There is no greater intellectual challenge hidden here, but we feel that it helps to have seen these bounds at least once. Note that, since the resulting bounds depend on the expectation, we require that the X_i take values in $[0,1]$. In other words, unlike the bounds presented so far in this subsection, the following bounds are not invariant under additive rescaling and are not symmetric for upper and lower tails.

Theorem 1.10.10 (equivalent to Theorem 1.10.1). *Let $X_1,\ldots,X_n$ be independent random variables taking values in $[0,1]$. Let $X=\sum_{i=1}^n X_i$. Let $\lambda \geq 0$. Then*

$$\begin{aligned}
&\Pr(X \geq E[X]+\lambda)\\
&\leq \left(\frac{E[X]}{E[X]+\lambda}\right)^{E[X]+\lambda}\left(\frac{n-E[X]}{n-E[X]-\lambda}\right)^{n-E[X]-\lambda} &(1.10.17)\\
&\leq e^{\lambda}\left(\frac{E[X]}{E[X]+\lambda}\right)^{E[X]+\lambda} = \exp\left(-(E[X]+\lambda)\ln\left(1+\frac{\lambda}{E[X]}\right)+\lambda\right) &(1.10.18)\\
&\leq \exp\left(-\frac{\lambda^2}{2E[X]+\frac{2}{3}\lambda}\right) &(1.10.19)\\
&\leq \exp\left(-\frac{1}{3}\min\left\{\frac{\lambda^2}{E[X]},\lambda\right\}\right), &(1.10.20)
\end{aligned}$$

where the bound in (1.10.17) *is read as* 0 *for $\lambda > n-E[X]$ and as $(\frac{E[X]}{n})^n$ for $\lambda = n-E[X]$. For $\lambda \leq E[X]$, equation* (1.10.20) *simplifies to*

$$\Pr[X \geq E[X]+\lambda] \leq \exp\left(-\frac{\lambda^2}{3E[X]}\right). \tag{1.10.21}$$

Theorem 1.10.11 (equivalent to Theorem 1.10.5). *Let $X_1,\ldots,X_n$ be independent random variables taking values in $[0,1]$. Let $X=\sum_{i=1}^n X_i$. Let $\lambda \geq 0$. Then*

$$\begin{aligned}
\Pr[X \leq E[X]-\lambda] &\leq \left(\frac{E[X]}{E[X]-\lambda}\right)^{E[X]-\lambda}\left(\frac{n-E[X]}{n-E[X]+\lambda}\right)^{n-E[X]+\lambda} &(1.10.22)\\
&\leq e^{-\lambda}\left(\frac{E[X]}{E[X]-\lambda}\right)^{E[X]-\lambda} &(1.10.23)\\
&\leq \exp\left(-\frac{\lambda^2}{2E[X]}\right). &(1.10.24)
\end{aligned}$$

1.10.1.4 Chernoff Bounds Using the Variance

There are several versions of Chernoff bounds that take into account the variance. In certain situations, they can give significantly stronger bounds than the estimates discussed so far. Hoeffding [54] proved essentially the following result.

Theorem 1.10.12. *Let* $X_1,\dots,X_n$ *be independent random variables such that* $X_i \le E[X_i]+1$ *for all* $i=1,\dots,n$. *Let* $X=\sum_{i=1}^n X_i$. *Let* $\sigma^2 = \sum_{i=1}^n \mathrm{Var}[X_i] = \mathrm{Var}[X]$. *Then, for all* $\lambda \ge 0$,

$$\begin{aligned}
&\Pr[X \ge E[X]+\lambda] \\
&\le \left(\left(1+\frac{\lambda}{\sigma^2}\right)^{-\left(1+\frac{\lambda}{\sigma^2}\right)\frac{\sigma^2}{n+\sigma^2}}\left(1-\frac{\lambda}{n}\right)^{-\left(1-\frac{\lambda}{n}\right)\frac{n}{n+\sigma^2}}\right)^n && (1.10.25)\\
&\le \exp\left(-\lambda\left(\left(1+\frac{\sigma^2}{\lambda}\right)\ln\left(1+\frac{\lambda}{\sigma^2}\right)-1\right)\right) && \\
&= \exp\left(-\sigma^2\left(\left(1+\frac{\lambda}{\sigma^2}\right)\ln\left(1+\frac{\lambda}{\sigma^2}\right)-\frac{\lambda}{\sigma^2}\right)\right) && (1.10.26)\\
&\le \exp\left(-\frac{\lambda^2}{2\sigma^2+\frac{2}{3}\lambda}\right) && (1.10.27)\\
&\le \exp\left(-\frac{1}{3}\min\left\{\frac{\lambda^2}{\sigma^2},\lambda\right\}\right), && (1.10.28)
\end{aligned}$$

where (1.10.25) *is understood to mean* 0 *when* $\lambda > n$ *and* $(\frac{\sigma^2}{n+\sigma^2})^n$ *when* $\lambda = n$.

Obtaining (1.10.26) from (1.10.25) is non-trivial. This estimate can be found, for example, in Hoeffding [54]. From (1.10.26), we derive (1.10.27) in the same way as we derived (1.10.3) from (1.10.2).

By replacing X_i with $-X_i$, we obtain the analogous bounds for the lower tail.

Corollary 1.10.13. *If the condition* $X_i \le E[X_i]+1$ *in Theorem 1.10.12 is replaced by* $X_i \ge E[X_i]-1$, *then* $\Pr[X \le E[X]-\lambda]$ *satisfies the estimates* (1.10.25) *to* (1.10.28).

As discussed in Hoeffding [54], the bound (1.10.26) is the same as the inequality (8b) in Bennett [5], which is stronger than the bound (1.10.27) due to Bernstein [6] and the bound of $\exp(-\frac{1}{2}\lambda \operatorname{arcsinh}(\frac{\lambda}{2\sigma^2}))$ due to Prokhorov [84].

In comparison with the additive version of the usual Chernoff bounds for the upper tail (Theorem 1.10.10), very roughly speaking, we see that the Chernoff bounds working with the variance allow us to replace the expectation of X by its variance. When the X_i are binary random variables with $\Pr[X_i=1]$ small, then $E[X] \approx \mathrm{Var}[X]$ and there is not much value in using Theorem 1.10.12. For this reason, Chernoff bounds taking into account the variance have not been used a lot in the theory of randomized search heuristics. They can, however, be convenient when we have random variables with $\Pr[X_i = 1]$ close to 1.

For example, assume that a search point $y \in \{0,1\}^n$ is obtained from a given $x \in \{0,1\}^n$ via standard bit mutation with mutation rate p. Assume for simplicity that we are interested in estimating the number of ones in y (the same argument would hold for the Hamming distance of y from some other search point $z \in \{0,1\}^n$, e.g., a unique optimum). Now, the number of ones in y is simply $X = \sum_{i=1}^n y_i$ and thus X is a sum of independent binary random variables. However, differently from, e.g., the situation in Lemma 1.10.3, the expectation of X may be big. If $x_i = 1$, then $E[y_i] = 1-p$. Hence, if x has many ones, then $E[Y]$ is large. However, since $\mathrm{Var}[y_i] = p(1-p)$ regardless of x_i, the variance $\mathrm{Var}[X] = np(1-p)$ is small (assuming that p is small). Consequently, here the Chernoff bounds in this subsection give better estimates than, e.g., Theorem 1.10.10. See, e.g., [28] for an example where this problem appeared in a recent research paper.

When not too precise bounds are needed, looking separately at the number of zeros and ones of x that flip (and bounding these via simple Chernoff bounds) is a way to circumvent the use of Chernoff bounds taking into account the variance. Several research studies follow this approach despite the computations often being more technical.

Chernoff bounds using the variance can also be useful in ant colony algorithms and estimation-of-distribution algorithms, where again pheromone values or frequencies close to 0 or 1 can lead to a small variance. See [76, 103] for examples.

The bounds of Theorem 1.10.12 can be written in a multiplicative form, for example,

$$\Pr[X \geq (1+\delta)E[X]]$$

$$\leq \left(\left(1+\frac{\delta E[X]}{\sigma^2}\right)^{-\left(1+\frac{\delta E[X]}{\sigma^2}\right)\frac{\sigma^2}{n+\sigma^2}} \left(1-\frac{\delta E[X]}{n}\right)^{-\left(1-\frac{\delta E[X]}{n}\right)\frac{n}{n+\sigma^2}} \right)^n \tag{1.10.29}$$

$$\leq \exp\left(-\frac{\delta^2 E[X]^2}{2\sigma^2 + \frac{2}{3}\delta E[X]}\right). \tag{1.10.30}$$

This is useful when working with relative errors, however, it seems that unlike for some previous bounds (compare, e.g., (1.10.2) and (1.10.18)) the multiplicative forms are not much simpler here.

Obviously, the case where all X_i satisfy $X_i \leq E[X_i] + b$ for some number b (instead of 1) can be reduced to the case $b = 1$ by dividing all random variables by b. For the reader's convenience, we state the resulting Chernoff bounds here.

Theorem 1.10.14 (equivalent to Theorem 1.10.12 and Corollary 1.10.13). *Let $X_1,\ldots,X_n$ be independent random variables. Let b be such that $X_i \le E[X_i]+b$ for all $i=1,\ldots,n$. Let $X=\sum_{i=1}^n X_i$. Let $\sigma^2=\sum_{i=1}^n \mathrm{Var}[X_i]=\mathrm{Var}[X]$. Then, for all $\lambda \ge 0$,*

$$\begin{aligned}
&\Pr[X \ge E[X]+\lambda] \\
&\le \left(\left(1+\frac{b\lambda}{\sigma^2}\right)^{-\left(1+\frac{b\lambda}{\sigma^2}\right)\frac{\sigma^2}{nb^2+\sigma^2}}\left(1-\frac{\lambda}{nb}\right)^{-\left(1-\frac{\lambda}{nb}\right)\frac{nb^2}{nb^2+\sigma^2}}\right)^n && (1.10.31)\\
&\le \exp\left(-\frac{\lambda}{b}\left(\left(1+\frac{\sigma^2}{b\lambda}\right)\ln\left(1+\frac{b\lambda}{\sigma^2}\right)-1\right)\right) && (1.10.32)\\
&\le \exp\left(-\frac{\lambda^2}{\sigma^2(2+\frac{2}{3}\frac{b\lambda}{\sigma^2})}\right) && (1.10.33)\\
&\le \exp\left(-\frac{1}{3}\min\left\{\frac{\lambda^2}{\sigma^2},\frac{\lambda}{b}\right\}\right), && (1.10.34)
\end{aligned}$$

where (1.10.31) is understood to mean 0 when $\lambda > nb$ and $(\frac{\sigma^2}{nb^2+\sigma^2})^n$ when $\lambda = nb$.

When we have $X_i \ge E[X_i]-b$ instead of $X_i \le E[X_i]+b$ for all $i=1,\ldots,n$, then the above estimates hold for $\Pr[X \le E[X]-\lambda]$.

1.10.1.5 Relation Between the Different Chernoff Bounds

We proceed by discussing how the bounds presented so far are related. The main finding will be that the Chernoff bounds depending on the variance imply all other bounds discussed so far with the exception of the additive Chernoff bound for random variables having different ranges (Theorem 1.10.9).

Surprisingly, this fact is not stated in Hoeffding's paper [54]. More precisely, in [54] the analogue of Theorems 1.10.12 and 1.10.14 uses the additional assumption that all X_i have the same expectation. Since this assumption is not made for the theorems not involving the variance, Hoeffding explicitly states that the latter are stronger in this respect (see the penultimate paragraph of Section 3 of [54]).

It is, however, quite obvious that the common-expectation assumption can be easily removed. From random variables with arbitrary means we can obtain random variables all having mean zero by subtracting their expectation. This operation does not change the variance and does not change the distribution of $X - E[X]$. Consequently, Hoeffding's result for variables with identical expectations immediately yields our version of this result (Theorems 1.10.12 and 1.10.14). Theorem 1.10.12 implies Theorem 1.10.1 via the equivalent version of Theorem 1.10.10 (see again the penultimate paragraph of Section 3 of [54]).

Consequently, the first (strongest) bound in Theorem 1.10.12 (or, equivalently the first bound in Theorem 1.10.14) implies the first (strongest) bound in Theorem 1.10.1, which is equivalent to the first (strongest) bound in Theorem 1.10.5. Essentially all of the other bounds presented so far can be derived from these main theorems via simple, sometimes tedious, estimates. The sole exception is Theorem 1.10.9, which can lead to significantly stronger estimates when the random variables have ranges of different size.

As an example, let $X_1,\dots,X_n$ be independent random variables such that $X_1,\dots,X_{n-1}$ take the values 0 and $(n-1)^{-1/2}$ with equal probability $\frac{1}{2}$ and such that X_n takes the values 0 and 1 with equal probability $\frac{1}{2}$. Let $X=\sum_{i=1}^n X_i$. Then $E[X]=\frac{1}{2}(\sqrt{n-1}+1)$. Theorem 1.10.9, taking $c_i=(n-1)^{-1/2}$ for $i\in[1..n-1]$ and $c_n=1$, yields the estimate

$$\Pr[X\geq E[X]+\lambda]\leq \exp\left(-\frac{2\lambda^2}{\sum_{i=1}^n c_i^2}\right)=\exp(-\lambda^2). \tag{1.10.35}$$

Note that $\mathrm{Var}[X]=\frac{1}{2}=:\sigma^2$. Consequently, the strongest Chernoff bound of Theorem 1.10.12, equation (1.10.25), gives an estimate larger than $(1+\frac{\lambda}{\sigma^2})^{-(1+\frac{\lambda}{\sigma^2})\frac{n\sigma^2}{n+\sigma}}=\exp(-\Theta(\lambda\log\lambda))$. Consequently, in this case Theorem 1.10.9 gives a significantly stronger estimate than Theorem 1.10.12.

1.10.1.6 Tightness of Chernoff Bounds, Lower Bounds for Deviations (Anti-Concentration)

As a very general and not at all precise rule of thumb, we can say that often the sharpest Chernoff bounds presented so far give an estimate of the tail probability that is near-tight. This is good to know from the perspective of proof design, since it indicates that failing to prove a desired statement usually cannot be overcome by trying to invent sharper Chernoff bounds. We shall not try to make this statement precise.

However, occasionally we also need lower bounds for the deviation from the expectation as a crucial argument in our analysis. For example, when generating several offspring independently in parallel, as, e.g., in a $(1+\lambda)$ EA, we expect the best of these to be significantly better than the expectation, and the efficiency of the algorithm relies on such desired deviations from the expectation.

Lower bounds for deviations from the expectation, occasionally called anti-concentration results, seem to be harder to work with. For this reason, we only briefly give some indications of how to handle them, and refer the reader to the literature. We note that there is a substantial body of mathematical literature on this topic (see, e.g., [73] and the references therein), which, however, is not always easy to use for algorithmic problems. We also note that for binomially distributed random variables, also the estimates in Theorem 1.10.39 can be used to derive lower bounds for tail probabilities.

Estimating Binomial Coefficients

For binomial distributions, estimating the (weighted) sum of binomial coefficients arising in the expression for the tail probability often works well (though the calculations may become tedious). In the theory of randomized search heuristics, this approach was used, among others, in the analysis of the $(1+\lambda)$ EA in [28, 40, 44, 51, 56] and the $(1+(\lambda,\lambda))$ GA in [23]. The following elementary bound was shown in [19, Lemma 3].

Lemma 1.10.15. *Let $n \in \mathbb{N}$ and $X \sim \mathrm{Bin}(n, \frac{1}{2})$. Then*

$$\Pr\left[X \geq E[X] + \tfrac{1}{2}\sqrt{E[X]}\right] \geq \tfrac{1}{8}, \tag{1.10.36}$$

$$\Pr\left[X \leq E[X] - \tfrac{1}{2}\sqrt{E[X]}\right] \geq \tfrac{1}{8}. \tag{1.10.37}$$

Two-Stage Rounding Trick

Estimating binomial coefficients works well for binomial distributions. However, a neat trick allows us to extend such results to sums of independent, non-identically distributed binary random variables. The rough idea is that we can sample a binary random variable X with $\Pr[X=1]=p$ by first sampling the unique random variable Y which takes values in $\{\frac{1}{2}, \lfloor p+\frac{1}{2}\rfloor\}$ and satisfies $E[Y] = E[X] = p$, and then, if $Y = \frac{1}{2}$, replacing Y with a uniform choice in $\{0,1\}$. If we view sampling X as rounding p randomly to 0 or 1 in such a way that the expectation is p, then this two-stage procedure consists of first rounding p to $\{0,\frac{1}{2}\}$ or $\{\frac{1}{2},1\}$ with expectation p and then (if necessary) rounding the result to $\{0,1\}$ without changing the expectation.

We use this trick below to show by elementary means two results which previously had been shown only via deeper methods. We first extend Lemma 1.10.15 above from fair coin flips to sums of independent binary random variables having different distributions. A similar result was shown in [81, first item of Lemma 6] for $X \sim \mathrm{Bin}(n,p)$, that is, for sums of identically distributed binary random variables (the result is stated without a lower bound on the variance, but by regarding, e.g., $\mathrm{Bin}(n,n^{-2})$, it becomes clear that a restriction such as $p \in [\frac{1}{n}, 1-\frac{1}{n}]$ is necessary). We have not found the general result of Lemma 1.10.16 in the literature, even though it is clear that such results can be shown via a normal approximation.

Lemma 1.10.16. *Let $v_0 > 0$. There are constants $c, C > 0$ such that the following is true. Let $n \in \mathbb{N}$. Let $p_1,\ldots,p_n \in [0,1]$. For all $i \in [1..n]$, let X_i be a binary random variable with $\Pr[X_i = 1] = p_i$. Assume that $X_1,\ldots,X_n$ are* *independent. Let* $X = \sum_{i=1}^n X_i$. *Assume that* $\mathrm{Var}[X] = \sum_{i=1}^n p_i(1-p_i) \geq v_0$. *Then*

$$\Pr\left[X \geq E[X] + c\sqrt{\mathrm{Var}[X]}\right] \geq C, \tag{1.10.38}$$

$$\Pr\left[X \leq E[X] - c\sqrt{\mathrm{Var}[X]}\right] \geq C. \tag{1.10.39}$$

Proof. Let us first assume that $p_i \leq \frac{1}{2}$ for all $i \in [1..n]$ and show the claim under the weaker assumption $\mathrm{Var}[X] \geq \frac{1}{2}v_0$. We define independent random variables Y_i such that

$$\Pr[Y_i = \tfrac{1}{2}] = 2p_i,$$
$$\Pr[Y_i = 0] = 1 - 2p_i.$$

Let $Y = \sum_{i=1}^n Y_i$ and note that $E[Y] = E[X]$.

Based on the Y_i, we define independent binary random variables Z_i as follows. If $Y_i = 0$, then $Z_i := 0$. Otherwise, that is, if $Y_i = \frac{1}{2}$, then we let Z_i be uniformly distributed in $\{0,1\}$. An elementary calculation shows that $\Pr[Z_i = 1] = p_i$, that is, the Z_i have the same distribution as the X_i. Hence it suffices to show our claim for $Z := \sum_{i=1}^n Z_i$.

Let c be a sufficiently small constant. Our main argument for the lower bound on the upper tail (1.10.38) will be that with constant probability we have the event

$$A := \text{“}Y \geq E[Y] - \tfrac{1}{2}c\sqrt{\mathrm{Var}[X]}\text{”}.$$

In this case, again with constant probability, we have $Z \geq E[Z \mid A] + c\sqrt{\mathrm{Var}[X]}$, which implies $Z \geq E[Y] - \frac{1}{2}c\sqrt{\mathrm{Var}[X]} + c\sqrt{\mathrm{Var}[X]} = E[X] + \frac{1}{2}c\sqrt{\mathrm{Var}[X]}$. In other words, we have

$$\Pr\left[X \geq E[X] + \tfrac{1}{2}c\sqrt{\mathrm{Var}[X]}\right] \geq \Pr[A] \cdot \Pr\left[(Z \mid A) \geq E[Z \mid A] + c\sqrt{\mathrm{Var}[X]}\right]$$

and we shall argue that both factors are at least constant.

For the first factor, we note that for all $i \in [1..n]$, we have $E[Y_i] = p_i$. An elementary calculation thus shows that $\mathrm{Var}[Y_i] = \frac{1}{2}p_i(1-2p_i) \leq \frac{1}{2}p_i(1-p_i) = \frac{1}{2}\mathrm{Var}[X_i]$ and hence $\mathrm{Var}[Y] \leq \frac{1}{2}\mathrm{Var}[X]$. With Cantelli's inequality (Lemma 1.6.10), we compute

$$\begin{aligned}
\Pr[A] &= \Pr\left[Y \geq E[Y] - \frac{1}{2}c\sqrt{\mathrm{Var}[X]}\right] \\
&\geq \Pr\left[Y \geq E[Y] - \frac{1}{\sqrt{2}}c\sqrt{\mathrm{Var}[Y]}\right] \\
&\geq 1 - \Pr\left[Y \leq E[Y] - \frac{1}{\sqrt{2}}c\sqrt{\mathrm{Var}[Y]}\right]
\end{aligned}$$

$$\ge 1 - \frac{1}{1+c^2/2} = \frac{c^2}{2+c^2}.$$

For the second factor, we note that once Y is determined, $Z \sim \text{Bin}(2Y, \frac{1}{2})$. We estimate

$$\begin{aligned} E[Y] - \tfrac{1}{2}c\sqrt{\text{Var}[X]} &\ge E[Y] - \tfrac{c}{\sqrt{2v_0}} \text{Var}[X] \\ &\ge E[Y] - \tfrac{c}{\sqrt{2v_0}} E[X] \ge (1 - \tfrac{c}{\sqrt{2v_0}})E[X] =: q, \end{aligned}$$

where we use the fact that $\text{Var}[X] \ge \frac{1}{2}v_0$ implies $\sqrt{\text{Var}[X]} \le \sqrt{2/v_0}\,\text{Var}[X]$. Hence, conditional on A, we have $Z \sim \text{Bin}(2\tilde{q}, \frac{1}{2})$ for some $\tilde{q} \ge q$, and thus

$$\Pr\left[(Z \mid A) \ge E[Z \mid A] + c\sqrt{\text{Var}[X]}\right] \ge \frac{1}{8}$$

by Lemma 1.10.15 and $c\sqrt{\text{Var}[X]} \le c\sqrt{E[X]} \le \frac{1}{2}\sqrt{(1 - \frac{c}{\sqrt{2v_0}})E[X]} \le \frac{1}{2}\sqrt{E[Z \mid A]}$, where the middle inequality assumes that c is sufficiently small.

To prove (1.10.39), we argue as follows. Let $K = E[X] + \frac{1}{2}c\sqrt{\text{Var}[X]}$. Then

$$\begin{aligned} &\Pr\left[X \le E[X] - \tfrac{1}{2}c\sqrt{\text{Var}[X]}\right] \\ &\qquad \ge \sum_{k=0}^{2K} \Pr[Y = \tfrac{k}{2}] \cdot \Pr\left[(Z \mid Y = \tfrac{k}{2}) \le E[X] - \tfrac{1}{2}c\sqrt{\text{Var}[X]}\right]. \end{aligned}$$

Now $(Z \mid Y = \frac{k}{2}) \sim \text{Bin}(k, \frac{1}{2})$, and hence Lemma 1.8.9(b) implies that the second factor is smallest for $k = 2K$. Consequently,

$$\begin{aligned} &\Pr\left[X \le E[X] - \tfrac{1}{2}c\sqrt{\text{Var}[X]}\right] \\ &\qquad \ge \Pr[Y \le K] \cdot \Pr\left[(Z \mid Y = K) \le E[X] - \tfrac{1}{2}c\sqrt{\text{Var}[X]}\right]. \end{aligned}$$

We estimate the two factors separately. For the first one, in an analogous fashion to that before, we obtain $\Pr[Y \le K] = \Pr[Y \le E[Y] + \frac{1}{2}c\sqrt{\text{Var}[X]}] \ge \frac{c^2}{2+c^2}$. For the second factor, we compute

$$\begin{aligned} &\Pr\left[(Z \mid Y = K) \le E[X] - \tfrac{1}{2}c\sqrt{\text{Var}[X]}\right] \\ &\quad = \Pr\left[\text{Bin}(2K, \tfrac{1}{2}) \le E[\text{Bin}(2K, \tfrac{1}{2})] - K + E[X] - \tfrac{1}{2}c\sqrt{\text{Var}[X]}\right] \\ &\quad = \Pr\left[\text{Bin}(2K, \tfrac{1}{2}) \le E[\text{Bin}(2K, \tfrac{1}{2})] - c\sqrt{\text{Var}[X]}\right]. \end{aligned}$$

For $c \le \frac{1}{2}$, we have $c\sqrt{\mathrm{Var}[X]} \le \frac{1}{2}\sqrt{E[X]} \le \frac{1}{2}\sqrt{E[\mathrm{Bin}(2K,\frac{1}{2})]}$ and Lemma 1.10.15 yields $\Pr\left[(Z \mid Y=K) \le E[X] - \frac{1}{2}c\sqrt{\mathrm{Var}[X]}\right] \ge \frac{1}{8}$.

Now assume that the p_i are not all in $[0,\frac{1}{2}]$. Let $I' = \{i \in [1..n] \mid p_i \le \frac{1}{2}\}$ and $I'' = [1..n] \setminus I'$. Let $X' = \sum_{i \in I'} X_i$ and $X'' = \sum_{i \in I''} X_i$. Since $\mathrm{Var}[X] = \mathrm{Var}[X'] + \mathrm{Var}[X'']$, by symmetry (possibly replacing the p_i by $1-p_i$), we can assume that $\mathrm{Var}[X'] \ge \frac{1}{2}\mathrm{Var}[X]$. Now $\mathrm{Var}[X] \ge v_0$ implies $\mathrm{Var}[X'] \ge \frac{1}{2}v_0$, and by the above we have $X' \ge E[X'] + \frac{1}{2}c\sqrt{\mathrm{Var}[X']} \ge E[X'] + \frac{1}{2\sqrt{2}}c\sqrt{\mathrm{Var}[X]}$ with constant probability. By Cantelli's inequality again, we have $X'' \ge E[X''] - \frac{c}{4}\sqrt{\mathrm{Var}[X'']} \ge E[X''] - \frac{c}{4\sqrt{2}}\sqrt{\mathrm{Var}[X]}$ with constant probability. Hence $X = X' + X'' \ge E[X'] + E[X''] + \frac{c}{4\sqrt{2}}\mathrm{Var}[X] = E[X] + \frac{c}{4\sqrt{2}}\mathrm{Var}[X]$ with constant probability. The proof that $X \le E[X] - \frac{c}{4\sqrt{2}}\mathrm{Var}[X]$ with constant probability is analogous. By replacing our original c by $4\sqrt{2}c$, we obtain the precise formulation of the claim. □

We now use the two-stage rounding trick to give an elementary proof of the following result.

Lemma 1.10.17. *Let $n \in \mathbb{N}$ and $p_1,\dots,p_n \in [0,1]$. For all $i \in [1..n]$, let X_i be a binary random variable with $\Pr[X_i = 1] = p_i$. Assume that $X_1,\dots,X_n$ are independent. Let $X = \sum_{i=1}^n X_i$. If $\mathrm{Var}[X] \ge 1$, then, for all $k \in [0..n]$,*

$$\Pr[X = k] \le \frac{2}{\sqrt{\mathrm{Var}[X]}}.$$

This result (without making the leading constant precise) was proven for the special case where all p_i are between $\frac{1}{6}$ and $\frac{5}{6}$ in [94, Lemma 9, arXiv version]. This proof uses several deep arguments from probability theory. In [63, Lemma 3], the result stated in [94] was minimally extended to the case where only a linear number of the p_i are between $\frac{1}{6}$ and $\frac{5}{6}$.

Proof (of Lemma 1.10.17). In a similar fashion to the proof of Lemma 1.10.16, we define independent random variables Y_i such that

$$\Pr[Y_i = \tfrac{1}{2}] = 2p_i,$$
$$\Pr[Y_i = 0] = 1 - 2p_i$$

when $p_i \le \frac{1}{2}$ and

$$\Pr[Y_i = 1] = 2(p_i - \tfrac{1}{2}) = 2p_i - 1,$$
$$\Pr[Y_i = \tfrac{1}{2}] = 1 - 2(p_i - \tfrac{1}{2}) = 2 - 2p_i$$

for $p_i > \frac{1}{2}$. If $Y_i \in \{0,1\}$, then $Z_i := Y_i$; otherwise (that is, when $Y_i = \frac{1}{2}$), we let Z_i be uniformly distributed on $\{0,1\}$. As before, the Z_i are just an alternative definition of the X_i. Hence $Z = \sum_{i=1}^n Z_i$ has the same distribution as X.

For $\ell \in \{0, \frac{1}{2}, 1\}$, let $I_\ell := |\{i \in [1..n] \mid Y_i = \ell\}|$. Since

$$\Pr[Y_i = \tfrac{1}{2}] = 2\min\{p_i, 1-p_i\} \geq 2\operatorname{Var}[X_i],$$

we have $E[I_{\frac{1}{2}}] \geq 2\operatorname{Var}[X]$. Since the Y_i are independent, we have $\Pr[I_{\frac{1}{2}} \leq 2(1-\delta)\operatorname{Var}[X]] \leq \exp(-\delta^2 \operatorname{Var}[X])$ for all $\delta \in [0,1]$, by (1.10.12).

Finally, note that by (1.4.18) we have $\Pr[\operatorname{Bin}(a, \frac{1}{2}) = k] \leq \sqrt{\frac{2}{\pi a}}$ for all $a \in \mathbb{N}$ and $k \in \mathbb{N}_0$.

Writing $a_0 = \lfloor 2(1-\delta)\operatorname{Var}[X] \rfloor$ and combining these arguments, we obtain

$$\begin{aligned}
\Pr[X = k] &= \Pr[Z = k] \\
&= \sum_{a=0}^{n} \Pr[I_{\frac{1}{2}} = a] \sum_{b=0}^{k} \Pr[I_1 = b] \Pr[Z = k \mid I_{\frac{1}{2}} = a \wedge I_1 = b] \\
&= \sum_{a=0}^{n} \Pr[I_{\frac{1}{2}} = a] \sum_{b=0}^{k} \Pr[I_1 = b] \Pr[\operatorname{Bin}(a, \tfrac{1}{2}) = k - b] \\
&\leq \Pr[I_{\frac{1}{2}} \leq a_0] \\
&\quad + \sum_{a=a_0+1}^{n} \Pr[I_{\frac{1}{2}} = a] \sum_{b=0}^{k} \Pr[I_1 = b] \Pr[\operatorname{Bin}(a, \tfrac{1}{2}) = k - b] \\
&\leq \exp(-\delta^2 \operatorname{Var}[X]) + \sum_{a=a_0+1}^{n} \Pr[I_{\frac{1}{2}} = a] \sum_{b=0}^{k} \Pr[I_1 = b] \sqrt{\frac{2}{\pi(a_0+1)}} \\
&\leq \exp(-\delta^2 \operatorname{Var}[X]) + \sqrt{\frac{1}{(1-\delta)\pi \operatorname{Var}[X]}}.
\end{aligned}$$

For $\operatorname{Var}[X] \geq 1$, by taking $\delta = 0.75$ and estimating $\exp(-\delta^2 \operatorname{Var}[X]) \leq \frac{1}{e\delta^2 \operatorname{Var}[X]} \leq \frac{1}{e\delta^2 \sqrt{\operatorname{Var}[X]}}$, where we have used the estimate $e^x \geq ex$, an alternative version of Lemma 1.4.1, we obtain the bound $2\operatorname{Var}[X]^{-1/2}$. □

We did not aim to optimize the implicit constants in the result above. We note that if we take $\delta = \operatorname{Var}[X]^{-1/4}$, the claimed probability becomes $(1+o(1))\frac{1}{\sqrt{\pi \operatorname{Var}[X]}}$ for $\operatorname{Var}[X] \to \infty$.

Approximation via the Normal Distribution

The generic approach of approximating binomial distributions via normal distributions is not often used in the theory of randomized search. In [81], the Berry–Esseen inequality was employed to prove a result similar to Lemma 1.10.16 for the special case of binomial distributions. Unlike many other proofs relying on the normal approximation, this proof is quite short and elegant.

In [64], the normal approximation was used to show that the best of $k \in \omega(1) \cap o(\sqrt{n})$ independent random initial search points in $\{0,1\}^n$ has with probability $1-o(1)$ a distance of $\frac{n}{2} - \sqrt{\frac{n}{2}(\ln k - \frac{1}{2}\ln\ln k \pm c_k)}$ from the optimum, where c_k is an arbitrary sequence tending to infinity.

In [94, Lemma 7, arXiv version], a very general result on how a sum of independent random variables with bounded expectation and variance can be approximated by a normal distribution was used to analyze the performance of an estimation-of-distribution algorithm. This analysis is highly technical.

Order Statistics

The result about the best of k independent initial individuals in [64] actually says something about the maximum order statistic of k independent $\mathrm{Bin}(n, \frac{1}{2})$ random variables. In general, the maximum order statistic is strongly related to lower bounds for tail probabilities, as the following elementary argument (more or less explicit in all work on the $(1+\lambda)$ EA) shows: Let $X_1, \ldots, X_\lambda$ be independent random variables following the same distribution. Let $X_{\max} = \max\{X_i \mid i \in [1..\lambda]\}$. Then

$$\begin{aligned}
&\Pr[X^* \geq D] \leq \lambda \Pr[X_1 \geq D], \\
&\Pr[X^* \geq D] = 1 - (1 - \Pr[X_1 \geq D])^\lambda \geq 1 - \exp(\lambda \Pr[X_1 \geq D]).
\end{aligned}$$

Consequently, $\Pr[X^* \geq D]$ is constant if and only if $\Pr[X_1 \geq D] = \Theta(\frac{1}{\lambda})$.

For the maximum order statistics of binomially distributed random variables with small success probability, Gießen and Witt [51, Lemma 4(3)] proved the following result and used it in the analysis of the $(1+\lambda)$ EA.

Lemma 1.10.18. *Let $\alpha \geq 0$ and $c > 0$ be constants. Let $n \in \mathbb{N}$, and let all of the following asymptotics be for $n \to \infty$. Let $k = n(\ln n)^{-\alpha}$ and $\lambda = \omega(1)$. Let $X_{\max}$ be the maximum of λ independent random variables with distribution $\mathrm{Bin}(k, \frac{c}{n})$. Then $E[X_{\max}] = (1 \pm o(1)) \frac{1}{1+\alpha} \frac{\ln \lambda}{\ln\ln \lambda}$.*

Extremal Situations

Occasionally, it is desirable to understand which situation gives the smallest or the largest deviations. For example, let $X_1, \ldots, X_n$ be independent binary random variables with expectations $E[X_i] = p_i$. Then it could be useful to know that $X = \sum_{i=1}^n X_i$ deviates most (in some suitable sense) from its expectation when all p_i are $\frac{1}{2}$. Such statements can be made formal and can be proven with the notions of majorization and Schur-convexity. We refer to [89] for a nice treatment of this topic. Such arguments have been used to analyze estimation-of-distribution algorithms in [94].

Staying on One Side of the Expectation, and Feige's Inequality

When it suffices to know that with reasonable probability we will stay (more or less) on one side of the expectation, then the following results can be useful.

A very general bound is Feige's inequality [48, Theorem 1], which has found applications in the analysis of randomized search heuristics, among others, in [15, 16, 65, 93].

Lemma 1.10.19 (Feige's inequality). *Let $X_1,\dots,X_n$ be independent non-negative random variables with expectations $\mu_i := E[X_i]$ satisfying $\mu_i \le 1$. Let $X = \sum_{i=1}^n X_i$. Then*

$$\Pr[X \le E[X] + \delta] \ge \min\{\tfrac{1}{13}, \tfrac{\delta}{\delta+1}\}.$$

For binomial distributions, we have stronger guarantees. Besides bounds comparing the binomial distribution with its normal approximation [90], the following specific bounds are known.

Lemma 1.10.20. *Let $n \in \mathbb{N}$, $p \in [0,1]$, and $k = \lfloor np \rfloor$. Let $X \sim \mathrm{Bin}(n,p)$.*

(a) If $\frac{1}{n} < p$, then $\Pr[X \ge E[X]] > \frac{1}{4}$.
(b) If $0.29/n \le p < 1$, then $\Pr[X > E[X]] \ge \frac{1}{4}$.
(c) If $\frac{1}{n} \le p \le 1 - \frac{1}{n}$, then $\Pr[X \ge E[X]] \ge \frac{1}{2\sqrt{2}} \frac{\sqrt{np(1-p)}}{\sqrt{np(1-p)+1}+1}$.
(d) If $\frac{1}{n} \le p < 1$, then $\Pr[X > E[X]] > \frac{1}{2} - \sqrt{\frac{n}{2\pi k(n-k)}}$.
(e) If $\frac{1}{n} \le p < 1 - \frac{1}{n}$, then $\Pr[X > E[X] + 1] \ge 0.037$.

Surprisingly, all these results are quite recent. The bound (a), from [52], appears to be the first general result of this type at all.[1] It was followed up by estimate (c), from [83], which gives stronger estimates when $np(1-p) > 8$. The result (d), from [21], is the only one to give a bound tending to $\frac{1}{2}$ for both np and $n(1-p)$ tending to infinity. The estimates (b) and (e) are also from [21]. A lower bound on the probability of exceeding the expectation by more than one, such as (e), was needed in an analysis of an evolutionary algorithm with self-adjusting mutation rate [28, Lemma 9].

[1] For $p \in [\frac{1}{n}, \frac{1}{2}]$, this result follows from the proof of Lemma 6.4 in [85]. The lemma itself only states the bound $\Pr[X \ge E[X]] \le \min\{p, \frac{1}{4}\}$ for $p \le \frac{1}{2}$. The assumption $p \le \frac{1}{2}$ appears to be crucial for the proof.

1.10.1.7 Proofs of the Chernoff Bounds

As discussed in Section 1.10.1.5, all Chernoff bounds stated so far can be derived from the strongest bounds of Theorem 1.10.9 or 1.10.12 via elementary estimates that have nothing to do with probability theory. We shall not detail these estimates – the reader can find them all in the literature, for example, in [54]. We shall, however, sketch how to prove the two central inequalities (1.10.15) and (1.10.25). One reason for this is that we can then argue in Section 1.10.2.2 that these proofs (and thus also all Chernoff bounds presented so far) hold not only for independent random variables, but also for negatively correlated ones.

A second reason is that, occasionally, it can be profitable to have this central argument ready to prove Chernoff bounds for particular distributions for which the classical bounds are not applicable or do not give sufficient results. This has been done, e.g., in [3, 27, 35, 67, 79, 80, 102].

The central step in almost all proofs of Chernoff bounds, going back to Bernstein [6], is the following one-line argument. Let $h > 0$. Then

$$\Pr[X \geq t] = \Pr[e^{hX} \geq e^{ht}] \leq \frac{E[e^{hX}]}{e^{ht}} = e^{-ht} \prod_{i=1}^{n} E[e^{hX_i}]. \tag{1.10.40}$$

Here, the first equality stems simply from the fact that the function $x \mapsto e^{hx}$ is monotonically increasing. The inequality in (1.10.40) is Markov's inequality (Lemma 1.6.7) applied to the (non-negative) random variable e^{hX}. The last equality exploits the independence of the X_i, which carries over to the e^{hX_i}.

It now remains to estimate $E[e^{hX_i}]$ and to choose h so as to minimize the resulting expression. We do this as an example for the case where all X_i take values in $[0,1]$ and $E[X] < t < n$. Since the exponential function is convex, $E[e^{hX_i}]$ is maximized (which is the worst case for our estimate) when X_i is concentrated on the values 0 and 1, that is, we have $\Pr[X_i = 1] = E[X_i]$ and $\Pr[X_i = 0] = 1 - E[X_i]$. In this case, $E[e^{hX_i}] = (1 - E[X_i])e^0 + E[X_i]e^h$. By the inequality of arithmetic and geometric means, we compute

$$\begin{aligned}\prod_{i=1}^{n} E[e^{hX_i}] &\leq \prod_{i=1}^{n} (1 - E[X_i] + E[X_i]e^h) \\ &\leq \left(\frac{1}{n}\sum_{i=1}^{n}(1 - E[X_i] + E[X_i]e^h)\right)^n \\ &\leq \left(\frac{1}{n}(n - E[X] + E[X]e^h)\right)^n.\end{aligned}$$

This gives the tail estimate $\Pr[X \geq t] \leq e^{-ht}(\frac{1}{n}(n - E[X] + E[X]e^h))^n$, which is minimized by taking

$$h = \ln\left(\frac{(n - E[X])t}{(n - t)E[X]}\right),$$

which then gives the strongest multiplicative Chernoff bound (1.10.1) if we rewrite it using $t = (1+\delta)E[X]$.

Since it may help reading the literature, we add that $E[e^X]$ is called the *exponential moment* of X and $h \mapsto E[e^{hX}]$ is called the *moment-generating function* of X.

From the above proof sketch together with the remark on the tightness of Markov's inequality following Lemma 1.6.7, we see that in almost all cases, our Chernoff bounds are not absolutely tight, that is, they hold with "$<$" instead of "$\le$." The sole exceptions are (i) when X takes only two values with positive probability, (ii) when the tail event consists of a single point, for example, when $X \ge n$ or $X \le 0$ if X is a sum of n binary random variables, or (iii) when the tail event is empty, for example, when $X \ge n+1$ if X is a sum of n binary random variables. Having a "$<$" in a Chernoff bound will not drastically change things, but can occasionally be nice for cosmetic reasons.

1.10.1.8 Chernoff Bounds with Estimates for the Expectation

Often we do not know the precise value of the expectation or it is tedious to compute it. In such cases, we can exploit the fact that *all Chernoff bounds discussed in this chapter are also valid when the expectation is replaced by an upper or lower bound on it.* This is obvious for many bounds; for example, from (1.10.13), (1.10.14), and (1.10.12) we immediately derive the estimates

$$\Pr[X \ge \mu^+ + \lambda] \le \Pr[X \ge \mu + \lambda] \le \exp\left(-\frac{2\lambda^2}{n}\right),$$
$$\Pr[X \le \mu^- - \lambda] \le \Pr[X \le \mu - \lambda] \le \exp\left(-\frac{2\lambda^2}{n}\right),$$
$$\Pr[X \le (1-\delta)\mu^-] \le \Pr[X \le (1-\delta)\mu] \le \exp\left(-\frac{\delta^2\mu}{2}\right) \le \exp\left(-\frac{\delta^2\mu^-}{2}\right)$$

for all $\mu^+ \ge E[X] =: \mu$ and $\mu^- \le E[X]$.

This is less obvious for a bound such as $\Pr[X \ge (1+\delta)\mu^+] \le \exp(-\frac{1}{3}\delta^2\mu^+)$, since now also the probability of the tail event decreases for increasing μ^+. However, for such bounds also we can replace $E[X]$ by an estimate, as the following argument shows.

Theorem 1.10.21. *(a) Upper tail: Let $X_1,\dots,X_n$ be independent random variables taking values in $[0,1]$. Let $X = \sum_{i=1}^n X_i$. Let $\mu^+ \ge E[X]$. Then, for all $\delta \ge 0$,*

$$\Pr[X \geq (1+\delta)\mu^+] \leq \left(\frac{1}{1+\delta}\right)^{(1+\delta)\mu^+} \left(\frac{n-\mu^+}{n-(1+\delta)\mu^+}\right)^{n-(1+\delta)\mu^+}, \tag{1.10.41}$$

where this bound is read as 0 *for* $\delta > \frac{n-\mu^+}{\mu^+}$ *and as* $(\frac{\mu^+}{n})^n$ *for* $\delta = \frac{n-\mu^+}{\mu^+}$. *Consequently, all Chernoff bounds of Theorem 1.10.1 (including* (1.10.7) *and* (1.10.8) *and Corollary 1.10.4) and those of Theorem 1.10.10 are valid when all occurrences of* $E[X]$ *are replaced by* μ^+. *The additive bounds* (1.10.13) *and* (1.10.15) *and those of Theorem 1.10.12 and 1.10.14 are trivially valid with the expectation replaced by an upper bound on it.*

(b) Lower tail: All Chernoff bounds of Theorems 1.10.5 and 1.10.11, the ones in (1.10.14) *and* (1.10.16), *and those of Corollary 1.10.13 are valid when all occurrences of* $E[X]$ *are replaced by* $\mu^- \leq E[X]$.

Proof. We first show (1.10.41). There is nothing to do when $(1+\delta)\mu^+ > n$, so let us assume that $(1+\delta)\mu^+ \leq n$. Let $\gamma = \frac{\mu^+ - E[X]}{n - E[X]}$. For all $i \in [1..n]$, define Y_i by $Y_i = X_i + \gamma(1 - X_i)$. Since $\gamma \leq 1$, $Y_i \leq 1$. By definition, $Y_i \geq X_i$, and thus also $Y \geq X$ for $Y := \sum_{i=1}^n Y_i$. Also, $\mu^+ = E[Y]$. Hence

$$\Pr[X \geq (1+\delta)\mu^+] \leq \Pr[Y \geq (1+\delta)\mu^+] = \Pr[Y \geq (1+\delta)E[Y]].$$

Now (1.10.41) follows immediately from Theorem 1.10.1, equation (1.10.1). Since (1.10.41) implies all other Chernoff bounds of Theorem 1.10.1 (including (1.10.7) and (1.10.8) and Corollary 1.10.4) via elementary estimates, all these bounds are valid with $E[X]$ replaced by μ^+ as well. This extends to Theorem 1.10.10, since that is just a reformulation of Theorem 1.10.1. For the remaining (additive) bounds, replacing $E[X]$ by an upper bound only decreases the probability of the tail event, so clearly these remain valid.

To prove our claim about lower tail bounds, it suffices to note that all bounds in Theorem 1.10.5 are monotonically decreasing in $E[X]$. So, replacing $E[X]$ by some $\mu^- < E[X]$ makes the tail event less likely and increases the probability in the statement. Similarly, the additive bounds are not affected when $E[X]$ is replaced by μ^-. □

We note without proof that the variance of the random variable Y constructed above is at most that of X. Since the tail bound in Theorem 1.10.12 is increasing in σ^2, the same argument as above also shows that multiplicative versions of Theorem 1.10.12 such as (1.10.29) remain valid when all occurrences of $E[X]$ are replaced by an upper bound $\mu^+ \geq E[X]$.

1.10.2 Chernoff Bounds for Sums of Dependent Random Variables

In the previous subsection, we discussed large-deviation bounds for the classical setting of sums of independent random variables. In the analysis of algorithms, often we cannot fully satisfy the assumption of independence. The dependencies may appear minor, maybe even in our favor in some sense, so we could hope for some good large-deviation bounds.

In this section, we discuss three such situations which all lead to (essentially) the known Chernoff bounds being applicable despite the absence of perfect independence. The first of these was already discussed in Section 1.8.3, so we just note here how it also implies the usual Chernoff bounds.

1.10.2.1 Unconditional Sequential Domination

In the analysis of sequential random processes such as iterative randomized algorithms, we rarely encounter situations where the events in different iterations are independent, simply because the actions of our algorithm depend on the results of the previous iterations. However, owing to the independent randomness used in each iteration, we can often say that, independent of what happened in iterations $1,\ldots,t-1$, in iteration t we have a particular event with at least some probability p.

This property was made precise in the definition of *unconditional sequential domination* before Lemma 1.8.8. The lemma then showed that unconditional sequential domination leads to domination by a sum of independent random variables. Any upper tail bound for this sum is naturally valid also for the sum of the original random variables. We make this elementary insight precise in the following lemma. This type of argument has been used in, among others, analyses of evolutionary algorithms for shortest-path problems [32, 33, 37]. There, one can show that, in each iteration, independent of the past, with at least a certain probability an extra edge of a desired path is found. This type of argument was also used in [34] to construct a monotonic function that is difficult to optimize.

Lemma 1.10.22. *Let $(X_1,\ldots,X_n)$ and $(X_1^*,\ldots,X_n^*)$ be finite sequences of discrete random variables. Assume that $X_1^*,\ldots,X_n^*$ are independent.*

(a) If $(X_1^*,\ldots,X_n^*)$ unconditionally sequentially dominates $(X_1,\ldots,X_n)$, then for all $\lambda \in \mathbb{R}$, we have $\Pr[\sum_{i=1}^n X_i \geq \lambda] \leq \Pr[\sum_{i=1}^n X_i^* \geq \lambda]$ and the latter expression can be bounded by Chernoff bounds for the upper tail of independent random variables.

(b) If $(X_1^*,\ldots,X_n^*)$ unconditionally sequentially subdominates $(X_1,\ldots,X_n)$, then for all $\lambda \in \mathbb{R}$, we have $\Pr[\sum_{i=1}^n X_i \leq \lambda] \leq \Pr[\sum_{i=1}^n X_i^* \leq \lambda]$ and the latter expression can be bounded by Chernoff bounds for the lower tail of independent random variables.

1.10.2.2 Negative Correlation

Occasionally, we encounter random variables that are not independent, but that display an intuitively even better negative-correlation behavior. Take as an example the situation where we do not flip bits independently with probability $\frac{k}{n}$, but we flip a set of exactly k bits chosen uniformly at random from all sets of k out of n bits. Let $X_1, \ldots, X_n$ be the indicator random variables for the events that bit $1, \ldots, n$ flips. Clearly, the X_i are not independent. If $X_1 = 1$, then $\Pr[X_2 = 1] = \frac{k-1}{n-1}$, which is different from the unconditional probability $\frac{k}{n}$. However, things feel even better than independent: knowing that $X_1 = 1$ actually reduces the probability that $X_2 = 1$. This intuition is made precise in the following notion of *negative correlation*.

Let $X_1, \ldots, X_n$ be binary random variables. We say that $X_1, \ldots, X_n$ are 1-*negatively correlated* if, for all $I \subseteq [1..n]$, we have

$$\Pr[\forall i \in I : X_i = 1] \le \prod_{i \in I} \Pr[X_i = 1].$$

We say that $X_1, \ldots, X_n$ are 0-*negatively correlated* if, for all $I \subseteq [1..n]$, we have

$$\Pr[\forall i \in I : X_i = 0] \le \prod_{i \in I} \Pr[X_i = 0].$$

Finally, we call $X_1, \ldots, X_n$ *negatively correlated* if they are both 0-negatively correlated and 1-negatively correlated.

In simple words, these conditions require that the event that a set of variables is all zero or all one is at most as likely as in the case of independent random variables. It seems natural that sums of such random variables are at least as strongly concentrated as independent random variables, and, in fact, Panconesi and Srinivasan [82] were able to prove that negatively correlated random variables admit Chernoff bounds. To be precise, they only proved that 1-negative correlation implies Chernoff bounds for the upper tail, but it is not too difficult to show (see below) that their main argument works for all bounds proven via Bernstein's exponential moments method. In particular, for sums of 1-negatively correlated random variables we obtain all Chernoff bounds for the upper tail that have been presented in this chapter for independent random variables (as far as they can be applied to binary random variables). We prove a slightly more general result, as this helps in arguing that we can also work with upper bounds for the expectation instead of the precise expectation. We then use a symmetry argument to argue that 0-negative correlation implies all lower tail bounds presented so far.

Theorem 1.10.23 (1-negative correlation implies upper tail bounds). *Let $X_1, \ldots, X_n$ be 1-negatively correlated binary random variables. Let $a_1, \ldots, a_n, b_1, \ldots, b_n \in \mathbb{R}$ with $a_i \le b_i$ for all $i \in [1..n]$. Let $Y_1, \ldots, Y_n$ be ran-*

dom variables with $\Pr[Y_i = a_i] = \Pr[X_i = 0]$ *and* $\Pr[Y_i = b_i] = \Pr[X_i = 1]$*. Let* $Y = \sum_{i=1}^n Y_i$*.*

(a) If $a_1, \ldots, a_n, b_1, \ldots, b_n \in [0,1]$*, then* Y *satisfies the Chernoff bounds given in* (1.10.1) *to* (1.10.5)*,* (1.10.7) *to* (1.10.9)*,* (1.10.13)*,* (1.10.17) *to* (1.10.21)*, and* (1.10.25) *to* (1.10.30)*, where in the latter we use* $\sigma^2 := \sum_{i=1}^n \mathrm{Var}[Y_i]$*.*

(b) Without the restriction to $[0,1]$ *specified in (a),* Y *satisfies the Chernoff bound of* (1.10.15) *with* $c_i := b_i - a_i$ *and the bounds of* (1.10.31) *to* (1.10.34) *with* $\sigma^2 := \sum_{i=1}^n \mathrm{Var}[Y_i]$*.*

Each of these results also holds when all occurrences of $E[Y]$ *are replaced by* μ^+ *for some* $\mu^+ \geq E[Y]$*.*

Proof. Let $X'_1, \ldots, X'_n$ be independent binary random variables such that for each $i \in [1..n]$, the random variables X_i and X'_i are identically distributed. Let $c_i := b_i - a_i$ for all $i \in [1..n]$. Note that $Y_i = a_i + c_i X_i$ for all $i \in [1..n]$. Let $Y'_i = a_i + c_i X'_i$. Let $Y' = \sum_{i=1}^n Y'_i$.

We first show that the 1-negative correlation of the X_i implies $E[Y^\ell] \leq E[(Y')^\ell]$ for all $\ell \in \mathbb{N}_0$. There is nothing to show for $\ell = 0$ and $\ell = 1$, so let $\ell \geq 2$. Since $Y = (\sum_{i=1}^n a_i) + (\sum_{i=1}^n c_i X_i)$, we have $Y^\ell = \sum_{k=0}^\ell \binom{\ell}{k} (\sum_{i=1}^n a_i)^{\ell-k} (\sum_{i=1}^n c_i X_i)^k$. By linearity of expectation, it suffices to show that $E[(\sum_{i=1}^n c_i X_i)^k] \leq E[(\sum_{i=1}^n c_i X'_i)^k]$. We have $(\sum_{i=1}^n c_i X_i)^k = \sum_{(i_1,\ldots,i_k) \in [1..n]^k} \prod_{j=1}^k c_{i_j} X_{i_j}$. Applying the definition of 1-negative correlation to the set $I = \{i_1, \ldots, i_k\}$, we compute

$$\begin{aligned} E\left[\prod_{j=1}^k c_{i_j} X_{i_j}\right] &= \left(\prod_{j=1}^k c_{i_j}\right) \Pr[\forall j \in [1..k] : X_{i_j} = 1] \\ &\leq \left(\prod_{j=1}^k c_{i_j}\right)\left(\prod_{i \in I} \Pr[X_i = 1]\right) \\ &= \left(\prod_{j=1}^k c_{i_j}\right)\left(\prod_{i \in I} \Pr[X'_i = 1]\right) = E\left[\prod_{j=1}^k c_{i_j} X'_{i_j}\right]. \end{aligned}$$

Consequently, by linearity of expectation, $E[(\sum_{i=1}^n c_i X_i)^k] \leq E[(\sum_{i=1}^n c_i X'_i)^k]$ for all $k \in \mathbb{N}$, and thus $E[Y^\ell] \leq E[(Y')^\ell]$.

We recall from Section 1.10.1.7 that essentially all large-deviation bounds are proven via upper bounds on the exponential moment $E[e^{hY}]$ of the random variable hY, where $h > 0$ is suitably chosen. Since the random variable Y is bounded, by Fubini's theorem we have

$$E[e^{hY}] = E\left[\sum_{\ell=0}^\infty \frac{h^\ell Y^\ell}{\ell!}\right] = \sum_{\ell=0}^\infty \frac{h^\ell E[Y^\ell]}{\ell!}. \tag{1.10.42}$$

Since $E[Y^\ell] \le E[(Y')^\ell]$, we have $E[e^{hY}] \le E[e^{hY'}]$. Consequently, we obtain for Y all Chernoff bounds which we could prove with the classical methods for Y'.

It remains to show that we can also work with an upper bound $\mu^+ \ge E[Y]$. For this, note that when we apply the construction of Theorem 1.10.21 to our random variables Y_i, that is, we define $Z_i = Y_i + \gamma(1 - Y_i)$ for a suitable $\gamma \in [0,1]$, then the resulting random variables Z_i have the same properties as the Y_i, that is, there are a_i' and c_i' such that $Z_i = a_i' + c_i' X_i$. Consequently, we have $\Pr[Y \ge (1+\delta)\mu^+] \le \Pr[Z \ge (1+\delta)\mu^+] = \Pr[Z \ge (1+\delta)E[Z]]$ for the sum $Z = \sum_{i=1}^n Z_i$, and the last expression can bounded via the results we have just proved. □

Theorem 1.10.24 (0-negative correlation implies lower tail bounds). *Let* $X_1,\dots,X_n$ *be 0-negatively correlated binary random variables. Let* $a_1,\dots,a_n,b_1,\dots,b_n \in \mathbb{R}$ *with* $a_i \le b_i$ *for all* $i \in [1..n]$. *Let* $Y_1,\dots,Y_n$ *be random variables with* $\Pr[Y_i = a_i] = \Pr[X_i = 0]$ *and* $\Pr[Y_i = b_i] = \Pr[X_i = 1]$. *Let* $Y = \sum_{i=1}^n Y_i$.

(a) If $a_1,\dots,a_n,b_1,\dots,b_n \in [0,1]$, *then* Y *satisfies the Chernoff bounds given in* (1.10.10) *to* (1.10.12), (1.10.14), (1.10.22) *to* (1.10.24), *and those in Corollary 1.10.13 with* $\sigma^2 := \sum_{i=1}^n \mathrm{Var}[Y_i]$.

(b) Without the restriction to $[0,1]$ *specified in (a),* Y *satisfies the Chernoff bound of* (1.10.16) *with* $c_i := b_i - a_i$ *and those of the last paragraph of Theorem 1.10.14 with* $\sigma^2 := \sum_{i=1}^n \mathrm{Var}[Y_i]$.

Each of these results also holds when all occurrences of $E[Y]$ *are replaced by* μ^- *for some* $\mu^- \le E[Y]$.

Proof. Let $\tilde{Y}_i := 1 - Y_i$. Then the $\tilde{Y}_i$ satisfy the assumptions of Theorem 1.10.23 (with $\tilde{a}_i = 1 - b_i$, $\tilde{b}_i = 1 - a_i$, and $\tilde{X}_i = 1 - X_i$; note that the latter are 1-negatively correlated, since the X_i are 0-negatively correlated; note further that $\tilde{a}_i, \tilde{b}_i \in [0,1]$ if $a_i, b_i \in [0,1]$). Hence Theorem 1.10.23 gives the usual Chernoff bounds for the $\tilde{Y}_i$. As in Section 1.10.1.2, these translate into the estimates (1.10.10) to (1.10.12) and these imply (1.10.22) to (1.10.24). The bound (1.10.13) for the $\tilde{Y}_i$ immediately translates to (1.10.14) for the Y_i. Finally, the results of Theorem 1.10.12 imply those of Corollary 1.10.13. All these results are obviously weaker when $E[Y]$ is replaced by some $\mu^- \le E[Y]$. □

1.10.2.3 Hypergeometric Distribution

It remains to point out some situations where we encounter negatively correlated random variables. One typical situation (but by far not the only one) is sampling without replacement, which leads to the *hypergeometric distribution*.

Say we choose n elements randomly from a given N-element set S *without replacement*. For a given m-element subset T of S, we wonder how many of its elements we have chosen. This random variable is said to be hypergeometrically distributed with parameters N, n, and m.

More formally, let S be any N-element set. Let $T \subseteq S$ have exactly m elements. Let U be a subset of S chosen uniformly from all n-element subsets of S. Then $X = |U \cap T|$ is a random variable with a hypergeometric distribution (with parameters N, n, and m). By definition,

$$\Pr[X = k] = \frac{\binom{m}{k}\binom{N-m}{n-k}}{\binom{N}{n}}$$

for all $k \in [\max\{0, n+m-N\}..\min\{n,m\}]$.

It is easy to see that $E[X] = \frac{|U||T|}{|S|} = \frac{mn}{N}$: Enumerate $T = \{t_1, \ldots, t_m\}$ in an arbitrary manner (before choosing U). For $i = 1, \ldots, m$, let X_i be the indicator random variable for the event $t_i \in U$. Clearly, $\Pr[X_i = 1] = \frac{|U|}{|S|} = \frac{n}{N}$. Since $X = \sum_{i=1}^{m} X_i$, we have $E[X] = \frac{mn}{N}$ by linearity of expectation (Lemma 1.6.4).

It is also obvious that the X_i are not independent. If $n < m$ and $X_1 = \ldots = X_n = 1$, then we necessarily have $X_i = 0$ for $i > n$. Fortunately, however, these dependencies are of the negative-correlation type. This is intuitively clear, but also straightforward to prove.

Let $I \subseteq [1..m]$, $W = \{t_i \mid i \in I\}$, and $w = |W| = |I|$. Then $\Pr[\forall i \in I : X_i = 1] = \Pr[W \subseteq U]$. Since U is uniformly chosen, it suffices to count the number of U that contain W, which is $\binom{|S \setminus W|}{|U \setminus W|}$, and to compare them with the total number of possible U. Hence

$$\begin{aligned}\Pr[W \subseteq U] &= \binom{N-w}{n-w} \Big/ \binom{N}{n} \\ &= \frac{n \cdot \ldots \cdot (n-w+1)}{N \cdot \ldots \cdot (N-w+1)} < \left(\frac{n}{N}\right)^w = \prod_{i \in I} \Pr[X_i = 1].\end{aligned}$$

In a similar fashion, we have

$$\begin{aligned}\Pr[\forall i \in I : X_i = 0] &= \Pr[U \cap W = \emptyset] \\ &= \binom{N-w}{n} \Big/ \binom{N}{n} \\ &= \frac{(N-n)\ldots(N-n-w+1)}{N\ldots(N-w+1)} \\ &\le \left(\frac{N-n}{N}\right)^w = \prod_{i \in I} \Pr[X_i = 0],\end{aligned}$$

where we read $\binom{N-w}{n} = 0$ when $n > N - w$.

Together with Theorems 1.10.23 and 1.10.24, we obtain the following theorem.

Theorem 1.10.25. *Let $N \in \mathbb{N}$. Let S be some set of cardinality N; for convenience, let $S = [1..N]$. Let $n \leq N$, and let U be a subset of S having cardinality n uniformly chosen from all such subsets. For $i \in [1..N]$, let X_i be the indicator random variable for the event $i \in U$. Then $X_1,\ldots,X_N$ are negatively correlated.*

Consequently, if X is a random variable having a hypergeometric distribution with parameters N, n, and m, then the usual Chernoff bounds for sums of n independent binary random variables (listed in Theorems 1.10.23 and 1.10.24) hold.

Note that for hypergeometric distributions we have symmetry in n and m, that is, the hypergeometric distribution with parameters N, n, and m is the same as the hypergeometric distribution with parameters N, m, and n. Hence, for Chernoff bounds depending on the number of random variables, for example those in Theorem 1.10.7, we can make this number $\min\{n,m\}$ by interpreting the random experiment in the right fashion.

That the hypergeometric distribution satisfies the Chernoff bounds of Theorem 1.10.7 has been attributed to Chvátal [14] in some recent publications, but this is not correct. As Chvátal writes, the aim of his note was solely to give an elementary proof of the fact that the hypergeometric distribution satisfies the strongest Chernoff bound of Theorem 1.10.1 (which implies the bounds of Theorem 1.10.7), whereas the result itself is from Hoeffding [54].

For a hypergeometric random variable X with parameters N, n, and m, [3, Lemma 2] showed that if $m < \frac{N}{2e}$ and $z \geq \frac{n}{2}$, then $\Pr[X = z] \leq (\frac{2em}{N})^z$. With Theorem 1.10.25, we can use the usual Chernoff bound (1.10.18) and obtain the stronger bound

$$\Pr[X \geq z] \leq e^{z-E[X]} \left(\frac{E[X]}{z}\right)^z \leq \left(\frac{eE[X]}{z}\right)^z = \left(\frac{enm}{zN}\right)^z, \qquad (1.10.43)$$

which is at most $(\frac{2em}{N})^z$ for $z \geq n/2$.

Theorem 1.10.25 can be extended to pointwise maxima of several families such as (X_i) in Theorem 1.10.25 if these are independent. This result was used in the analysis of a population-based genetic algorithm in [23], but might be useful also in other areas of discrete algorithmics.

Lemma 1.10.26. *Let $k, N \in \mathbb{N}$. For all $j \in [1..k]$, let $n_j \in [1..N]$. Let S be some set of cardinality N; for convenience, let $S = [1..N]$. For all $j \in [1..k]$, let U_j be a subset of S having cardinality n_j uniformly chosen from all such subsets. Let the U_j be stochastically independent. For all $i \in S$, let X_i be the indicator random variable for the event that $i \in U_j$ for some $j \in [1..k]$. Then the random variables $X_1,\ldots,X_N$ are negatively correlated.*

Note that the situation in the lemma above can be seen as sampling with partial replacement. We sample a total of $\sum_j n_j$ elements, but we replace the elements chosen after rounds $n_1, n_1+n_2, \ldots$ only. We expect that other partial replacement scenarios will also lead to negatively correlated random variables, and thus to the usual Chernoff bounds.

We recall that negative correlation can also be useful without Chernoff bounds. For example, in Section 1.9 we used the lemma above to prove a lower bound on the coupon collector time (or, equivalently, on the runtime of the randomized local search heuristic on monotonic functions).

1.10.3 Chernoff Bounds for Functions of Independent Variables, Martingales, and Bounds for Maxima

So far, we have discussed tail bounds for random variables which can be written as a sum of (more or less) independent random variables. Sometimes, the random variable we are interested in is determined by the outcomes of many independent random variables, but not simply as a sum of these. Nevertheless, if each of the independent random variables has only a limited influence on the result, then bounds similar to those of Theorem 1.10.9 can be proven. Such bounds can be found under the names of *Azuma's inequality*, *martingale inequalities*, and the *method of bounded differences*.

As far as possible, we shall try to avoid the use of martingales. The following two bounds due to McDiarmid [70] need martingales in their proof, but not in their statement.

Theorem 1.10.27 (method of bounded differences). *Let $X_1, \ldots, X_n$ be independent random variables taking values in the sets $\Omega_1, \ldots, \Omega_n$, respectively. Let $\Omega := \Omega_1 \times \ldots \times \Omega_n$. Let $f : \Omega \to \mathbb{R}$. For all $i \in [1..n]$ let $c_i > 0$ be such that for all $\omega, \bar{\omega} \in \Omega$ we have that if for all $j \neq i$, $\omega_j = \bar{\omega}_j$, then $|f(\omega) - f(\bar{\omega})| \le c_i$.*

Let $X = f(X_1, \ldots, X_n)$. Then, for all $\lambda \ge 0$,

$$\Pr[X \ge E[X] + \lambda] \le \exp\left(-\frac{2\lambda^2}{\sum_{i=1}^n c_i^2}\right),$$

$$\Pr[X \le E[X] - \lambda] \le \exp\left(-\frac{2\lambda^2}{\sum_{i=1}^n c_i^2}\right).$$

The version of Azuma's inequality given above is due to McDiarmid [70] and is stronger than the bound $\exp(-\lambda^2/2\sum_{i=1}^n c_i^2)$ given by several other authors.

Theorem 1.10.27 has found numerous applications in discrete mathematics and computer science, but only a few in the analysis of randomized search heuristics (the only one we are aware of is [11]). All other analyses of randomized search heuristics that need Chernoff-type bounds for random variables that are determined by independent random variables, but in a way other than as a simple sum, have resorted to the use of martingales.

One reason for this might be that the bounded-differences assumption is easily proven in discrete mathematics problems such as the analysis of random graphs, whereas in algorithms the sequential nature of the use of randomness makes it hard to argue that a particular random variable sampled now has a bounded influence on the final result regardless of how we condition on all future random variables. A more natural condition might be that the outcome of the current random variable has only a limited influence on the expected result determined by the future random variables. For this reason, we are optimistic that the following result might become useful in the analysis of randomized search heuristics. This result is a weak version of Theorem 3.7 in [70].

Theorem 1.10.28 (method of bounded conditional expectations). *Let $X_1,\dots,X_n$ be independent random variables taking values in the sets $\Omega_1,\dots,\Omega_n$, respectively. Let $\Omega := \Omega_1 \times \dots \times \Omega_n$. Let $f : \Omega \to \mathbb{R}$. For all $i \in [1..n]$ let $c_i > 0$ be such that for all $\omega_1 \in \Omega_1,\dots,\omega_{i-1} \in \Omega_{i-1}$ and all $\omega_i, \bar{\omega}_i \in \Omega_i$ we have*

$$|E[f(\omega_1,\dots,\omega_{i-1},\omega_i,X_{i+1},\dots,X_n)] - E[f(\omega_1,\dots,\omega_{i-1},\bar{\omega}_i,X_{i+1},\dots,X_n)]| \\ \le c_i.$$

Let $X = f(X_1,\dots,X_n)$. Then, for all $\lambda \ge 0$,

$$\Pr[X \ge E[X] + \lambda] \le \exp\left(-\frac{2\lambda^2}{\sum_{i=1}^n c_i^2}\right),$$
$$\Pr[X \le E[X] - \lambda] \le \exp\left(-\frac{2\lambda^2}{\sum_{i=1}^n c_i^2}\right).$$

Here is an example of how the new theorem can be helpful. The *compact genetic algorithm (cGA)* without frequency boundaries maximizes a function $f : \{0,1\}^n \to \mathbb{R}$ as follows. There is a (hypothetical) population size $K \in \mathbb{N}$, which we assume to be an even integer. The cGA sets the initial *frequency vector* $\tau^{(0)} \in [0,1]^n$ to $\tau^{(0)} = (\frac{1}{2},\dots,\frac{1}{2})$. Then, in each iteration $t = 1,2,\dots$ it generates two search points $x^{(t,1)}, x^{(t,2)} \in \{0,1\}^n$ randomly such that, independently for all $j \in \{1,2\}$ and $i \in [1..n]$, we have $\Pr[x_i^{(t,j)} = 1] = \tau_i^{(t)}$. If $f(x^{(t,1)}) < f(x^{(t,2)})$, then we swap the two variables, that is, we set $(x^{(t,1)}, x^{(t,2)}) \leftarrow (x^{(t,2)}, x^{(t,1)})$. Finally, in this iteration, we update the frequency vector by setting $\tau^{(t+1)} \leftarrow \tau^{(t)} + \frac{1}{K}(x^{(t,1)} - x^{(t,2)})$.

Let us analyze the behavior of the frequency $\tau_i^{(t)}$ of a *neutral bit* $i \in [1..n]$, that is, one that has property that $f(x) = f(y)$ for all x and y which differ only in the i-th bit.

Lemma 1.10.29. *Let K be an even integer. Consider a run of the cGA with hypothetical population size K on an objective function $f:\{0,1\}^n \to \mathbb{R}$ having a neutral bit i. For all $T \in \mathbb{N}$, the probability that within the first T iterations the frequency of the i-th bit has converged to one of the absorbing states 0 or 1 is at most $2\exp\left(-\frac{K^2}{32T}\right)$.*

Proof. To ease reading, let $X_t := \tau_i^{(t)}$. We have $X_0 = \frac{1}{2}$ with probability one. Once X_t is determined, we have

$$\begin{aligned}\Pr[X_{t+1} = X_t + \tfrac{1}{K}] &= X_t(1-X_t),\\ \Pr[X_{t+1} = X_t - \tfrac{1}{K}] &= X_t(1-X_t),\\ \Pr[X_{t+1} = X_t] &= 1-2X_t(1-X_t).\end{aligned}$$

In particular, we have $E[X_{t+1} \mid X_0,\dots,X_t] = E[X_{t+1} \mid X_t] = X_t$. By induction, we have $E[X_T \mid X_t] = X_t$ for all $T > t$.

Our aim is to show that, with probability at least $1-2\exp\left(-\frac{K^2}{32T}\right)$, X_T has not yet reached to one of the absorbing states 0 and 1. We first write the frequencies as results obtained from independent random variables. For convenience, these will be continuous random variables, but it is easy to see that we could have used discrete ones instead. For all $t = 1,2,\dots$, let R_t be a random number uniformly distributed in the interval $[0,1]$. Define $Y_0, Y_1, \dots$ as follows. We have $Y_0 = \frac{1}{2}$ with probability one. For $t \in \mathbb{N}_0$, we set

$$\begin{aligned}\Pr[Y_{t+1} = Y_t + \tfrac{1}{K}] &\text{ if } R_t \ge 1 - Y_t(1-Y_t),\\ \Pr[Y_{t+1} = Y_t - \tfrac{1}{K}] &\text{ if } R_t \le Y_t(1-Y_t),\\ \Pr[Y_{t+1} = Y_t] &\text{ otherwise.}\end{aligned}$$

It is easy to see that $(X_0, X_1, \dots)$ and $(Y_0, Y_1, \dots)$ are identically distributed. Note that Y_T is a function g of $(R_1, \dots, R_T)$. For concrete values $r_1, \dots, r_t \in [0,1]$, we have $E[g(r_1,\dots,r_t,R_{t+1},\dots,R_T)] = E[Y_T \mid Y_t] = Y_t$. Consequently, for all $\overline{r}_t \in [0,1]$, the two expectations $E[g(r_1,\dots,r_{t-1},r_t,R_{t+1},\dots,R_T)]$ and $E[g(r_1,\dots,r_{t-1},\overline{r}_t,R_{t+1},\dots,R_T)]$ are two possible outcomes of Y_t given a common value for Y_{t-1} (which is determined by $r_1,\dots,r_{t-1}$), and hence differ by at most $c_t = \frac{2}{K}$. We can thus apply Theorem 1.10.28 as follows.

$$\begin{aligned}\Pr[Y_T \in \{0,1\}] &= \Pr\left[|Y_T - \tfrac{1}{2}| \ge \tfrac{1}{2}\right]\\ &= \Pr\left[|g(R_1,\dots,R_T) - E[g(R_1,\dots,R_T)]| \ge \tfrac{1}{2}\right]\\ &\le 2\exp\left(-\frac{(\frac{1}{2})^2}{2T(\frac{2}{K})^2}\right) = 2\exp\left(-\frac{K^2}{32T}\right).\end{aligned}$$

□

Note that it is not obvious how to obtain this result with the classical method of bounded differences (Theorem 1.10.27). In particular, the above construction does not satisfy the bounded-differences condition, that is, there are values $r_1,\dots,r_T$ and $\overline{r}_t$ such that $g(r_1,\dots,r_T)$ and $g(r_1,\dots,r_{t-1},\overline{r}_t,r_{t+1},\dots,r_T)$ differ by significantly more than $\frac{2}{K}$. To see this, consider the following example. Let $r_i = 1$ for even i and $r_i = \frac{1}{4}$ for odd i. Then $g(r_1,\dots,r_T) = \frac{1}{2}$ for even T and $g(r_1,\dots,r_T) = \frac{1}{2} - \frac{1}{K}$ for odd T. However, for all even T we have $g(\frac{1}{2},r_2,\dots,r_T) = g(\frac{1}{2},r_2,\dots,r_{T+1}) = \min\{1,\frac{1}{2}+\frac{T}{2}\cdot\frac{1}{K}\}$, showing that a change in the first variable leads to a drastic change in the g-values for larger T.

This example shows that our stochastic modeling of the process cannot be analyzed via the method of bounded differences. We cannot rule out the possibility that a different modeling might admit an analysis via the method of bounded differences, but nevertheless this example suggests that Theorem 1.10.28 is a useful tool in the theory of randomized search heuristics.

Without going into details (and, in particular, without defining the notion of a martingale), we note that both Theorem 1.10.27 and Theorem 1.10.28 are special cases of the following martingale result, which is often attributed to Azuma [2] despite the fact that it had already been proposed by Hoeffding [54]. Readers familiar with martingales may find it more natural to use this result rather than the previous two theorems in their work; however, it has to be said that not all researchers in the theory of algorithms are familiar with martingales.

Theorem 1.10.30 (Azuma–Hoeffding inequality). *Let $X_0,X_1,\dots,X_n$ be a martingale. Let $c_1,\dots,c_n > 0$ with $|X_i - X_{i-1}| \le c_i$ for all $i \in [1..n]$. Then, for any $\lambda \ge 0$,*

$$\Pr[X_n - X_0 \ge \lambda] \le \exp\left(-\frac{\lambda^2}{2\sum_{i=1}^n c_i^2}\right).$$

This result has found several applications in the theory of randomized search heuristics, for example, in [25, 40, 62].

We observe that the theorem above is a direct extension of Theorem 1.10.9 to martingales (note that the c_i there are twice as large as here, which explains the different location of the 2 in the bounds). In a similar vein, there are martingale versions of most other Chernoff bounds presented in this chapter. We refer to McDiarmid [70] for more details.

1.10.3.1 Tail Bounds for Maxima and Minima of Partial Sums

We end this section with a gem already contained in Hoeffding's work. It builds on the following elementary observation: if $X_0,X_1,\dots,X_n$ form a martingale, then $Y_0,Y_1,\dots,Y_n$ defined as follows also form a martingale. Let $\lambda \in \mathbb{R}$. Let $i \in [0..n]$ be minimal with $X_i \ge \lambda$, if such an X_i exists, and $i = n+1$ otherwise. Let $Y_j = X_j$ for $j \le i$ and $Y_j = X_i$ for $j > i$. Then $Y_n \ge \lambda$ if and only if $\max_{i\in[1..n]} X_i \ge \lambda$. Since $Y_0,\dots,Y_n$ is a martingale with martingale differences bounded at least as well as for $X_0,\dots,X_n$ (and also all other variation measures at least as good as for $X_0,\dots,X_n$), all large-deviation bounds provable for the martingale $X_0,\dots,X_n$ via the Bernstein method are valid also for $Y_0,\dots,Y_n$, that is, for $\max_{i\in[1..n]} X_i$. Since we have not introduced martingales here, we omit the details and state only some implications of this observation. The reader will find more details in [54, end of Section 2] and [70, end of Section 3.5]. It seems that the authors of [54] and [70] do not see this extension as very important (see also the comment at the end of Section 2 in [70]). We feel that this might be different for randomized search heuristics. For example, to prove that a randomized search heuristic has at least some optimization time T, we need to show that the distance of each of the first $T-1$ solutions from the optimum is positive, that is, that the minimum of these differences is positive.

Theorem 1.10.31 (tail bounds for maxima and minima). *Let $X_1,\dots,X_n$ be independent random variables. For all $i \in [1..n]$, let $S_i = \sum_{j=1}^{i} X_j$. Assume that one of the results in Section 1.10.1 yields the tail bound $\Pr[S_n \geq E[S_n]+\lambda] \leq p$. Then we also have*

$$\Pr[\exists i \in [1..n] : S_i \geq E[S_i]+\lambda] \leq p. \tag{1.10.44}$$

In an analogous manner, each tail bound $\Pr[S_n \leq E[S_n]-\lambda] \leq p$ derivable from Section 1.10.1 can be strengthened to $\Pr[\exists i \in [1..n] : S_i \leq E[S_i]-\lambda] \leq p$.

Note that if the X_i in the theorem are non-negative, then, trivially, (1.10.44) implies the uniform bound

$$\Pr[\exists i \in [1..n] : S_i \geq E[S_n]+\lambda] \leq p. \tag{1.10.45}$$

Note also that the deviation parameter λ does not scale with i. In particular, a bound like $\Pr[\exists i \in [1..n] : S_i \geq (1+\delta)E[S_i]] \leq p$ cannot be derived.

1.10.4 Chernoff Bounds for Geometric Random Variables

As is visible from Lemma 1.10.36 below, sums of independent geometric random variables occur frequently in the analysis of randomized search heuristics. Surprisingly, it was only in 2007 that a Chernoff-type bound was used to analyze such sums in the theory of randomized search heuristics [31] (for subsequent uses see, e.g., [4, 23, 32, 35, 104]). Even more surprisingly, Witt [102] only recently proved good tail bounds for sums of geometric random variables having significantly different success probabilities. Note that geometric random variables are unbounded. Hence the Chernoff bounds presented so far cannot be applied directly.

We start this subsection with simple Chernoff bounds for sums of identically distributed geometric random variables, as these can be derived from the Chernoff bounds for sums of independent 0,1 random variables discussed so far. We remark that a sum X of n independent geometric distributions with success probability $p > 0$ is closely related to the *negative binomial distribution* $\mathrm{NB}(n, 1-p)$ with parameters n and $1-p$: we have $X \sim \mathrm{NB}(n, 1-p) + n$.

Theorem 1.10.32. *Let $X_1, \ldots, X_n$ be independent geometric random variables with common success probability $p > 0$. Let $X := \sum_{i=1}^n X_i$ and $\mu := E[X] = \frac{n}{p}$.*

(a) For all $\delta \geq 0$,

$$\Pr[X \geq (1+\delta)\mu] \leq \exp\left(-\frac{\delta^2}{2}\frac{n-1}{1+\delta}\right) \leq \exp\left(-\frac{1}{4}\min\{\delta^2, \delta\}(n-1)\right). \tag{1.10.46}$$

(b) For all $0 \leq \delta \leq 1$,

$$\Pr[X \leq (1-\delta)\mu] \leq (1-\delta)^n \left(\frac{(1-\delta)(\mu-n)}{(1-\delta)\mu - n}\right)^{(1-\delta)\mu - n} \tag{1.10.47}$$

$$\leq (1-\delta)^n \exp(\delta n) \tag{1.10.48}$$

$$\leq \exp\left(-\frac{\delta^2 n}{2-\frac{4}{3}\delta}\right), \tag{1.10.49}$$

where the first bound reads as p^n for $(1-\delta)\mu = n$ and as 0 for $(1-\delta)\mu < n$. For $0 \leq \delta < 1$ and $\lambda \geq 0$, we also have

$$\Pr[X \leq (1-\delta)\mu] \leq \exp\left(-\frac{2\delta^2 pn}{1-\delta}\right), \tag{1.10.50}$$

$$\Pr[X \leq \mu - \lambda] \leq \exp\left(-\frac{2p^3\lambda^2}{n}\right). \tag{1.10.51}$$

The bounds (1.10.50) and (1.10.51) are interesting only for relatively large values of p. Since part (a) has been proven in [4], we prove only (b). The main idea in both cases is exploiting the natural relation between a sum of independent identically distributed geometric random variables and a sequence of Bernoulli events.

Proof. Let $Z_1, Z_2, \ldots$ be independent binary random variables with $\Pr[Z_i = 1] = p$ for all $i \in \mathbb{N}$. Let $n \leq K \leq \frac{n}{p}$. Let $Y_K = \sum_{i=1}^K Z_i$. Then $X \leq K$ if and only if $Y_K \geq n$. Consequently, by Theorem 1.10.1,

$$\begin{aligned}\Pr[X \leq K] &= \Pr[Y_K \geq n] \\ &= \Pr\left[Y_K \geq \left(1 + \left(\frac{n}{Kp} - 1\right)\right) E[Y_K]\right]\end{aligned}$$

$$
\begin{aligned}
&\leq \left(\frac{Kp}{n}\right)^n \left(\frac{K-Kp}{K-n}\right)^{K-n} \\
&\leq \left(\frac{Kp}{n}\right)^n \exp(n-Kp) \\
&\leq \exp\left(\frac{p\lambda^2}{2K+\frac{2}{3}\lambda}\right),
\end{aligned}
$$

where we have used the shorthand $\lambda := \mu - K$ for the absolute deviation. From Theorem 1.10.7, we derive

$$
\begin{aligned}
\Pr[X \leq K] &= \Pr[Y_K \geq n] \\
&= \Pr[Y_K \geq E[Y_K] + (n-Kp)] \\
&\leq \exp\left(-\frac{2(n-Kp)^2}{K}\right) \\
&= \exp\left(-\frac{2p^2\lambda^2}{\mu-\lambda}\right) \leq \exp\left(-\frac{2p^2\lambda^2}{\mu}\right).
\end{aligned}
$$

Replacing K by $(1-\delta)\mu$ and λ by $\delta\mu$ in these equations gives the claim. □

When the geometric random variables have different success probabilities, the following bounds can be employed.

Theorem 1.10.33. *Let $X_1,\ldots,X_n$ be independent geometric random variables with success probabilities $p_1,\ldots,p_n > 0$. Let $p_{\min} := \min\{p_i \mid i \in [1..n]\}$. Let $X := \sum_{i=1}^n X_i$ and $\mu = E[X] = \sum_{i=1}^n \frac{1}{p_i}$.*

(a) For all $\delta \geq 0$,

$$
\begin{aligned}
\Pr[X \geq (1+\delta)\mu] &\leq \frac{1}{1+\delta}(1-p_{\min})^{\mu(\delta-\ln(1+\delta))} && (1.10.52)\\
&\leq \exp(-p_{\min}\mu(\delta-\ln(1+\delta))) && (1.10.53)\\
&\leq \left(1+\frac{\delta\mu p_{\min}}{n}\right)^n \exp(-\delta\mu p_{\min}) && (1.10.54)\\
&\leq \exp\left(-\frac{(\delta\mu p_{\min})^2}{2n(1+\frac{\delta\mu p_{\min}}{n})}\right). && (1.10.55)
\end{aligned}
$$

(b) For all $0 \leq \delta \leq 1$,

$$
\begin{aligned}
\Pr[X \leq (1-\delta)\mu] &\leq (1-\delta)^{p_{\min}\mu} \exp(-\delta p_{\min}\mu) && (1.10.56)\\
&\leq \exp\left(-\frac{\delta^2\mu p_{\min}}{2-\frac{4}{3}\delta}\right) && (1.10.57)\\
&\leq \exp(-\tfrac{1}{2}\delta^2\mu p_{\min}). && (1.10.58)
\end{aligned}
$$

The estimates (1.10.52) and (1.10.53) are from [60], the bound (1.10.54) is from [89], and (1.10.55) follows from the previous bound by standard estimates. This last bound, when applied to identically distributed random variables, is essentially the same as (1.10.46).

For the lower tail bounds, (1.10.56) from [60] is identical to (1.10.48) for identically distributed variables. Hence (1.10.47) is the strongest estimate for identically distributed geometric random variables. Equation (1.10.56) gives (1.10.57) via the same estimate that gives (1.10.49) from (1.10.48). Estimate (1.10.58) appeared earlier in [89].

Overall, it remains surprising that such useful bounds have been proven only relatively late and have not yet appeared in a scientific journal.

The bounds of Theorem 1.10.33 allow the geometric random variables to have different success probabilities; however, the tail probability depends only on the smallest of them. This is partially justified by the fact that the corresponding geometric random variable has the largest variance, and thus might be most detrimental to the desired strong concentration. If the success probabilities vary significantly, however, then this result gives overly pessimistic tail bounds, and the following result of Witt [102] can lead to stronger estimates.

Theorem 1.10.34. *Let $X_1,\dots,X_n$ be independent geometric random variables with success probabilities $p_1,\dots,p_n > 0$. Let $X = \sum_{i=1}^n X_i$, $s = \sum_{i=1}^n (\frac{1}{p_i})^2$, and $p_{\min} := \min\{p_i \mid i \in [1..n]\}$. Then, for all $\lambda \geq 0$,*

$$\Pr[X \geq E[X] + \lambda] \leq \exp\left(-\frac{1}{4}\min\left\{\frac{\lambda^2}{s}, \lambda p_{\min}\right\}\right), \tag{1.10.59}$$

$$\Pr[X \leq E[X] - \lambda] \leq \exp\left(-\frac{\lambda^2}{2s}\right). \tag{1.10.60}$$

In the analysis of randomized search heuristics, it appears that we often encounter sums of independent geometrically distributed random variables $X_1,\dots,X_n$ with success probabilities p_i proportional to i. For this case, the following result from [23, Lemma 4] gives stronger tail bounds than the previous result. See Section 1.4.2 for the definition of the harmonic number H_n.

Theorem 1.10.35. *Let $X_1,\dots,X_n$ be independent geometric random variables with success probabilities $p_1,\dots,p_n$. Assume that there is a number $C \leq 1$ such that $p_i \geq C\frac{i}{n}$ for all $i \in [1..n]$. Let $X = \sum_{i=1}^n X_i$. Then*

$$E[X] \leq \tfrac{1}{C} n H_n \leq \tfrac{1}{C} n(1 + \ln n), \tag{1.10.61}$$

$$\Pr[X \geq (1+\delta)\tfrac{1}{C} n \ln n] \leq n^{-\delta} \text{ for all } \delta \geq 0. \tag{1.10.62}$$

As announced in Section 1.8.2, we now present a few examples where the existing literature gives only an upper bound on the expected runtime, but where a closer look at the proofs easily gives more details about the distribution, which in particular allows one to obtain tail bounds on the runtime. We note that similar results have previously (and before [20]) only been presented for the (1+1) EA optimizing the LEADINGONES test function [35] and for RLS optimizing the ONEMAX test function [102]. Zhou et al. [104] implicitly gave several results of this type; however, the resulting runtime guarantees are not optimal, owing to the use of an inferior Chernoff bound for geometric random variables.

Lemma 1.10.36. *(a) The runtime T of the (1+1) EA on the* ONEMAX *function is dominated by the independent sum $\sum_{i=1}^{n} \mathrm{Geom}(\frac{i}{en})$ [47]. Hence $E[T] \le enH_n$ and $\Pr[T \ge (1+\delta)en\ln n] \le n^{-\delta}$ for all $\delta \ge 0$.*

(b) For any function $f : \{0,1\}^n \to \mathbb{R}$, the runtime T of the (1+1) EA is dominated by $\mathrm{Geom}(n^{-n})$ [47]. Hence $E[T] \le n^n$ and $\Pr[T \ge \gamma n^n] \le (1-n^{-n})^{\gamma n^n} \le e^{-\gamma}$ for all $\gamma \ge 0$.

(c) The runtime T of the (1+1) EA for finding Eulerian cycles in undirected graphs using perfect matchings in the adjacency lists as the genotype and using the edge-based mutation operator is dominated by the independent sum $\sum_{i=1}^{m/3} \mathrm{Geom}(\frac{i}{2em})$ [36]. Hence $E[T] \le 2emH_{m/3}$ and $\Pr[T \ge 2(1+\delta)em\ln\frac{m}{3}] \le (\frac{m}{3})^{-\delta}$ for all $\delta \ge 0$.

(d) The runtime of the (1+1) EA for sorting an array of length n by minimizing the number of inversions is dominated by the independent sum $\sum_{i=1}^{\binom{n}{2}} \mathrm{Geom}(\frac{3i}{4e\binom{n}{2}})$ [88]. Hence $E[T] \le \frac{4e}{3}\binom{n}{2}H_{\binom{n}{2}} \le \frac{2e}{3}n^2(1+2\ln n)$ and $\Pr[T \ge (1+\delta)\frac{4e}{3}n^2\ln n] \le \binom{n}{2}^{-\delta}$.
Similarly, the runtime of the (1+1) EA using a tree-based representation for the sorting problem [30] has a runtime satisfying $T \preceq \sum_{i=1}^{\binom{n}{2}} \mathrm{Geom}(\frac{1}{2e})$. Hence the expected optimization time is $E[T] = 2e\binom{n}{2}$ and we have the tail bound $\Pr[T \ge (1+\delta)E[T]] \le \exp(-\delta^2 n/(2+2\delta))$. This example shows that a superior representation can not only improve the expected runtime, but also lead to significantly lighter tails (negative exponential vs. inverse polynomial).

(e) The runtime of the multi-criteria (1+1) EA for the single-source shortest-path problem in a graph G can be described as follows. Let ℓ be such that there is a shortest path from the source to any vertex having at most ℓ edges. Then there are random variables G_{ij}, $i \in [1..\ell]$, $j \in [1..n-1]$, such that (i) $G_{ij} \sim \mathrm{Geom}(\frac{1}{en^2})$ for all $i \in [1..\ell]$ and $j \in [1..n-1]$, (ii) for all $j \in [1..n-1]$, the variables $G_{1j},\dots,G_{\ell j}$ are independent, and (iii) T is dominated by $\max\{\sum_{i=1}^{\ell} G_{ij} \mid j \in [1..n-1]\}$ [32, 88]. Consequently, for $\delta = \max\{\frac{4\ln(n-1)}{\ell-1}, \sqrt{\frac{4\ln(n-1)}{\ell-1}}\}$ and $T_0 := (1+\delta)\frac{\ell}{p}$, we have $E[T] \le (1+\frac{1}{\ln(n-1)})T_0$ and $\Pr[T \ge (1+\varepsilon)T_0] \le (n-1)^{-\varepsilon}$ for all $\varepsilon \ge 0$.

Proof. We shall not show the domination statements, as these can be easily derived from the original analyses cited in the theorem. Given the domination result, parts (a), (c), and (d) follow immediately from Theorem 1.10.35. Part (b) follows directly from the law of the geometric distribution.

To prove part (e), let $X_1,\dots,X_\ell$ be independent geometrically distributed random variables with parameter $p=\frac{1}{en^2}$. Let $X=\sum_{i=1}^{\ell} X_i$. Let $\delta=\max\{\frac{4\ln(n-1)}{\ell-1},\sqrt{\frac{4\ln(n-1)}{\ell-1}}\}$. Then, by (1.10.46), $\Pr[X\geq(1+\delta)E[X]]\leq \exp(-\frac{1}{2}\frac{\delta^2}{1+\delta}(\ell-1))\leq\exp(-\frac{1}{4}\min\{\delta^2,\delta\}(\ell-1))\leq\exp(-\frac{1}{4}\frac{4\ln(n-1)}{\ell-1}(\ell-1))=\frac{1}{n-1}$. For all $\varepsilon>0$, again by (1.10.46), we compute

$$\begin{aligned}\Pr[X\geq(1+\varepsilon)(1+\delta)E[X]]&\leq\exp\left(-\frac{1}{2}\frac{(\delta+\varepsilon+\delta\varepsilon)^2}{(1+\delta)(1+\varepsilon)}(\ell-1)\right)\\&\leq\exp\left(-\frac{1}{2}\frac{\delta^2(1+\varepsilon)^2}{(1+\delta)(1+\varepsilon)}(\ell-1)\right)\\&\leq\exp\left(-\frac{1}{2}\frac{\delta^2}{1+\delta}(\ell-1)\right)^{1+\varepsilon}\leq(n-1)^{-(1+\varepsilon)}.\end{aligned}$$

Let $Y_1,\dots,Y_{n-1}$ be random variables with a distribution equal to that of X. We do not make any assumptions about the correlation of the Y_i; in particular, they do not need to be independent. Let $Y=\max\{Y_i\mid i\in[1..n-1]\}$, and recall that the runtime T is dominated by Y. Let $T_0=(1+\delta)E[X]=(1+\delta)\frac{\ell}{p}$. Then $\Pr[Y\geq(1+\varepsilon)T_0]\leq(n-1)\Pr[X\geq(1+\varepsilon)T_0]\leq(n-1)^{-\varepsilon}$ by the union bound (Lemma 1.5.1). By Corollary 1.6.2,

$$E[Y]\leq\left(1+\frac{1}{\ln(n-1)}\right)T_0.$$

□

We note that not all classical proofs reveal details about the distribution. For results obtained via random walk arguments, for example, the optimization of the short path function SPC_n [57], monotone polynomials [98], or vertex covers on path-like graphs [77], as well as for results proven via additive drift [53], the proofs often give little information about the runtime distribution (an exception is the analysis of the needle function and the ONEMAX function in [50]).

For results obtained via the average weight decrease method [75] or via multiplicative drift analysis [38], the proofs also do not give information about the runtime distribution. However, the probabilistic runtime bound of type $\Pr[T\geq T_0+\lambda]\leq(1-\delta)^\lambda$ obtained from the tail bound in [29] implies that the runtime is dominated by $T\preceq T_0-1+\mathrm{Geom}(1-\delta)$.

1.10.5 Tail Bounds for the Binomial Distribution

For binomially distributed random variables, tail bounds exist which are slightly stronger than the bounds for general sums of independent 0,1 random variables. The difference are small, but since they have been used in the analysis of randomized search heuristics, we briefly describe them here.

In this section, let X always be a binomially distributed random variable with parameters n and p, that is, $X = \sum_{i=1}^{n} X_i$ with independent X_i satisfying $\Pr[X_i = 1] = p$ and $\Pr[X_i = 0] = 1 - p$. The following estimate seems well known (e.g., it was used in [56] without proof or reference). Gießen and Witt [51, Lemma 3] gave an elementary proof via estimates of binomial coefficients and the binomial identity. We find the proof below more intuitive.

Lemma 1.10.37. *Let* $X \sim \mathrm{Bin}(n,p)$*. Let* $k \in [0..n]$*. Then*

$$\Pr[X \geq k] \leq \binom{n}{k} p^k.$$

Proof. For all $T \subseteq [1..n]$ with $|T| = k$, let A_T be the event that $X_i = 1$ for all $i \in T$. Clearly, $\Pr[A_T] = p^k$. The event "$X \geq k$" is the union of the events A_T, with T as above. Hence $\Pr[X \geq k] \leq \sum_T \Pr[A_T] = \binom{n}{k} p^k$ by the union bound (Lemma 1.5.1). □

When the binomial coefficient is estimated by $\binom{n}{k} \leq (\frac{en}{k})^k$, which is often an appropriate way to derive more understandable expressions, the above bound reverts to (1.10.7), a slightly weaker version of the classical multiplicative bound (1.10.2). Since we are not aware of an application of Lemma 1.10.37 that does not estimate the binomial coefficient in this way, its main value might be its simplicity.

The following tail bound for the binomial distribution was shown by Klar [61], again with elementary arguments. In many cases, it is significantly stronger than Lemma 1.10.37. However, again we do not see an example where this tail bound would have improved an existing analysis of a randomized search heuristic.

Lemma 1.10.38. *Let* $X \sim \mathrm{Bin}(n,p)$ *and* $k \in [np..n]$*. Then*

$$\Pr[X \geq k] \leq \frac{(k+1)(1-p)}{k+1-(n+1)p} \Pr[X = k].$$

Note that, trivially, $\Pr[X = k] \leq \Pr[X \geq k]$, so it is immediately clear that this estimate is quite tight (the gap is at most the factor $\frac{(k+1)(1-p)}{k+1-(n+1)p}$). With elementary arguments, Lemma 1.10.38 gives the slightly weaker estimate

$$\Pr[X \geq k] \leq \frac{k - kp}{k - np} \Pr[X = k], \tag{1.10.63}$$

which has appeared also in [49, equation (VI.3.4)]. For $p = \frac{1}{n}$, the typical mutation rate in standard bit mutation, Lemma 1.10.38 gives

$$\Pr[X \geq k] \leq \frac{k+1}{k} \Pr[X = k]. \tag{1.10.64}$$

Writing Lemma 1.10.37 in the equivalent form $\Pr[X \geq k] \leq (\frac{1}{1-p})^{n-k} \Pr[X = k]$ and noting that $(\frac{1}{1-p})^{n-k} \geq \exp(p(n-k))$, we see that in many cases Lemma 1.10.38 gives substantially better estimates.

Finally, we mention the following estimates for the probability function of the binomial distribution stemming from [8]. By summing over all values for $k' \geq k$, upper and lower bounds for tail probabilities can be be derived.

Theorem 1.10.39. *Let* $X \sim \mathrm{Bin}(n,p)$ *with* $np \geq 1$. *Let* $h > 0$ *such that* $k = np + h \in \mathbb{N}$. *Let* $q = 1 - p$.

(a) If $hqn/3 \geq 1$, *then*

$$\Pr[X = k] < \frac{1}{\sqrt{2\pi pqn}} \exp\left(-\frac{h^2}{2pqn} + \frac{h}{qn} + \frac{h^3}{p^2n^2}\right).$$

(b) If $k < n$, *then*

$$\Pr[X = k] > \frac{1}{\sqrt{2\pi pqn}} \exp\left(-\frac{h^2}{2pqn} - \frac{h^3}{2q^2n^2} - \frac{h^4}{3p^3n^3} - \frac{h}{2pn} - \frac{1}{12k} - \frac{1}{12(n-k)}\right).$$

References

[1] Antipov, D., Doerr, B., Fang, J., Hetet, T.: Runtime analysis for the $(\mu+\lambda)$ EA optimizing OneMax. In: Genetic and Evolutionary Computation Conference, GECCO 2018, pp. 1459–1466. ACM (2018)
[2] Azuma, K.: Weighted sums of certain dependent variables. Tohoku Mathematical Journal **19**, 357–367 (1967)
[3] Badkobeh, G., Lehre, P.K., Sudholt, D.: Unbiased black-box complexity of parallel search. In: Parallel Problem Solving from Nature, PPSN 2014, pp. 892–901. Springer (2014)
[4] Baswana, S., Biswas, S., Doerr, B., Friedrich, T., Kurur, P.P., Neumann, F.: Computing single source shortest paths using single-objective fitness. In: 10th Workshop on Foundations of Genetic Algorithms, FOGA 2009, pp. 59–66. ACM (2009)
[5] Bennett, G.: Probability inequalities for the sum of independent random variables. Journal of the American Statistical Association **57**, 33–45 (1962)

[6] Bernstein, S.N.: On a modification of Chebyshev's inequality and of the error formula of Laplace. Ann. Sci. Inst. Sav. Ukraine, Sect. Math. 1 **4**, 38–49 (1924)

[7] Bienaymé, I.J.: Considérations à l'appui de la découverte de Laplace. Comptes Rendus de l'Académie des Sciences **37**, 309–324 (1853)

[8] Bollobás, B.: Random Graphs. Cambridge University Press (2001)

[9] Borisovsky, P.A., Eremeev, A.V.: Comparing evolutionary algorithms to the (1+1)-EA. Theoretical Computer Science **403**, 33–41 (2008)

[10] Böttcher, S., Doerr, B., Neumann, F.: Optimal fixed and adaptive mutation rates for the LeadingOnes problem. In: Parallel Problem Solving from Nature, PPSN 2010, pp. 1–10. Springer (2010)

[11] Buzdalov, M., Doerr, B.: Runtime analysis of the $(1+(\lambda,\lambda))$ genetic algorithm on random satisfiable 3-CNF formulas. In: Genetic and Evolutionary Computation Conference, GECCO 2017, pp. 1343–1350. ACM (2017)

[12] Buzdalov, M., Doerr, B., Kever, M.: The unrestricted black-box complexity of jump functions. Evolutionary Computation **24**, 719–744 (2016)

[13] Chernoff, H.: A measure of asymptotic efficiency for tests of a hypothesis based on the sum of observations. Annals of Mathematical Statistics **23**, 493–507 (1952)

[14] Chvátal, V.: The tail of the hypergeometric distribution. Discrete Mathematics **25**, 285–287 (1979)

[15] Corus, D., Dang, D., Eremeev, A.V., Lehre, P.K.: Level-based analysis of genetic algorithms and other search processes. IEEE Transactions on Evolutionary Computation **22**, 707–719 (2018)

[16] Dang, D., Lehre, P.K.: Simplified runtime analysis of estimation of distribution algorithms. In: Genetic and Evolutionary Computation Conference, GECCO 2015, pp. 513–518. ACM (2015)

[17] Dang, D.C., Lehre, P.K.: Self-adaptation of mutation rates in non-elitist populations. In: Parallel Problem Solving from Nature, PPSN 2016, pp. 803–813. Springer (2016)

[18] Doerr, B.: Analyzing randomized search heuristics: Tools from probability theory. In: A. Auger, B. Doerr (eds.) Theory of Randomized Search Heuristics, pp. 1–20. World Scientific (2011)

[19] Doerr, B.: A lower bound for the discrepancy of a random point set. Journal of Complexity **30**, 16–20 (2014)

[20] Doerr, B.: Better runtime guarantees via stochastic domination. In: Evolutionary Computation in Combinatorial Optimization, EvoCOP 2018, pp. 1–17. Springer (2018)

[21] Doerr, B.: An elementary analysis of the probability that a binomial random variable exceeds its expectation. Statistics and Probability Letters **139**, 67–74 (2018)

[22] Doerr, B., Doerr, C.: The impact of random initialization on the runtime of randomized search heuristics. Algorithmica **75**, 529–553 (2016)

[23] Doerr, B., Doerr, C.: Optimal static and self-adjusting parameter choices for the $(1+(\lambda,\lambda))$ genetic algorithm. Algorithmica **80**, 1658–1709 (2018)

[24] Doerr, B., Doerr, C., Yang, J.: k-bit mutation with self-adjusting k outperforms standard bit mutation. In: Parallel Problem Solving from Nature, PPSN 2016, pp. 824–834. Springer (2016)

[25] Doerr, B., Doerr, C., Yang, J.: Optimal parameter choices via precise black-box analysis. In: Genetic and Evolutionary Computation Conference, GECCO 2016, pp. 1123–1130. ACM (2016)

[26] Doerr, B., Fouz, M., Witt, C.: Quasirandom evolutionary algorithms. In: Genetic and Evolutionary Computation Conference, GECCO 2010, pp. 1457–1464. ACM (2010)

[27] Doerr, B., Fouz, M., Witt, C.: Sharp bounds by probability-generating functions and variable drift. In: Genetic and Evolutionary Computation Conference, GECCO 2011, pp. 2083–2090. ACM (2011)

[28] Doerr, B., Gießen, C., Witt, C., Yang, J.: The $(1+\lambda)$ evolutionary algorithm with self-adjusting mutation rate. Algorithmica **81**, 593–631 (2019)

[29] Doerr, B., Goldberg, L.A.: Adaptive drift analysis. Algorithmica **65**, 224–250 (2013)

[30] Doerr, B., Happ, E.: Directed trees: A powerful representation for sorting and ordering problems. In: Congress on Evolutionary Computation, CEC 2008, pp. 3606–3613. IEEE (2008)

[31] Doerr, B., Happ, E., Klein, C.: A tight bound for the (1 + 1)-EA for the single source shortest path problem. In: Congress on Evolutionary Computation, CEC 2007, pp. 1890–1895. IEEE (2007)

[32] Doerr, B., Happ, E., Klein, C.: Tight analysis of the (1+1)-EA for the single source shortest path problem. Evolutionary Computation **19**, 673–691 (2011)

[33] Doerr, B., Happ, E., Klein, C.: Crossover can provably be useful in evolutionary computation. Theoretical Computer Science **425**, 17–33 (2012)

[34] Doerr, B., Jansen, T., Sudholt, D., Winzen, C., Zarges, C.: Mutation rate matters even when optimizing monotone functions. Evolutionary Computation **21**, 1–21 (2013)

[35] Doerr, B., Jansen, T., Witt, C., Zarges, C.: A method to derive fixed budget results from expected optimisation times. In: Genetic and Evolutionary Computation Conference, GECCO 2013, pp. 1581–1588. ACM (2013)

[36] Doerr, B., Johannsen, D.: Adjacency list matchings: an ideal genotype for cycle covers. In: Genetic and Evolutionary Computation Conference, GECCO 2007, pp. 1203–1210. ACM (2007)

[37] Doerr, B., Johannsen, D.: Edge-based representation beats vertex-based representation in shortest path problems. In: Genetic and Evo-

lutionary Computation Conference, GECCO 2010, pp. 759–766. ACM (2010)
[38] Doerr, B., Johannsen, D., Winzen, C.: Multiplicative drift analysis. Algorithmica **64**, 673–697 (2012)
[39] Doerr, B., Kodric, B., Voigt, M.: Lower bounds for the runtime of a global multi-objective evolutionary algorithm. In: Congress on Evolutionary Computation, CEC 2013, pp. 432–439. IEEE (2013)
[40] Doerr, B., Künnemann, M.: Optimizing linear functions with the $(1+\lambda)$ evolutionary algorithm—different asymptotic runtimes for different instances. Theoretical Computer Science **561**, 3–23 (2015)
[41] Doerr, B., Le, H.P., Makhmara, R., Nguyen, T.D.: Fast genetic algorithms. In: Genetic and Evolutionary Computation Conference, GECCO 2017, pp. 777–784. ACM (2017)
[42] Doerr, B., Sudholt, D., Witt, C.: When do evolutionary algorithms optimize separable functions in parallel? In: Foundations of Genetic Algorithms, FOGA 2013, pp. 48–59. ACM (2013)
[43] Doerr, B., Winzen, C.: Ranking-based black-box complexity. Algorithmica **68**, 571–609 (2014)
[44] Doerr, B., Witt, C., Yang, J.: Runtime analysis for self-adaptive mutation rates. In: Genetic and Evolutionary Computation Conference, GECCO 2018, pp. 1475–1482. ACM (2018)
[45] Droste, S.: Analysis of the (1+1) EA for a dynamically bitwise changing OneMax. In: Genetic and Evolutionary Computation Conference, GECCO 2003, pp. 909–921. Springer (2003)
[46] Droste, S.: Analysis of the (1+1) EA for a noisy OneMax. In: Genetic and Evolutionary Computation Conference, GECCO 2004, pp. 1088–1099. Springer (2004)
[47] Droste, S., Jansen, T., Wegener, I.: On the analysis of the (1+1) evolutionary algorithm. Theoretical Computer Science **276**, 51–81 (2002)
[48] Feige, U.: On sums of independent random variables with unbounded variance and estimating the average degree in a graph. SIAM Journal of Computing **35**, 964–984 (2006)
[49] Feller, W.: An Introduction to Probability Theory and Its Applications, vol. I, third edn. Wiley (1968)
[50] Garnier, J., Kallel, L., Schoenauer, M.: Rigorous hitting times for binary mutations. Evolutionary Computation **7**, 173–203 (1999)
[51] Gießen, C., Witt, C.: The interplay of population size and mutation probability in the $(1 + \lambda)$ EA on OneMax. Algorithmica **78**, 587–609 (2017)
[52] Greenberg, S., Mohri, M.: Tight lower bound on the probability of a binomial exceeding its expectation. Statistics and Probability Letters **86**, 91–98 (2014)
[53] He, J., Yao, X.: Drift analysis and average time complexity of evolutionary algorithms. Artificial Intelligence **127**, 51–81 (2001)

[54] Hoeffding, W.: Probability inequalities for sums of bounded random variables. Journal of the American Statistical Association **58**, 13–30 (1963)

[55] Hwang, H., Panholzer, A., Rolin, N., Tsai, T., Chen, W.: Probabilistic analysis of the (1+1)-evolutionary algorithm. Evolutionary Computation **26**, 299–345 (2018)

[56] Jansen, T., De Jong, K.A., Wegener, I.: On the choice of the offspring population size in evolutionary algorithms. Evolutionary Computation **13**, 413–440 (2005)

[57] Jansen, T., Wegener, I.: Evolutionary algorithms - how to cope with plateaus of constant fitness and when to reject strings of the same fitness. IEEE Transactions on Evolutionary Computation **5**, 589–599 (2001)

[58] Jansen, T., Wegener, I.: On the analysis of a dynamic evolutionary algorithm. Journal of Discrete Algorithms **4**, 181–199 (2006)

[59] Jansen, T., Zarges, C.: On benefits and drawbacks of aging strategies for randomized search heuristics. Theoretical Computer Science **412**, 543–559 (2011)

[60] Janson, S.: Tail bounds for sums of geometric and exponential variables. ArXiv e-prints **arXiv:1709.08157** (2017)

[61] Klar, B.: Bounds on tail probabilities of discrete distributions. Probability in the Engineering and Informational Sciences **14**, 161–171 (2000)

[62] Kötzing, T.: Concentration of first hitting times under additive drift. Algorithmica **75**, 490–506 (2016)

[63] Krejca, M.S., Witt, C.: Lower bounds on the run time of the univariate marginal distribution algorithm on OneMax. In: Foundations of Genetic Algorithms, FOGA 2017, pp. 65–79. ACM (2017)

[64] de Perthuis de Laillevault, A., Doerr, B., Doerr, C.: Money for nothing: Speeding up evolutionary algorithms through better initialization. In: Genetic and Evolutionary Computation Conference, GECCO 2015, pp. 815–822. ACM (2015)

[65] Lehre, P.K., Nguyen, P.T.H.: Improved runtime bounds for the univariate marginal distribution algorithm via anti-concentration. In: Genetic and Evolutionary Computation Conference, GECCO 2017, pp. 1383–1390. ACM (2017)

[66] Lehre, P.K., Witt, C.: Black-box search by unbiased variation. Algorithmica **64**, 623–642 (2012)

[67] Lehre, P.K., Witt, C.: Concentrated hitting times of randomized search heuristics with variable drift. In: 25th International Symposium on Algorithms and Computation, ISAAC 2014, pp. 686–697. Springer (2014)

[68] Lehre, P.K., Yao, X.: On the impact of mutation-selection balance on the runtime of evolutionary algorithms. IEEE Transactions on Evolutionary Computation **16**, 225–241 (2012)

[69] Lissovoi, A., Oliveto, P.S., Warwicker, J.A.: On the runtime analysis of generalised selection hyper-heuristics for pseudo-boolean optimisation.

In: Genetic and Evolutionary Computation Conference, GECCO 2017, pp. 849–856. ACM (2017)
[70] McDiarmid, C.: Concentration. In: Probabilistic Methods for Algorithmic Discrete Mathematics, vol. 16, pp. 195–248. Springer, Berlin (1998)
[71] Mitzenmacher, M., Upfal, E.: Probability and Computing—Randomized Algorithms and Probabilistic Analysis. Cambridge University Press (2005)
[72] Müller, A., Stoyan, D.: Comparison Methods for Stochastic Models and Risks. Wiley (2002)
[73] Nagaev, S.V.: Lower bounds on large deviation probabilities for sums of independent random variables. Theory of Probability and Its Applications **46**, 79–102 (2001)
[74] Neumann, F., Sudholt, D., Witt, C.: A few ants are enough: ACO with iteration-best update. In: Genetic and Evolutionary Computation Conference, GECCO 2010, pp. 63–70. ACM (2010)
[75] Neumann, F., Wegener, I.: Randomized local search, evolutionary algorithms, and the minimum spanning tree problem. Theoretical Computer Science **378**, 32–40 (2007)
[76] Neumann, F., Witt, C.: Runtime analysis of a simple ant colony optimization algorithm. Algorithmica **54**, 243–255 (2009)
[77] Oliveto, P.S., He, J., Yao, X.: Analysis of the (1+1)-EA for finding approximate solutions to vertex cover problems. IEEE Transactions on Evolutionary Computation **13**, 1006–1029 (2009)
[78] Oliveto, P.S., Lehre, P.K., Neumann, F.: Theoretical analysis of rank-based mutation - combining exploration and exploitation. In: Congress on Evolutionary Computation, CEC 2009, pp. 1455–1462. IEEE (2009)
[79] Oliveto, P.S., Witt, C.: Simplified drift analysis for proving lower bounds in evolutionary computation. Algorithmica **59**, 369–386 (2011)
[80] Oliveto, P.S., Witt, C.: Erratum: Simplified drift analysis for proving lower bounds in evolutionary computation. CoRR **abs/1211.7184** (2012)
[81] Oliveto, P.S., Witt, C.: Improved time complexity analysis of the simple genetic algorithm. Theoretical Computer Science **605**, 21–41 (2015)
[82] Panconesi, A., Srinivasan, A.: Randomized distributed edge coloring via an extension of the Chernoff–Hoeffding bounds. SIAM Journal on Computing **26**, 350–368 (1997)
[83] Pelekis, C., Ramon, J.: A lower bound on the probability that a binomial random variable is exceeding its mean. Statistics and Probability Letters **119**, 305–309 (2016)
[84] Prokhorov, Y.: Convergence of random processes and limit theorems in probability theory. Theory of Probability and Its Applications **1**, 157–214 (1956)
[85] Rigollet, P., Tong, X.: Neyman-Pearson classification, convexity and stochastic constraints. Journal of Machine Learning Research **12**, 2831–2855 (2011)

[86] Robbins, H.: A remark on Stirling's formula. American Mathematical Monthly **62**, 26–29 (1955)
[87] Rowe, J.E., Sudholt, D.: The choice of the offspring population size in the $(1, \lambda)$ evolutionary algorithm. Theoretical Computer Science **545**, 20–38 (2014)
[88] Scharnow, J., Tinnefeld, K., Wegener, I.: The analysis of evolutionary algorithms on sorting and shortest paths problems. Journal of Mathematical Modelling and Algorithms **3**, 349–366 (2004)
[89] Scheideler, C.: Probabilistic Methods for Coordination Problems. University of Paderborn (2000). Habilitation thesis. Available at `http://citeseerx.ist.psu.edu/viewdoc/summary?doi=10.1.1.70.1319`
[90] Slud, E.V.: Distribution inequalities for the binomial law. Annals of Probability **5**, 404–412 (1977)
[91] Sudholt, D.: The impact of parametrization in memetic evolutionary algorithms. Theoretical Computer Science **410**, 2511–2528 (2009)
[92] Sudholt, D.: A new method for lower bounds on the running time of evolutionary algorithms. IEEE Transactions on Evolutionary Computation **17**, 418–435 (2013)
[93] Sudholt, D., Thyssen, C.: A simple ant colony optimizer for stochastic shortest path problems. Algorithmica **64**, 643–672 (2012)
[94] Sudholt, D., Witt, C.: Update strength in EDAs and ACO: how to avoid genetic drift. In: Genetic and Evolutionary Computation Conference, GECCO 2016, pp. 61–68. ACM (2016). Extended version *CoRR* abs/1607.04063
[95] Tchebichef, P.: Des valeurs moyennes. Journal de mathématiques pures et appliquées, série 2 **12**, 177–184 (1867)
[96] Tchebichef, P.: Sur les valeurs limites des intégrales. Journal de mathématiques pures et appliquées, série 2 **19**, 157–160 (1874)
[97] Wegener, I.: Theoretical aspects of evolutionary algorithms. In: International Colloquium on Automata, Languages and Programming, ICALP 2001, pp. 64–78. Springer (2001)
[98] Wegener, I., Witt, C.: On the optimization of monotone polynomials by simple randomized search heuristics. Combinatorics, Probability & Computing **14**, 225–247 (2005)
[99] Witt, C.: Runtime analysis of the $(\mu + 1)$ EA on simple pseudo-Boolean functions. Evolutionary Computation **14**, 65–86 (2006)
[100] Witt, C.: Population size versus runtime of a simple evolutionary algorithm. Theoretical Computer Science **403**, 104–120 (2008)
[101] Witt, C.: Tight bounds on the optimization time of a randomized search heuristic on linear functions. Combinatorics, Probability & Computing **22**, 294–318 (2013)
[102] Witt, C.: Fitness levels with tail bounds for the analysis of randomized search heuristics. Information Processing Letters **114**, 38–41 (2014)
[103] Witt, C.: Upper bounds on the running time of the univariate marginal distribution algorithm on OneMax. Algorithmica **81**, 632–667 (2019)

[104] Zhou, D., Luo, D., Lu, R., Han, Z.: The use of tail inequalities on the probable computational time of randomized search heuristics. Theoretical Computer Science **436**, 106–117 (2012)

Chapter 2
Drift Analysis

Johannes Lengler

Abstract Drift analysis is one of the major tools for analysing evolutionary algorithms and nature-inspired search heuristics. In this chapter we give an introduction to drift analysis and give some examples of how to use it for the analysis of evolutionary algorithms.

2.1 Introduction

Drift analysis goes back to the seminal paper of Hajek [37], and has since become ubiquitous in the analysis of evolutionary algorithms (EAs). Google Scholar lists more than 100,000 hits for the phrases 'Drift' and 'Evolutionary Algorithm', so a comprehensive review of all applications or even just all existing drift theorems is far beyond the scope of this chapter. Instead, the chapter serves two purposes.

Firstly, it provides a self-contained introduction to drift analysis (Section 2.3), which has so far been absent in the literature.[1] This introduction is suitable for graduate students and for theory-affine researchers who have not yet encountered drift analysis. This first part of the chapter will contain illustrative examples, and will discuss in detail the different requirements of the most basic drift theorems, specifically on additive drift, variable drift and multiplicative drift. Counterexamples are given to point out when some drift theorems are not applicable or give poor results.

Secondly, Section 2.4 provides an overview of the most important recent developments in drift analysis, including lower and tail bounds, weak drift, negative drift, and population drift. This section is much more concise, and may serve also as a quick reference for the expert reader.

Johannes Lengler
Department of Computer Science, ETH Zürich, Switzerland

[1] A briefer introduction can be found in [51].

B. Doerr, F. Neumann (eds.), *Theory of Evolutionary Computation*,
Natural Computing Series, https://doi.org/10.1007/978-3-030-29414-4_2

2.2 Basics of Drift Analysis

2.2.1 Motivation

To analyse the runtime of an evolutionary algorithm (or, more generally, any randomised algorithm), one of the most common and successful approaches consists of the following three steps.

1. Identify a quantity X_t, the *potential* (also called the *drift function* or *distance function*), that adequately measures the progress that the algorithm has made after t steps.
2. For any value of X_t, understand the nature of the random variable $X_t - X_{t+1}$, the one-step change of the potential.
3. Translate the data from step 2 into information about the runtime T of the algorithm, i.e. the number of steps until the algorithm achieves its goal.

Drift analysis is concerned with step 3. Generally, good drift theorems require as little information as possible about the potential X_{t+1}, and give as much information as possible about T. In the basic theorems, we only require (bounds on) the expectation $E[X_t - X_{t+1} \mid X_t = s]$ for all s, which is called the *drift*, in order to derive (bounds on) the expectation $E[T]$. Drift analysis has become a successful theory because the framework above is very general, and good tools for step 3 exist, which apply to a multitude of situations. In contrast, steps 1 and 2 often do not generalise from one problem to another. Frequently, step 1 is the part of a runtime analysis that carries the key insight, and it usually requires much more ingenuity than the other steps. On the other hand, step 2, the analysis of $X_t - X_{t+1}$, requires arguably less insight. However, step 2 is usually the most lengthy and technical part of a runtime analysis. Therefore, the complexity of a proof can often be substantially reduced if only some basic information such as the expectation $E[X_t - X_{t+1} \mid X_t = x]$ is needed in step 2.

For evolutionary algorithms, a natural candidate for the potential X_t is the fitness $f(x^{(t)})$ of the best individual in the current population, especially so if the population consists only of a single individual, as for example in (1+1) EAs. In a sense, this fitness measures the 'progress' up to time t, since it would correspond exactly to the quality of the output if the algorithm terminated with this generation. However, it is not necessarily the best choice to measure the progress that the algorithm has made towards finding a global optimum. For example, consider the linear fitness function[2] $f : \{0,1\}^n$ with $f(x) = (n-1) \cdot x_1 + \sum_{i=2}^{n} x_i$, which puts a very large emphasis on the first bit. The optimum (for maximisation) is the string $x_{\mathrm{OPT}} = (1,\ldots,1)$, but the two strings $x_1 = (1,0,0,\ldots,0)$ and $x_2 = (0,1,1,\ldots,1)$ have the same fitness $f(x_1) = f(x_2) = n-1$. However, the string x_2 is much more similar to x_{OPT}

[2] We follow the standard convention that for an n-dimensional vector x, we denote its components by $x_1,\ldots,x_n$.

than x_1 is, so we should choose a potential that assigns a higher rating to x_2 than to x_1. We will see later (Example 2.3.12) some good choices for the potential in this example.

Historically, before drift analysis was fully developed in the EA community, it was preceded by the *fitness level method* [74]. In retrospect this method may be regarded as a special case of the variable drift theorem, which we will introduce in Section 2.3.2. Likewise, the *method of expected weight decrease* [64] may be regarded as a predecessor of the multiplicative drift theorem, presented in Section 2.3.3. It is fair to say that the development of drift analysis has boosted our understanding of evolutionary algorithms, either by simplifying existing results, by achieving greater precision, or as a means to obtain qualitatively new results that might not have been achievable with the old techniques. For example, the original proof by Droste, Jansen, and Wegener that the $(1+1)$ EA takes time $O(n \log n)$ on all linear functions needed 7 pages [26], while Doerr, Johannsen, and Winzen could reduce the proof to a single page [18]. To obtain the leading constant with the fitness level method would have been quite challenging and perhaps out of reach. With drift analysis, in a groundbreaking paper, Witt [75] could derive the leading constant not only for the standard mutation rate $1/n$ but for any mutation rate c/n, where c is a constant, in a proof of 2–3 pages!

2.2.2 General Set-Up

Throughout this chapter, we will assume that $(X_t)_{t \geq 0}$ is a sequence of non-negative random variables with a finite state space $\mathcal{S} \subseteq \mathbb{R}_0^+$ such that $0 \in \mathcal{S}$. We will denote the minimum positive state by $s_{\min} := \min(\mathcal{S} \setminus \{0\})$. The *stopping time*, or *hitting time of* 0, of $(X_t)_{t \geq 0}$ is defined as the smallest t such that $X_t = 0$. We are generally interested in the *drift* $\Delta_t(s) := E[X_t - X_{t+1} \mid X_t = s]$, where $t \geq 0$ and $s \in \mathcal{S}$.

As with all conditional expectations, $\Delta_t(s)$ is not well-defined if $\Pr[X_t = s] = 0$. So in other words, $\Delta_t(s)$ is undefined for situations that never occur. Obviously, this is not a practical issue, and it is convenient (and common in the community) to be sloppy about such cases. So we will use phrases such as '$\Delta_t(s) \leq 1$ for all $t \geq 0$' as a shorthand for '$\Delta_t(s) \leq 1$ for all $t \geq 0$ for which the conditional expectation $\Delta_t(s)$ is well-defined'.

In Section 2.4 we will often need to work with pointwise drift and filtrations, i.e. we need to condition on the complete history (or at least the current state) of the algorithm, instead of just conditioning on the value of X_t. In these cases, we will denote the filtration associated with the algorithm's history up to time t by $\mathcal{F}_t$. Moreover, tail bounds will be formulated for a fixed initial search point $X_0 = s_0$. For details and an explanation of the technical terms 'pointwise drift' and 'filtration', see the corresponding paragraph in Section 2.2.3 below.

Throughout the chapter, f will denote a fitness function to be optimised, either maximised or minimised. For a $(1+\lambda)$ algorithm, we will use the convention that $x^{(t)}$ is the search point after t generations.

2.2.3 Variants

In the literature, terminology may vary between different authors, and there are often slightly different set-ups considered. We highlight some variants which occur frequently. A reader who is new to drift analysis may skip this section on first reading.

1. *Signs.* We consider the change $X_t - X_{t+1}$. In the literature, the difference is sometimes considered with opposite signs, $X_{t+1} - X_t$, which is arguably a more natural choice. However, since we consider drift towards zero, with our choice the drift is usually positive instead of negative. Moreover, our choice is more consistent with the established term 'negative drift', which refers to a drift that points away from the target.
2. *Markov chains.* Instead of *any* sequence of random variables, the sequence X_t is sometimes assumed to be a Markov chain, i.e. the state X_t should completely determine the distribution of X_{t+1}. While this is a mathematically appealing scenario, it usually does not apply in the context of evolutionary algorithms. For instance, in the example in Section 2.2.1 above, the information $X_t = n-1$ would tell us that the current fitness is $n-1$, but the two search points x_1 and x_2 differ in nature. Thus, the subsequent trajectory of search points depends on more information than is contained in X_t, and so do the subsequent potentials $X_{t+1}, X_{t+2}, \ldots$. So, even in this very simple example, we do not have a Markov chain.

   There are several papers on the theory of EAs which ignore this point, either accidentally or perhaps consciously for the sake of exposition, since Markov chains are an easily accessible concept. These papers contain drift theorems for a Markov chain X_t, but use them for runtime analyses in which X_t is not a Markov chain. So, technically speaking, the proofs are not correct. However, this is a purely technical issue: since the Markov property is not really needed for drift theorems, the results derived are still correct. An alternative was used in [54], where the authors assumed an underlying Markov process Y_t with arbitrary state space $\mathcal{S}$, and a function $\alpha : \mathcal{S} \to \mathbb{R}$. Then they formulated drift theorems for $X_t := \alpha(Y_t)$. This is a more precise description of randomised algorithms, where the internal state (e.g., the current population) is described by Y_t, and the real-valued potential is described by X_t. It has the advantage that expressions such as $E[X_t - X_{t+1} \mid Y_t = s]$ are still well-defined even if $\Pr[Y_t = s] = 0$. This is especially relevant in continuous domains. For example, assume that Y_0 is a real number drawn uniformly at random from $[0,1]$. Then $\Pr[Y_0 = s] = 0$ for all $s \in [0,1]$.

3. *Filtrations and pointwise drift.* We have defined the drift as a random variable that is conditioned on the value of X_t, i.e. $\Delta_t(x) = E[X_t - X_{t+1} \mid X_t = s]$. Instead, it is also possible to condition on the whole history of X_t, or even on the whole history of the algorithm. (Recall that, in general, the potential X_t does not completely describe the state of the algorithm at time t). In mathematical terms, the set of such histories is described by a *filtration* of σ-algebras $\mathcal{F}_0 \subseteq \mathcal{F}_1 \subseteq \ldots$, where, intuitively, the σ-algebra $\mathcal{F}_t$ contains all the information that is available after the first t steps of the algorithm.[3] For example, instead of requiring that $E[X_t - X_{t+1} \mid X_t = s] \leq 1$ for all $t \geq 0$, we would ask that $E[X_t - X_{t+1} \mid F_t] \leq 1$ for all $t \geq 0$ and all histories F_t up to time t such that $X_t = s$ in F_t.[4] In this case, we also speak of *pointwise drift*, and we will write[5] $E[X_t - X_{t+1} \mid \mathcal{F}_t, X_t = s] \leq 1$ to mean that, for every history F of the algorithm up to time t with the property $X_t = s$, we have $E[X_t - X_{t+1} \mid F] \leq 1$.
   Obviously, pointwise drift is a much stronger condition, and requiring such a strong condition in a drift theorem gives a priori a weaker theorem. However, for most applications it does not make a big difference to consider either version. Intellectually, it is arguably easier to imagine a fixed history of the algorithm, and to think about the next step in this fixed setting. Therefore, it is not uncommon in the EA community to formulate drift theorems using filtrations. However, we will also see examples (Examples 2.3.2 and 2.3.12) where the weaker condition '$X_t = s$' is beneficial.
   The basic drift theorems concerned with the expected runtime $E[T]$ can be formulated with either form of conditioning, and in this chapter we choose the stronger form (i.e. with weaker requirements), conditioning on $X_t = s$. However, once the drift theorems include tail bounds, things become more subtle, and it becomes essential to condition on every possible history. Therefore, we will switch to using filtrations and pointwise drift in the last part of the chapter.
4. *Infinite search spaces.* We assume in this chapter that the state space $\mathcal{S}$ is finite. This makes sense in the context of this book, since in discrete optimisation the search spaces, and also the state spaces of the algorithms, tend to be finite (although they may be huge). However, there are problems, especially in continuous optimisation, in which infinite state spaces are more natural. Generally, all drift theorems mentioned in this chapter still hold if the state space $\mathcal{S} \subseteq \mathbb{R}_0^+$ is infinite, but bounded.[6] For unbounded search

---

[3] Mathematically speaking, it is the coarsest σ-algebra which makes all random choices of the algorithm up to time t measurable.

[4] This is sometimes sloppily described by $E[X_t - X_{t+1} \mid X_0, \ldots, X_t]$. However, note that this is not quite correct, since it conditions only on the past values of X_t, and not on the history of the algorithm. In particular, conditioning on $X_0, \ldots, X_t$ usually does not determine the current state of the algorithm (e.g. the current search point or population).

[5] By abuse of notation, for brevity.

[6] Some statements, such as Theorems 2.3.3 and 2.3.11, additionally require that the infimum $s_{\min} := \inf(S \setminus \{0\})$ is strictly positive.

spaces, things become more complicated. The upper bounds on $E[T]$ in the drift theorems still hold in these cases, while the lower bounds on $E[T]$ fail in general [54], as we will discuss briefly after Theorem 2.3.1. Collections of drift theorems for unbounded spaces can be found in [45, 54].

5. *Drift versus expected drift.* Unfortunately, the meaning of the term 'drift' is somewhat inconsistent in the literature. We have defined it as the *expected* change $E[X_t - X_{t+1} \mid X_t = s]$. However, some authors also use 'drift' to refer to the conditional random variable $X_t - X_{t+1} \mid X_t = s$, and our definition would be the 'expected drift' in their terminology. Some authors would also call the conditional expectation $E[X_t - X_{t+1} \mid \mathcal{F}_t]$ 'drift', which is itself a random variable (by the randomness in the history of the algorithm). Again, our notion of drift would be the expected drift $E_{\mathcal{F}_t}\left[E[X_t - X_{t+1} \mid \mathcal{F}_t]\right]$, also called the 'average drift' in this terminology [40]. Yet another notion uses 'drift' to refer to the conditional random variable $X_t - X_{t+1} \mid \mathcal{F}_t$. Fortunately, the heterogeneous nomenclature usually does not lead to confusion, except for some minor notational irritations.

2.3 Elementary Introduction to Drift Analysis

We start with an elementary introduction to drift analysis. We will discuss the three main workhorses, The additive drift theorem (Theorem 2.3.1), the variable drift theorem (Theorem 2.3.3), and the multiplicative drift theorem (Theorem 2.3.11). All of them give upper bounds on the expected hitting time $E[T]$, the additive drift theorem also giving matching lower bounds.[7]

2.3.1 Additive Drift

The simplest possible drift is additive drift, i.e. X_{t+1} differs from X_t in expectation by an additive constant. The theorem in its modern form dates back to He and Yao [38, 39], who built on work by Hajek [37],[8] which they stripped

[7] Note that the expectation of a random variable may not always give the full story. There are even cases where the value of $E[T]$ may be misleading. We will discuss such examples in Section 2.4.2, where we consider drift theorems that give tail bounds on T.

[8] They were apparently all unaware that the result had been proven even earlier by Tweedie [73, Theorem 6], and a yet earlier proof in Russian has been attributed to Menshikov [58, Bibliographical Notes on Section 2.6]. The aditive drift theorem has been proven and rediscovered many times, and it is known under various names. For example, in stability theory it is considered a special case of Dynkin's formula [59, Theorem 11.3.1], or a generalisation of Foster's criterion [2, Proposition 4.5]. In these contexts, drift analysis is often called the *Lyapunov function method*, e.g. [58, Theorem 2.6.2]. However, the hitting time is often only a side aspect in these areas.

of substantial technical overhead that was due to the fact that Hajek's focus was more on deciding whether hitting times actually exist for unbounded state spaces. He and Yao proved their theorem using (without explicit reference) the *optional stopping theorem* for martingales [36]. Here we give an elementary proof taken from [54], since this proof gives some insight into the differences between upper and lower bounds.

Theorem 2.3.1 (Additive Drift Theorem [39]). *Let $(X_t)_{t \geq 0}$ be a sequence of non-negative random variables with a finite state space $\mathcal{S} \subseteq \mathbb{R}_0^+$ such that $0 \in \mathcal{S}$. Let $T := \inf\{t \geq 0 \mid X_t = 0\}$.*

(a) If there exists $\delta > 0$ such that for all $s \in \mathcal{S} \setminus \{0\}$ and for all $t \geq 0$,

$$\Delta_t(s) := E[X_t - X_{t+1} \mid X_t = s] \geq \delta, \tag{2.1}$$

then

$$E[T] \leq \frac{E[X_0]}{\delta}. \tag{2.2}$$

(b) If there exists $\delta > 0$ such that for all $s \in \mathcal{S} \setminus \{0\}$ and for all $t \geq 0$,

$$\Delta_t(s) := E[X_t - X_{t+1} \mid X_t = s] \leq \delta, \tag{2.3}$$

then

$$E[T] \geq \frac{E[X_0]}{\delta}. \tag{2.4}$$

Proof.
(a). As we are interested only in the hitting time T of zero, we may assume without loss of generality that $X_{T+1} = X_{T+2} = \ldots = 0$.

We may rewrite the condition (2.1) as $E[X_{t+1} \mid X_t = s] \leq E[X_t \mid X_t = s] - \delta$. Since this holds for all $s \in \mathcal{S} \setminus \{0\}$, and since $T > t$ if and only if $X_t > 0$, we conclude

$$E[X_{t+1} \mid T > t] \leq E[X_t \mid T > t] - \delta. \tag{2.5}$$

By the law of total probability, we have

$$\begin{aligned} E[X_t] &= \Pr[T > t] \cdot E[X_t \mid T > t] + \Pr[T \leq t] \cdot \underbrace{E[X_t \mid T \leq t]}_{=0} \\ &= \Pr[T > t] \cdot E[X_t \mid T > t]. \end{aligned} \tag{2.6}$$

Proceeding similarly for X_{t+1}, we obtain

$$\begin{aligned} E[X_{t+1}] &= \Pr[T > t] \cdot E[X_{t+1} \mid T > t] + \Pr[T \leq t] \cdot \underbrace{E[X_{t+1} \mid T \leq t]}_{=0} \\ &\overset{(2.5)}{\leq} \Pr[T > t] \cdot (E[X_t \mid T > t] - \delta) \end{aligned}$$

$$\overset{(2.6)}{=} E[X_t] - \delta \cdot \Pr[T > t]. \tag{2.7}$$

Since T is a random variable that takes values in $\mathbb{N}_0$, we may write $E[T] = \sum_{t=0}^{\infty} \Pr[T > t]$. Thus

$$\begin{aligned} \delta \cdot E[T] \overset{\tau \to \infty}{\longleftarrow} \sum_{t=0}^{\tau} \delta \Pr[T > t] &\overset{(2.7)}{\leq} \sum_{t=0}^{\tau} (E[X_t] - E[X_{t+1}]) \\ &= E[X_0] - \underbrace{E[X_{\tau+1}]}_{\geq 0} \leq E[X_0], \end{aligned} \tag{2.8}$$

which proves (a).

(b) Analogously to (a), the calculations (2.5), (2.6), (2.7) and (2.8) hold with reversed inequalities, except for the very last step in (2.8). So, (2.8) becomes

$$\delta \cdot E[T] \overset{\tau \to \infty}{\longleftarrow} \sum_{t=0}^{\tau} \delta \Pr[T > t] \geq E[X_0] - E[X_{\tau+1}]. \tag{2.9}$$

There are only two possible cases. The first case is that $\Pr[T > t]$, which is a non-increasing sequence, does *not* converge to 0. In this case, $E[T] = \sum_{t=0}^{\infty} \Pr[T > t] = \infty$, in which case (b) holds trivially. The second is that $\Pr[T > t] \to 0$, and by (2.6) we also have

$$E[X_{\tau+1}] = \underbrace{\Pr[T > t]}_{\to 0} \cdot \underbrace{E[X_{\tau+1} \mid T > t]}_{\leq \max S < \infty} \to 0. \tag{2.10}$$

Now (b) follows from (2.9) and (2.10). □

The proof also shows what can generally go wrong for infinite search spaces. The proof of (a) goes through unmodified. For (b), the inequality (2.9) is generally true. Moreover, it is *tight* if the condition (2.3) is tight. The problem is that $E[X_{\tau+1}]$ may not go to zero. For example, consider the Markov chain where X_{t+1} is either 0 or $2X_t$, both with probability 1/2. Here $E[T] = 2$, but $E[X_t - X_{t+1}] = 0$ for all $t \geq 0$. In particular, the condition (2.3) is satisfied with $\delta = 1$ (or any other $\delta > 0$), but the conclusion of (b) does not hold. On the other hand, for the tight choice $\delta = 0$, we see that we have equality in (2.9) since $E[X_{\tau+1}] = E[X_0]$.

Note that if the drift in Theorem 2.3.1 is *exactly* δ in each step, then the upper and lower bounds match. In this case, Theorem 2.3.1 can be seen as an *invariance theorem*, which states that the expected hitting time of 0 is *independent* of the exact distribution of the progress, as long as the expectation of the progress (i.e. the drift) remains fixed. In particular, if X_0 is an integer multiple of δ, this includes the deterministic case in which X_t decreases in each step by exactly δ, with probability 1. Thus a process of constant drift cannot be accelerated (or slowed down) by redistributing the probability mass. We will resume this point in Section 2.3.2 when we discuss why other drift theorems are *not* tight in general.

We conclude this section on additive drift with an application.

Example 2.3.2 (RLS on LEADINGONES*).* Consider *random local search* (RLS) on the n-dimensional hypercube $\{0,1\}^n$. RLS is a (1+1) algorithm (i.e. it has population size one and generates only one offspring in each generation). The mutation operator flips exactly one bit, which is chosen uniformly at random. RLS has elitist selection, i.e. the offspring replaces the parent if and only if its fitness is at least as large as the parent's fitness. A pseudocode description is given in Algorithm 2.1.

Algorithm 2.1: Random Local Search (RLS) maximising a fitness function $f:\{0,1\}^n \to \mathbb{R}$

```
1 Choose x^(0) ∈ {0,1}^n uniformly at random;
2 for t = 0,1,2,... do
3     Pick i ∈ {1,...,n} uniformly at random, and create y^(t) by flipping the i-th
        bit in x^(t);
4     if f(y^(t)) ≥ f(x^(t)) then
5         x^(t+1) ← y^(t);
6     else
7         x^(t+1) ← x^(t);
```

We study RLS on the LEADINGONES fitness function, which returns the number of initial one-bits before the first zero bit. Formally,

$$\text{LEADINGONES}(x) = \sum_{k=1}^{n}\prod_{i=1}^{k} x_i = \max\{i \in \{1,\dots,n\} \mid \underbrace{11\dots1}_{i\text{ times}} \text{ is a prefix of } x\}.$$

The LEADINGONES problem is a classical benchmark problem for evolutionary algorithms, and RLS on LEADINGONES has been studied in much greater detail than we can present here, with methods and results that go far beyond drift analysis [10, 49]. We examine the bounds that we obtain from the additive drift theorem for different potential functions.

Naive potential. We first choose as the potential $X_t := n - f(x^{(t)})$, the distance in fitness from the optimum. The state space is $\mathcal{S} = \{0,\dots,n\}$. We need to compute the drift $\Delta_t(s) := E[X_t - X_{t+1} \mid X_t = s]$ for every state $s \in \mathcal{S}\setminus\{0\}$, so we fix such an s. For convenience, we write $k := n - s \in \{0,\dots,n-1\}$ for the fitness in this case. Note that $X_t = s$ implies that the first k bits of $x^{(t)}$ are one-bits, but the $k+1$-st bit is a zero-bit. Obviously, the potential changes if and only if we flip the $k+1$-st bit, so let us denote this event by $\mathcal{E}$. Since the flipped bit is chosen uniformly, we have $\Pr[\mathcal{E}] = 1/n$. Hence the drift is

$$\Delta_t(s) = \Pr[\mathcal{E}] \cdot \underbrace{E[X_t - X_{t+1} \mid X_t = s \text{ and } \mathcal{E}]}_{=:E(s)} = \frac{1}{n} \cdot E(s). \tag{2.11}$$

So, it remains to bound the conditional expectation $E(s)$. Such conditional expectations occur quite frequently when a drift is computed. Assume that $X_t = s$ (i.e. $f(x^{(t)}) = k = n - s$), and that $\mathcal{E}$ occurs. Obviously, $E(s) \geq 1$, since we improve at least the $k+1$-st bit. On the other hand, we improve the fitness by at least 2 if and only if the $k+2$-nd bit happens to be a one-bit. Note that since the algorithm is elitist and has fitness $f(x_t) = k$, the $k+2$-nd bit has had no influence on the fitness of previous search points. Therefore, by symmetry, it has probability 1/2 of being a one-bit[9] and we obtain

$$\Pr[X_t - X_{t+1} \geq 2 \mid X_t = s \text{ and } \mathcal{E}] = \Pr[x^{(t)}_{k+2} = 1 \mid X_t = s \text{ and } \mathcal{E}] = \frac{1}{2}.$$

Analogously, $X_t - X_{t+1} \geq i$ if and only if the bits with indices $k+2, \ldots, k+i$ are all one-bits, which happens with probability 2^{-i+1}. Since $X_t - X_{t+1}$ is an integer non-negative random variable, we may sandwich it as follows.

$$\begin{aligned} 1 \leq E(s) &= \sum_{i=1}^{s} \Pr[X_t - X_{t+1} \geq i \mid X_t = s \text{ and } \mathcal{E}] \\ &= 1 + \sum_{i=2}^{s} 2^{-i+1} < 1 + \sum_{i=1}^{\infty} 2^{-i} = 2. \end{aligned} \tag{2.12}$$

Hence, by (2.11),

$$\frac{1}{n} \leq \Delta_t(k) \leq \frac{2}{n}, \tag{2.13}$$

and Theorem 2.3.1 implies that

$$\frac{n}{2} E[X_0] \leq E[T] \leq n E[X_0]. \tag{2.14}$$

To estimate $E[X_0] = n - E[f(x^{(0)})]$, we observe that $f(x^{(0)}) \geq i$ happens if only if the first i bits are all one-bits, which happens with probability 2^{-i}. Hence, a similar calculation to that before shows

$$E[f(x^{(0)})] = \sum_{i=1}^{n} \Pr[f(x^{(0)}) \geq i] = \sum_{i=1}^{n} 2^{-i} = 1 - 2^{-n} \in [0,1], \tag{2.15}$$

and thus $n - 1 \leq E[X_0] \leq n$. Hence, by (2.14) we get $(n-1)n/2 \leq E[T] \leq n^2$, and thus $E[T] = \Theta(n^2)$.

[9] Note that such an argument would not be true if we were to condition on one particular history of the algorithm, compare the discussion on filtrations in Section 2.2.2.

Translated potential. The analysis so far gives the asymptotics of $E[T]$, but it is not tight up to constant factors. The problem is, as (2.12) shows, that the inequality $E(k) \geq 1$ is rather coarse except for the few exceptional cases where k is almost n. In fact, in the border case $k = n-1$ we have equality, $E(k) = 1$. Hence, we do not have a perfectly constant drift, which is one reason for the discrepancy between the upper and lower bounds. Such border effects can often be remedied by *translating the potential function.* In this case, we consider

$$Y_t := \begin{cases} X_t + 1 & \text{if } X_t \geq 1, \\ 0 & \text{otherwise.} \end{cases} \tag{2.16}$$

The effect is that the drift increases when there is a substantial chance of reaching 0 in the next step. In our case, we get an additional term for $i = n-k+1$ in (2.12), which equals the term for $i = n-k$. Intuitively, the term for $i = n-k$ counts double since in this case the potential drops from 2 to 0, rather than from 1 to 0. Consequently, we get the following for the potential $Y_t = s+1$, which corresponds as before to a fitness $f(x^{(t)}) = k = n-s$:

$$\begin{aligned} E[Y_t - Y_{t+1} \mid Y_t = s+1 \text{ and } \mathcal{E}] &= \sum_{i=1}^{n-k} \Pr[Y_t - Y_{t+1} \geq i \mid Y_t = s+1 \text{ and } \mathcal{E}] \\ &= 1 + \sum_{i=2}^{n-k} 2^{-i+1} + 2^{n-k+1} = 2. \end{aligned} \tag{2.17}$$

Hence, the drift with respect to Y_t is exactly $2/n$, and Theorem 2.3.1 gives a tight result:

$$E[T] = \frac{n}{2} E[Y_0]. \tag{2.18}$$

From (2.16) it is easy to compute $E[Y_0]$ exactly as

$$E[Y_0] = n - E[f(x^{(0)})] + 1 \cdot \Pr[Y_0 > 0] \overset{(2.15)}{=} n - (1 - 2^{-n}) + 1 - 2^{-n} = n.$$

Together with (2.18), the additive drift theorem (Theorem 2.3.1) now implies $E[T] = n^2/2$.

The above example illustrates how important it is for Theorem 2.3.1 that the drift be as uniform as possible, to get matching upper and lower bounds. The example also shows that rescaling of the potential function may be a way to smooth out inhomogeneities. Following this approach systematically leads to the variable drift theorem, which we will discuss in the next section.

2.3.2 Variable Drift

2.3.2.1 The Variable Drift Theorem

The additive drift theorem is useful because it is tight, but it requires us to find a potential function that has constant drift. Is this even always possible? The perhaps surprising answer is 'Yes', as we will discuss in Section 2.5. Unfortunately, it can be rather hard to find a good potential. However, there are helpful tools. Even if we start with a potential functions with the 'wrong' scaling, Mitavskiy, Rowe and Cannings [60], and Johannsen in his PhD thesis [42] developed a theorem which *automatically* rescales the drift in the right way. A similar result was obtained independently (and earlier) by Baritompa and Steel [1].

Theorem 2.3.3 (Variable Drift Theorem [42, 70]). *Let $(X_t)_{t\geq 0}$ be a sequence of non-negative random variables with a finite state space $\mathcal{S} \subseteq \mathbb{R}_0^+$ such that $0 \in \mathcal{S}$. Let $s_{\min} := \min(\mathcal{S} \setminus \{0\})$, let $T := \inf\{t \geq 0 \mid X_t = 0\}$, and for $t \geq 0$ and $s \in \mathcal{S}$ let $\Delta_t(s) := E[X_t - X_{t+1} \mid X_t = s]$. If there is an increasing function[10] $h : \mathbb{R}^+ \to \mathbb{R}^+$ such that for all $s \in \mathcal{S} \setminus \{0\}$ and all $t \geq 0$,*

$$\Delta_t(s) \geq h(s), \tag{2.19}$$

then

$$E[T] \leq \frac{s_{\min}}{h(s_{\min})} + E\left[\int_{s_{\min}}^{X_0} \frac{1}{h(\sigma)} d\sigma\right], \tag{2.20}$$

where the expectation in the latter term is over the random choice of X_0.

We remark that the condition that h be increasing is usually satisfied, since progress typically becomes harder as the algorithm approaches an optimum. We will see in the proof why the condition is necessary, and an example showing that it is necessary can be found in [45]. However, variants of the theorem for non-decreasing drift functions do exist [16, 28].

We present a proof of the variable drift theorem, for two reasons. Firstly, the theorem is so central that it deserves to come with a proof. Secondly, we will gain valuable insights from the proof. In particular, it will enable us to understand when the upper bound on $E[T]$ is tight, and realise when the upper bound may be misleading. A reader who is completely new to drift analysis may first skip ahead to some examples, and return to the proof when we discuss tightness of the variable drift theorem.

Proof (of Theorem 2.3.3, adapted from [42]). The main insight of the proof lies in an appropriate rescaling of X_t by the function

[10] Some formulations in the literature require h to be integrable. However, since we assume $\mathcal{S}$ to be finite, the interval $[s_{\min}, X_0]$ is a compact interval, on which every monotone function is integrable.

$$g(s) := \begin{cases} \frac{s_{\min}}{h(s_{\min})} + \int_{s_{\min}}^{s} \frac{1}{h(\sigma)} d\sigma, & s \geq s_{\min}, \\ \frac{s}{h(s_{\min})}, & 0 \leq s \leq s_{\min}. \end{cases} \tag{2.21}$$

The integral is well-defined since h is increasing. Note that g is strictly increasing. We claim that for all $s \in \mathcal{S} \setminus \{0\}$ and all $r \geq 0$,

$$g(s) - g(r) \geq \frac{s - r}{h(s)}. \tag{2.22}$$

To prove the claim, we distinguish three cases. First, assume that $s \geq r \geq s_{\min}$. Then

$$g(s) - g(r) = \int_{r}^{s} \frac{1}{h(\sigma)} d\sigma \geq \int_{r}^{s} \frac{1}{h(s)} d\sigma = \frac{s - r}{h(s)}. \tag{2.23}$$

Similarly, if $r \geq s \geq s_{\min}$, then

$$g(r) - g(s) = \int_{s}^{r} \frac{1}{h(\sigma)} d\sigma \leq \int_{s}^{r} \frac{1}{h(s)} d\sigma = \frac{r - s}{h(s)}, \tag{2.24}$$

and multiplication by -1 yields the claim. The only remaining case is $s \geq s_{\min} > r \geq 0$ (since we have assumed $s \in \mathcal{S} \setminus \{0\}$), and in this case

$$\begin{aligned} g(s) - g(r) &= \frac{s_{\min}}{h(s_{\min})} + \int_{s_{\min}}^{s} \frac{1}{h(\sigma)} d\sigma - \frac{r}{h(s_{\min})} \geq \frac{s_{\min} - r}{h(s_{\min})} + \frac{s - s_{\min}}{h(s)} \\ &\geq \frac{s - r}{h(s)}. \end{aligned} \tag{2.25}$$

Now let us consider the rescaled random variable $Y_t := g(X_t)$. This random variable takes values of the form $g(s)$, where $s \in S$. For all $s \in \mathcal{S} \setminus \{0\}$,

$$\begin{aligned} E[Y_t - Y_{t+1} \mid Y_t = g(s)] &= E[g(X_t) - g(X_{t+1}) \mid g(X_t) = g(s)] \\ &\overset{(2.22)}{\geq} E\left[\frac{X_t - X_{t+1}}{h(X_t)} \,\middle|\, X_t = s\right] = \frac{\Delta_t(s)}{h(s)} \overset{(2.19)}{\geq} 1. \end{aligned} \tag{2.26}$$

Hence Y_t has at least a constant drift. The theorem follows by applying the additive drift theorem (Theorem 2.3.1) to Y_t. □

Example 2.3.4 (Coupon Collector, RLS on ONEMAX*).* The most classical example of variable drift is the *Coupon Collector Process* (CCP): there are n types of coupons, and a collector wants to have at least one coupon of each type. However, the coupons are sold in opaque wrappings, so she cannot see the type of a coupon before buying it. If each type occurs with the same frequency $1/n$, how many coupons does she need to buy before she has every type at least once?

The CCP and its variants appear in various contexts in the study of EAs. The most basic example is the runtime of RLS (Algorithm 2.1 on page 97) for maximising the OneMax fitness function, which counts the number of one-bits in a bit string. Formally, for $x \in \{0,1\}^n$,

$$\textsc{OneMax}(x) = \sum_{i=1}^{n} x_i. \tag{2.27}$$

The one-bits correspond to the coupons in the CCP that the collector has already obtained. Since RLS flips exactly one bit in each round, and a one-bit stays a one-bit forever, a round of RLS corresponds exactly to the purchase of a coupon. Thus the number of rounds of RLS on OneMax is equivalent to the number of purchases in the CCP.[11]

To analyse the CCP, we let X_t be the number of missing coupons after t purchases, and as usual we denote by T the hitting time of 0. Then, for $X_t = s$, the probability of obtaining a new type with the next purchase is s/n. In this case X_t decreases by one, so X_t has a drift of $\Delta_t(s) = s/n$. The minimum positive value of X_t is $s_{\min} = 1$. Hence, the variable drift theorem with function $h(s) = s/n$ gives the upper bound

$$E[T] \le \frac{1}{h(1)} + E\left[\int_1^{X_0} \frac{n}{\sigma} d\sigma\right] = n(1 + E[\ln(X_0)]) \le n\ln n + n. \tag{2.28}$$

The drift in Example 2.3.4 was *multiplicative*, i.e. $\Delta_t(s)$ was proportional to s. This is by far the most important special case of the variable drift theorem, important enough that in Section 2.3.3 we will provide it with a theorem of its own, the multiplicative drift theorem. Any reader who is eager to see some more cute examples of a similar type is invited to peek ahead.

The upper bound in Example 2.3.4 is remarkably tight. The expected runtime is indeed $E[T] = n\ln n + \Theta(n)$, both for the CCP [61] and for RLS on OneMax [10]. We will discuss in the next section when we can expect the bounds from the variable drift theorem to be tight, and see situations in which they are rather inaccurate. Before that, we give a more serious example coming from applications.

Example 2.3.5 (Genetic Programming). Genetic programming (GP) uses evolutionary principles to automatically generate programs which match some desired input–output scheme. The programs are typically represented as syntax trees [48], where the leaves correspond to variables $x_1, \ldots, x_n$, and the inner nodes correspond to operators such as AND, OR or NOT. Here we restrict ourselves to the Boolean domain, for simplicity. Then each syntax tree

[11] Except for the initial conditions: for the CCP, the collector usually starts with no coupons, while RLS starts with a random bit string and thus with a random initial number of ones/coupons.

τ represents a Boolean term, and thus defines a pseudo-Boolean function $f_\tau : \{0,1\}^n \to \{0,1\}$. Doerr, Lissovoi and Oliveto [25] studied the problem of learning the AND function $\text{AND}(x_1,\ldots,x_n) = x_1 \wedge \ldots \wedge x_n$, if the inner nodes may be either AND or OR. To turn this into an optimisation problem, we assign to each syntax tree τ the number $F(\tau)$ of inputs $x = (x_1,\ldots,x_n) \in \{0,1\}^n$ for which $f_\tau(x_1,\ldots,x_n) \neq \text{AND}(x_1,\ldots,X_n)$. So the goal is to reduce the potential F to zero. The search procedure considered in [25] uses a mutation operator which adds, substitutes or deletes nodes, or which deletes whole subtrees of the current syntax tree. The actual algorithm is rather complicated, and we refer the reader to [25] for more details.

We define $X_t := F(\tau_t)$, where τ_t is the syntax tree after t steps. The authors of [25] showed that X_t has the following drift:

$$E[X_t - X_{t+1} \mid X_t = s] \geq h(s) := \begin{cases} \frac{\delta s \ln s}{\ln n} & \text{if } s \geq n, \\ \delta s & \text{if } s < n, \end{cases} \tag{2.29}$$

where $\delta = \Theta(1/n^2)$ depends on the number of variables, but is independent of s. Note that h is increasing and that $X_0 \leq 2^n$. Therefore, the variable drift theorem immediately gives the following upper bound on the expected optimisation time T:

$$E[T] \leq \frac{1}{h(1)} + \int_1^{2^n} \frac{1}{h(\sigma)} d\sigma = \frac{1}{h(1)} + \int_1^n \frac{1}{\delta\sigma} d\sigma + \int_n^{2^n} \frac{\ln n}{\delta\sigma \ln \sigma} d\sigma. \tag{2.30}$$

To compute the integral, we note that the inverse derivative of $1/\sigma$ is $\ln \sigma$, and the inverse derivative of $1/(\sigma \ln \sigma)$ is $\ln \ln \sigma$. Hence,

$$E[T] \leq \frac{1}{\delta} + \frac{\ln n}{\delta} + \frac{\ln n}{\delta}(\ln \ln 2^n - \ln \ln n) = O\left(\frac{\log^2 n}{\delta}\right) = O(n^2 \log^2 n). \tag{2.31}$$

So once we have found the drift as in (2.29), the drift theorems make it an easy task to compute the expected runtime. Of course, the main contribution of the authors of [25] was to actually compute the drift.

2.3.2.2 Tightness of the Variable Drift Theorem

In general, the bound in the variable drift theorem does not need to be tight, even if we assume that $h(s)$ is a tight lower bound on the drift (i.e. if (2.19) is an equality). However, in many situations the bound *is* tight, especially if the potential X_t does not jump around too much. Let us unravel the proof of Theorem 2.3.3 to understand this phenomenom better.

We first note that the proof is a reduction to the additive drift theorem, which *is* tight (see the discussion after Theorem 2.3.1). So, the only possible problem is the estimate (2.26) of the drift. This estimate may not be tight if (2.22), the inequality $g(s)-g(r) \geq (s-r)/h(s)$, is too coarse. Note that to estimate the drift, we use (2.22) specifically for $s = X_t$ and $r = X_{t+1}$. These are not arbitrary values; for example, for RLS on ONEMAX, they differ by at most one. We proved (2.22) by case distinction, so let us inspect one of the cases for illustration. For convenience, we restate the argument for $s > r > s_{\min}$:

$$g(s)-g(r) = \int_r^s \frac{1}{h(\sigma)} d\sigma \geq \int_r^s \frac{1}{h(s)} d\sigma = \frac{s-r}{h(s)}. \tag{2.23}$$

The crucial step is to use $1/h(\sigma) \geq 1/h(s)$ for the range $r \leq \sigma \leq s$. In general, this may be a bad estimate. However, if $s = X_t$ and $r = X_{t+1}$ are close to each other, then σ runs through a small range, and $1/h(\sigma)$ may not vary too much. For example, s and r differ at most by one for RLS on ONEMAX, and the function $1/h(\sigma) = n/\sigma$ does not vary much in such a small range, especially if r and s are large. We will see in Section 2.4.1 that large jumps are still tolerable if they occur with sufficiently small probability. The following artificial example from [33] illustrates how large jumps can lead to bad upper bounds. The idea of the construction is similar to the initial example on page 96.

Example 2.3.6 (RLS with shortcuts). Consider a (1+1) algorithm that in each step creates the optimum with probability $1/n$, and with probability $1-1/n$ does an RLS step as in Algorithm 2.1. To minimise ONEMAX, we may naively try the fitness as the potential, $X_t := \text{ONEMAX}(x^{(t)})$. For $X_t = s > 0$, there is a probability of $1/n$ of jumping directly to the optimum, thus decreasing the potential by s. On the other hand, there is a probability of $(1-1/n)\cdot i/n$ of decreasing the potential by 1 with a normal RLS step. Together, the drift is

$$\Delta_t(s) = h(s) := \frac{1}{n}\cdot s + \left(1-\frac{1}{n}\right)\frac{s}{n} = \frac{2s}{n} - \frac{s}{n^2} = (1 \pm o(1))\frac{2s}{n}. \tag{2.32}$$

Thus, the variable drift theorem (Theorem 2.3.3) yields

$$E[T] \leq \frac{1}{h(1)} + E\left[\int_1^{X_0} (1 \pm o(1))\frac{n}{2\sigma} d\sigma\right] = \Theta(n \log n). \tag{2.33}$$

However, since in each step we have a probability of at least $1/n$ of jumping directly to the optimum, the expected runtime is at most $E[T] \leq n$, so (2.33) is not tight. The problem can be understood by inspecting the transformed variable $Y_t := g(X_t)$ considered in the proof of the variable drift theorem, see (2.21). For simplicity, we ignore the factor $(1+o(1))$ in (2.32) and obtain

$$Y_t := \begin{cases} \frac{n}{2}(1+\ln X_t) & \text{if } X_t \geq 1, \\ 0 & \text{if } X_t = 0. \end{cases} \tag{2.34}$$

Computing the drift of Y_t directly, we obtain the following for $X_t = s$, i.e. for $Y_t = n/2 \cdot (1+\ln s)$:

$$\begin{aligned} E[Y_t - Y_{t+1} \mid X_t = s] &= \frac{1}{n} \cdot \frac{n}{2}(1+\ln s) + \left(1 - \frac{1}{n}\right) \frac{s}{n} \cdot \frac{n}{2} (\ln s - \ln(s-1)) \\ &= \frac{\ln s}{2} \pm O(1). \end{aligned} \tag{2.35}$$

Thus, we do not have constant drift in the scaled potential. However, in the proof of the variable drift theorem, we bound the drift by 1 (see (2.26)), which is the reason for the additional $\log n$ factor.

Fortunately, it is quite common that there are no large jumps in the fitness value. Mutation-based evolutionary algorithms tend to take small steps, and other nature-based search heuristics such as ant colony optimisation and estimation-of-distribution algorithms tend to make rather small updates to reasonable functions. However, note that this is not necessarily true for crossover operations. Also, depending on the fitness function, a small (genotypic) change may cause a large (phenotypic) jump in the fitness, as the next example shows.

Example 2.3.7 (RLS on BINVAL*).* We consider RLS (Algorithm 2.1 on page 97) for minimising the BINVAL function given by

$$\text{BINVAL}(x) = \sum_{i=1}^{n} 2^{n-i} x_i. \tag{2.36}$$

If we choose the potential $X_t := \text{BINVAL}(x^{(t)})$ to be identical to the fitness, then we observe that each one-bit has probability $1/n$ of being flipped. If the i-th bit is flipped from one to zero, this reduces the potential by 2^i. Hence, at a search point x with potential $s := \text{BINVAL}(x)$ the drift is

$$E[X_t - X_{t+1} \mid x^{(t)} = x] = \sum_{1 \leq i \leq n,\ x_i^{(t)}=1} \frac{1}{n} \cdot 2^{n-i} = \frac{1}{n} \sum_{i=1}^{n} 2^{n-i} x_i = \frac{s}{n}. \tag{2.37}$$

In particular, since the latter term depends only on s, we can write

$$E[X_t - X_{t+1} \mid X_t = s] = \frac{s}{n}. \tag{2.38}$$

Therefore we can apply the variable drift theorem (Theorem 2.3.3) with $h(s) = s/n$ and $s_{\min} = 1$, and obtain

$$E[T] \leq \frac{1}{1/n} + E\left[\int_1^{X_0} \frac{n}{\sigma} d\sigma\right] = n + n \cdot E[\ln X_0] = \Theta(n^2), \tag{2.39}$$

where the last equality follows since $X_0 \leq 2^{n+1}$, and since with probability at least 1/2 the first bit in X_0 is a one-bit, which implies $E[X_0] \geq 2^{n-1}$.

However, the bound (2.39) is far from tight. In fact, if we use the *OneMax potential* $\text{ONEMAX}(x) := \sum_{i=1}^n x_i$, then the drift with respect to ONEMAX is still $\Delta_t^{\text{ONEMAX}}(s) = s/n$, which leads to a runtime bound of $E[T] \leq n + n \cdot E[\text{ONEMAX}(x^{(0)})] \leq n \ln n + n$.[12]

The reason why (2.39) is not tight is that there may be some very large jumps in the potential (see the discussion before this example). For example, consider the situation where only a single one-bit is left. RLS operates symmetrically on BINVAL, so this one-bit is at a random position.[13] In particular, with probability at least 1/2, the bit is in the first half, and thus $X_t \geq 2^{n/2}$. Therefore, in (2.25) we estimate $h(\sigma) \leq h(s)$ for values of σ that range at least between $s_{\min} = 1$ and $2^{n/2}$. Thus the estimate is off by an exponential factor. Consequently, the rescaled potential $Y_t = g(X_t) = n(1 + \ln X_t)$ does not have constant drift. While the drift is always at least 1 by (2.26), if there is only a single one-bit left in the first half of the string, the rescaled potential decreases with probability $1/n$ from $Y_t \geq n(1 + \ln 2^{n/2}) = \Omega(n^2)$ to 0. Hence, the drift of Y_t in this situation is $1/n \cdot \Omega(n^2) = \Omega(n)$, causing the runtime bound to be almost a factor n too large.

2.3.2.3 When Rescaling Beats the Variable Drift Theorem

We have seen an example which illustrates why the variable drift theorem does not always give tight results. Unfortunately, a common reason is that the potential does not represent very well the progress the algorithm has made, in which case a truly new insight is needed. However, sometimes the problem can be solved by directly considering the rescaled potential. We illustrate this by an artificial example taken from [54].

Example 2.3.8 (Random Decline). Let $a > 0$ be a constant, let $n \in \mathbb{N}^+$ and consider the following Markov chain on $\mathcal{S} = \{0, \ldots, N\}$, where N is a sufficiently large integer compared with n. For this exposition, we will assume that N is so large that the process never hits the right border. We start with $X_0 = n$, and for each $t \geq 0$ we draw X_{t+1} uniformly at random from $\{0, 1, 2, \ldots, \min\{\lfloor aX_t \rfloor, N\}\}$.

If $a < 2$, then for $\mathcal{S} \in S \setminus \{0\}$ and all $t \geq 0$ we have a drift of

[12] Alternatively, we could observe that RLS behaves in exactly the same way on BINVAL and on ONEMAX, so the runtimes are the same.

[13] Note that this is specific to RLS, which uses only one-bit flips. An algorithm which flips two or more bits per step would not operate symmetrically, since it would trade a one-bit of large weight for a zero-bit of small weight, but not vice versa.

$$\Delta_t(s) \geq s - \frac{a}{2}s = \frac{2-a}{2} \cdot s. \tag{2.40}$$

Therefore, by the variable drift theorem, $E[T] = O(\log n)$. However, the theorem does not make any statement for $a \geq 2$.[14] Nevertheless, let us inspect the rescaled potential $Y_t := 1 + \ln(X_t)$. We give only an estimate; the full calculation, including error terms, can be found in [54]. For every $s \in \mathcal{S} \setminus \{0\}$ that is smaller than N/a,

$$\begin{aligned} E[Y_t - Y_{t+1} \mid Y_t = 1 + \ln s] &= 1 + \ln(s) - \frac{1}{\lfloor as+1 \rfloor} \sum_{k=1}^{\lfloor as \rfloor} (1 + \ln k) \\ &\approx \ln(s) - \frac{1}{as}\left(\int_1^{as} \ln \sigma \, d\sigma\right) \\ &= \ln(x) - \frac{1}{as}[\sigma \ln(\sigma) - \sigma]_{\sigma=1}^{as} \\ &\approx \ln(s) - (\ln(as) - 1) = 1 - \ln a. \end{aligned} \tag{2.41}$$

Thus we see that if $a < e = 2.71\ldots$ is a constant, then the drift of Y_t is also constant. Hence, by the additive drift theorem (Theorem 2.3.1) we get $E[T] = O(E[Y_0]) = O(\ln n)$. So, the analysis of the rescaled random variable applies to a wider range than the variable drift theorem does. In fact, the condition $a < e$ is tight for logarithmic runtime, since for $a \geq e$ the expected runtime is $\omega(\ln n)$ [54].

We have seen that once we try out the rescaling $Y_t = 1 + \ln(X_t)$, the rest is very simple and mostly calculations. We will discuss in Section 2.5 how to see that this particular rescaling is worth trying.

2.3.2.4 Further Applications of the Variable Drift Theorem

We conclude this section with some more applications of the variable drift theorem. They illustrate the fact that even if the drift is a highly complicated function, the variable drift theorem gives us an explicit expression for the expected runtime, which we can evaluate by elementary calculus. The impatient reader is free to skip this section.

Example 2.3.9 ($(1+\lambda)$ EA on ONEMAX*).* In 2017, Gießen and Witt [34] analysed the $(1+\lambda)$ EA (Algorithm 2.2) for minimising the ONEMAX function (see (2.27)).

The potential was identical to the fitness, $X_t = \text{ONEMAX}(x^{(t)})$. To bound the drift $\Delta_t(s)$, the authors of [34] used order statistics of the binomial distribution. They showed that $\Delta_t(s) \geq h(s)$, where[15]

---

[14] Worse: the statement could be applied for a non-constant a such as $a = 2(1 - 1/n)$, and would lead to the misleading bound $E[T] = O(n \log n)$.

[15] For the case $\lambda = \omega(1)$. The other case, $\lambda = O(1)$, is similar.

Algorithm 2.2: The $(1+\lambda)$ EA with offspring population size λ and mutation rate c/n, minimising a fitness function $f:\{0,1\}^n \to \mathbb{R}$

```
1 Choose x^(0) ∈ {0,1}^n uniformly at random;
2 for t = 0,1,2,... do
3     for i = 1,...,λ do
4         Create y^(t,i) by flipping each bit of x^(t) independently with probability
          c/n;
5         y^(t) ← argmin{f(y^(t,i))} (breaking ties randomly);
6         if f(y^(t)) ≤ f(x^(t)) then
7             x^(t+1) ← y^(t);
8         else
9             x^(t+1) ← x^(t);
```

$$h(s) := \begin{cases} (1-o(1))\frac{\ln\lambda}{\ln\ln\lambda} & \text{if } s \geq \frac{n}{(\ln\lambda)^{1/(\ln\ln\ln\lambda)}}, \\ (1/2-o(1))e^{-c}\frac{\ln\lambda}{\ln\ln\lambda} & \text{if } s \geq \frac{n}{\ln\lambda}, \\ (1-o(1))e^{-c}\min\{c,1\}/2 & \text{if } s \geq \frac{n}{\lambda}, \\ (1-o(1))e^{-c}\frac{c}{\sqrt{\ln n}} & \text{if } s \geq \frac{n}{\lambda\sqrt{\ln n}}, \\ (1-o(1))ce^{-c}\lambda\frac{s}{n} & \text{if } s < \frac{n}{\lambda\sqrt{\ln n}}. \end{cases} \tag{2.42}$$

Obviously, computing the drift is non-trivial, and this was the major contribution of the paper. Despite the complexity of the formula, once we know it we can easily obtain a runtime bound by the variable drift theorem:

$$E[T] \leq \frac{1}{h(1)} + E\left[\int_1^{X_{\max}} \frac{1}{h(\sigma)} d\sigma\right]. \tag{2.43}$$

The integral can now be computed by splitting it into six ranges, and evaluating it with elementary calculus. Actually, $h(\sigma)$ is constant for all ranges except for the last one, which gives one of the leading terms:

$$\int_1^{n/(\lambda\sqrt{\ln n})} (1+o(1))\frac{e^c n}{c\lambda\sigma} d\sigma = (1+o(1))\frac{e^c n \ln(n/(\lambda\sqrt{\ln n}))}{c\lambda}. \tag{2.44}$$

Proceeding like this for all six ranges, the authors of [34] obtained the final result

$$E[T] \leq (1+o(1))\left(\frac{e^c}{c}\cdot\frac{n\ln n}{\lambda} + \frac{1}{2}\cdot\frac{n\ln\ln\lambda}{\ln\lambda}\right). \tag{2.45}$$

The authors also proved a matching lower bound by the techniques discussed in Section 2.4.1.

Example 2.3.10 (Island Model on ONEMAX*).* Doerr, Fischbeck, Frahnow, Friedrich, Kötzing and Schirneck [13] studied island models in various topologies. For the complete graph as the migration topology, the algorithm consists of λ independent $(1+1)$ EAs, except that every τ rounds all individuals are updated by the current best search point; see Algorithm 2.3.

Algorithm 2.3: Island model with λ islands and migration interval τ for minimising $f:\{0,1\}^n \to \mathbb{R}$

```
1  Choose x^(0,1),...,x^(0,λ) ∈ {0,1}^n uniformly at random;
2  for t = 0,1,2,... do
3      for i = 1,...,λ do
4          Create y^(t,i) by flipping each bit of x^(t,i) independently with probability
           1/n;
5          if f(y^(t,i)) ≤ f(x^(t,i)) then
6              x^(t+1,i) ← y^(t,i);
7          else
8              x^(t+1,i) ← x^(t,i);
9      if (t+1 mod τ) = 0 then
10         for i = 1,...,λ do
11             y ← argmin{f(y^(t+1,i))} (breaking ties randomly);
12             x^(t+1,i) ← y;
```

For minimising the ONEMAX function, the most interesting phase[16] turns out to be the phase when the current best search point has its fitness in some interval $[s_0, s_1]$, where $s_0 = \min\{n, n\ln\lambda/(2\tau)\}$ and $s_1 = n/(\tau\ln\lambda)$. The authors of [13] defined X_t to be the fitness after t migrations, i.e. $X_t = \text{ONEMAX}(x^{(t\tau,i)})$ holds for every $1 \le i \le \lambda$. To identify the end of the phase, we truncate X_t, i.e. we define $X_t := 0$ if $\text{ONEMAX}(x^{(t\tau,i)}) < s_0$. Note that the minimum non-zero value of X_t is thus $s_{\min} = s_0$. The drift of X_t for all $t \ge 0$ and all $s \in [s_0, s_1]$ turns out to be

$$\Delta_t(s) \ge h(s) := \frac{c\ln\lambda}{\ln(n\ln\lambda/(\tau s))}, \tag{2.46}$$

for some constant $c > 0$. Note that the function $h(s)$ is increasing. Thus, by the variable drift theorem (Theorem 2.3.3), we may bound the expected number of migrations T_0 before a fitness of less than s_0 is achieved by

$$E[T_0] \le \frac{s_0}{h(s_0)} + \frac{1}{c\ln\lambda}\int_{s_0}^{s_1} \ln\left(\frac{n\ln\lambda}{\tau\sigma}\right) d\sigma, \tag{2.47}$$

[16] For some parameter regimes.

where we have used $X_0 \leq s_1$. The latter integral can now be evaluated by elementary analysis, and yields

$$\int_{s_0}^{s_1} \ln\left(\frac{n\ln\lambda}{\tau\sigma}\right) d\sigma = \frac{\tau}{n\ln\lambda}\Big[\sigma(1-\ln\sigma)\Big]_{\tau s_0/(n\ln\lambda)}^{\tau s_1/(n\ln\lambda)}, \tag{2.48}$$

from which the authors of [13] computed their runtime bounds. We refrain from stating the final result, since it involves several case distinctions with respect to τ and λ.

2.3.3 Multiplicative Drift

A very important special case of variable drift is *multiplicative drift*, where the drift is proportional to the potential. Introduced in [15, 19, 20], it has become the most widely used variant of drift analysis in the field of evolutionary algorithms. In fact, all of Examples 2.3.4, 2.3.6, 2.3.7 and 2.3.8 had multiplicative drift. In particular, Examples 2.3.6, 2.3.7 and 2.3.8 show that the same limitations as for variable drift apply.

Theorem 2.3.11 (Multiplicative Drift [20], special case of Theorem 2.3.3). *Let $(X_t)_{t\geq 0}$ be a sequence of non-negative random variables with a finite state space $\mathcal{S} \subseteq \mathbb{R}_0^+$ such that $0 \in \mathcal{S}$. Let $s_{\min} := \min(\mathcal{S}\setminus\{0\})$, let $T := \inf\{t \geq 0 \mid X_t = 0\}$ and, for $t \geq 0$ and $s \in \mathcal{S}$, let $\Delta_t(s) := E[X_t - X_{t+1} \mid X_t = s]$. Suppose there exists $\delta > 0$ such that for all $s \in \mathcal{S}\setminus\{0\}$ and all $t \geq 0$ the drift is*

$$\Delta_t(s) \geq \delta s. \tag{2.49}$$

Then

$$E[T] \leq \frac{1 + E[\ln(X_0/s_{\min})]}{\delta}. \tag{2.50}$$

We conclude this section by giving some applications of the multiplicative drift theorem.

Example 2.3.12 ((1+1) EA on Linear Functions). One of the cornerstones in the theory of evolutionary algorithms is the analysis of linear pseudo-Boolean functions $f : \{0,1\}^n \to \mathbb{R}$, i.e. functions of the form $f(x) = \sum_{i=1}^n w_i x_i$, where the w_i are constants. To avoid trivialities, we assume that the weights are non-zero, and by the symmetry of the search space we may assume that they are non-negative and sorted, $w_1 \geq w_2 \geq \ldots \geq w_n > 0$. We have already seen two examples of such functions: OneMax in Example 2.3.4 and BinVal in Example 2.3.7.

To analyse how the $(1+1)$ EA with mutation rate $c = 1/n$ (Algorithm 2.2 with offspring population size $\lambda = 1$) minimises a linear function, a naive approach would be to use the fitness as the potential, $X_t := f(x^{(t)})$. Similarly to the analysis of RLS on BINVAL, this yields a multiplicative drift of at least

$$\Delta_t(s) \geq \Omega(s/n), \tag{2.51}$$

since the (1+1) EA has at least a constant probability of performing an RLS step, i.e. of flipping exactly one bit. Therefore, the multiplicative drift theorem gives the bound

$$E[T] \leq O\left(\frac{1+E[\ln(X_0/w_n)]}{\delta}\right). \tag{2.52}$$

For ONEMAX-like functions where all weights are similar, this bound is $O(n \ln n)$, which turns out to be tight. However, for other linear functions such as BINVAL, the bound is not tight, for the same reason as for RLS on BINVAL (Example 2.3.7). Rather, the expected runtime is $\Theta(n \ln n)$, as was first shown by Droste, Jansen and Wegener [26].

For the ONEMAX potential $\mathrm{OM}_t := \mathrm{ONEMAX}(x^{(t)})$, the situation is rather interesting. For functions such as BINVAL, there are search points (e.g. the search point $(1,0,\ldots,0)$, where only the highest-valued bit has not yet been optimised) for which the drift is negative, i.e. $E[\mathrm{OM}_t - \mathrm{OM}_{t+1} \mid x^{(t)} = (1,0,\ldots,0)] < 0$. Nevertheless, Jägersküpper [41] showed by a coupling argument that bits of larger weight are more likely to be optimised, so that we still have a multiplicative drift [18] for all $t \geq 0$ and all $s \in \{1,\ldots,n\}$,

$$\Delta_t(s) = E[\mathrm{OM}_t - \mathrm{OM}_{t+1} \mid \mathrm{OM}_t = s] = \Omega(s/n), \tag{2.53}$$

from which a runtime bound $E[T] = O(n \ln n)$ follows. So this is one of the cases where it is beneficial to avoid filtrations and pointwise drift, see also the paragraph 'Drift versus expected drift' in Section 2.2.3.

The results can be tightened if one considers more carefully crafted potentials. Doerr, Johannsen and Winzen showed [19], building on ideas from [39], that the drift function $\varphi(x) := \sum_{i=1}^{\lfloor n/2 \rfloor} \frac{5}{4} x_i + \sum_{i=\lfloor n/2 \rfloor+1}^{n} x_i$ even has *pointwise* multiplicative drift, i.e. for all $t \geq 0$ and all search points $x \in \{0,1\}^n$,

$$E[\varphi(x^{(t)}) - \varphi(x^{(t+1)}) \mid x^{(t)} = x] = \Omega(\varphi(x)/n). \tag{2.54}$$

This yields again the runtime bound $E[T] = O(n \ln n)$. Pointwise multiplicative drift giving similar runtime bounds can also be achieved by other potential functions [20].

Similar techniques can also be used to show that the (1+1) EA still has runtime $\Theta(n \ln n)$ on every linear function if the mutation rate is c/n for an arbitrary constant c [15, 54, 75]. However, this requires a considerably more complicated potential function which must necessarily depend on the mutation rate [21].

Example 2.3.13 (Minimum Spanning Trees). Consider the following *minimum spanning tree (MST)* problem proposed in [64]. Let $G=(V,E)$ be a connected graph with n vertices, m edges $e_1,\dots,e_m$ and positive integer edge weights $w_1,\dots,w_m$. We denote by $w_{\max}:=\max_i w_i$ the maximum weight. A bit string $x\in\{0,1\}^m$ represents a subgraph of G with vertex set V, where the edge e_i is present if and only if $x_i=1$. The fitness of a bit string is given by $f(x)=\sum_{i=1}^n w_i x_i + p(x)$, where $p(x)$ is a punishment term for non-trees that ensures we find a spanning tree quickly, and stay within the set of spanning trees afterwards.

We consider the performance of the $(1+1)$ EA on this problem. In [64] it was shown that the algorithm quickly finds a spanning tree, so we assume for simplicity that the initial search point $x^{(0)}$ represents such a tree. We consider the potential function $X_t := \sum_{i=1}^n w_i x_i^{(t)} - w_{\text{opt}}$, where w_{opt} is the weight of a minimum spanning tree. Then, relying on results from [64], it was shown in [20] that the potential function has a multiplicative drift of

$$\Delta_t(s) = E[X_t - X_{t+1} \mid X_t = s] \geq \frac{s}{em^2}. \tag{2.55}$$

Hence, by the multiplicative drift theorem (Theorem 2.3.11) the expected runtime (starting from a spanning tree) is at most

$$E[T] \leq em^2(1+\ln(mw_{\max})), \tag{2.56}$$

since the minimum potential of a non-optimal search point is at least $s_{\min}\geq 1$, and since $mw_{\max}$ is an upper bound on X_0. It is an open question whether (2.56) is tight, since the best lower bound is $\Omega(m^2 \ln m)$ [64], which is a *tight* bound for RLS [69].

There are numerous other applications of the multiplicative drift theorem, including in evolutionary algorithms on other problems [17, 20, 24, 32], ant colony optimisation [30], island models [57], genetic programming [23] and estimation-of-distribution algorithms [31].

2.4 Advanced Drift Theorems

In this section we review the most important developments in drift analysis in recent years, in particular lower and tail bounds, weak drift, negative drift, and population drift. Note that unlike those in the previous section, many advanced theorems, especially about tail bounds, make assumptions about the *pointwise drift* (see Section 2.2.2).

2.4.1 Lower Bounds

As discussed in Section 2.3.2, the variable drift theorem and the multiplicative drift theorem only have a chance to give tight results if we have some restriction on the probability of making large jumps. From the earlier discussion on pages 103ff, we know that the critical estimates in the proof of the variable drift theorem are (2.23), (2.24) and (2.25). If $h(X_{t+1})/h(X_t)$ is always close to one, then these estimates are tight. For example, (2.23) tells us that, for the potential function g from the proof, $g(X_t)-g(X_{t+1}) \geq (X_t - X_{t+1})/h(X_t)$ if $X_t \geq X_{t+1} \geq s_{\min}$. But the same argument also shows that $g(X_t)-g(X_{t+1}) \leq (X_t - X_{t+1})/h(X_{t+1})$ in this case. If $h(X_t)$ and $h(X_{t+1})$ differ at most by a factor $c > 1$, then the upper and lower bound also differ at most by a factor c. Following this idea, we get the following lower bound.

Theorem 2.4.1 (Variable Drift Theorem, Lower Bound 1). *Let $(X_t)_{t\geq 0}$ be a sequence of non-negative random variables with a finite state space $\mathcal{S} \subseteq \mathbb{R}_0^+$ such that $0 \in \mathcal{S}$. Let $s_{\min} := \min(\mathcal{S} \setminus \{0\})$, let $T := \inf\{t \geq 0 \mid X_t = 0\}$ and, for $t \geq 0$ and $s \in \mathcal{S}$, let $\Delta_t(s) := E[X_t - X_{t+1} \mid X_t = s]$. Suppose there is an increasing function $h: \mathbb{R}^+ \to \mathbb{R}^+$ and a constant $c \geq 1$ such that for all $s \in \mathcal{S} \setminus \{0\}$ and all $t \geq 0$ the following conditions hold:*

$$X_{t+1} \leq X_t, \tag{2.1}$$

$$\Delta_t(s) \leq h(s), \tag{2.2}$$

$$\frac{h(\max\{X_{t+1}, s_{\min}\})}{h(X_t)} \geq \frac{1}{c}. \tag{2.3}$$

Then

$$E[T] \geq \frac{1}{c} \cdot \left(\frac{s_{\min}}{h(s_{\min})} + E\left[\int_{s_{\min}}^{X_0} \frac{1}{h(\sigma)} d\sigma \right] \right), \tag{2.4}$$

where the expectation on the latter term is over the random choice of X_0.

Note that the theorem has the rather strong assumption that the sequence X_t is non-increasing, see also the discussion after Theorem 2.4.3. This is necessary because otherwise, even if positive and negative contributions to the drift are known up to constant factors, the relative error may increase due to cancellation effects. However, for non-increasing X_t, the upper bound of Theorem 2.3.1 and the lower bound of Theorem 2.4.1 are directly comparable: they differ exactly by a factor c.

Despite its arguably natural form, it seems that Theorem 2.4.1 has never been formulated in this version in the literature,[17] perhaps because it usually

[17] Although Feldmann and Kötzing [28] gave bounds following the same ideas.

does not give tight leading constants. For example, consider RLS on OneMax as in Example 2.3.4. There X_t is given by the fitness, and $h(s) = s/n$. The largest jump occurs when X_t decreases from 2 to 1, in which case $h(X_{t+1})/h(X_t) = 1/2$. Thus the lower bound is a factor 2 from the upper bound.

Doerr, Fouz and Witt [14] have given a variant which usually gives a tighter lower bound. In fact, it gives a matching lower bound in many applications.

Theorem 2.4.2 (Variable Drift Theorem, Lower Bound 2 [14]). *Let $(X_t)_{t\geq 0}$ be a sequence of non-negative random variables with a finite state space $\mathcal{S} \subseteq \mathbb{R}_0^+$ such that $0 \in \mathcal{S}$, and with associated filtration $\mathcal{F}_t$. Let $s_{\min} := \min(\mathcal{S} \setminus \{0\})$, and let $T := \inf\{t \geq 0 \mid X_t = 0\}$. Suppose there are two functions $\xi, h : \mathbb{R}_0^+ \to \mathbb{R}^+$ such that h is monotone increasing, and such that for all $s \in \mathcal{S} \setminus \{0\}$ and for all $t \geq 0$ the following three conditions hold:*

$$X_{t+1} \leq X_t, \tag{2.5}$$

$$X_{t+1} \geq \xi(X_t), \tag{2.6}$$

$$E[X_t - X_{t+1} \mid \mathcal{F}_t, X_t = s] \leq h(\xi(s)). \tag{2.7}$$

Then

$$E[T] \geq \frac{s_{\min}}{h(s_{\min})} + E\left[\int_{s_{\min}}^{X_0} \frac{1}{h(\sigma)} d\sigma\right], \tag{2.8}$$

where the expectation in the latter term is over the random choice of X_0.

To apply Theorem 2.4.2, one should first choose ξ such that (2.6) is satisfied, and afterwards choose h in such a way that the composition $h \circ \xi$ is the drift, cf. Example 2.4.4 below. In particular, the function h in Theorem 2.4.2 is not identical to the function h in the upper-bound version, Theorem 2.3.3. A formulation of Theorem 2.4.3 in which the function h corresponds directly to the same function as in Theorem 2.3.3 can be found in [12].

We remark that Gießen and Witt [33] have developed a version in which the deterministic condition (2.6) is replaced by a probabilistic condition. The exact formulation is rather technical. For the special case where (2.6) holds with some fixed probability p independent of X_t, a simplified version was developed by Doerr, Doerr and Yang [12]. Moreover, the theorem in [33] simplifies for multiplicative drift [75]. We give here the version in [52], which assumes bounds on the probability that X_t drops by more than a multiplicative factor. A version in which an *additive* bound on $|X_t - X_{t+1}|$ is assumed can be found in [23].

Theorem 2.4.3 (Multiplicative Drift Theorem, Lower Bound [52, 75]). *Let $(X_t)_{t\geq 0}$ be a sequence of non-negative random variables with a* *finite state space* $\mathcal{S} \subseteq \mathbb{R}_0^+$ *such that* $0 \in \mathcal{S}$*, and with associated filtration* $\mathcal{F}_t$*. Let* $s_{\min} := \min(\mathcal{S} \setminus \{0\})$*, and let* $T := \inf\{t \geq 0 \mid X_t = 0\}$*. Suppose there are two constants* $0 < \beta, \delta \leq 1$ *such that for all* $s \in \mathcal{S} \setminus \{0\}$ *and all* $t \geq 0$ *the following conditions hold:*

$$X_{t+1} \leq X_t, \tag{2.9}$$

$$\Pr[X_t - X_{t+1} \geq \beta X_t \mid \mathcal{F}_t, X_t = s] \leq \frac{\beta\delta}{1 + \ln(s/s_{\min})}, \tag{2.10}$$

$$E[X_t - X_{t+1} \mid \mathcal{F}_t, X_t = s] \leq \delta s. \tag{2.11}$$

Then

$$E[T] \geq \frac{1-\beta}{1+\beta} \cdot \frac{1 + E[\ln(X_0/s_{\min})]}{\delta}. \tag{2.12}$$

Recently, Doerr, Doerr and Kötzing [11] showed that the monotonicity condition (2.9) can be completely removed if (2.11) is replaced by the condition that, for all $s, s' \in \mathcal{S} \setminus \{0\}$ with $s' \leq s$,

$$E[\max\{s' - X_{t+1}, 0\} \mid \mathcal{F}_t, X_t = s] \leq \delta s'. \tag{2.13}$$

The authors of [11] showed that this condition is satisfied for very natural processes. In particular, it is satisfied for processes with multiplicative drift if the jump probability $p(s) := \Pr[X_{t+1} \leq s' \mid \mathcal{F}_t, X_t = s]$ is a decreasing function of s, whenever $s' \leq s$.[18] This modification extends the scope of Theorem 2.4.3 considerably, since many evolutionary algorithms are non-monotone processes. Moreover, it seems likely that the proof in [11] can be extended to generalise related lower bounds, in particular the lower bounds for variable drift in Theorems 2.4.1 and 2.4.2.

We conclude the discussion of lower bounds with an easy example to demonstrate how to apply Theorem 2.4.2 and 2.4.3.

Example 2.4.4 (RLS on OneMax*, Lower Bound).* Consider once more RLS on OneMax as in Example 2.3.4. We want to apply Theorem 2.4.2. Since X_t decreases by at most one, we choose $\xi(s) := s - 1$ to satisfy (2.6) as tightly as possible. Since the drift is $\Delta_t(s) = s/n$, we choose $h(s) := (s+1)/n$ so that $h(\xi(s)) = \Delta_t(s)$. Thus we obtain the lower bound

$$E[T] \geq \frac{s_{\min}}{h(s_{\min})} + E\left[\int_{s_{\min}}^{X_0} \frac{1}{h(\sigma)} d\sigma\right] = \frac{1}{2/n} + E\left[\int_1^{X_0} \frac{n}{\sigma+1} d\sigma\right]$$

[18] In other words, it should more likely to jump into the interval $[0, s']$ if you start closer to it.

$$= \frac{n}{2} + n \cdot E[\ln(X_0+1) - \ln 2], \tag{2.14}$$

which is easily seen to be at least $n \ln n - O(n)$.

Note that Theorem 2.4.3 would give a less tight bound if naively applied. To satisfy (2.10) for $s=2$, it would be necessary to choose $\beta \geq 1/2$, and for $s=1$ we even need $\beta \geq 1$, which renders the bound useless. However, this problem can be overcome by truncating the search space; see [11] for details.

2.4.2 Tail Bounds

In some cases, we would also like to understand T beyond its expectation. In particular, we may want T to be concentrated, i.e. we want bounds on the probability that T deviates substantially from its expectation. This is desirable for at least two reasons. Firstly, it gives more concrete guarantees about T, for example that the algorithm will converge in a certain number of steps with 99% probability. Secondly, it might also happen that the expectation is misleading. For example, consider the following variant of the gambler's ruin problem. A gambler starts with \$1, and with each game she either wins or loses \$1, but the probability of losing is $1/2+1/n$, so slightly larger than the probability $1/2-1/n$ of winning. Let T be the time until she is broke, i.e. the number of games until she has no money left. Then the drift towards 0 is $2/n$, and therefore $E[T] = n/2$ by the additive drift theorem. However, it can be computed that $\Pr[T \leq 27] \geq 70\%$, which holds even for the fair game where winning and losing are equally likely. Therefore, for large n the expectation $n/2$ is rather misleading, since *typical values* of T are very different. Such discrepancies can be ruled out by concentration results.

For the standard drift theorems, we need additional assumptions about X_t for such concentration results to hold, with one notable exception. The following *upper tail bound for multiplicative drift* holds without any further requirements, as pointed out by Doerr and Goldberg [15]. We give the simplified formulation presented in [20]. We also present the proof of Doerr and Goldberg, which is remarkably short and elegant.

Theorem 2.4.5 (Multiplicative Drift, Upper Tail Bound [15, 20]). *Let $(X_t)_{t\geq 0}$ be a sequence of non-negative random variables with a finite state space $\mathcal{S} \subseteq \mathbb{R}_0^+$ such that $0 \in \mathcal{S}$. Let $s_{\min} := \min(\mathcal{S} \setminus \{0\})$, and let $T := \inf\{t \geq 0 \mid X_t = 0\}$. Suppose that $X_0 = s_0$, and that there exists $\delta > 0$ such that, for all $s \in \mathcal{S} \setminus \{0\}$ and all $t \geq 0$,*

$$E[X_t - X_{t+1} \mid X_t = s] \geq \delta s. \tag{2.15}$$

Then, for all $r \geq 0$,

$$\Pr\left[T > \left\lceil \frac{r+\ln(s_0/s_{\min})}{\delta} \right\rceil\right] \le e^{-r}. \tag{2.16}$$

Proof. For every fixed $\rho = \lceil (r+\ln(s_0/s_{\min}))/\delta \rceil \in \mathbb{N}$, by Markov's inequality,

$$\Pr[T > \rho] = \Pr[X_\rho > 0] \le \frac{E[X_\rho]}{s_{\min}} \overset{(*)}{\le} (1-\delta)^\rho \frac{s_0}{s_{\min}}, \tag{2.17}$$

where (*) comes from applying (2.15) and linearity of expectation ρ times. Since $(1-x) \le e^{-x}$ for all $x \in \mathbb{R}$, we obtain $\Pr[T > \rho] \le e^{-\rho\delta} s_0/s_{\min} \le e^{-r}$. □

For all of the other main drift theorems, including additive drift, variable drift and lower tails for multiplicative drift, we need assumptions on the probability of large jumps. For example, consider the process on $\mathcal{S} = \{0, n\}$ in which $X_t = n$ has probability $1/n$ of jumping to zero, and stays at n otherwise. Then X_t has drift one towards 0, but the hitting time T is geometrically distributed. In particular, T is not concentrated.[19] So, we need to make some assumption about the distribution of $|X_t - X_{t+1}|$.

The easiest assumption is that large jumps do not occur at all, i.e. $|X_{t+1} - X_t| < c$ for some parameter c. This case occurs in various situations, for example for RLS, for some ant colony optimisation algorithms such as the max–min ant system (MMAS) and for the compact genetic algorithm (cGA). We refer the reader to Kötzing [43] for a large collection of additive drift theorems with this assumption.

While there are situations without large jumps, there are even more cases in which large jumps may occur, but are unlikely. Thus research has focused on drift theorems with assumptions about the jump probability, usually some type of exponentially falling bounds, i.e. $\Pr[|X_{t+1} - X_t| > j] \le c \cdot (1+\eta)^{-j}$ for some parameters $c, \eta > 0$. In this chapter we stick with this type of condition, although generalisations are possible. Kötzing has made the point that exponentially falling jump probabilities imply a sub-Gaussian distribution of $X_t - \varepsilon t$, which is sufficient to derive most known tail bounds [44].[20] Lehre and Witt have given a very general framework for drift theorems [52, 53], in which only weak conditions on the exponential probability-generating function $e^{\lambda(X_t - X_{t+1})}$ are needed.[21] Most major drift theorems, including concentration bounds, can be derived from this framework, so that it arguably renders the other drift theorems unnecessary [52]. However, researchers have continued to use specialised drift theorems, possibly because the framework of Lehre and Witt comes with a substantial technical overhead. We give their main theorem at the end of the section for quick reference, but discussing its relation to the other drift theorems is beyond the scope of this chapter, and we refer the reader to the very nice exposition in [52].

[19] For example, $\Pr[T > 2E[T]] = (1-1/n)^{2n} \approx e^{-2}$.

[20] And arguably more natural, using the Azuma–Hoeffding inequality.

[21] More precisely, only the *expectation* of this function needs to be bounded.

Even with bounds on the probability of making jumps, lower tail bounds remain rather delicate. Unfortunately, it is *not* true in general that the runtime is concentrated around the expectation. This problem occurs when the drift is too weak, as the following counterexample shows.

Example 2.4.6 (Runtime is Not Concentrated Around Mean for Weak Drift). We consider the following artificial random walk on the set $\{0,1,\ldots,N\}$ for some (very large) constant N. We start at $X_0 = n$, where n is much smaller than N. For $X_t = s$, with probability $1/n^4$ we take a step to the left, $X_{t+1} := X_t - 1$, and otherwise we flip an unbiased coin to see whether we take a step to the left or to the right. We say that we take a *biased step* in the first case, and an *unbiased step* in the second.[22] Effectively, this process can be summarised as

$$X_{t+1} = \begin{cases} X_t - 1 & \text{with probability } \frac{1}{2}(1+1/n^4), \\ X_t + 1 & \text{with probability } \frac{1}{2}(1-1/n^4). \end{cases} \tag{2.18}$$

Then the drift is easily seen to be

$$\Delta_t(s) = \frac{1}{n^4}, \tag{2.19}$$

so that by the additive drift theorem (Theorem 2.3.1) we obtain

$$E[T] = n^4. \tag{2.20}$$

So, in terms of expectation, drift analysis can handle the problem quite well. However, it turns out that the expectation is completely misleading. Consider the first n^3 steps of the algorithm. By a union bound, with probability $1 - O(1/n)$ all of these steps are unbiased. Hence, with high probability the first n^3 steps are given by an unbiased random walk, also known as a *gambler's ruin process.* This process is well studied, and it is known that the probability of walking from n to 0 in at most αn^2 steps is $1 - O(\alpha^{-1/2})$ for all $\alpha > 1$ [36]. In particular, with $\alpha = n$, the probability that an unbiased random walk starting at n hits 0 in at most n^3 steps is $1 - O(n^{-1/2})$. Thus, with high probability the stopping time T of our process satisfies $T = O(n^3)$.[23] Hence, with high probability, T is asymptotically much smaller than its expectation $E[T] = n^4$.

[22] We have neglected the border case $X_t = N$ in the description. However, if N is large enough, e.g. $N = e^n$, then we cannot hit the right border in $o(N)$ steps, so the arguments are unaffected by the right border. For (2.20) we require that the drift is also $1/n^4$ at the border.

[23] In fact, if we are mathematically sloppy, the 'typical case' is $T = \Theta(n^2)$.

This example is rather prototypical for situations with weak drift. In fact, it was shown in [22] that in general[24] for weak additive drift the value of $E[T]$ is not dominated by 'typical' cases, but at least a constant proportion of $E[T]$ comes from exceptional cases in which T is much larger than $E[T]$. We also remark that Example 2.4.6 above can easily be adapted to multiplicative drift, for example by making the probability of an unbiased step X_t/n^{10}. Since X_t changes in each step by at most one, by Theorem 2.4.3 the bound $E[T] = O(n^{10} \log n)$ given by the multiplicative drift theorem is tight up to constant factors. However, as before, the runtime is $O(n^3)$ with high probability, so that with high probability the runtime is much smaller than the expected runtime.

Despite this problem, good tail bounds for additive drift have been developed. The following theorem follows from combining Theorems 10, 12 and 13 in [44].[25]

Theorem 2.4.7 (Additive Drift, Tail Bounds, following [44]). *Let $(X_t)_{t\geq 0}$ be a sequence of non-negative random variables with a finite state space $\mathcal{S} \subseteq \mathbb{R}_0^+$ such that $0 \in \mathcal{S}$, and with associated filtration $\mathcal{F}_t$. Let $s_{\min} := \min(\mathcal{S} \setminus \{0\})$, and let $T := \inf\{t \geq 0 \mid X_t = 0\}$. Suppose that $X_0 = s_0$, and that there exist $\delta, \eta, r > 0$ such that for all $s \in \mathcal{S} \setminus \{0\}$, all $j \in \mathbb{N}_0$ and all $t \geq 0$ the following conditions hold:*

$$\Pr[|X_{t+1} - X_t| > j \mid \mathcal{F}_t] \leq \frac{r}{(1+\eta)^j}, \tag{2.21}$$

$$E[X_t - X_{t+1} \mid \mathcal{F}_t, X_t = s] \leq \delta. \tag{2.22}$$

Then, for all $x \geq 0$,

$$\Pr\left[T \leq \frac{s_0 - x}{\delta}\right] \leq \exp\left\{-\frac{\eta x}{8} \cdot \min\left\{1, \frac{\eta^2 \delta x}{32 r s_0}\right\}\right\}. \tag{2.23}$$

If, instead of (2.22), *we have*

$$E[X_t - X_{t+1} \mid \mathcal{F}_t, X_t = s] \geq \delta, \tag{2.24}$$

then

$$\Pr\left[T \geq \frac{s_0 + x}{\delta}\right] \leq \exp\left\{-\frac{\eta x}{8} \cdot \min\left\{1, \frac{\eta^2 \delta x}{32 r s_0}\right\}\right\}. \tag{2.25}$$

---

[24] Under some weak assumptions, in particular assuming that large step sizes are unlikely as in (2.21) below.

[25] Actually, the statement in [44] is stronger, since it states that *at no point* during the whole process does X_t deviate substantially from its expectation, whereas we consider only values of X_t that are relevant for the runtime.

Note that the bounds in (2.23) and (2.25) give concentration only if the right hand side is of the form $\exp\{-\Phi\}$ for a large term Φ. In particular, consider the case where δ and r are constants and $x = \Theta(s_0)$. Then $\Phi = \omega(1)$ if and only if the bound s_0/δ on the expected runtime satisfies $s_0/\delta = o(x^2) = o(s_0^2)$. On the other hand, for $s_0/\delta = \omega(s_0^2)$ the runtime bound from the drift is larger than the time that an unbiased random walk would need to hit 0; see also Example 2.4.6. So, it is not surprising that Theorem 2.4.7 does not give concentration in this regime. Tight concentration bounds for the regime of weak drift can be found in [44].

We conclude the section with a consideration of the tail bounds in the general framework of Lehre and Witt [52, 53]. Note that [52, 53] both contain also several corollaries that correspond to simplified special cases, in particular some cases which resemble more closely our variant of the variable drift theorem.

Theorem 2.4.8 (General Drift Theorem, Tail Bounds [53]). *Let $a \geq 0$, and let $(X_t)_{t\geq 0}$ be a sequence of random variables with a finite state space $\mathcal{S} \subseteq \mathbb{R}_0^+$ such that the interval $[0,a] \cap S$ is absorbing, and with associated filtration $\mathcal{F}_t$. Let $T_a := \inf\{t \geq 0 \mid X_t \leq a\}$, and assume that $X_0 = s_0 > a$. Moreover, let $\lambda > 0$, let $g : \mathbb{R}_0^+ \to \mathbb{R}_0^+$ be a function such that $g(0) = 0$ and $g(s) \geq g(a)$ for all $s > a$, and let $\beta : \mathbb{N} \to \mathbb{R}^+$.*

(a) If, for all $t \geq 0$,

$$E[e^{-\lambda(g(X_t)-g(X_{t+1}))} \mid \mathcal{F}_t, X_t > 0] \leq \beta(t), \tag{2.26}$$

then for all $t \geq 0$,

$$\Pr[T_a > t] < \left(\prod_{r=0}^{t-1} \beta(r)\right) e^{\lambda(g(s_0)-g(a))}. \tag{2.27}$$

(b) If, for all $t \geq 0$,

$$E[e^{\lambda(g(X_t)-g(X_{t+1}))} \mid \mathcal{F}_t, X_t > 0] \geq \beta(t), \tag{2.28}$$

then for all $t \geq 0$,

$$\Pr[T_a < t] \leq \left(\prod_{r=0}^{t-1} \beta(r)\right) e^{-\lambda(g(s_0)-g(a))}. \tag{2.29}$$

In general, in order to obtain tail bounds for variable drift, we can either apply Theorem 2.4.8; or we can rescale X_t, as discussed in Section 2.3.2, to turn variable drift into additive drift, and then apply Theorem 2.4.7. Unfortunately, both approaches tend to be very technical. The most important case is obtaining tight lower tail bounds for multiplicative drift. Even with the framework of Lehre and Witt, in order to derive lower tail bounds for the

(1+1) EA on OneMax, it is still necessary to split the process into phases of relatively constant drift [52]. An easy and comprehensive lower tail bound for multiplicative drift is still absent in the literature.

2.4.3 Negative Drift

If the drift does not point towards zero, but instead points with a constant rate away from zero, then it takes exponential time to cross an interval. The first theorem of this type was proven by Oliveto and Witt [65, 66], following Hajek's classical work [37]. We give a formulation close to [54, 71] because it avoids o-notation for the length of the interval. Explicit constants can be found in [44, 67, 76].

Theorem 2.4.9 (Negative Drift, following [54, 65, 66, 71]). *For all $a,b,\delta,\eta,r>0$, with $a<b$, there is a $c>0, n_0 \in \mathbb{N}$ such that the following holds for all $n \geq n_0$. Suppose $(X_t)_{t\geq 0}$ is a sequence of random variables with a finite state space $\mathcal{S} \subseteq \mathbb{R}_0^+$, and with associated filtration $\mathcal{F}_t$. Assume that $X_0 \geq bn$, and let $T_a := \min\{t \geq 0 \mid X_t \leq an\}$ be the hitting time of $\mathcal{S} \cap [0,an]$. Assume further that for all $s \in \mathcal{S}$ with $s > an$, for all $j \in \mathbb{N}_0$ and for all $t \geq 0$ the following conditions hold:*

$$E[X_t - X_{t+1} \mid \mathcal{F}_t, X_t = s] \leq -\delta, \tag{2.30}$$

$$\Pr[|X_t - X_{t+1}| \geq j \mid \mathcal{F}_t, X_t = s] \leq \frac{r}{(1+\eta)^j}. \tag{2.31}$$

Then

$$\Pr[T_a \leq e^{cn}] \leq e^{-cn}. \tag{2.32}$$

Negative drift is helpful for proving lower bounds [54, 67, 71], but not only so. It may also be used to show that an algorithm stays in a desired parameter regime. For example, Neumann, Sudholt and Witt used it to show that an ant colony optimisation (ACO) algorithm has good runtime because all pheromone values stay in a desirable range [63]. Similarly, Kötzing and Molter [47], as well as subsequent authors [30, 55, 56], used negative drift to show that ACO algorithms tend to stay close to the optimum, thus enabling the algorithm to follow the optimum in a dynamically changing environment. In a different setting, Sudholt and Witt [72] showed that the compact genetic algorithm (cGA) is efficient on OneMax[26] because, for each position, the probability of sampling a one-bit never becomes too low. Similar ideas

[26] In some parameter regimes.

have been applied for population-based non-elitist algorithms in the strong-selection weak-mutation (SSWM) regime [68].

2.4.4 Populations

If the algorithm uses population sizes larger than one, or if it does not work at all with populations, as in the case of ant colony optimisation (ACO) or estimation-of-distribution algorithms (EDAs),[27] then it is often challenging to find a single potential X_t which captures well the quality of the current population. As before, *if* such a potential can be found, then drift analysis can take care of the rest. In some cases, it suffices to consider the current best optimum as the potential (see Example 2.3.10 and [13]), or some average quality [29, 72]. A systematic approach was developed by Corus, Dang, Eremeev and Lehre [3, 4], who gave the so-called level-based theorem for population-based algorithms. With this theorem, they identified a generic situation in which a good potential can be found automatically. A population-based algorithm in their sense[28] is any algorithm of the following form. In each round it maintains a population of size λ, and from this population it generates some probability distribution $\mathcal{D}$. For the next round, it produces λ samples independently from $\mathcal{D}$, which form the next generation.

This framework of population-based algorithms applies to many situations, often with a twist to the usual algorithm description. Firstly, it does include all (μ, λ) evolutionary or genetic algorithms if the λ offspring are generated independently of each other. In this case, let P_i be the i-th offspring population.[29] Then, from P_i, a complex process determines some probability distribution $\mathcal{D}$ from which the next offspring is sampled. This process subsumes selection and mutation/crossover. Other population-based algorithms include simulated annealing, and, surprisingly, EDAs [7]. While these latter algorithms conceptually maintain a probability distribution rather than a population, they do produce a sample population in each round, from which the next distribution is computed. This offspring population makes them fit into the framework of population-based algorithms.

The level-based theorem assumes a partitioning of the search space into fitness levels that need to be climbed by the population. It gives an upper bound on the expected runtime if certain conditions are satisfied. The exact formulation is rather technical, so we refer the reader to [4]. Qualitatively, three ingredients are required:

---

[27] ACO algorithms maintain pheromone values; EDAs maintain a probability distribution, rather than a population of search points.

[28] Conflicting terminology exists.

[29] *Not* the parent population, since the next parents are not sampled independently from these. Rather, the parents of the next generation need to compete with each other in the selection step.

1. If part of the population has at least fitness level i, then the probability of sampling an offspring at level $i+1$ is sufficiently large.
2. The fraction of the population which has fitness level at least i increases in expectation.
3. The population size is large enough.

Although it has only recently been developed, the level-based theorem has already found quite a number of applications, including the analysis of genetic algorithms with a multitude of selection mechanisms and benchmark functions [4], of EDAs [7, 50], of self-adaptive algorithms [9] and of algorithms in situations that are dynamic [5] or noisy [6], or provide only partial information [8].

2.5 Finding the Potential Function

At the very beginning of the chapter, we listed three ingredients for runtime analysis via drift theory: finding a good potential function, computing the drift and transferring the knowledge about the drift into information about the runtime. In this chapter, we have discussed the third point, because it is based on a universal technique that applies to many settings. In contrast, the first two points are highly problem-dependent, and cannot be generalised well. As mentioned before, the second point is usually not the hardest part, though it is often the most technical part and sometimes tedious. On the other hand, the first task – finding a good potential function – is often the hardest part, and it requires a lot of insight into the problem. Unfortunately, it is difficult to give general advice on how to find an appropriate fitness function for a given problem. Nevertheless, we will try to give some approaches which may be helpful.

A first question may be whether drift analysis is always applicable, or whether there are cases where the method fails completely. More concretely, is there always a good potential function, ideally one with constant drift? The answer is pleasantly clear: 'Yes'. In theory, there is even a surprisingly simple answer to the question of what this potential may look like. We may always choose the *canonical potential* $X_t := E[T \mid \mathcal{F}_t] - t$, where $\mathcal{F}_t$ is the history of the algorithm up to time t. Note that T is, as usual, the total number of steps in the process; it is not just the number of steps remaining. For the canonical potential, we always get a drift of exactly 1, for rather trivial reasons [20, 39]. The canonical potential does not look very helpful, since it seemingly only helps in finding the runtime if we already know the runtime. However, the canonical potential gives us a natural candidate for the right potential function if we have any *guess* as to what the runtime might be. The guess may come from heuristic considerations or from simulations. The situation resembles induction, where finding the right induction hypothesis is sometimes much harder than actually proving the inductive step. With the random decline in Example 2.3.8 we have already seen a case where the situation was obscure, but after the right scaling, $Y_t = 1 + \ln(X_t)$, it was rather easy to check that the drift was constant. How do we get to such a scaling? Reinspecting the example, we find that it is very natural to guess that the runtime is logarithmic, so a scaling of the form $Y_t = c_1 + c_2 \log(X_t)$ is a natural candidate. Indeed, any scaling of this form would have been sufficient. Choosing $c_1 = c_2 = 1$ was just the most convenient choice, owing to the fact that then $Y_t = 1$ if and only if $X_t = 1$. We have seen other examples of the rescaling technique in Example 2.3.2 and in the variable drift theorem.

Note that the canonical potential is more than 'just' a rescaling technique, since it defines X_t from scratch. In particular, we can theoretically compute the expected runtime for *every random process*[30] by drift analysis, by using the canonical potential. In practice, the main problem is that the history F_t (or even the current state) is too complicated to work with, and likewise the canonical potential is often too complex too handle. Therefore, the art of drift analysis lies in finding a potential which is simple and manageable, but which still resembles the canonical potential.

Let us consider next the (quite realistic) scenario where we already have some candidate for a potential function, but this candidate is still not good enough. Let us first discuss what it means that a potential function is 'not good'. If we want additive drift, it means that there are different states s_1, s_2 of the algorithm with very different drift. If we want multiplicative drift, it means that the ratio between the drift and the potential is very different for some states s_1, s_2, because we want the drift to be proportional to the potential. So, the first task is to look for states with such discrepancies. Then we can try to repair this defect: if the drift at s_1 is too large (compared with the drift at s_2) then we must try to decrease the difference between the potential of s_1 and the potentials of typical successor states of s_1. We can do this either by decreasing the potential of s_1 or by increasing the potentials of successor states. Hopefully, this will improve the accuracy of the potential function. We may iterate this procedure until we arrive at a good potential function.

For concreteness, let us study this approach with an example. Consider the (1+1) EA with standard mutation rate $1/n$ for minimising BinVal, where $\text{BinVal}(x) = \sum_{i=1}^{n} 2^{n-i} x_i$. Our first guess is to use the fitness function as the potential, $X_t := \text{BinVal}(x^{(t)})$. Our hope is that we will get multiplicative drift, as we got for RLS in Example 2.3.7. However, with this potential we have two problems. Firstly, the potential may make huge jumps (e.g. decrease from 2^{n-1} to 0), so we need to be careful when applying the multiplicative drift theorem, as we saw in Examples 2.3.6 and 2.3.7. Secondly, the drift is not very close to multiplicative. For example, consider the search points $s_1 := (0,\ldots,0,1)$ and $s_2 := (1,0,\ldots,0)$. The potentials are $x_1 := 1$ and $x_2 :=$

[30] If the expected runtime is finite. However, the process does not need to have finite, bounded or discrete search spaces.

2^{n-1}, respectively. A mutation of s_1 is accepted only if it flips the last bit, and no other bit, which happens with probability $\approx 1/(en) = x_1/(en)$. On the other hand, for s_2 we accept every mutation that flips the first bit, which happens with probability $1/n$. We may also flip a few other bits, but the potential still goes down by $(1-o(1))2^{n-1}$ in expectation if we flip the first bit. Therefore the drift is $\approx 2^{n-1}/n = x_2/n$. So the drift for s_2 is by factor e larger than desired, if we compare it with s_1. Hence, we must try to decrease the potential for s_2 and/or increase it for s_1. A natural way to do this is to reduce the weight of the higher-order bits in computing the potential. This might also alleviate the effects of large jumps.

How much should we reduce the weight? In the extreme case, we would make all weights equal, i.e. we would use the OneMax potential. This works well on strings where the higher-order bits are all zero. For example, for s_2 we get a drift of $1/n$, and the ratio between drift and fitness is generally very close to $1/n$ if all one-bits are in the last, say, 10% of the string. However, there is a problem for s_1. Here we accept an offspring whenever we flip the first bit, and in this case we flip an expected number of $(n-1)/n$ other bits. Therefore, the drift for s_2 is $1/n \cdot (1-(n-1)/n) = 1/n^2$, so it is too small. Hence we need to increase the potential of s_2 compared with the potential of its typical offspring, i.e. we need to increase the weight of higher-order bits. It takes some fiddling to get the right trade-off, but Doerr, Johannsen and Winzen figured out that a good choice is a weight of 5/4 for the first half of the bits, and 1 for the second half [19]. This choice works not only for BinVal but for all linear functions (cf. Example 2.3.12). In principle, the same approach can also be used for mutation rates other than $1/n$. In one of the most important results on the theory of evolutionary algorithms, for a mutation rate of the form c/n for any constant $c > 0$ Witt [75] managed to find weights which lead to a good potential. In this way, he could prove in just 2–3 pages that the runtime of the (1+1) EA is $(1 \pm o(1))e^c/c \cdot n \ln n$, settling a question that had been open for years.

We should keep in mind that the methods discussed above are only guidelines, which may be helpful in some situations but fruitless in others. Finding the right drift function often requires ingenuity, and cannot be reduced to a simple cooking recipe. Thus it is still one of the most challenging, but also most rewarding tasks in runtime analysis.

2.6 Conclusion

We have seen how drift analysis can be applied to transform knowledge about the drift into knowledge about the runtime of an algorithm. In this chapter we have restricted ourselves to applications in the analysis of evolutionary algorithms, but drift analysis can be applied to other randomised algorithms or random processes. We refer the reader to [35], which contains a nice variety of applications of drift analysis, including algorithms for approximate vertex cover, 2-SAT and random sorting, and applications to processes such as the Moran process.

In theory, it is always possible to apply drift analysis to obtain matching upper and lower bounds on the expected runtime. However, in practice there are many situations which are still difficult to handle because we do not know a good potential function. In particular, the more complex the state space and the behaviour of the algorithm are, the more difficult it is to find a single real-valued function which is a sufficiently good measure of progress. For example, in genetic programming (GP) the states are trees instead of strings, which makes the situation considerably more complex. In the few cases where theoretical results exist, this is mostly because the tree structure is unimportant for the problem [23, 27, 46, 62], with the notable exception of [25]. Similarly, while there have been impressive advances for large population sizes, especially through the level-based theorem (see Section 2.4.4), these techniques are still limited to some special cases of population dynamics. In particular, they consider only the number of individuals on each fitness level. This limits the complexity of the interactions that we can understand with this method – for example, the approach is blind to beneficial crossovers that happen between search points on the same fitness level. In general, it remains a major challenge to apply drift analysis to complex state spaces, and to algorithms which maintain and utilise a large diversity within their population, for example through crossover.

References

[1] Baritompa, B., Steel, M.: Bounds on absorption times of directionally biased random sequences. Random Structures & Algorithms **9**(3), 279–293 (1996)
[2] Bramson, M.: Stability of queueing networks. Probability Surveys **5**, 169–345 (2008)
[3] Corus, D., Dang, D.C., Eremeev, A.V., Lehre, P.K.: Level-based analysis of genetic algorithms and other search processes. In: Proceedings of the International Conference on Parallel Problem Solving from Nature (PPSN 2014), pp. 912–921. Springer (2014)
[4] Corus, D., Dang, D.C., Eremeev, A.V., Lehre, P.K.: Level-based analysis of genetic algorithms and other search processes. IEEE Transactions on Evolutionary Computation (2017)
[5] Dang, D.C., Jansen, T., Lehre, P.K.: Populations can be essential in tracking dynamic optima. Algorithmica **78**(2), 660–680 (2017)
[6] Dang, D.C., Lehre, P.K.: Efficient optimisation of noisy fitness functions with population-based evolutionary algorithms. In: Proceedings of the

International Workshop on Foundations of Genetic Algorithms (FOGA 2015), pp. 62–68. ACM (2015)
[7] Dang, D.C., Lehre, P.K.: Simplified runtime analysis of estimation of distribution algorithms. In: Proceedings of the Genetic and Evolutionary Computation Conference (GECCO 2015), pp. 513–518. ACM (2015)
[8] Dang, D.C., Lehre, P.K.: Runtime analysis of non-elitist populations: From classical optimisation to partial information. Algorithmica **75**(3), 428–461 (2016)
[9] Dang, D.C., Lehre, P.K.: Self-adaptation of mutation rates in non-elitist populations. In: Proceedings of the International Conference on Parallel Problem Solving from Nature (PPSN 2016), pp. 803–813. Springer (2016)
[10] Doerr, B., Doerr, C.: The impact of random initialization on the runtime of randomized search heuristics. Algorithmica **75**(3), 529–553 (2016)
[11] Doerr, B., Doerr, C., Kötzing, T.: Static and self-adjusting mutation strengths for multi-valued decision variables. Algorithmica pp. 1–37 (2017)
[12] Doerr, B., Doerr, C., Yang, J.: Optimal Parameter Choices via Precise Black-Box Analysis. arXiv preprint arXiv:1807.03403 (2018)
[13] Doerr, B., Fischbeck, P., Frahnow, C., Friedrich, T., Kötzing, T., Schirneck, M.: Island models meet rumor spreading. In: Proceedings of the Genetic and Evolutionary Computation Conference (GECCO 2017), pp. 1359–1366. ACM (2017)
[14] Doerr, B., Fouz, M., Witt, C.: Sharp bounds by probability-generating functions and variable drift. In: Proceedings of the Genetic and Evolutionary Computation Conference (GECCO 2011), pp. 2083–2090. ACM (2011)
[15] Doerr, B., Goldberg, L.A.: Adaptive drift analysis. Algorithmica **65**(1), 1–27 (2013)
[16] Doerr, B., Hota, A., Kötzing, T.: Ants easily solve stochastic shortest path problems. In: Proceedings of the Genetic and Evolutionary Computation Conference (GECCO 2012), pp. 17–24. ACM (2012)
[17] Doerr, B., Johannsen, D.: Edge-based representation beats vertex-based representation in shortest path problems. In: Proceedings of the Genetic and Evolutionary Computation Conference (GECCO 2010), pp. 758–766. ACM (2010)
[18] Doerr, B., Johannsen, D., Winzen, C.: Drift analysis and linear functions revisited. In: Proceedings of the Congress on Evolutionary Computation (CEC 2010), pp. 1–8. IEEE (2010)
[19] Doerr, B., Johannsen, D., Winzen, C.: Multiplicative drift analysis. In: Proceedings of the Genetic and Evolutionary Computation Conference (GECCO 2010), pp. 1449–1456. ACM (2010)
[20] Doerr, B., Johannsen, D., Winzen, C.: Multiplicative drift analysis. Algorithmica **64**, 673–697 (2012)

[21] Doerr, B., Johannsen, D., Winzen, C.: Non-existence of linear universal drift functions. Theoretical Computer Science **436**, 71–86 (2012)
[22] Doerr, B., Kötzing, T., Lagodzinski, G., Lengler, J.: Bounding bloat in genetic programming. arXiv preprint arXiv:2287.2831 (2018)
[23] Doerr, B., Kötzing, T., Lagodzinski, J., Lengler, J.: Bounding bloat in genetic programming. In: Proceedings of the Genetic and Evolutionary Computation Conference (GECCO 2017), pp. 921–928. ACM (2017)
[24] Doerr, B., Künnemann, M.: Optimizing linear functions with the (1+ λ) evolutionary algorithm - different asymptotic runtimes for different instances. Theoretical Computer Science **561**, 3–23 (2015)
[25] Doerr, B., Lissovoi, A., Oliveto, P.S.: On the evolution of boolean functions with conjunctions and disjunctions via genetic programming. arXiv preprint arXiv:1903.11936 (2018)
[26] Droste, S., Jansen, T., Wegener, I.: On the analysis of the (1+1) evolutionary algorithm. Theoretical Computer Science **276**, 51–81 (2002)
[27] Durrett, G., Neumann, F., O'Reilly, U.M.: Computational complexity analysis of simple genetic programming on two problems modeling isolated program semantics. In: Proceedings of the International Workshop on Foundations of Genetic Algorithms (FOGA 2011), pp. 69–80. ACM (2011)
[28] Feldmann, M., Kötzing, T.: Optimizing expected path lengths with ant colony optimization using fitness proportional update. In: Proceedings of the International Workshop on Foundations of Genetic Algorithms (FOGA 2013), pp. 65–74. ACM (2013)
[29] Friedrich, T., Kötzing, T., Krejca, M.S., Sutton, A.M.: The benefit of recombination in noisy evolutionary search. In: International Symposium on Algorithms and Computation (ISAAC 2015), pp. 140–150. Springer (2015)
[30] Friedrich, T., Kötzing, T., Krejca, M.S., Sutton, A.M.: Robustness of ant colony optimization to noise. Evolutionary computation **24**(2), 237–254 (2016)
[31] Friedrich, T., Kötzing, T., Krejca, M.S., Sutton, A.M.: The compact genetic algorithm is efficient under extreme gaussian noise. IEEE Transactions on Evolutionary Computation **21**(3), 477–490 (2017)
[32] Gießen, C., Kötzing, T.: Robustness of populations in stochastic environments. Algorithmica **75**(3), 462–489 (2016)
[33] Gießen, C., Witt, C.: Optimal Mutation Rates for the (1+λ) EA on OneMax Through Asymptotically Tight Drift Analysis. Algorithmica pp. 1–22 (2017)
[34] Gießen, C., Witt, C.: The Interplay of Population Size and Mutation Probability in the (1+λ) EA on OneMax. Algorithmica **78**(2), 587–609 (2017)
[35] Göbel, A., Kötzing, T., Krejca, M.S.: Intuitive analyses via drift theory. arXiv preprint arXiv:1806.01919 (2018)

[36] Grimmett, G., Stirzaker, D.: Probability and random processes. Oxford University Press (2001)

[37] Hajek, B.: Hitting-time and occupation-time bounds implied by drift analysis with applications. Advances in Applied probability **14**(3), 502–525 (1982)

[38] He, J., Yao, X.: Drift analysis and average time complexity of evolutionary algorithms. Artificial Intelligence **127**, 57–85 (2001)

[39] He, J., Yao, X.: A study of drift analysis for estimating computation time of evolutionary algorithms. Natural Computing **3**, 21–35 (2004)

[40] He, J., Yao, X.: Average drift analysis and population scalability. IEEE Transactions on Evolutionary Computation **21**(3), 426–439 (2017)

[41] Jägersküpper, J.: A Blend of Markov-Chain and Drift Analysis. In: Proceedings of the International Conference on Parallel Problem Solving from Nature (PPSN 2008), pp. 41–51. Springer (2008)

[42] Johannsen, D.: Random combinatorial structures and randomized search heuristics. Ph.D. thesis, Universität des Saarlandes (2010)

[43] Kötzing, T.: Concentration of first hitting times under additive drift. In: Proceedings of the Genetic and Evolutionary Computation Conference (GECCO 2014), pp. 1391–1398. ACM (2014)

[44] Kötzing, T.: Concentration of first hitting times under additive drift. Algorithmica **75**(3), 490–506 (2016)

[45] Kötzing, T., Krejca, M.S.: First-hitting times under additive drift. In: Proceedings of the International Conference on Parallel Problem Solving from Nature (PPSN 2018). pp. 92–104. Springer (2018).

[46] Kötzing, T., Lagodzinski, G., Lengler, J., Melnichenko, A.: Destructiveness of lexicographic parsimony pressure and alleviation by a concatenation crossover in genetic programming. In: Proceedings of the International Conference on Parallel Problem Solving from Nature (PPSN 2018). pp. 42–54. Springer (2018).

[47] Kötzing, T., Molter, H.: ACO beats EA on a dynamic pseudo-boolean function. In: Proceedings of the International Conference on Parallel Problem Solving from Nature (PPSN 2012), pp. 113–122. Springer (2012)

[48] Koza, J.R.: Genetic programming as a means for programming computers by natural selection. MIT Press (1992)

[49] Ladret, V.: Asymptotic hitting time for a simple evolutionary model of protein folding. Journal of Applied Probability **42**(1), 39–51 (2005)

[50] Lehre, P.K., Nguyen, P.T.H.: Improved runtime bounds for the univariate marginal distribution algorithm via anti-concentration. In: Proceedings of the Genetic and Evolutionary Computation Conference (GECCO 2017), pp. 1383–1390. ACM (2017)

[51] Lehre, P.K., Oliveto, P.S.: Theoretical analysis of stochastic search algorithms. arXiv preprint arXiv:1709.00890 (2017)

[52] Lehre, P.K., Witt, C.: General drift analysis with tail bounds. arXiv preprint arXiv:1307.2559 (2013)

[53] Lehre, P.K., Witt, C.: Concentrated hitting times of randomized search heuristics with variable drift. In: International Symposium on Algorithms and Computation (ISAAC 2014), pp. 686–697. Springer (2014)
[54] Lengler, J., Steger, A.: Drift analysis and evolutionary algorithms revisited. Combinatorics, Probability and Computing **27**(4), 643–666 (2018)
[55] Lissovoi, A., Witt, C.: Runtime analysis of ant colony optimization on dynamic shortest path problems. Theoretical Computer Science **561**, 73–85 (2015)
[56] Lissovoi, A., Witt, C.: MMAS versus population-based EA on a family of dynamic fitness functions. Algorithmica **75**(3), 554–576 (2016)
[57] Lissovoi, A., Witt, C.: A runtime analysis of parallel evolutionary algorithms in dynamic optimization. Algorithmica **78**(2), 641–659 (2017)
[58] Menshikov, M., Popov, S., Wade, A.: Non-homogeneous Random Walks: Lyapunov Function Methods for Near-Critical Stochastic Systems, vol. 209. Cambridge University Press (2016)
[59] Meyn, S.P., Tweedie, R.L.: Markov chains and stochastic stability. Springer Science & Business Media (2012)
[60] Mitavskiy, B., Rowe, J., Cannings, C.: Theoretical analysis of local search strategies to optimize network communication subject to preserving the total number of links. International Journal of Intelligent Computing and Cybernetics pp. 243–284 (2009)
[61] Motwani, R., Raghavan, P.: Randomized Algorithms. Cambridge University Press (1995)
[62] Neumann, F.: Computational complexity analysis of multi-objective genetic programming. In: Proceedings of the Genetic and Evolutionary Computation Conference (GECCO 2012), pp. 799–806. ACM (2012)
[63] Neumann, F., Sudholt, D., Witt, C.: A few ants are enough: ACO with iteration-best update. In: Proceedings of the Genetic and Evolutionary Computation Conference (GECCO 2010), pp. 63–70. ACM (2010)
[64] Neumann, F., Wegener, I.: Randomized local search, evolutionary algorithms, and the minimum spanning tree problem. Theoretical Computer Science **378**(1), 32–40 (2007)
[65] Oliveto, P.S., Witt, C.: Simplified drift analysis for proving lower bounds in evolutionary computation. Algorithmica **59**(3), 369–386 (2011)
[66] Oliveto, P.S., Witt, C.: Erratum: Simplified drift analysis for proving lower bounds in evolutionary computation. arXiv preprint arXiv:1211.7184 (2012)
[67] Oliveto, P.S., Witt, C.: Improved time complexity analysis of the simple genetic algorithm. Theoretical Computer Science **605**, 21–41 (2015)
[68] Paixão, T., Heredia, J.P., Sudholt, D., Trubenová, B.: Towards a runtime comparison of natural and artificial evolution. Algorithmica **78**(2), 681–713 (2017)
[69] Reichel, J., Skutella, M.: Evolutionary algorithms and matroid optimization problems. Algorithmica **57**(1), 187–206 (2010)

[70] Rowe, J.E., Sudholt, D.: The choice of the offspring population size in the $(1,\lambda)$ EA. In: Proceedings of the Genetic and Evolutionary Computation Conference (GECCO 2012), pp. 1349–1356 (2012)
[71] Rowe, J.E., Sudholt, D.: The choice of the offspring population size in the $(1, \lambda)$ evolutionary algorithm. Theoretical Computer Science **545**, 20–38 (2014)
[72] Sudholt, D., Witt, C.: Update strength in EDAs and ACO: How to avoid genetic drift. In: Proceedings of the Genetic and Evolutionary Computation Conference (GECCO 2016), pp. 61–68. ACM (2016)
[73] Tweedie, R.: Criteria for classifying general Markov chains. Advances in Applied Probability **8**(4), 737–771 (1976)
[74] Wegener, I.: Methods for the analysis of evolutionary algorithms on pseudo-boolean functions. In: Evolutionary Optimization, pp. 349–369. Springer (2003)
[75] Witt, C.: Tight bounds on the optimization time of a randomized search heuristic on linear functions. Combinatorics, Probability and Computing **22**(2), 294–318 (2013)
[76] Witt, C.: Upper bounds on the runtime of the univariate marginal distribution algorithm on Onemax. In: Proceedings of the Genetic and Evolutionary Computation Conference (GECCO 2017), GECCO, pp. 1415–1422. ACM (2017)

Chapter 3
Complexity Theory for Discrete Black-Box Optimization Heuristics

Carola Doerr

Abstract A predominant topic in the theory of evolutionary algorithms and, more generally, theory of randomized black-box optimization techniques is *running-time analysis.* Running-time analysis is aimed at understanding the performance of a given heuristic on a given problem by bounding the number of function evaluations that are needed by the heuristic to identify a solution of a desired quality. As in general algorithms theory, this running-time perspective is most useful when it is complemented by a meaningful *complexity theory* that studies the limits of algorithmic solutions.

In the context of discrete black-box optimization, several *black-box complexity models* have been developed to analyze the best possible performance that a black-box optimization algorithm can achieve on a given problem. The models differ in the classes of algorithms to which these lower bounds apply. This way, black-box complexity contributes to a better understanding of how certain algorithmic choices (such as the amount of memory used by a heuristic, its selective pressure, or properties of the strategies that it uses to create new solution candidates) influence performance.

In this chapter we review the different black-box complexity models that have been proposed in the literature, survey the bounds that have been obtained for these models, and discuss how the interplay of running-time analysis and black-box complexity can inspire new algorithmic solutions to well-researched problems in evolutionary computation. We also discuss in this chapter several interesting open questions for future work.

Carola Doerr
Sorbonne Université, CNRS, LIP6, Paris, France

B. Doerr, F. Neumann (eds.), *Theory of Evolutionary Computation*,
Natural Computing Series, https://doi.org/10.1007/978-3-030-29414-4_3

3.1 Introduction and Historical Remarks

One of the driving forces in theoretical computer science is the fruitful interplay between *complexity theory* and the *theory of algorithms*. While the former measures the minimum computational effort that is needed to solve a given problem, the latter is aimed at designing and analyzing efficient algorithmic solutions which prove that a problem can be solved with a certain computational effort. When, for a problem, the lower bounds on the resources needed to solve it are identical to (or not much smaller than) the upper bounds attained by some specific algorithm, we can be certain that we have an (almost) optimal algorithmic solution to this problem. Big gaps between lower and upper bounds, in contrast, indicate that more research effort is needed to understand the problem: it may be that more efficient algorithms for the problem exist, or that the problem is indeed "harder" than what the lower bound suggests.

Many different complexity models coexist in the theoretical computer science literature. The arguably most classical one measures the number of arithmetic operations that an algorithm needs to perform on the problem data until it obtains a solution for the problem. A solution can be a "yes/no" answer (a decision problem), a classification of a problem instance according to some criteria (a classification problem), a vector of decision variables that maximize or minimize some objective function (an optimization problem), etc. In the optimization context, we are typically interested only in algorithms that satisfy some minimal quality requirements such as a guarantee that the suggested solutions ("the output" of the algorithm) are always optimal or are optimal with some large enough probability, or that they are not worse than an optimal solution by more than some additive or multiplicative factor C, etc.

In the *white-box* setting, in which the algorithms have full access to the data describing the problem instance, complexity theory is a well-established and very intensively studied research objective. In *black-box optimization*, where the algorithms do not have access to the problem data and can learn about the problem at hand only through the evaluation of potential solution candidates, complexity theory is a much less present topic, with rather large fluctuations in the number of publications. In the context of heuristic solutions to black-box optimization problems, which is the topic of this book, complexity theory has been systematically studied only since 2010, using the notion of *black-box complexity*. Luckily, black-box complexity theory can build on results in related research domains such as information theory, discrete mathematics, cryptography, and others.

In this chapter, we review the state of the art in this currently very active area of research, which is concerned with bounding the best possible performance that an optimization algorithm can achieve in a black-box setting.

3.1.1 Black-Box vs. White-Box Complexity

Most of the traditional complexity measures assume that the algorithms have access to the problem data, and count the number of steps that are needed until the algorithm outputs a solution. In the black-box setting, these complexity measures are not very meaningful, as the algorithms are asked to optimize a problem without having direct access to it. As a consequence, the performance of a black-box optimization algorithm is therefore traditionally measured by the number of function evaluations that the algorithm does until it queries for the first time a solution that satisfies some specific performance criteria. In this book, we are mostly interested in the expected number of evaluations needed until an *optimal* solution is evaluated for the first time. It is therefore natural to define black-box complexity as the *minimum number of function evaluations that any black-box algorithm needs to perform, on average, until it queries an optimal solution for the first time.*

We typically consider classes of problems, for example, the set of traveling salesperson instances of planar graphs with integer edge weights. For such a class $\mathcal{F} \subseteq \{f : S \to \mathbb{R}\}$ of problem instances, we take a worst-case view and measure the expected number of function evaluations that an algorithm needs to optimize any instance $f \in \mathcal{F}$. That is, the black-box complexity of a problem $\mathcal{F}$ is $\inf_A \sup_{f \in \mathcal{F}} E[T(A,f)]$, the best (among all algorithms A) worst-case (among all problem instances f) expected number $E[T(A,f)]$ of function evaluations that are needed to optimize any $f \in \mathcal{F}$. A formal definition will be given in Section 3.2.

The black-box complexity of a problem can be very different from its white-box counterpart. We will discuss, for example, in Sections 3.2.4 and 3.6.3.5 the fact that there are a number of NP-hard problems whose black-box complexity is of small polynomial order.

3.1.2 Motivation and Objectives

The ultimate objective of black-box complexity is to support the investigation and design of efficient black-box optimization techniques. This is achieved in several complementary ways.

A first benefit of black-box complexity is that it enables the above-mentioned evaluation of how well we have understood a black-box optimization problem, and how suitable the state-of-the-art heuristics are. Where large gaps between lower and upper bounds exist, we may want to explore alternative algorithmic solutions, in the hope of identifying more efficient solvers. Where the lower and upper bounds match or are close, we can stop striving for more efficient algorithms.

Another advantage of black-box complexity studies is that they allow us to investigate how certain algorithmic choices influence the performance: By restricting the class of algorithms under consideration, we can judge how these restrictions increase the complexity of a black-box optimization problem. In the context of evolutionary computation, interesting restrictions include the amount of memory that is available to the algorithms, the number of solutions that are sampled in every iteration, the way new solution candidates are generated, the selection principles according to which it is decided which search points to keep for future reference, etc. Comparing the unrestricted with the restricted black-box complexity of a problem (i.e. its black-box complexity with respect to *all* versus that with respect to a *subclass* of all algorithms) quantifies the performance loss caused by these restrictions. This way, we can understand, for example, the effects of not storing the set of all previously evaluated solution candidates, but only a small subset.

The black-box complexity of a problem can be significantly smaller than the performance of a best known 'standard' heuristic. In such cases, the small complexity is often attained by a problem-tailored black-box algorithm, which is not representative of common black-box heuristics. Interestingly, it turns out that we can nevertheless learn from such highly specific algorithms, as they often incorporate some ideas that could be beneficial far beyond the particular problem at hand. As we shall demonstrate in Section 3.9, even for very well-researched optimization problems, such ideas can give rise to the design of novel heuristics which are provably more efficient than standard solutions. This way, black-box complexity serves as a source of inspiration for the development of novel algorithmic ideas that lead to the design of better search heuristics.

3.1.3 Relationship to Query Complexity

As indicated above, black-box complexity is studied in several different contexts, which reach far beyond evolutionary computation. In the 1960s and 1970s, for example, this complexity measure was very popular in the context of combinatorial games, such as coin-weighing problems of the type "given n coins of two different types, what is the minimum number of weighings that is needed to classify the coins according to their weight?" Interpreting a weighing as a function evaluation, we see that such questions can be formulated as black-box optimization problems.

Black-box complexity also plays an important role in cryptography, where a common research question concerns the minimum amount of information that suffices to break a secret code. Quantum computing, communication complexity, and information theory are other research areas where (variants of) black-box complexity are intensively studied performance measures. While in these settings the precise model is often not exactly identical to a model of the kind we are faced with in black-box optimization, some of the tools developed in these related areas can be useful in our context.

A significant part of the literature studies the performance of deterministic algorithms. Randomized black-box complexities are much less understood. They can be much smaller than their deterministic counterparts. Since deterministic algorithms form a subclass of randomized ones, any lower bound proven for the randomized black-box complexity of a problem also applies to any deterministic algorithm. In some cases, a strict separation between deterministic and randomized black-box complexities can be proven. This is the case for the LEADINGONES function, as we shall briefly discuss in Section 3.3.6. For other problems, the deterministic and randomized black-box complexities coincide. Characterizing those problems for which access to random bits can provably decrease the complexity is a wide-open research question.

In several contexts, in particular the research domains mentioned above, black-box complexity is typically referred to as *query* or *oracle complexity*, with the idea that the algorithms do not evaluate the function values of the solution candidates themselves but rather query them from an oracle. This interpretation is mostly identical to the black-box scenario classically considered in evolutionary computation.

3.1.4 Scope of This Chapter

In this chapter, as in the remainder of this book, we restrict our attention to discrete optimization problems, i.e., the maximization or minimization of functions $f : S \to \mathbb{R}$ that are defined over finite search spaces S. As in the previous chapters, we will mostly deal with the optimization of pseudo-Boolean functions $f : \{0,1\}^n \to \mathbb{R}$, permutation problems $f : S_n \to \mathbb{R}$, and functions $f : [0..r-1]^n \to \mathbb{R}$ defined for strings over an alphabet of bounded size, where, here and in the following, we use the following abbreviations: $[0..r-1] := \{0,1,\ldots,r-1\}$ represents the set of non-negative integers smaller than r, $[n] := \{1,2,\ldots,n\}$, and S_n represents the set of all permutations (one-to-one maps) $\sigma : [n] \to [n]$.

We point out that black-box complexity notions are also studied for infinite search spaces S. In the context of continuous optimization problems, studies of black-box complexity are aimed at bounding the best possible *convergence rates* that a derivative-free black-box optimization algorithm can achieve, see [57, 86] for examples.

3.1.5 Target Audience and Complementary Material

This chapter is written with a reader in mind who is familiar with black-box optimization, and who brings with them some background in theoretical running-time analysis. We will give an exhaustive survey of existing results. Where appropriate, we provide proof ideas and discuss some historical developments. Readers interested in a more gentle introduction to the basic concepts of black-box complexity are referred to [61]. A slide presentation on selected aspects of black-box complexity, along with a summary of complexity bounds known back in spring 2014, can be found in the tutorial [20].

3.1.6 Overview of the Content

Black-box complexity is formally defined in Section 3.2. We also provide there a summary of useful tools. In Section 3.2.4 we discuss why classical complexity statements such as NP-hardness results do not necessarily imply hardness in the black-box complexity model.

In Sections 3.3-3.7 we review the different black-box complexity models that have been proposed in the literature. For each model, we discuss the main results that have been achieved for it. For several benchmark problems, including most notably OneMax, LeadingOnes, and Jump, but also combinatorial problems such as the minimum spanning tree problem and shortest-paths problems, bounds have been derived for various complexity models. For OneMax and LeadingOnes, we compare these different bounds in Section 3.8, to summarize where gaps between upper and lower bounds exist, and to highlight the increasing complexities imposed by the restrictive models.

We will demonstrate in Section 3.9 that the complexity-theoretic view of black-box optimization can inspire the design of more efficient optimization heuristics. This is made possible by questioning some of the state-of-the-art choices that are made in evolutionary computation and neighboring disciplines.

Finally, we show in Section 3.10 that research efforts originally motivated by the study of black-box complexity have yielded improved bounds for long-standing open problems in classical computer science.

In Section 3.11, we conclude this chapter with a summary of open questions and problems in discrete black-box complexity and directions for future work.

3.2 The Unrestricted Black-Box Model

In this section we introduce the most basic black-box model, which is the *unrestricted* one. This model contains all black-box optimization algorithms. Any lower bound in this model therefore immediately applies to any of the restricted models which we discuss in Sections 3.4-3.7. We also discuss in this section some useful tools for the analysis of black-box complexity and

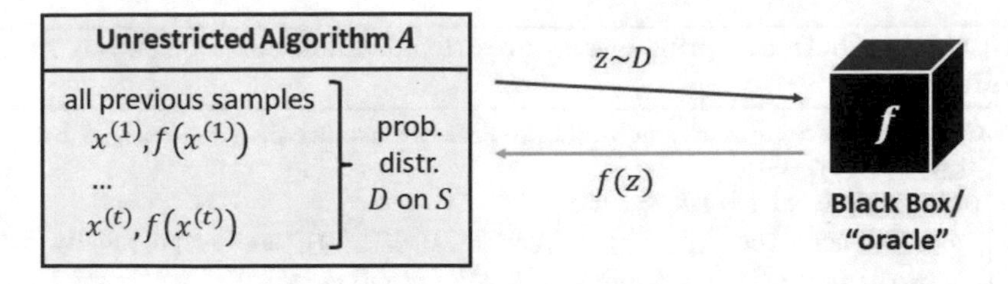

Fig. 3.1 In the unrestricted black-box model, the algorithm can store the full history of previously queried search points. For each of these already evaluated candidate solutions x, the algorithm has access to its absolute function value $f(x) \in \mathbb{R}$. There are no restrictions on the structure of the distributions D from which new solution candidates are sampled.

demonstrate that the black-box complexity of a problem can be very different from its classical white-box complexity.

The unrestricted black-box model was introduced by Droste, Jansen, Wegener in [54]. The only assumption that it makes is that the algorithms do not have any information about the problem at hand other than the fact that it stems from some function class $\mathcal{F} \subseteq \{f : S \to \mathbb{R}\}$. The only way an unrestricted black-box algorithm can learn about the instance f is by evaluating the function values $f(x)$ of potential solution candidates $x \in S$. We can assume that the evaluation is done by some oracle, from which $f(x)$ is queried. In the unrestricted model, the algorithms can update after any such query the strategy by which the next search point(s) are generated. In this book, we are mostly interested in the performance of *randomized black-box heuristics*, so that these strategies are often *probability distributions* over the search space from which the next solution candidates are sampled. This process continues until an optimal search point $x \in \arg\max f$ is queried for the first time.

The algorithms that we are interested in are thus those that maintain a probability distribution D over the search space S. In every iteration, a new solution candidate x is sampled from this distribution and the function value $f(x)$ of this search point is evaluated. After this evaluation, the probability distribution D is updated according to the information gathered through the sample $(x, f(x))$. The next iteration starts again by sampling a search point from this updated distribution D, and so on. This structure is summarized in Algorithm 3.1, which models *unrestricted randomized black-box algorithms*. A visualization is provided in Fig. 3.1.

Note that in Algorithm 3.1, in every iteration only one new solution candidate is sampled. In contrast, many evolutionary algorithms and other black-box optimization techniques generate and evaluate several search points *in parallel*. It is not difficult to see that lower bounds obtained for the unrestricted black-box complexity described here apply immediately to such *population-based* heuristics, since an unrestricted algorithm is free to ignore

Algorithm 3.1: Blueprint of an unrestricted randomized black-box algorithm

1 **Initialization:** Sample $x^{(0)}$ according to some probability distribution $D^{(0)}$ over S and query $f(x^{(0)})$;
2 **Optimization: for** $t = 1,2,3,\ldots$ **do**
3 Depending on $\big((x^{(0)}, f(x^{(0)})), \ldots, (x^{(t-1)}, f(x^{(t-1)}))\big)$ choose a probability distribution $D^{(t)}$ over S and sample $x^{(t)}$ according to $D^{(t)}$;
4 Query $f(x^{(t)})$;

information obtained from previous iterations. As will be commented on in Section 3.7.2, the *parallel* black-box complexity of a function can be (much) larger than its sequential variant. Taking this idea to the extreme, i.e., requiring the algorithm to neglect information obtained through previous queries yields so-called *nonadaptive* black-box algorithms. A prime example for a nonadaptive black-box algorithm is random sampling (with and without repetitions). Nonadaptive algorithms play only a marginal role in evolutionary computation. From a complexity point of view, however, it can be interesting to study how much adaptation is needed for an efficient optimization; see also the discussions in Sections 3.3.2 and 3.10. For most problems, the adaptive and nonadaptive complexity differ by large factors. For some other problems, however, the two complexity notions coincide; see Section 3.3.1 for an example.

Note also that unrestricted black-box algorithms have access to the full history of previously evaluated solutions. The effects of restricting the available memory to a *population* of a certain size will be the focus of the memory-restricted black-box models discussed in Section 3.4.

In line 3 of Algorithm 3.1 we do not specify how the probability distribution $D^{(t)}$ is chosen. Thus, in principle, the algorithm can spend significant time on choosing this distribution. This can result in small polynomial black-box complexities for NP-hard problems; see Section 3.2.4. Droste, Jansen, and Wegener [54] therefore suggested restricting the set of algorithms to those that execute the choice of the distributions $D^{(t)}$ in a polynomial number of algebraic steps (i.e., polynomial time in the input length, where "time" refers to the classically considered complexity measure). They called this model the *time-restricted model.* In this chapter, we will not study this time-restricted model. That is, we allow the algorithms to spend arbitrary time on the choice of the distributions $D^{(t)}$. This way, we obtain very general lower bounds. Almost all upper bounds stated in this chapter nevertheless apply also to the time-restricted model. The polynomial bounds for NP-hard problems form, of course, an exception to this rule.

We comment, finally, on the fact that Algorithm 3.1 runs forever. As we have seen in previous chapters in this book, the pseudocode in Algorithm 3.1 is a common representation of black-box algorithms in the theory of heuristic optimization. Not specifying the termination criterion is justified by our performance measure, which is the expected number of function evaluations that an algorithm performs until (and including) the first iteration in which an optimal solution is evaluated; see Definition 3.2.1 below. Other performance measures for black-box heuristics have been discussed in the literature [11, 34, 65], but in the realm of black-box complexity, the *average optimization time* is still the predominant performance indicator. See Section 3.11 for a discussion of the possibility of extending existing results to other, possibly more complex performance measures.

3.2.1 Formal Definition of Black-Box Complexity

In this section, we give a very general definition of black-box complexity. More precisely, we formally define the black-box complexity of a class $\mathcal{F}$ of functions with respect to some class $\mathcal{A}$ of algorithms. The unrestricted black-box complexity will be the complexity of $\mathcal{F}$ with respect to all black-box algorithms that follow the blueprint provided in Algorithm 3.1.

For a black-box optimization algorithm A and a function $f : S \to \mathbb{R}$, let $T(A,f) \in \mathbb{R} \cup \{\infty\}$ be the number of function evaluations that algorithm A does until and including the evaluation in which it evaluates for the first time an optimal search point $x \in \arg\max f$. As in previous chapters, we call $T(A,f)$ the *running time of A for f* or, synonymously, the *optimization time of A for f*. When A is a randomized algorithm, $T(A,f)$ is a random variable that depends on the random decisions made by A. We are mostly interested in its expected value $E[T(A,f)]$.

With this performance measure in place, the definition of the black-box complexity of a class $\mathcal{F}$ of functions $S \to \mathbb{R}$ with respect to some class $\mathcal{A}$ of algorithms now follows the usual approach in complexity theory.

Definition 3.2.1. For a given black-box algorithm A, the *A-black-box complexity of $\mathcal{F}$* is

$$E[T(A,\mathcal{F})] := \sup_{f \in \mathcal{F}} E[T(A,f)],$$

the worst-case expected running time of A on $\mathcal{F}$.

The *$\mathcal{A}$-black-box complexity of $\mathcal{F}$* is

$$E[T(\mathcal{A},\mathcal{F})] := \inf_{A \in \mathcal{A}} E[T(A,\mathcal{F})],$$

the minimum ("best") complexity among all $A \in \mathcal{A}$ for $\mathcal{F}$.

Thus, formally, the *unrestricted black-box complexity* of a problem class $\mathcal{F}$ is $E[T(\mathcal{A},\mathcal{F})]$, where $\mathcal{A}$ is the collection of all unrestricted black-box algorithms, i.e., all algorithms that can be expressed in the framework of Algorithm 3.1.

The following lemma formalizes the intuition that every lower bound for the unrestricted black-box model also applies to any restricted black-box model.

Lemma 3.2.2. *Let $\mathcal{F} \subseteq \{f : S \to \mathbb{R}\}$. For every collection $\mathcal{A}'$ of black-box optimization algorithms for $\mathcal{F}$, the $\mathcal{A}'$-black-box complexity of $\mathcal{F}$ is at least as large as its unrestricted black-box complexity.*

Formally, this lemma holds because $\mathcal{A}'$ is a subclass of the set $\mathcal{A}$ of all unrestricted black-box algorithms. The infimum in the definition of $E[T(\mathcal{A}',\mathcal{F})]$ is therefore taken over a smaller class, thus giving values that are at least as large as $E[T(\mathcal{A},\mathcal{F})]$.

3.2.2 Tools for Proving Lower Bounds

Lemma 3.2.2 shows that the unrestricted black-box complexity of a class $\mathcal{F}$ of functions is a lower bound for the performance of any black-box algorithm on $\mathcal{F}$. In other words, no black-box algorithm can optimize $\mathcal{F}$ more efficiently than what the unrestricted black-box complexity of $\mathcal{F}$ indicates. We are therefore particularly interested in proving *lower bounds* for the black-box complexity of a problem. This is the topic of this section.

To date, the most powerful tool to prove lower bounds for randomized query complexity models such as our unrestricted black-box model is the so-called *minimax principle* of Yao [88]. In order to discuss this principle, we first need to recall that we can interpret every randomized unrestricted black-box algorithm as a probability distribution over deterministic algorithms. In fact, randomized black-box algorithms are often defined this way.

Deterministic black-box algorithms are those for which the probability distributions in line 3 of Algorithm 3.1 are one-point distributions. That is, for every t and for every sequence $((x^{(0)},f(x^{(0)})),\ldots,(x^{(t-1)},f(x^{(t-1)})))$ of previous queries, there exists a search point $s \in S$ such that $D^{(t)}\Big(((x^{(0)},f(x^{(0)})),\ldots,(x^{(t-1)},f(x^{(t-1)})))\Big)(s) = 1$ and $D^{(t)}\Big(((x^{(0)},f(x^{(0)})),\ldots,(x^{(t-1)},f(x^{(t-1)})))\Big)(y) = 0$ for all $y \neq s$. In other words, we can interpret deterministic black-box algorithms as *decision trees*. A decision tree for a class $\mathcal{F}$ of functions is a rooted tree in which the nodes are labeled by the search points that the algorithm queries. The first query is the label of the root node, say $x^{(0)}$. The edges from the root node to its neighbors are labeled with the possible objective values $\{g(x^{(0)}) \mid g \in \mathcal{F}\}$. After evaluating $f(x^{(0)})$, the algorithm follows the (unique) edge $\{x^{(0)},x^{(1)}\}$ which is labeled with the value $f(x^{(0)})$. The next query is the label of the endpoint $x^{(1)}$ of this edge. We call $x^{(1)}$ a level-1 node. The level-2 neighbors of $x^{(0)}$ (i.e., all neighbors of $x^{(1)}$ except the root node $x^{(0)}$) are the potential search points to be queried in the next iteration. As before, the algorithm chooses as the next query the neighbor $x^{(2)}$ of $x^{(1)}$ to which the unique edge labeled with the value $f(x^{(1)})$ leads. This process continues until an optimal search point has been queried. The optimization time $T(A,f)$ of the algorithm A on the function f equals the depth of this node plus one (the "plus one" accounts for the evaluation of the root node).

We can easily see that, in this model, it does not make sense to query the same search point twice. Such a query would not reveal any new information about the objective function f. For this reason, on every rooted path in the decision tree, every search point appears at most once. This shows that the *depth of the decision tree* is bounded by $|S|-1$. The *width* of the tree, however, can be as large as the size of the set $\mathcal{F}(S) := \{g(s) \mid g \in \mathcal{F}, s \in S\}$, which can be infinite or even uncountable, for example, if $\mathcal{F}$ equals the set of all linear or monotone functions $f : \{0,1\}^n \to \mathbb{R}$. As we shall see below, Yao's minimax principle can only be applied to problems for which $\mathcal{F}(S)$ is finite. Luckily, it is often possible to identify subclasses $\mathcal{F}'$ of $\mathcal{F}$ for which $\mathcal{F}'(S)$ is finite and whose complexity is identical to or not much smaller than that of the whole class $\mathcal{F}$.

When S and $\mathcal{F}(S)$ are finite, the number of (nonrepetitive) deterministic decision trees, and hence the number of deterministic black-box algorithms for $\mathcal{F}$, is finite. In this case, we can apply Yao's minimax principle. This theorem, intuitively speaking, allows us to restrict our attention to bounding the expected running time $E[T(A,f)]$ of a best possible *deterministic* algorithm A on a *random* instance f taken from $\mathcal{F}$ according to some probability distribution p. By Yao's minimax principle, this best possible expected running time is a lower bound for the expected performance of a best possible *randomized* algorithm on an *arbitrary* input. In our words, it is thus a lower bound on the unrestricted black-box complexity of the class $\mathcal{F}$.

Analyzing deterministic black-box algorithms is often considerably easier than directly bounding the performance of any possible randomized algorithm. An a priori challenge in applying this theorem is the identification of a probability distribution p on $\mathcal{F}$ for which the expected optimization time of a best possible deterministic algorithm is large. Luckily, for many applications some rather simple distributions on the inputs suffice, for example the uniform distribution, which assigns equal probability to each problem instance $f \in \mathcal{F}$. Another difficulty in the application of the theorem is the above-mentioned identification of subclasses $\mathcal{F}'$ of $\mathcal{F}$ for which $\mathcal{F}'(S)$ is finite.

Formally, Yao's minimax principle reads as follows.

Theorem 3.2.3 (Yao's minimax principle). *Let Π be a problem with a finite set $\mathcal{I}$ of input instances (of a fixed size) permitting a finite set $\mathcal{A}$ of deterministic algorithms. Let p be a probability distribution over $\mathcal{I}$ and let q be a probability distribution over $\mathcal{A}$. Then,*

$$\min_{A \in \mathcal{A}} E[T(I_p, A)] \leq \max_{I \in \mathcal{I}} E[T(I, A_q)],$$

where I_p denotes a random input chosen from $\mathcal{I}$ according to p, A_q denotes a random algorithm chosen from $\mathcal{A}$ according to q, and $T(I,A)$ denotes the running time of algorithm A on input I.

The formulation of Theorem 3.2.3 is taken from the book by Motwani and Raghavan [77], where an extended discussion of this principle can be found.

A straightforward but still quite handy application of Yao's minimax principle gives the following lower bound.

Theorem 3.2.4 (simple information-theoretic lower bound, Theorem 2 in [54]). *Let S be finite. Let $\mathcal{F}$ be a set of functions $\{f : S \to \mathbb{R}\}$ such that for every $s \in S$ there exists a function $f_s \in \mathcal{F}$ for which the size of $f_s(S) := \{f_s(x) \mid x \in S\}$ is bounded by k and for which s is a unique optimum, i.e.,* $\arg\max f_s = \{s\}$ *and $|f_s(S)| \leq k$. The unrestricted black-box complexity of $\mathcal{F}$ is at least $\lceil \log_k(|S|) \rceil - 1$.*

To prove Theorem 3.2.4 it suffices to select for every $s \in S$ one function f_s as in the statement and to consider the uniform distribution over the set $\{f_s \mid s \in S\}$. Every deterministic black-box algorithm that eventually solves any instance f_s has to have at least one node labeled s. We therefore need to distribute all $|S|$ potential optima on the decision tree that corresponds to this deterministic black-box algorithm. Since the outdegree of every node is bounded from above by k, the average distance from a node to the root is at least $\lceil \log_k(|S|) \rceil - 2$.

An informal interpretation of Theorem 3.2.4, which in addition ignores the rounding of the logarithms, is as follows. In the setting of Theorem 3.2.4, optimizing a function f_s corresponds to *learning* s. A binary encoding of the optimum s requires $\log_2(|S|)$ bits. With every query, we obtain at most $\log_2(k)$ bits of information, namely, the number of bits needed to encode which of the at most k possible objective values is assigned to the queried search point. We therefore need to query at least $\log_2(|S|)/\log_2(k) = \log_k(|S|)$ search points to obtain the information that is required to decode s. This "hand-wavy" interpretation often gives a good first idea of the lower bounds that can be proven by Theorem 3.2.4.

This intuitive proof for Theorem 3.2.4 shows that it works best if at every search point *exactly* k answers are possible, and each of them is equally likely. This situation, however, is not typical for black-box optimization processes, where usually only a (possibly small) subset of function values are likely to appear next. As a rule of thumb, the larger the difference of the potential function value from the function value of the current best solution, the less likely an algorithm is to obtain it in the next iteration. Such *transition probabilities* are not taken into account in Theorem 3.2.4. The theorem also does not cover very well the situation in which, at a certain step, fewer than k answers are possible. Even for fully symmetric problem classes, this situation is likely to appear in the later parts of the optimization process, where those problem instances that are still aligned with all previously evaluated function values all map the next query to one out of fewer than k possible function values. Covering these two shortcomings of Theorem 3.2.4 is one of the main challenges in black-box complexity. One step in this direction is the *matrix lower bound theorem* presented in [9] and the subsequent publication [7]. As also acknowledged there, however, the verification of the conditions under which these two generalizations apply is often quite tedious, so that the two methods are unfortunately not yet easily and very generally applicable. So far, they have been used to derive lower bounds for the black-box complexity of the OneMax and the Jump benchmark functions; see Sections 3.3.2 and 3.3.7.

Another tool that will be very useful in the subsequent sections is the following theorem, which allows us to transfer lower bounds proven for a simpler problem to a problem that is derived from it by a composition with another function. Most notably, it allows us to bound the black-box complexity of functions of unitation (i.e; functions for which the function value depends only on the number of ones in the string) by that of the OneMax problems. We will apply this theorem to show that the black-box complexity of the jump functions is at least as large as that of OneMax; see Section 3.3.7.

Theorem 3.2.5 (generalization of Theorem 2 in [28]). *For all problem classes $\mathcal{F}$, all classes of algorithms $\mathcal{A}$, and all maps $g : \mathbb{R} \to \mathbb{R}$ that are such that for all $f \in \mathcal{F}$ it holds that $\{x \mid g(f(x))$ optimal$\} = \{x \mid f(x)$ optimal$\}$ the $\mathcal{A}$-black-box complexity of $g(\mathcal{F}) := \{g \circ f \mid f \in \mathcal{F}\}$ is at least as large as that of $\mathcal{F}$.*

The intuition behind Theorem 3.2.5 is that with a knowledge of $f(x)$, we can compute $g(f(x))$, so that every algorithm that optimizes $g(\mathcal{F})$ can also be used to optimize $\mathcal{F}$, by evaluating the $f(x)$ values, feeding $g(f(x))$ to the algorithm, and querying the solution candidates that this algorithm suggests.

3.2.3 Tools to Prove Upper Bounds

We now present general upper bounds for the black-box complexity of a problem. We recall that, by definition, a small upper bound for the black-box complexity of a problem $\mathcal{F}$ shows that there exists an algorithm which solves every problem instance $f \in \mathcal{F}$ efficiently. When the upper bound for a problem is smaller than the expected performance of well-understood search heuristics, the question of whether these state-of-the-art heuristics can be improved or whether the unrestricted black-box model is too generous arises.

The simplest upper bound for the black-box complexity of a class $\mathcal{F}$ of functions is the expected performance of random sampling without repetitions.

Lemma 3.2.6. *For every finite set S and every class $\mathcal{F} \subset \{f : S \to \mathbb{R}\}$ of real-valued functions over S, the unrestricted black-box complexity of $\mathcal{F}$ is at most $(|S|+1)/2$.*

This simple bound can be tight, as we shall discuss in Section 3.3.1. A similarly simple upper bound is presented in the next subsection.

3.2.3.1 Function Classes vs. Individual Instances

In all of the above we have discussed the black-box complexity of a *class* of functions, and not of individual problem instances. This is justified by the following observation, which also explains why in the following we will usually consider generalizations of the benchmark problems typically studied in the theory of randomized black-box optimization.

Lemma 3.2.7. *For every function $f : S \to \mathbb{R}$, the unrestricted black-box complexity of the class $\{f\}$ that consists only of f is one. The same holds for any class $\mathcal{F}$ of functions that all have their optimum at the same point, i.e., for which there exists a search point $x \in S$ such that, for all $f \in \mathcal{F}$, $x \in \arg\max f$ holds.*

More generally, if $\mathcal{F}$ is a collection of functions $f : S \to \mathbb{R}$ and $X \subseteq S$ is such that for all $f \in \mathcal{F}$ there exists at least one point $x \in X$ such that $x \in \arg\max f$, the unrestricted black-box complexity of $\mathcal{F}$ is at most $(|X|+1)/2$.

For every finite set $\mathcal{F}$ of functions, the unrestricted black-box complexity is bounded from above by $(|\mathcal{F}|+1)/2$.

The proof of this lemma is quite straightforward. For the first statement, the algorithm which queries any point in $\arg\max f$ in the first query certifies this bound. Similarly, the second statement is certified by the algorithm that queries x in the first iteration. The algorithm which queries the points in X in random order proves the third statement. Finally, note that the third statement implies the fourth by letting X be the set that contains, for each function $f \in \mathcal{F}$, one optimal solution $x_f \in \arg\max f$.

Lemma 3.2.7 indicates that function classes $\mathcal{F}$ for which $\cup_{f \in \mathcal{F}} \arg\max f$ or, more precisely, for which a small set X as in the third statement of Lemma 3.2.7 exists are not very interesting research objects in the unrestricted black-box model. We therefore typically choose generalizations of the benchmark problems in such a way that any set X which contains for each objective function $f \in \mathcal{F}$ at least one optimal search point has to be large. We shall often even have $|X| = |\mathcal{F}|$, i.e., the optima of any two functions in $\mathcal{F}$ are pairwise different.

We will see in Section 3.6 that Lemma 3.2.7 does not apply to all of the restricted black-box models. In fact, in the unary unbiased black-box model considered there, the black-box complexity of a single function can be of order $n \log n$. That is, even if the algorithm "knows" where the optimum is, it may still need $\Omega(n \log n)$ steps to generate it.

3.2.3.2 Upper Bounds via Restarts

In several situations, rather than bounding the expected optimization time of a black-box heuristic, it can be easier to show that the probability that it solves a given problem within s iterations is at least p. If p is large enough (for an asymptotic bound, it suffices that this success probability is constant), then a restarting strategy can be used to obtain upper bounds on the black-box complexity of the problem. Either the algorithm is finished after at most s steps, or it is initialized from scratch, independently of all previous runs. This way, we obtain the following lemma.

Lemma 3.2.8 (Remark 7 in [37]). *Suppose for a problem $\mathcal{F}$ that there exists an unrestricted black-box algorithm A that, with constant success probability, solves any instance $f \in \mathcal{F}$ in s iterations (that is, it queries an optimal solution within s queries). Then the unrestricted black-box complexity of $\mathcal{F}$ is at most $O(s)$.*

Lemma 3.2.8 also applies to almost all of the restricted black-box models that we will discuss in Sections 3.4-3.7. In general, it applies to all black-box models in which restarts are allowed. It does not apply to the (strict version of the) elitist black-box model, which we discuss in Section 3.7.4.

3.2.4 Polynomial Bounds for NP-Hard Problems

Our discussion in Section 3.1.1 indicates that the classical complexity notions developed for white-box optimization and decision problems are not very meaningful in the black-box setting. This is impressively demonstrated by a number of NP-hard problems that have a small polynomial black-box complexity. We present such an example here, taken from [54, Section 3].

One of the best-known NP-complete problems is MaxClique. For a given graph $G=(V,E)$ of $|V|=n$ nodes and for a given parameter k, it asks whether there exists a complete subgraph $G' = (V' \subseteq V, E' := E \cap \{\{u,v\} \in E \mid u,v \in V'\})$ of size $|V'| \geq k$. A complete graph is a graph in which every two vertices are connected by a direct edge between them. The optimization version of MaxClique asks us to find a complete subgraph of the largest possible size. A polynomial-time optimization algorithm for this problem implies P=NP.

The unrestricted black-box complexity of MaxClique is, however, only of order n^2. This bound can be achieved as follows. In the first $\binom{n}{2}$ queries, the algorithm queries the presence of individual edges. This way, it learns the structure of the problem instance. From this information, all future solution candidates can be evaluated without any oracle queries. That is, a black-box algorithm can now compute an optimal solution *offline*, i.e., without the need for further function evaluations. This offline computation may take exponential time, but in the black-box complexity model, we do not charge the algorithm for the time needed between two queries. The optimal solution of the MaxClique instance can be queried in the $(\binom{n}{2}+1)$-st query.

Theorem 3.2.9 (Section 3 in [54]). *The unrestricted black-box complexity of* MaxClique *is at most* $\binom{n}{2}+1$ *and thus* $O(n^2)$.

Several similar results can be obtained. For most of the restricted black-box complexity models this has been explicitly done; see also Section 3.6.3.5.

One way to avoid such small complexities would be to restrict the time that an algorithm can spend between any two queries. This suggestion was made in [54]. In our opinion, this requirement would, however, carry a few disadvantages such as a mixture of different complexity measures. We will therefore, in this chapter, not explicitly verify that the algorithms run in polynomial time. Most upper bounds are nevertheless easily seen to be obtained by polynomial-time algorithms. Where polynomial bounds are proven for NP-hard problems, there must be at least one iteration for which the respective algorithm, according to today's knowledge, needs excessive time.

3.3 Known Black-Box Complexities in the Unrestricted Model

We survey existing results for the unrestricted black-box model, and proceed by problem type. For each benchmark problem considered, we first introduce its generalization to classes of similar problem instances. We discuss which of characteristics of the original problem are maintained in these generalizations. We will see that for some classical benchmark problems, different generalizations have been proposed in the literature.

3.3.1 Needle

Our first benchmark problem is an example that shows that the simple upper bound given in Lemma 3.2.6 can be tight. The function that we generalize is the Needle function, which assigns 0 to all search points $s \in S$ except for one distinguished optimum, which has a function value of one. In order to obtain the above-mentioned property that every function in the generalized class has a different optimum than any other function (see the discussion after Lemma 3.2.7), while at the same time maintaining the characteristics of the problem, the following generalization is made. For every $s \in S$, we let $f_s : S \to \mathbb{R}$ be the function which assigns the function value 1 to the unique optimum $s \in S$ and 0 to all other search points $x \neq s$. We let $\text{Needle}(S) := \{f_s \mid s \in S\}$ be the set of all such functions.

Confronted with such a function f_s, we do not learn anything about the target string s until we have found it. It seems quite intuitive that the best we can do in such a case is to query search points at random, without repetitions. That this is indeed optimal is the statement of the following theorem, which can be easily proven by Yao's minimax principle applied to NEEDLE(S) with the uniform distribution.

Theorem 3.3.1 (Theorem 1 in [54]). *For every finite set S, the unrestricted black-box complexity of* NEEDLE(S) *is* $(|S|+1)/2$.

3.3.2 OneMax

The best-studied benchmark function in the theory of randomized black-box optimization is certainly ONEMAX. ONEMAX assigns to each bit string x of length n the number $\sum_{i=1}^{n} x_i$ of ones in it. The natural generalization of this particular function to a nontrivial class of functions is as follows.

Definition 3.3.2 (OneMax). For all $n \in \mathbb{N}$ and all $z \in \{0,1\}^n$ let

$$\mathrm{OM}_z : \{0,1\}^n \to [0..n], x \mapsto \mathrm{OM}_z(x) = |\{i \in [n] \mid x_i = z_i\}|,$$

the function that assigns to each length-n bit string x the number of bits in which x and z agree. Being the unique optimum of OM_z, the string z is called its *target string*.

We refer to $\text{ONEMAX}_n := \{\mathrm{OM}_z \mid z \in \{0,1\}^n\}$ as the set of all (generalized) ONEMAX functions. We will often omit the subscript n.

We easily observe that, for every n, the original ONEMAX function OM counting the number of ones corresponds to $\mathrm{OM}_{(1,\dots,1)}$. It is, furthermore, not difficult to prove that, for every $z \in \{0,1\}^n$, the *fitness landscape* of OM_z is *isomorphic* to that of OM. This can be seen by observing that $\mathrm{OM}_z(x) = \mathrm{OM}(x \oplus z \oplus (1,\dots,1))$ for all $x, z \in \{0,1\}^n$, which shows that $\mathrm{OM}_z = \mathrm{OM} \circ \alpha_z$ for the Hamming automorphism $\alpha_z : \{0,1\}^n \to \{0,1\}^n, x \mapsto x \oplus z \oplus (1,\dots,1)$. As we shall discuss in Section 3.6, a Hamming automorphism is a one-to-one map $\alpha : \{0,1\}^n \to \{0,1\}^n$ such that for all x and all z the Hamming distance between x and z is identical to that between $\alpha(x)$ and $\alpha(z)$. This shows that the generalization of OM to functions OM_z preserves its problem characteristics. In essence, the generalization is just a "relabeling" of the search points.

3.3.2.1 The Unrestricted Black-Box Complexity of OneMax

With Definition 3.3.2 at hand, we can study the unrestricted black-box complexity of this important class of benchmark functions.

Interestingly, it turns out that the black-box complexity of $\textsc{OneMax}_n$ has been studied in several different contexts, long before Droste, Jansen, and Wegener introduced black-box complexity. In fact, Erdős and Rényi [55] as well as several other authors studied it in the early 1960s, inspired by a question about so-called *coin-weighing problems*.

In our terminology, Erdős and Rényi [55] showed that the unrestricted black-box complexity of $\textsc{OneMax}$ is at least $(1-o(1))n/\log_2(n)$ and at most $(1+o(1))\log_2(9)n/\log_2(n)$. The upper bound was improved to $(1+o(1))2n/\log_2(n)$ in [10, 73, 74]. Identical or weaker bounds have been proven several times in the literature. Some publications appeared at the same time as the work of Erdős and Rényi (see the discussion in [6]), and some much later [2, 6, 54].

Theorem 3.3.3 ([10, 55, 73, 74]). *The unrestricted black-box complexity of* $\textsc{OneMax}$ *is at least* $(1-o(1))n/\log_2(n)$ *and at most* $(1+o(1))2n/\log_2(n)$. *It is thus* $\Theta(n/\log n)$.

The lower bound in Theorem 3.3.3 follows from Yao's minimax principle, applied to $\textsc{OneMax}_n$ with the uniform distribution. Informally, we can use the arguments given after Theorem 3.2.4: since the optimum can be anywhere in $\{0,1\}^n$, we need to learn the n bits of the target string z. With each function evaluation, we receive at most $\log_2(n+1)$ bits of information, namely the objective value, which is an integer between 0 and n. We therefore need at least (roughly) $n/\log_2(n+1)$ iterations. Using Theorem 3.2.4, this reasoning can be turned into a formal proof.

The upper bound given in Theorem 3.2.4 is quite interesting because it is obtained by a very simple strategy. Erdős and Rényi showed that $O(n/\log n)$ bit strings sampled independently and uniformly at random from the hypercube $\{0,1\}^n$ have a high probability of revealing the target string. That is, an asymptotically optimal unrestricted black-box algorithm for $\textsc{OneMax}$ can just sample $O(n/\log n)$ random samples. From these samples and the corresponding objective values, the target string can be identified without further queries. Its computation, however, may not be possible in polynomial time. The fact that $\textsc{OneMax}_n$ can be optimized in $O(n/\log n)$ queries also in polynomial time was proven in [6].[1] The reader interested in a formal analysis of the strategy used by Erdős and Rényi may refer to Section 3 of [35], where a detailed proof of the $O(n/\log n)$ random sampling strategy is presented.

In the context of *learning*, it is interesting to note that the random sampling strategy of Erdős and Rényi is *nonadaptive*, i.e., the t-th search point does not depend on the previous $t-1$ evaluations. In the black-box context, a last query, in which the optimal solution is evaluated, is needed. This query certainly depends on the previous $O(n/\log n)$ evaluations, but note

[1] Bshouty [6] mentions that also the constructions of Lindström [73, 74] and Cantor and Mills [10] can be done in polynomial time. But this was not explicitly mentioned in the latter publications. The method of Bshouty also has the advantage that it generalizes to $\textsc{OneMax}$ functions over alphabets larger than $\{0,1\}$; see also Section 3.10.

that here we know the answer to this evaluation already (with high probability). For nonadaptive strategies, learning z with $(1+o(1))2n/\log n$ queries is optimal [55]. The intuitive reason for this lower bound is that a random guess typically has an objective value close to $n/2$. More precisely, instead of using the whole range of $n+1$ possible answers, almost all function values are in an $O(\sqrt{n})$ range around $n/2$, giving, very informally, the lower bound $\log_2(n)/\log_2(O(\sqrt{n})) = \Omega(2n/\log n)$.

Using the probabilistic method (or the constructive result of Bshouty [6]), the random sampling strategy can be derandomized. This derandomization says that for every n, there is a sequence of $t = \Theta(n/\log n)$ strings $x^{(1)},\dots,x^{(t)}$ such that the objective values $\mathrm{OM}_z(x^{(1)}),\dots,\mathrm{OM}_z(x^{(t)})$ uniquely determine the target string z. Such a derandomized version will be used in later parts of this chapter, for example, in the context of the k-ary unbiased black-box complexity of OneMax studied in Section 3.6.3.2.

Theorem 3.3.4 (from [55] and others). *For every n there is a sequence $x^{(1)},\dots,x^{(t)}$ of $t = \Theta(n/\log n)$ bit strings such that for every two length-n bit strings $y \neq z$ there exists an index i with $\mathrm{OM}_z(x^{(i)}) \neq \mathrm{OM}_y(x^{(i)})$.*

For some very concrete OneMax instances, i.e., for instances of bounded dimension n, very precise bounds for the black-box complexity are known; see [7] and the pointers in [29, Section 1.4] for details. Here, in this chapter, we are only concerned with the asymptotic complexity of OneMax_n with respect to the problem dimension n. Unsurprisingly, this benchmark problem will also be studied in almost all of the restricted black-box models that we describe in the subsequent sections. A summary of known results can be found in Section 3.8.

3.3.3 BinaryValue

Another intensively studied benchmark function is the binary-value function $\mathrm{BV}(x) := \sum_{i=1}^{n} 2^{i-1} x_i$, which assigns to each bit string the value of the binary number it represents. As $2^i > \sum_{j=1}^{i} 2^{j-1}$, the bit value of the bit $i+1$ dominates the effect of all bits $1,\dots,i$ on the function value.

Two straightforward generalizations of BV to function classes exist. The first one is the collection of all functions

$$\mathrm{BV}_z : \{0,1\}^n \to [0..2^n], x \mapsto \sum_{i=1}^{n} 2^{i-1} \mathbb{1}(x_i, z_i),$$

where $\mathbb{1}(a,b) := 1$ if and only if $a = b$, and $\mathbb{1}(a,b) := 0$ otherwise. In light of Definition 3.3.2, this may seem like a natural extension of BV to a class of functions. It also satisfies our sought condition that for any two functions $\mathrm{BV}_z \neq \mathrm{BV}_{z'}$ the respective optima z and z' differ, so that the smallest set containing its optimum for each function is the full n-dimensional hypercube $\{0,1\}^n$. However, we can easily see that the unrestricted black-box complexity of the set $\text{BINARYVALUE}^*_n := \{\text{BV}_z \mid z \in \{0,1\}^n\}$ so defined is very small.

Theorem 3.3.5 (Theorem 4 in [54]). *The unrestricted black-box complexity of* BINARYVALUE^*_n *is* $2-2^{-n}$.

Proof. The lower bound follows from observing that, for an instance BV_z for which z is chosen uniformly at random, the probability of querying the optimum z in the first query is 2^{-n}. In all other cases, at least two queries are needed.

For the upper bound, we only need to observe that for any two target strings $z \neq z'$ and for every search point $x \in \{0,1\}^n$ we have $\text{BV}_z(x) \neq \text{BV}_{z'}(x)$. More precisely, it is easy to see that from $\text{BV}_z(x)$ we can easily determine for which bits $i \in [n]$ the bit value of x_i is identical to z_i. This shows that by querying the objective value of a random string in the first query we can compute the optimum z, which we query in the second iteration if the first value is not already optimal. □

Theorem 3.3.5 is possible because the objective values disclose a lot of information about the target string. A second generalization of BV has therefore been suggested in the literature. In light of the typical behavior of black-box heuristics, which do not discriminate between bit positions, and in particular with respect to the unbiased black-box model defined in Section 3.6, this variant seems to be the more "natural" choice in the context of evolutionary algorithms. This second generalization of BV collects together all functions $\text{BV}_{z,\sigma}$, defined as

$$\text{BV}_{z,\sigma} : \{0,1\}^n \to \mathbb{N}_0, x \mapsto \sum_{i=1}^{n} 2^{i-1} \delta(x_{\sigma(i)}, z_{\sigma(i)}).$$

Denting by $\sigma(x)$ the string $(x_{\sigma(1)} \cdots x_{\sigma(n)})$, we easily see that $\text{BV}_{z,\sigma}(x) = \text{BV}(\sigma(x \oplus z \oplus (1,\ldots,1)))$, thus showing that the class $\{\text{BV}_{z,\sigma} \mid z \in \{0,1\}^n, \sigma \in S_n\}$ can be obtained from BV by composing it with an $\oplus$-shift of the bit values and a permutation of the indices $i \in [n]$. Since $z = \arg\max \text{BV}_{z,\sigma}$, we call z the *target string* of $\text{BV}_{z,\sigma}$. Similarly, we call σ the *target permutation* of $\text{BV}_{z,\sigma}$.

Going through the bit string one by one, i.e., flipping one bit at a time, shows that at most $n+1$ function evaluations are needed to optimize any $\text{BV}_{z,\sigma}$ instance. This simple upper bound can be improved by observing that for each query x and for each $i \in [n]$ we can derive from $\text{BV}_{z,\sigma}(x)$ whether or not $x_{\sigma(i)} = z_{\sigma(i)}$, even if we cannot yet locate $\sigma(i)$. Hence, all we need to do is to identify the target permutation σ. This can be done by a binary search, which gives the following result.

Theorem 3.3.6 (Theorem 16 in [44]). *The unrestricted black-box complexity of* $\text{BINARYVALUE}_n := \{\text{BV}_{z,\sigma} \mid z \in \{0,1\}^n, \sigma \in S_n\}$ *is at most* $\lceil \log_2 n \rceil + 2$.

In a learning-related sense, in which we want to *learn* both z and σ, the bound in Theorem 3.3.6 is tight, as, informally, the identification of σ requires us to learn $\Theta(\log(n!)) = \Theta(n \log n)$ bits, while with every query we obtain $\log_2(2^n) = n$ bits of information. In our optimization context, however, we do not necessarily need to learn σ in order to optimize $\text{BV}_{z,\sigma}$. A similar situation will be discussed in Section 3.3.6, where we study the unrestricted black-box complexity of LEADINGONES. For LEADINGONES, it can be formally proven that the complexities of optimization and learning are identical (up to at most n queries). We are not aware of any formal statement showing whether or not a similar argument holds for the class BINARYVALUE_n.

3.3.4 Linear Functions

OM and BV are representatives of the class of linear functions $f: \{0,1\}^n \to \mathbb{R}, x \mapsto \sum_{i=1}^n f_i x_i$. We can generalize this class in the same way as above to obtain the collection

$$\text{LINEAR}_n := \left\{ f_z : \{0,1\}^n \to \mathbb{R}, x \mapsto \sum_{i=1}^n f_i \mathbb{1}(x_i, z_i) \mid z \in \{0,1\}^n \right\}$$

of generalized linear functions. ONEMAX_n and BINARYVALUE_n are both contained in this class.

Not much is known about the black-box complexity of this class. The only known bounds are summarized by the following theorem.

Theorem 3.3.7 (Theorem 3.3.3 above and Theorem 4 in [54]). *The unrestricted black-box complexity of the class* LINEAR_n *is at most* $n+1$ *and at least* $(1-o(1))n/\log_2 n$.

The upper bound is attained by an algorithm that starts with a random or a fixed bit string x and flips one bit at a time, using the better of the parent and the offspring as the starting point for the next iteration. A linear lower bound seems likely, but has not been formally proven.

3.3.5 Monotone and Unimodal Functions

For the sake of completeness, we mention that the class LINEAR_n is a subclass of the class of generalized monotone functions.

Definition 3.3.8 (monotone functions). Let $n \in \mathbb{N}$ and let $z \in \{0,1\}^n$. A function $f: \{0,1\}^n \to \mathbb{R}$ is said to be *monotone with respect to* z if for all $y, y' \in \{0,1\}^n$ with $\{i \in [n] \mid y_i = z_i\} \subsetneq \{i \in [n] \mid y'_i = z_i\}$ it holds that

$f(y) < f(y')$. The class MONOTONE$_n$ contains all such functions that are monotone with respect to some $z \in \{0,1\}^n$.

The above-mentioned algorithm which flips one bit at a time (see the discussion after Theorem 3.3.7) solves any of these instances in at most $n+1$ queries, giving the following theorem.

Theorem 3.3.9. *The unrestricted black-box complexity of the class* MONOTONE$_n$ *is at most* $n+1$ *and at least* $(1-o(1))n/\log_2 n$.

Monotone functions are instances of so-called *unimodal functions.* A function f is unimodal if and only if for every nonoptimal search point x there exists a direct neighbor y of x with $f(y) > f(x)$. The unrestricted black-box complexity of this class of unimodal functions was studied in [54, Section 8], where a lower bound that depends on the number of different function values that the objective functions can map to was presented.

3.3.6 LeadingOnes

After ONEMAX, probably the second most investigated function in the theory of discrete black-box optimization is the leading-ones function $\text{LO} : \{0,1\}^n \to [0..n]$, which assigns to each bit string x the length of the longest prefix of ones, i.e., $\text{LO}(x) := \max\{i \in [0..n] \mid \forall j \in [i] : x_j = 1\}$. Like for BINARYVALUE, two generalizations have been studied, an $\oplus$-invariant version and an $\oplus$- and permutation-invariant version. As a consequence of the unbiased black-box complexity model which we will discuss in Section 3.6, the latter is the more frequently studied.

Definition 3.3.10 (LeadingOnes function classes). Let $n \in \mathbb{N}$. For any $z \in \{0,1\}^n$, let

$$\text{LO}_z : \{0,1\}^n \to \mathbb{N}, x \mapsto \max\{i \in [0..n] \mid \forall j \in [i] : x_j = z_j\},$$

the length of the maximal joint prefix of x and z. Let $\text{LEADINGONES}^*_n := \{\text{LO}_z \mid z \in \{0,1\}^n\}$.

For $z \in \{0,1\}^n$ and $\sigma \in S_n$, let

$$\text{LO}_{z,\sigma} : \{0,1\}^n \to \mathbb{N}, x \mapsto \max\{i \in [0..n] \mid \forall j \in [i] : x_{\sigma(j)} = z_{\sigma(j)}\},$$

the maximal joint prefix of x and z with respect to σ. The set LEADINGONES$_n$ is the collection of all such functions, i.e.,

$$\text{LEADINGONES}_n := \{\text{LO}_{z,\sigma} \mid z \in \{0,1\}^n, \sigma \in S_n\}.$$

The unrestricted black-box complexity of the set LEADINGONES$^*_n$ is easily seen to be around $n/2$. This is the complexity of the algorithm which starts with a random string x and, given an objective value of $\mathrm{LO}_z(x)$, replaces x by the string that is obtained from x by flipping the $\mathrm{LO}_z(x)+1$-st bit in x. The lower bound is a simple application of Yao's minimax principle to the uniform distribution over all possible problem instances. It is crucial here to note that the algorithms do not have any information about the "tail" $(z_j \dots z_n)$ until they have seen for the first time a search point of function value at least $j-1$.

Theorem 3.3.11 (Theorem 6 in [54]). *The unrestricted black-box complexity of the set* LEADINGONES^*_n *is* $n/2 \pm o(n)$. *The same holds for the set* $\{\mathrm{LO}_{z,\sigma} \mid z \in \{0,1\}^n\}$, *for any fixed permutation* $\sigma \in S_n$.

The unrestricted black-box complexity of LEADINGONES_n is also quite well understood.

Theorem 3.3.12 (Theorem 4 in [1]). *The unrestricted black-box complexity of* LEADINGONES_n *is* $\Theta(n \log\log n)$.

Both the upper and the lower bounds in Theorem 3.3.12 are quite involved. For the lower bound, Yao's minimax principle is applied to the uniform distribution over the instances $\mathrm{LO}_{z,\sigma}$ with $z_{\sigma(i)} := (i \mod 2), i = 1, \dots, n$. Informally, this choice indicates that the complexity of the LEADINGONES problem originates in the difficulty of identifying the target permutation. Indeed, as soon as we know the permutation, we need at most $n+1$ queries to identify the target string z (and only around $n/2$ on average, by Theorem 3.3.11). To measure the amount of information that an algorithm can have about the target permutation σ, a potential function is designed that maps each search point x to a real number. To prove the lower bound in Theorem 3.3.12, it is necessary to show that, for every query x, the expected increase in this potential is not very large. Using drift analysis, this can be used to bound the expected time needed to accumulate the amount of information needed to uniquely determine the target permutation.

The proof of the upper bound will be sketched in Section 3.6.3.3, in the context of the unbiased black-box complexity of LEADINGONES_n.

It may be interesting to note that the $O(n \log\log n)$ bound in Theorem 3.3.12 cannot be achieved by deterministic algorithms. In fact, Theorem 3 in [1] states that the *deterministic unrestricted black-box complexity* of LEADINGONES_n is $\Theta(n \log n)$.

3.3.7 Classes of Jump Functions

Another class of popular pseudo-Boolean benchmark functions is that of so-called "jump" functions. In black-box complexity, this class is currently one of the most intensively studied problems, with a number of surprising results, which in addition carry some interesting ideas for potential refinements of state-of-the-art heuristics. For this reason, we discuss this class in more detail, and compare the known complexity bounds with running-time bounds for some standard and recently developed heuristics.

For a nonnegative integer ℓ, the function $\textsc{Jump}_{\ell,z}$ is derived from the $\textsc{OneMax}$ function OM_z with target string $z \in \{0,1\}^n$ by "blanking out" any useful information within the strict ℓ-neighborhood of the optimum z and its bitwise complement $\bar{z}$, by giving all these search points a fitness value of 0. In other words,

$$\textsc{Jump}_{\ell,z}(x) := \begin{cases} n & \text{if } \text{OM}_z(x) = n, \\ \text{OM}_z(x) & \text{if } \ell < \text{OM}_z(x) < n - \ell, \\ 0 & \text{otherwise.} \end{cases} \tag{3.3.1}$$

This definition is mostly similar to the two definitions used in [53, 71] that we shall discuss below, which do not fully agree.

3.3.7.1 Known Running-Time Bounds for Jump Functions

We summarize here the known running-time results for the optimization of jump functions via randomized optimization heuristics. The reader interested only in black-box complexity results can skip this section.

Droste, Jansen, and Wegener [53] analyzed the optimization time of the (1+1) EA on jump functions. From this work, it is not difficult to see that the expected running time of the (1+1) EA on $\textsc{Jump}_{\ell,z}$ is $\Theta(n^{\ell+1})$, for all $\ell \in \{1, \ldots, \lfloor n/2 \rfloor - 1\}$ and all $z \in \{0,1\}^n$. This running time is dominated by the time needed to "jump" from a local optimum x with function value $\text{OM}_z(x) = n - \ell - 1$ to the unique global optimum z.

The *fast genetic algorithm* proposed in [39] significantly reduces this running time by using a generalized variant of standard bit mutation, which goes through its input and flips each bit independently with probability c/n. By choosing in every iteration the expected step size c in this mutation rate c/n from a power-law distribution with exponent β (more precisely, in every iteration, c is chosen independently of all previous iterations, and independently of the current state of the optimization process), an expected running time of $O(\ell^{\beta-0.5}((1+o(1))e/\ell)^{\ell} n^{\ell})$ on $\textsc{Jump}_{\ell,z}$ is achieved, uniformly for all jump sizes $\ell > \beta - 1$. This is only a polynomial factor worse than the $((1+o(1))e/\ell)^{\ell} n^{\ell}$ expected running time of a (1+1) EA which for every jump size ℓ uses a bit flip probability of ℓ/n, which is the optimal static choice.

Dang and Lehre [17] showed an expected running time of $O((n/c)^{\ell+1})$ for a large class of nonelitist population-based evolutionary algorithms with mutation rate c/n, where c is supposed to be a constant.

Jump functions were originally studied to investigate the usefulness of crossover, i.e., the recombination of two or more search points into a new solution candidate. In [64], a $(\mu+1)$ genetic algorithm using crossover was shown to optimize any $\text{JUMP}_{\ell,z}$ function in an expected number $O(\mu n^2(\ell-1)^3+4^{\ell-1}/p_c)$ of function evaluations, where $p_c < 1/(c(\ell-1)n)$ denotes the (artificially small) probability of doing a crossover. In [18] it was shown that for more "natural" parameter settings, most notably those with a nonvanishing probability of crossover, a standard $(\mu+1)$ genetic algorithm which uses crossover followed by mutation has an expected $O(n^\ell \log n)$ optimization time on any $\text{JUMP}_{\ell,z}$ function, which is a gain by a factor $O(n/\log n)$ over the above-mentioned bound for the standard (1+1) EA. In [16] it was shown that significant performance gains can be achieved by the usage of diversity mechanisms. We refer the interested reader to Chapter 7.6 of this book for a more detailed description of these mechanisms and running-time statements; see in particular Sections 8.4.4 and 8.4.5.

3.3.7.2 The Unrestricted Black-Box Complexity of Jump Functions

From the definition in (3.3.1), we can easily see that for every $n \in \mathbb{N}$ and for all $\ell \in [0..n/2]$ there exists a function $f : [0..n] \to [0..n]$ such that $\text{JUMP}_{\ell,z}(x) = f(\text{OM}_z(x))$ for all $z,x \in \{0,1\}^n$. By Theorem 3.2.5, we therefore obtain the result that for every class $\mathcal{A}$ of algorithms and for all ℓ, the $\mathcal{A}$-black-box complexity of $\text{JUMP}_{\ell,n} := \{\text{JUMP}_{\ell,z} \mid z \in \{0,1\}^n\}$ is at least as large as that of ONEMAX_n. Quite surprisingly, it turns out that this bound can be met for a broad range of jump sizes ℓ. Building on work [28] on the unbiased black-box complexity of jump functions (see Section 3.6.3.4 for a detailed description of the results proven in [27]), the following bounds were obtained in [9].

Theorem 3.3.13 ([9]). *For $\ell < n/2 - \sqrt{n}\log_2 n$, the unrestricted black-box complexity of $\text{JUMP}_{\ell,n}$ is at most $(1+o(1))2n/\log_2 n$, while for $n/2 - \sqrt{n}\log_2 n \le \ell < \lfloor n/2 \rfloor - \omega(1)$ it is at most $(1+o(1))n/\log_2(n-2\ell)$ (where, in this latter bound, $\omega(1)$ and $o(1)$ refer to $n-2\ell \to \infty$).*

For the extreme case of $\ell = \lfloor n/2 \rfloor - 1$, the unrestricted black-box complexity of $\text{JUMP}_{\ell,n}$ is $n+\Theta(\sqrt{n})$.

For all ℓ and every odd n, the unrestricted black-box complexity of $\text{JUMP}_{\ell,n}$ is at least $\lfloor \log_{\frac{n-2\ell+1}{2}} (2^{n-2}(n-2\ell-1)+1) \rfloor - \frac{2}{n-2\ell-1}$. For even n, it is at least $\lfloor \log_{\frac{n-2\ell+2}{2}} (1+2^{n-1}\frac{(n-2\ell)^2}{n-2\ell-1}) \rfloor - \frac{2}{n-2\ell}$.

The proofs of the results in Theorem 3.3.13 are built to a large extent on the techniques used in [28], which we shall discuss in Section 3.6.3.4. In addition to these techniques, [9] introduced a matrix lower-bound method, which allows one to prove stronger lower bounds than the simple information-theoretic result presented in Theorem 3.2.4 by taking into account the fact that the "typical" information obtained through a function evaluation can be much smaller than what the whole range $\{f(s) \mid s \in S\}$ of possible f-values suggests.

Note that even for the case of "small" $\ell < n/2 - \sqrt{n}\log_2$, the region around the optimum in which the function values are zero is actually quite large. This plateau contains $2^{(1-o(1))n}$ points and has a diameter that is linear in n.

For the case of the extreme jump functions, note also that, apart from the optimum, only the points x with $\mathrm{OM}_z(x) = \lfloor n/2 \rfloor$ and $\mathrm{OM}_z(x) = \lceil n/2 \rceil$ have a nonzero function value. It is thus quite surprising that these functions can nevertheless be optimized so efficiently. We shall see in Section 3.6.3.4 how this is possible.

One may wonder why, in the definition of $\textsc{Jump}_{\ell,n}$, we have fixed the jump size ℓ, as in this way it is "known" to the algorithm. It has been argued in [38] that the algorithms can learn ℓ efficiently, if this is needed; in some cases, including those of small ℓ-values, knowing ℓ may not be needed to achieve the above-mentioned optimization times. Whether or not knowledge of ℓ is needed can be decided adaptively.

3.3.7.3 Alternative Definitions of Jump Functions

Following up on results for the so-called unbiased black-box complexity of jump functions [28] (see Section 3.6.3.4), Jansen [62] proposed an alternative generalization of the classical jump function considered in [53]. To discuss this extension, we recall that the jump function analyzed in [53, Definition 24] is the $(1,\ldots,1)$ version of the maps $f_{\ell,z}$ that assign to every length-n bit string x the function value

$$f_{\ell,z}(x) := \begin{cases} \ell + n & \text{if } \mathrm{OM}_z(x) = n, \\ \ell + \mathrm{OM}_z(x) & \text{if } \mathrm{OM}_z(x) \leq n - \ell, \\ n - \mathrm{OM}_z(x) & \text{otherwise.} \end{cases}$$

We first describe the motivation behind the extension considered in the definition given by Equation (3.3.1). To this end, we first note that in the unrestricted black-box complexity model, $f_{\ell,z}$ can be very efficiently optimized by searching for the bitwise complement $\bar{z}$ of z and then inverting this string to the optimal search point z. Note also that, in this definition, the region around the optimum z provides more information than the functions $\textsc{Jump}_{\ell,z}$ defined via (3.3.1). When we are interested in bounding the expected optimization time of classical black-box heuristics, this additional information most often does not pose any problems. But, for our sought black-box complexity studies, this information can make a crucial difference. Lehre and Witt [71] therefore designed a different set of jump functions consisting of maps $g_{\ell,z}$ that assign to each x the function value

$$g_{\ell,z}(x) := \begin{cases} n & \text{if } \mathrm{OM}_z(x) = n, \\ \mathrm{OM}_z(x) & \text{if } \ell < \mathrm{OM}_z(x) \le n-\ell, \\ 0 & \text{otherwise.} \end{cases}$$

The definition in Equation (3.3.1) is mostly similar to this definition of Lehre and witt, with the only difference being the function values for bit strings x with $\mathrm{OM}_z(x) = n - \ell$. Note that in (3.3.1) the sizes of the "blanked-out areas" around the optimum and its complement are equal, while for $g_{\ell,z}$ the area around the complement is larger than that around the optimum.

As mentioned, Jansen [62] introduced yet another version of the jump function. His motivation was that the spirit of the jump function is to "[locate] an unknown target string that is hidden in some distance to points a search heuristic can find easily". Jansen's definition also has black-box complexity analysis in mind. For a given $z \in \{0,1\}^n$ and a search point x^* with $\mathrm{OM}_z(x^*) > n-\ell$, his jump function h_{ℓ,z,x^*} assigns to every bit string x the function value

$$h_{\ell,z,x^*}(x) := \begin{cases} n+1 & \text{if } x = x^*, \\ n - \mathrm{OM}_z(x) & \text{if } n-\ell < \mathrm{OM}_z(x) \le n \text{ and } x \ne x^*, \\ \ell + \mathrm{OM}_z(x) & \text{otherwise.} \end{cases}$$

Since these functions do not reveal information about the optimum other than its ℓ-neighborhood, the unrestricted black-box complexity of the class $\{h_{\ell,z,x^*} \mid z \in \{0,1\}^n, \mathrm{OM}_z(x^*) > n-\ell\}$ is $\left(\sum_{i=0}^{\ell-1}\binom{n}{i}+1\right)/2$ [62, Theorem 4]. This bound also holds if z is fixed to be the all-ones string, i.e., if we consider the unrestricted black-box complexity of the class $\{h_{\ell,(1,\ldots,1),x^*} \mid \mathrm{OM}_z(x^*) > n-\ell\}$. For constant ℓ, the black-box complexity of this class of jump functions is thus $\Theta(n^{\ell-1})$, very different from the results for the unrestricted black-box complexity of the $\textsc{Jump}_{\ell,z}$ functions considered above. In contrast to the latter, the expected optimization times stated for crossover-based algorithms in Section 3.3.7.1 above do not necessarily apply to the functions h_{ℓ,z,x^*}, as for these functions the optimum x^* is not located in the middle of the ℓ-neighborhood of z.

3.3.8 Combinatorial Problems

The results described above are mostly concerned with benchmark functions that were introduced to study some particular features of typical black-box optimization techniques, for example, their hill-climbing capabilities (via the OneMax function) or their ability to jump a plateau of a certain size (this is the focus of the jump functions). Running-time analysis, of course, also studies more "natural" combinatorial problems, such as satisfiability problems, partition problems, scheduling, and graph-based problems such as routing, vertex cover, and MaxCut, see [79] for a survey of running-time results for combinatorial optimization problems.

Apart from a few results for combinatorial problems derived in [54],[2] the first publication to present a systematic investigation of the black-box complexities of combinatorial optimization problems was [37]. In that publication, the two well-known problems of finding a *minimum spanning tree* (MST) in a graph and the *single-source shortest-paths problem* (SSSP) were considered. The study revealed that, for combinatorial optimization problems, the precise formulation of the problem can make a decisive difference. Modeling such problems therefore needs be done with care.

We will not be able to summarize all results proven in [37], but the following summarizes the most relevant ones for the unrestricted black-box model. [37] also studies the MST and the SSSP problem in various restricted black-box models: more precisely, in the unbiased black-box model (see Section 3.6), the ranking-based model (Section 3.5) and combinations thereof. We will briefly discuss results for the unbiased case in Sections 3.6.3.6 and 3.6.4.1.

3.3.8.1 Minimum Spanning Trees

For a given graph $G = (V,E)$ with edge weights $w : E \to \mathbb{R}$, the minimum spanning tree problem asks us to find a connected subgraph $G' = (V,E')$ of G such that the sum $\sum_{e \in E'} w(e)$ of the edge weights in G' is minimized. This problem has a natural bit string representation. Letting $m := |E|$, we can fix an enumeration $\nu : [m] \to$ e. In this way, we can identify a length-m bit string $x = (x_1,\dots,x_m)$ with the subgraph $G_x := (V,E_x)$, where $E_x := \{\nu(i) \mid i \in [m] \text{ with } x_i = 1\}$ is the set of edges i for which $x_i = 1$. Using this interpretation, the MST problem can be modeled as a pseudo-Boolean optimization problem $f : \{0,1\}^m \to \mathbb{R}$; see [79] for details. This formulation is one of the two most common formulations of the MST problem in the evolutionary computation literature. The other common formulation uses a biobjective fitness function $f : \{0,1\}^m \to \mathbb{R}^2$; the first component maps each subgraph to its number of connected components, while the second component measures the total weight of the edges in this subgraph. In the unrestricted black-box model, the two formulations are almost equivalent.[3]

Theorem 3.3.14 (Theorems 10 and 12 in [37]). *For the biobjective and the single-objective formulation of the MST problem on an arbitrary connected*

[2] More precisely, the following combinatorial problems were studied in [54]: MaxClique (Section 3), Sorting (Section 4), and the single-source shortest-paths problem (Sections 4 and 9).

[3] Note that this is not the case for the *ranking-based model* discussed in Section 3.5, since here it can make a decisive difference whether two rankings for the two components of the biobjective function are reported or whether this information is condensed further into one single ranking.

graph of n nodes and m edges, the unrestricted black-box complexity is strictly larger than $n-2$ and at most $2m+1$.

These bounds also apply if, instead of absolute function values $f(x)$, only their rankings are revealed; in other words, the ranking-based black-box complexity (which will be introduced in Section 3.5) of the MST problem is also at most $2m+1$.

The upper bound is shown by first learning the order of the edge weights and then testing, in increasing order of the edge weights, whether or not the inclusion of the corresponding edge forms a cycle or not. This way, the algorithm imitates the well-known MST algorithm of Kruskal.

The lower bound of Theorem 3.3.14 is obtained by applying Yao's minimax principle with a probability distribution on the problem instances that samples uniformly at random a spanning tree, gives weight 1 to all of its edges, and gives weight 2 to all other edges. By Cayley's formula, the number of spanning trees on n vertices is n^{n-2}. In intuitive terms, a black-box algorithm solving the MST problem therefore needs to learn $(n-2)\log_2 n$ bits. Since each query reveals a number between $2k-n+1$ and $2k$ (k being the number of edges included in the corresponding graph), it provides at most $\log_2 n$ bits of information. Hence, in the worst case, we get a running time of at least $n-2$ iterations. To turn this intuition into a formal proof, a drift theorem is used to show that in each iteration the number of consistent possible trees decreases by factor of at most $1/n$.

3.3.8.2 Single-Source Shortest Paths

For the SSSP problem, which asks us to connect all vertices of an edge-weighted graph to a distinguished source node through a path of smallest total weight, several formulations coexist. The first one, which was also considered in [54], uses the following multicriteria objective function. An algorithm can query arbitrary trees on $[n]$ and the objective value of any such tree is an $(n-1)$-tuple of the distances of the $n-1$ nonsource vertices to the source $s=1$ (if an edge is traversed which does not exist in the input graph, the entry in the tuple is ∞).

Theorem 3.3.15 (Theorems 16 and 17 in [37]). *For arbitrary connected graphs with n vertices and m edges, the unrestricted black-box complexity of the multiobjective formulation of the SSSP problem is $n-1$. For complete graphs, it is at least $n/4$ and at most $\lfloor (n+1)/2 \rfloor + 1$.*

Theorem 3.3.15 improves on the previous $n/2$ lower and the $2n-3$ upper bound given in [54, Theorem 9]. For the general case, the proof of the upper bound imitates Dijkstra's algorithm by first connecting all vertices to the source, then all but one of the vertices to the vertex of lowest distance to the source, then all but the two of lowest distance to the vertex of second lowest distance, and so on, fixing one vertex with each query. The lower bound is an application of Yao's minimax principle to a bound on deterministic algorithms, which is obtained through an additive drift analysis.

For complete graphs, it is essentially shown that the problem instance can be learned with $\lfloor (n+1)/2 \rfloor$ queries, while the lower bound is again a consequence of Yao's minimax principle.

The bound in Theorem 3.3.15 is not very satisfactory as already the size of the input is $\Omega(m)$. The discrepancy is due to the large amount of information that the objective function reveals about the problem instance. To avoid such low black-box complexities, and to shed a better light on the complexity of the SSSP problem, [37] considered also an alternative model for the SSSP problem, in which a representation of a candidate solution is a vector $(\rho(2),\ldots,\rho(n)) \in [n]^{n-1}$. Such a vector is interpreted such that the predecessor of node i is $\rho(i)$ (the indices run from 2 to n to match the indices with the labels of the nodes - node 1 is the source node to which a shortest path is sought). With this interpretation, the search space becomes the set $S_{[2..n]}$ of permutations of $[2..n]$, i.e., $S_{[2..n]}$ is the set of all one-to-one maps $\sigma : [2..n] \to [2..n]$. For a given graph G, the single-criterion objective function f_G is defined by assigning to each candidate solution $(\rho(2),\ldots,\rho(n))$ the function value $\sum_{i=2}^{n} d_i$, where d_i is the distance of the i-th node to the source node. If an edge - including loops - is traversed which does not exist in the input graph, d_i is set to n times the maximum weight $w_{\max}$ of any edge in the graph.

Theorem 3.3.16 (Theorem 18 in [37]). *The unrestricted black-box complexity of the SSSP problem with the single-criterion objective function is at most* $\sum_{i=1}^{n-1} i = n(n-1)/2$.

As in the multiobjective case, the bound in Theorem 3.3.16 is obtained by imitating Dijkstra's algorithm. In the single-objective setting, adding the i-th node to the shortest-path tree comes at a cost of at most $n-i$ function evaluations.

3.4 Memory-Restricted Black-Box Complexity

As mentioned in the previous section, as early as in the first publication on black-box complexity [54] it was noted that the unrestricted model can be too generous in the sense that it includes black-box algorithms that are highly problem-tailored and whose expected running time is much smaller than that of typical black-box algorithms. One potential reason for such a discrepancy is the fact that unrestricted algorithms are allowed to store and to access the full search history, while typical heuristics store only a subset ("population" in evolutionary computation) of previously evaluated samples and their corresponding objective values. Droste, Jansen, and Wegener [54] therefore

Algorithm 3.2: The $(\mu+\lambda)$ memory-restricted black-box algorithm for optimizing an unknown function $f: S \to \mathbb{R}$

1 **Initialization:**
2 $X \leftarrow \emptyset$;
3 Choose a probability distribution $D^{(0)}$ over S^{μ}, sample from it $x^{(1)},\dots,x^{(\mu)} \in S$, and query $f(x^{(1)}),\dots,f(x^{(\mu)})$;
4 $X \leftarrow \left\{\left(x^{(1)},f(x^{(1)})\right),\dots,\left(x^{(\mu)},f(x^{(\mu)})\right)\right\}$;
5 **Optimization: for** $t=1,2,3,\dots$ **do**
6 Depending only on the multiset X choose a probability distribution $D^{(t)}$ over S^{λ}, sample from it $y^{(1)},\dots,y^{(\lambda)} \in S$, and query $f(y^{(1)}),\dots,f(y^{(\mu)})$;
7 Set $X \leftarrow X \cup \left\{\left(y^{(1)},f(y^{(1)})\right),\dots,\left(y^{(\mu)},f(y^{(\lambda)})\right)\right\}$;
8 **for** $i=1,\dots,\lambda$ **do** Select $(x,f(x)) \in X$ and update $X \leftarrow X \setminus \{(x,f(x))\}$;

suggested adjusting the unrestricted black-box model to reflect this behavior. In their *memory-restricted model of size* μ,[4] the algorithms can store up to μ pairs $(x,f(x))$ of previous samples. Based only on this information, they decide on the probability distribution D from which the next solution candidate is sampled. Note that this also implies that the algorithm does not have any iteration counter or any other information about the time elapsed so far. Regardless of how many samples have been evaluated already, the sampling distribution D depends only on the μ pairs $\left(x^{(1)},f(x^{(1)})\right),\dots,\left(x^{(\mu)},f(x^{(\mu)})\right)$ stored in the memory.

We extend this model to a $(\mu+\lambda)$ *memory-restricted black-box model*, in which the algorithms have to query λ solution candidates in every round; see Algorithm 3.2 and Fig. 3.2. Following Definition 3.2.1, the $(\mu+\lambda)$ *memory-restricted black-box complexity* of a function class $\mathcal{F}$ is the black-box complexity with respect to the class $\mathcal{A}$ of all $(\mu+\lambda)$ memory-restricted black-box algorithms.

The memory-restricted model of size μ corresponds to the $(\mu+1)$ memory-restricted one, in which only one search point needs to be evaluated per round. Since this variant with $\lambda=1$ allows the highest degree of adaptation, it is not difficult to verify that for all $\mu \in \mathbb{N}$ and for all $\lambda > 1$ the $(\mu+\lambda)$ memory-restricted black-box complexity of a problem $\mathcal{F}$ is at least as large as its $(\mu+1)$ black-box complexity. The effects of larger λ have been studied in a *parallel black-box complexity model*, which we will discuss in Section 3.7.2.

While it seems intuitive that larger memory sizes yield smaller optimization times, this is not necessarily true for all functions. Indeed, the following discussion shows that memory is not needed for the efficient optimization of OneMax.

[4] In the original publication [54], a memory-restricted algorithm of size μ was called a black-box algorithm with *size bound* μ.

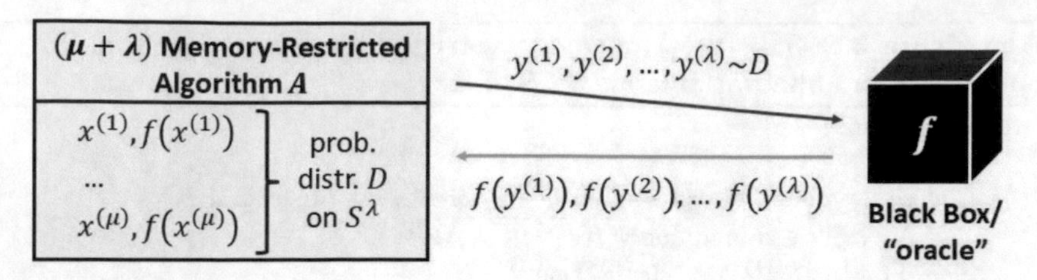

Fig. 3.2 In the $(\mu+\lambda)$ memory-restricted black-box model, the algorithm can store up to μ previously evaluated search points and their absolute function values. In each iteration, it queries the function values of λ new solution candidates. It then has to decide which of the $\mu+\lambda$ search points to keep in the memory for the next iteration.

3.4.1 OneMax

Droste, Jansen, and Wegener conjectured in [54, Section 6] that the $(1+1)$ memory-restricted black-box complexity of OneMax is $O(n \log n)$, in the belief that Randomized Local Search and the (1+1) EA are asymptotically optimal representatives of this class. This conjecture was refuted in [42], where a linear upper bound was presented. This bound was reduced further to $O(n/\log n)$ in [43], showing that even the most restrictive version of the memory-restricted black-box model does not increase the asymptotic complexity of OneMax. By the lower bound presented in Theorem 3.3.3, the $O(n/\log n)$ bound is asymptotically best possible.

Theorem 3.4.1 (Theorem 1 in [43]). *The* $(1+1)$ *memory-restricted black-box complexity of* OneMax *is* $\Theta(n/\log n)$.

The proof of Theorem 3.4.1 makes use of the $O(n/\log n)$ unrestricted strategy of Erdős and Rényi. To respect the memory restriction, the algorithm achieving the $O(n/\log n)$ expected optimization time works in rounds. In every round, a substring of size $s := \sqrt{n}$ of the target string z is identified, using $O(s/\log s)$ queries. The algorithm alternates between querying a string to obtain new information and queries which are used only to store the function value of the last query in the current memory. This works only if sufficiently many bits are available in the guessing string to store this information. It was shown that $O(n/\log n)$ bits suffice. These last remaining $O(n/\log n)$ bits of z are then identified with a constant number of guesses per position, giving an overall expected optimization time of $O(n/s)O(s/\log s)+O(n/\log n)=O(n/\log n)$.

3.4.2 Difference with Respect to the Unrestricted Model

In light of Theorem 3.4.1, it is interesting to find examples for which the $(\mu+1)$ memory-restricted black-box complexity is strictly (and potentially much) larger than its $((\mu+1)+1)$ memory-restricted complexity. This question was addressed in [82].

In the first step, it was shown that having a memory of one can make a decisive difference compared with not being able to store any information at all. In fact, it is easily seen that without any memory, for every function class $\mathcal{F}$ that for every $s \in S$ contains a function f_s such that s is the only optimal solution of f_s, the best one can do without any memory is random sampling, resulting in an expected optimization time of $|S|$. Assume now that there is a (fixed) search point $h \in S$ where a hint is given, in the sense that for all $s \in S$ the objective value $f_s(h)$ uniquely determines where the optimum s is located. Then, clearly, the $(1+1)$ memory-restricted algorithm which first queries h and then, based on $(h, f_s(h))$, queries s solves any problem instance f_s in at most two queries.

This idea can be generalized to a class of functions with two hints hidden in two different distinguished search points h_1 and h_2. Only the combination of $(h_1, f_s(h_1))$ with $(h_2, f_s(h_2))$ defines where the optimum s is located. This way, the $(2+1)$ memory-restricted black-box complexity of this class $\mathcal{F}(h_1,h_2)$ is at most three, while its $(1+1)$ memory-restricted complexity is at least $(S+1)/2$. For, say, $S = \{0,1\}^n$ we thus see that the discrepancies between the $(0+1)$ memory-restricted black-box complexity of a problem $\mathcal{F}$ and its $(1+1)$ memory-restricted complexity can be exponential, and so can be the difference between the $(1+1)$ memory-restricted black-box complexity and the $(2+1)$ memory-restricted complexity. We are not aware of any generalization of this result to arbitrary values of μ.

Theorem 3.4.2 ([82]). *There are classes of functions $\mathcal{F}(h) \subset \{f \mid f : \{0,1\}^n \to \mathbb{R}\}$ and $\mathcal{F}(h_1,h_2) \subset \{f \mid f : \{0,1\}^n \to \mathbb{R}\}$ such that*

- *the $(0+1)$ memory-restricted black-box complexity of $\mathcal{F}(h)$ is exponential in n, while its $(1+1)$ memory-restricted black-box complexity is at most two, and*
- *the $(1+1)$ memory-restricted black-box complexity of $\mathcal{F}(h_1,h_2)$ is exponential in n, while its $(2+1)$ memory-restricted black-box complexity is at most three.*

Storch [82] also presented a class of functions that is efficiently optimized by a standard (2+1) genetic algorithm, which is a $(2+1)$ memory-restricted black-box algorithm, in $O(n^2)$ queries on average, while its $(1+1)$ memory-restricted black-box complexity is exponential in n. This function class is built around so-called royal road functions, the main idea being that the genetic algorithm is guided towards the two "hints" between which the unique global optimum is located.

3.5 Comparison- and Ranking-Based Black-Box Complexity

Many standard black-box heuristics do not take advantage of knowing *exact* objective values. Instead, they use these function values only to rank the search points. This ranking determines the next steps, so that the absolute function values are not needed. Such algorithms are often referred to as *comparison-based* or *ranking-based.* To understand their efficiency *comparison-based* and *ranking-based* black-box complexity models were suggested in [44, 57, 86].

3.5.1 The Ranking-Based Black-Box Model

In the ranking-based black-box model, the algorithms receive a ranking of the search points currently stored in the memory of the population. This ranking is defined by the objective values of these points.

Definition 3.5.1. Let S be a finite set, let $f: S \to \mathbb{R}$ be a function, and let $\mathcal{C}$ be a subset of S. The *ranking* ρ of $\mathcal{C}$ with respect to f assigns to each element $c \in \mathcal{C}$ the number of elements in $\mathcal{C}$ with a smaller f-value plus 1, formally, $\rho(c) := 1 + |\{c' \in \mathcal{C} \mid f(c') < f(c)\}|$.

Note that two elements with the same f-value are assigned the same ranking.

In the ranking-based black-box model without memory restriction, an algorithm thus receives with every query a ranking of *all* previously evaluated solution candidates, while, in the memory-restricted case, naturally, only the ranking of those search points currently stored in the memory is revealed. To be more precise, the ranking-based black-box model without memory restriction subsumes all algorithms that can be described via the scheme of Algorithm 3.3. Fig. 3.3 illustrates these ranking-based black-box algorithms.

Likewise, the $(\mu + \lambda)$ memory-restricted ranking-based model contains those $(\mu + \lambda)$ memory-restricted algorithms that follow the blueprint in Algorithm 3.4; Fig. 3.4 illustrates this pseudocode.

These ranking-based black-box models capture many common search heuristics, such as $(\mu + \lambda)$ evolutionary algorithms, some ant colony optimization algorithms (e.g., simple versions of the Max-Min Ant Systems analyzed in [68, 78]), and Randomized Local Search. They do not include algorithms

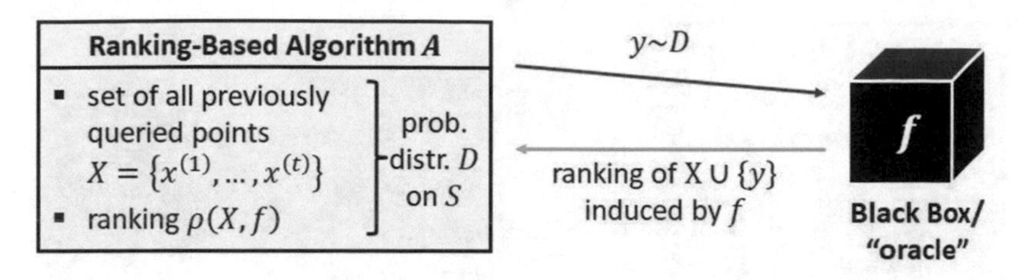

Fig. 3.3 A ranking-based black-box algorithm without memory restriction can store all previously evaluated search points. Instead of knowing their function values, it only has access to the ranking of the search points induced by the objective function f. Based on this, it decides upon a distribution D from which the next search point is sampled.

Algorithm 3.3: Blueprint of a ranking-based black-box algorithm without memory restriction

1 **Initialization:**
2 Sample $x^{(0)}$ according to some probability distribution $D^{(0)}$ over S;
3 $X \leftarrow \{x^{(0)}\}$;
4 **Optimization: for** $t = 1, 2, 3, \ldots$ **do**
5 Depending on $\{x^{(0)}, \ldots, x^{(t-1)}\}$ and its ranking $\rho(X, f)$ with respect to f, choose a probability distribution $D^{(t)}$ on S and sample from it $x^{(t)}$;
6 $X \leftarrow X \cup \{x^{(t)}\}$;
7 Query the ranking $\rho(X, f)$ of X induced by f;

Algorithm 3.4: The $(\mu+\lambda)$ memory-restricted ranking-based black-box algorithm for maximizing an unknown function $f : \{0,1\}^n \to \mathbb{R}$

1 **Initialization:**
2 $X \leftarrow \emptyset$;
3 Choose a probability distribution $D^{(0)}$ over S^μ and sample from it $X = \{x^{(1)}, \ldots, x^{(\mu)}\} \subseteq S$;
4 Query the ranking $\rho(X, f)$ of X induced by f;
5 **Optimization: for** $t = 1, 2, 3, \ldots$ **do**
6 Depending only on the multiset X and the ranking $\rho(X, f)$ of X induced by f choose a probability distribution $D^{(t)}$ on S^λ and sample from it $y^{(1)}, \ldots, y^{(\lambda)}$;
7 Set $X \leftarrow X \cup \{y^{(1)}, \ldots, y^{(\lambda)}\}$ and query the ranking $\rho(X, f)$ of X induced by f;
8 **for** $i = 1, \ldots, \lambda$ **do** Based on X and $\rho(X, f)$ select a (multi)subset Y of X of size μ and update $X \leftarrow Y$;

with fitness-dependent parameter choices, such as fitness-proportional mutation rates or fitness-dependent selection schemes.

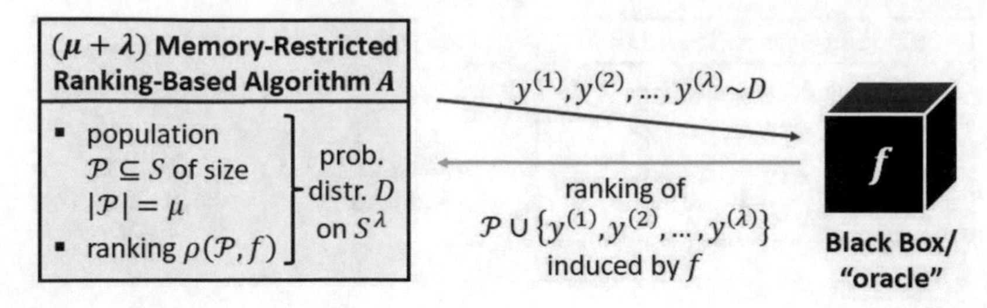

Fig. 3.4 A $(\mu+\lambda)$ memory-restricted ranking-based black-box algorithm can store up to μ previously evaluated search points and the ranking of this population induced by the objective function f. Using only this information, λ new solution candidates are sampled in each iteration and the ranking of the $(\mu+\lambda)$ points is revealed. Based on this ranking, the algorithm needs to select which μ points to keep in the memory.

Surprisingly, the unrestricted and the nonmemory-restricted ranking-based black-box complexities of OneMax coincide in asymptotic terms; the leading constants may be different.

Theorem 3.5.2 (Theorem 6 in [44]). *The ranking-based black-box complexity of* OneMax *without memory restriction is* $\Theta(n/\log n)$.

The upper bound for OneMax is obtained by showing that, for a sufficiently large sample base, a median search point x (i.e., a search point for which half of the search points have a ranking that is at most as large as that of x and the other half of the search points have a ranking that is at least as large as that of x) is very likely to have $n/2$ correct bits. It was shown furthermore that with $O(n/\log n)$ random queries, each of the function values in the interval $[n/2-\kappa\sqrt{n}, n/2+\kappa\sqrt{n}]$ appears at least once. This information is used to translate the ranking of the random queries into absolute function values, for those solution candidates y for which $\mathrm{OM}_z(y)$ lies in the interval $[n/2-\kappa\sqrt{n}, n/2+\kappa\sqrt{n}]$. The proof is then concluded by showing that it suffices to consider only these samples in order to identify the target string z of the problem instance OM_z.

For BinaryValue, in contrast, it makes a substantial difference whether absolute or relative objective values are available.

Theorem 3.5.3 (Theorem 17 in [44]). *The ranking-based black-box complexity of* BinaryValue_n *and* BinaryValue^*_n *is strictly larger than* $n-1$, *even when the memory is not bounded.*

This lower bound of $n-1$ is almost tight. In fact, an $n+1$ ranking-based algorithm is easily obtained by starting with a random initial search point and then, from left to right, flipping exactly one bit in each iteration. The ranking uniquely determines the permutation σ and the string z of the problem instance $\mathrm{BV}_{z,\sigma}$.

Theorem 3.5.3 can be shown with Yao's minimax principle applied to the uniform distribution over the problem instances. The crucial observation is that when optimizing $\mathrm{BV}_{z,\sigma}$ with a ranking-based algorithm, from t samples we can learn at most $t-1$ bits of the hidden bit string z, and not $\Theta(t \log t)$ bits as one might guess from the fact that there are $t!$ permutations of the set $[t]$.

This last intuition, however, gives a very general lower bound. Intuitively, if $\mathcal{F}$ is such that every $z \in \{0,1\}^n$ is the unique optimum for a function $f_z \in \mathcal{F}$, and we learn only the ranking of the search points evaluated so far, then for the t-th query we learn at most $\log_2(t!) = \Theta(t \log t)$ bits of information. Since we need to learn n bits in total, the ranking-based black-box complexity of $\mathcal{F}$ is of order at least $n/\log n$.

Theorem 3.5.4 (Theorem 21 in [44]). *Let $\mathcal{F}$ be a class of functions such that each $f \in \mathcal{F}$ has a unique global optimum and such that for all $z \in \{0,1\}^n$ there exists a function $f_z \in \mathcal{F}$ with $\{z\} = \arg\max f_z$. Then the unrestricted ranking-based black-box complexity of $\mathcal{F}$ is $\Omega(n/\log n)$.*

Results for the ranking-based black-box complexity of the two combinatorial problems MST and SSSP have been derived in [37]. Some of these bounds were mentioned in Section 3.3.8.

3.5.2 The Comparison-Based Black-Box Model

In the ranking-based model, the algorithms receive for every query quite a lot of information, namely the full ranking of the current population and its offspring. One may argue that some evolutionary algorithms use even less information. Instead of considering the full ranking, they base their decisions on a few selected points only. This idea is captured in the *comparison-based black-box model*. In contrast to the ranking-based model, here only the ranking of the queried points is revealed. In this model it can therefore make sense to query a search point more than once, to compare it with a different offspring, for example. Fig. 3.5 illustrates the $(\mu+\lambda)$ memory-restricted comparison-based black-box model. A comparison-based model without memory restriction is obtained by setting $\mu = \infty$.

We will not detail this model further, as it has received only marginal attention so far in the black-box complexity literature. We note, however, that Teytaud and co-authors [57, 86] have presented some very general lower bounds for the convergence rate of comparison-based and ranking-based evolutionary strategies in continuous domains. From these studies, results for the comparison-based black-box complexity of problems defined over discrete domains can be obtained. These bounds, however, seem to coincide with the information-theoretic ones that can be obtained through Theorem 3.2.4.

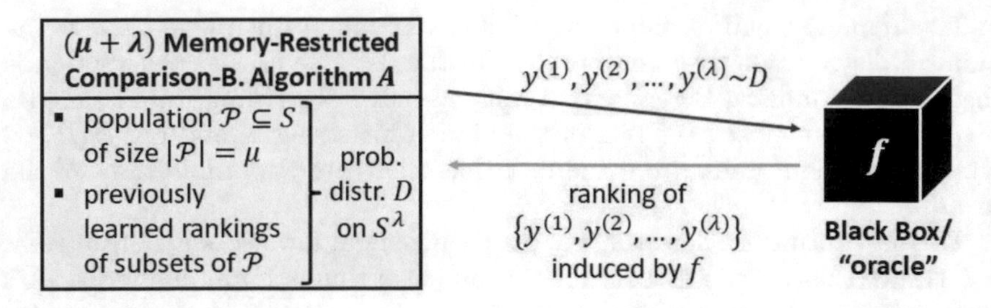

Fig. 3.5 A $(\mu+\lambda)$ memory-restricted comparison-based black-box algorithm can store up to μ previously evaluated search points and the comparison of these points that have been learned through previous queries. In the next iteration, λ solution candidates are queried, possibly containing some of the current population. Only the ranking of the μ queried points is revealed. Based on this ranking and the previous information about the relative fitness values, the algorithm needs to select which μ points to keep in the memory.

3.6 Unbiased Black-Box Complexity

As previously commented, the quest to develop a meaningful complexity theory for evolutionary algorithms and other black-box optimization heuristics seemed to have come to an early end after 2006, the only publication which picked up on this topic being that of Anil and Wiegand on the unrestricted black-box complexity of OneMax [2] (see Section 3.3.2). In 2010, the situation changed drastically. Black-box complexity was revived by Lehre and Witt in [71] (a journal version appeared as [72]). To overcome the drawbacks of the previous unrestricted black-box model, they restricted the class of black-box optimization algorithms in a natural way that still covers a large class of the classically used algorithms.

In their *unbiased black-box complexity model*, Lehre and Witt considered pseudo-Boolean optimization problems $\mathcal{F} \subseteq \{f : \{0,1\}^n \to \mathbb{R}\}$. The unbiased black-box model requires that all solution candidates must be sampled from distributions that are *unbiased*. In the context of pseudo-Boolean optimization, unbiasedness means that the distribution cannot discriminate between bit positions $1, 2, \ldots, n$ nor between the bit entries 0 and 1; a formal definition will be given in Sections 3.6.1 and 3.6.2. The unbiased black-box model also admits a notion of arity. A k-ary unbiased black-box algorithm is one that employs only such variation operators that take up to k arguments. This allows one, for example, to talk about mutation-only algorithms (unary unbiased algorithms) and to study the potential benefits of recombining previously sampled search points through distributions of higher arity.

In a crucial difference from the memory-restricted model, in the pure version of the unbiased black-box model the memory is not restricted. That is, the k search points that form the input for the k-ary variation operator can be any random or previously evaluated solution candidate. As in the case of the comparison- and the ranking-based black-box models, combined unbiased memory-restricted models have also been studied; see Section 3.7.

Before we formally introduce unbiased black-box models for pseudo-Boolean optimization problems in Section 3.6.2, we define and discuss in Section 3.6.1 the concept of unbiased variation operators. Known black-box complexities in the unbiased black-box models are surveyed in Section 3.6.3. In Section 3.6.4 we present extensions of the unbiased black-box models to search spaces different from $\{0,1\}^n$.

3.6.1 Unbiased Variation Operators

In order to formally define the unbiased black-box model, we first introduce the notion of *k-ary unbiased variation operators.* Informally, a k-ary unbiased variation operator takes as input up to k search points. It samples a new point $z \in \{0,1\}^n$ by applying some procedure to these previously evaluated solution candidates that treats all bit positions and the two bit values in an equal way.

Definition 3.6.1 (k-ary unbiased variation operator). Let $k \in \mathbb{N}$. A *k-ary unbiased distribution* $(D(. \mid y^{(1)},\ldots,y^{(k)}))_{y^{(1)},\ldots,y^{(k)} \in \{0,1\}^n}$ is a family of probability distributions over $\{0,1\}^n$ such that for all inputs $y^{(1)},\ldots,y^{(k)} \in \{0,1\}^n$ the following two conditions hold:

$$(i)\ \forall x,z \in \{0,1\}^n : D(x \mid y^{(1)},\ldots,y^{(k)}) = D(x \oplus z \mid y^{(1)} \oplus z,\ldots,y^{(k)} \oplus z)\,,$$

$$(ii)\ \forall x \in \{0,1\}^n\, \forall \sigma \in S_n : D(x \mid y^{(1)},\ldots,y^{(k)}) = D(\sigma(x) \mid \sigma(y^{(1)}),\ldots,\sigma(y^{(k)}))\,.$$

We refer to the first condition as *$\oplus$-invariance* and to the second as *permutation invariance.* A variation operator that creates an offspring by sampling from a k-ary unbiased distribution is called a *k-ary unbiased variation operator.*

To get some intuition about unbiased variation operators, we now summarize a few characterizations and consequences of Definition 3.6.1.

We first note that the combination of $\oplus$- and permutation invariance can be characterized as invariance under Hamming automorphisms. A Hamming automorphism is a one-to-one map $\alpha : \{0,1\}^n \to \{0,1\}^n$ that satisfies the condition that for any two points $x,y \in \{0,1\}^n$ their Hamming distance $H(x,y)$ is equal to the Hamming distance $H(\alpha(x),\alpha(y))$ of their images. A formal proof of the following lemma can be found in [37, Lemma 3].

Lemma 3.6.2. *A distribution $D(\cdot \mid x^1,\ldots,x^k)$ is unbiased if and only if, for all Hamming automorphisms $\alpha : \{0,1\}^n \to \{0,1\}^n$ and for all bit strings $y \in \{0,1\}^n$, the probability $D(y \mid x^1,\ldots,x^k)$ of sampling y from $(x^1,\ldots,x^k)$ equals the probability $D(\alpha(y) \mid \alpha(x^1),\ldots,\alpha(x^k))$ of sampling $\alpha(y)$ from $(\alpha(x^1),\ldots,\alpha(x^k))$.*

It is not difficult to see that the only 0-ary unbiased distribution over $\{0,1\}^n$ is the uniform one.

1-ary operators, also called *unary* operators, are sometimes referred to as *mutation operators,* in particular in the field of evolutionary computation. Standard bit mutation, as used in several $(\mu+\lambda)$ EAs and $(\mu+\lambda)$ EAs, is a unary unbiased variation operator. The random bit flip operation used by RLS, which chooses at random a bit position $i \in [n]$ and replaces the entry x_i by the value $1-x_i$, is also unbiased. In fact, all unary unbiased variation operators are of a very similar type, as the following definition and lemma, taken from [31], show. These results can be derived from a more general description of unbiased operators offered in [37], which characterizes unbiased operations on arbitrary search spaces. When restricted to pseudo-Boolean optimization, we obtain the following geometric interpretation.

Definition 3.6.3. Let $n \in \mathbb{N}$ and $r \in [0..n]$. For every $x \in \{0,1\}^n$, let flip_r be the variation operator that creates an offspring y from x by selecting r positions $i_1,\ldots,i_r$ in $[n]$ uniformly at random (without replacement), setting $y_i := 1-x_i$ for $i \in \{i_1,\ldots,i_r\}$, and copying $y_i := x_i$ for all other bit positions $i \in [n] \setminus \{i_1,\ldots,i_r\}$.

Using this definition, unary unbiased variation operators can be characterized as follows.

Lemma 3.6.4 (Lemma 1 in [31]). *For every unary unbiased variation operator $(p(\cdot|x))_{x\in\{0,1\}^n}$, there exists a family of probability distributions $(r_{p,x})_{x\in\{0,1\}^n}$ on $[0..n]$ such that for all $x,y \in \{0,1\}^n$ the probability $p(y|x)$ that $(p(\cdot|x))_{x\in\{0,1\}^n}$ samples y from x equals the probability that the routine first samples a random number r from $r_{p,x}$ and then obtains y by applying flip_r to x. On the other hand, each such family of distributions $(r_{p,x})_{x\in\{0,1\}^n}$ on $[0..n]$ induces a unary unbiased variation operator.*

From this characterization, we can easily see that neither the somatic contiguous hypermutation operator used in artificial immune systems (which selects a random position $i \in [n]$ and a random length $\ell \in [n]$ and flips the ℓ consecutive bits in positions $i, i+1 \mod n, \ldots, i+\ell \mod n$; see [15, Algorithm 3]), nor the asymmetric nor the position-dependent mutation operators considered in [63] and [12, 27], respectively, are unbiased.

2-ary operators, also called *binary* operators, are often referred to as *crossover operators.* A prime example of a binary unbiased variation operator is *uniform crossover*. Given two search points x and y, the uniform crossover operator creates an offspring z from x and y by choosing, independently for each index $i \in [n]$, the entry $z_i \in \{x_i, y_i\}$ uniformly at random. In contrast, the standard *one-point crossover operator* - which, given two search points $x, y \in \{0,1\}^n$ picks uniformly at random an index $k \in [n]$ and outputs from x and y one or both of the two offspring $x' := x_1 \ldots x_k y_{k+1} \ldots y_n$ and $y' := y_1 \ldots y_k x_{k+1} \ldots x_n$ - publicationsis not permutation-invariant, and therefore not an unbiased operator.

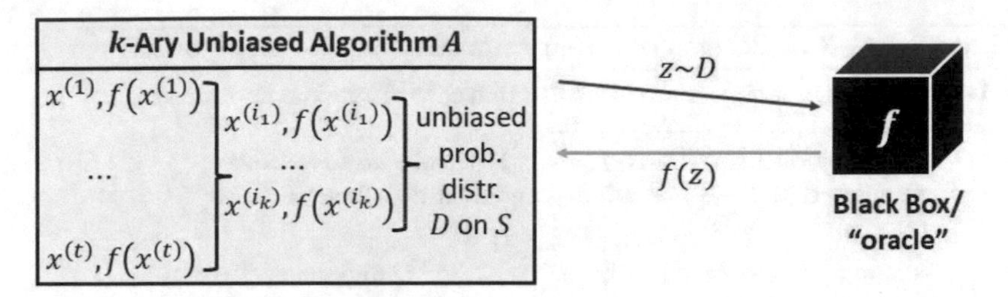

Fig. 3.6 In the k-ary unbiased black-box model, the algorithm can store the full query history. For every already evaluated search point x, the algorithm has access to the absolute function value $f(x) \in \mathbb{R}$. The distributions D from which new solution candidates are sampled have to be unbiased. They can depend on up to k previously evaluated solution candidates.

Some publications refer to the unbiased black-box model allowing variation operators of arbitrary arity as the *$*$-ary unbiased black-box model.* Black-box complexities in the $*$-ary unbiased black-box model are of the same asymptotic order as those in the unrestricted model. This has been formally shown in [81], for a general notion of unbiasedness that is not restricted to pseudo-Boolean optimization problems (see Definition 3.6.16).

Theorem 3.6.5 (Corollary 1 in [81]). *The $*$-ary unbiased black-box complexity of a problem class $\mathcal{F}$ is the same as its unrestricted black-box complexity.*

Apart from [81], most research on the unbiased black-box model assumes a restriction on the arity of the variation operators. We therefore concentrate in the remainder of this chapter on such restricted settings.

3.6.2 The Unbiased Black-Box Model for Pseudo-Boolean Optimization

With Definition 3.6.1 and its characterizations at hand, we can now introduce the unbiased black-box models. The k-ary unbiased black-box model covers all algorithms that follow the blueprint of Algorithm 3.5. Fig. 3.6 illustrates these algorithms. As in previous sections, the *k-ary unbiased black-box complexity* of some class of functions $\mathcal{F}$ is the complexity of $\mathcal{F}$ with respect to all k-ary unbiased black-box algorithms.

As Fig. 3.6 indicates, it is important to note that in line 3 of Algorithm 3.5 the k selected previously evaluated search points $x^{(i_1)},\ldots,x^{(i_k)}$ do not necessarily have to be the k *immediately previously* queried ones. That is, the algorithm can store and is allowed to choose from *all* previously sampled search points.

Algorithm 3.5: Scheme of a k-ary unbiased black-box algorithm

1 **Initialization:** Sample $x^{(0)} \in \{0,1\}^n$ uniformly at random and query $f(x^{(0)})$;
2 **Optimization: for** $t = 1,2,3,\dots$ **do**
3   Depending on $\left(f(x^{(0)}),\dots,f(x^{(t-1)})\right)$ choose up to k indices $i_1,\dots,i_k \in [0..t-1]$ and a k-ary unbiased distribution $(D(\cdot \mid y^{(1)},\dots,y^{(k)}))_{y^{(1)},\dots,y^{(k)} \in \{0,1\}^n}$;
4   Sample $x^{(t)}$ according to $D(\cdot \mid x^{(i_1)},\dots,x^{(i_k)})$ and query $f(x^{(t)})$;

Note further that for all $k \le \ell$, each k-ary unbiased black-box algorithm is contained in the ℓ-ary unbiased black-box model. This is due to the fact that we do not require the indices to be pairwise distinct.

The unary unbiased black-box model captures most of the commonly used mutation-based algorithms, such as the $(\mu+\lambda)$ EA and the (μ,λ) EA, Simulated Annealing, the Metropolis algorithm, and Randomized Local Search. The binary unbiased model subsumes many traditional genetic algorithms, such as the $(\mu+\lambda)$ GAs and the (μ,λ) GAs using uniform crossover. As we shall discuss in Section 3.9, the $(1+(\lambda,\lambda))$ GA introduced in [25] is also binary unbiased.

As a word of warning, we note that in [85] and [87] lower bounds are proven for what the authors of those publications call *mutation-based algorithms*. Their definitions are more restrictive than what Algorithm 3.5 proposes. The lower bounds proven in [85, 87] therefore do not (immediately) apply to the unary unbiased black-box model. A comparison of Theorem 12 in [85] and Theorem 3.1(5) in [87] with Theorem 9 in [31] shows that there can be substantial differences (in this case, a multiplicative factor $\approx e$ in the lower bound for the complexity of OneMax with respect to all mutation-based and all unary unbiased black-box algorithms, respectively). One of the main differences between the different models is that in [85, 87] only algorithms using standard bit mutation are considered. This definition excludes algorithms such as RLS for which the radius r fed into the variation operator flip_r is deterministic and is thus not sampled from a binomial distribution $\text{Bin}(n,p)$. When using the term "mutation-based algorithms," we should therefore always make precise which algorithmic framework we are referring to. Here, in this chapter, we will exclusively refer to the unary unbiased black-box algorithms defined via Algorithm 3.5.

3.6.3 Existing Results for Pseudo-Boolean Problems

In this section we survey existing bounds for the unbiased black-box complexity of several classical benchmark functions. As in previous sections, we proceed by function class, and not in historical order.

3.6.3.1 Functions with a Unique Global Optimum

As discussed in Section 3.2.3.1, the unrestricted black-box complexity of every function class $\mathcal{F} = \{f\}$ containing only one function f is one, certified by the algorithm that simply queries a point $x \in \arg\max f$ in the first step. The situation is different in the unbiased black-box model, as the following theorem reveals.

Theorem 3.6.6 (Theorem 6 in [71]). *Let $f : \{0,1\}^n \to \mathbb{R}$ be a function that has a single global optimum (i.e., in the case of maximization, the size of the set $\arg\max f$ is one). The unary unbiased black-box complexity of f is $\Omega(n \log n)$.*

Theorem 3.6.6 gives an $\Omega(n \log n)$ lower bound on the unary unbiased black-box complexity of several standard benchmark functions, such as OneMax and LeadingOnes. We shall see below that for some of these classes, including OneMax, this bound is tight, since it is met by different unary unbiased heuristics, such as the (1+1) EA or RLS. For other classes, including LeadingOnes, the lower bound can be improved through problem-specific arguments.

The proof of Theorem 3.6.6 uses multiplicative drift analysis. To this end, the potential $P(t)$ of an algorithm at time t is defined as the smallest Hamming distance from any of the previously queried search points $x^{(1)},\dots,x^{(t)}$ to the unique global optimum z or its bitwise complement $\bar{z}$. The algorithm has identified z (or its complement) if and only if $P(t) = 0$. The distance to $\bar{z}$ needs to be considered, as the algorithm that first identifies $\bar{z}$ and then flips all bits obtains z from $\bar{z}$ in only one additional query. As we have discussed for jump functions in Section 3.3.7, for some functions it can be substantially easier to identify $\bar{z}$ than to identify z itself. This is true in particular if there are paths leading to $\bar{z}$, such as in the original jump functions $f_{\ell,z}$ discussed in Section 3.3.7.3. The key step in the proof of Theorem 3.6.6 is to show that in one iteration $P(t)$ decreases by at most $200P(t)/n$, in expectation, provided that $P(t)$ is between $c \log\log n$ (for some positive constant $c > 0$) and $n/5$. Put differently, in this case $E[P(t) - P(t+1) \mid P(t)] \le \delta P(t)$ for $\delta := 200/n$. It can be shown, furthermore, that the probability of making very large gains in potential is very small. These two statements allow the application of a multiplicative drift theorem, which bounds the total expected optimization time by $\Omega((\log(n/10) - \log\log(n))/\delta) = \Omega(n \log n)$, provided that the algorithm reaches a state t with $P(t) \in (n/10, n/5]$. A short proof that every unary unbiased black-box algorithm reaches such a state with probability $1 - e^{-\Omega(n)}$ then concludes the proof of Theorem 3.6.6.

3.6.3.2 OneMax

The Unary Unbiased Black-Box Complexity of OneMax

Since OneMax is a unimodal function, the lower bound of Theorem 3.6.6 certainly applies to that function, thus showing that no unary unbiased black-box optimization can optimize OneMax faster than in expected time $\Omega(n \log n)$. This bound is attained by several classical heuristics, such as RLS, the (1+1) EA, and others. While the (1+1) EA has an expected optimization time of $(1 \pm o(1))en\ln(n)$ [32, 85], that of RLS is only $(1 \pm o(1))n\ln(n)$. More precisely, it is $n\ln(n/2) + \gamma n \pm o(1)$ [23], where $\gamma \approx 0.5772156649\ldots$ is the Euler-Mascheroni constant. The unary unbiased black-box complexity of OneMax is just slightly smaller than this term. It was slightly improved by an additive $\sqrt{n \log n}$ term in [69] through iterated initial sampling. In [31], the following very precise bound for the unary unbiased black-box complexity of OneMax was shown. It is smaller than the expected running time of RLS by an additive term that is between $0.138n \pm o(n)$ and $0.151n \pm o(n)$. It was also proven in [31] that a variant of RLS that uses fitness-dependent neighborhood structures attains this optimal bound, up to additive $o(n)$ lower-order terms.

Theorem 3.6.7 (Theorem 9 in [31]). *The unary unbiased black-box complexity of* OneMax *is* $n\ln(n) - cn \pm o(n)$ *for a constant* c *between* 0.2539 *and* 0.2665.

The Binary Unbiased Black-Box Complexity of OneMax

When Lehre and Witt initially defined the unbiased black-box model, they conjectured that also the binary black-box complexity of OneMax was $\Omega(n \log n)$ [P.K. Lehre, personal communication in 2010]. In light of our understanding of the role and usefulness of crossover in black-box optimization, such a bound would have indicated that crossover cannot be beneficial for simple hill-climbing tasks. Given that in 2010 all results seemed to indicate that it was at least very difficult, if not impossible, to rigorously prove any advantages of crossover for problems with smooth fitness landscapes, this conjecture came along very naturally. It was, however, soon refuted. In [35], a binary unbiased algorithm was presented that achieves linear expected running time on OneMax.

Theorem 3.6.8 (Theorem 9 in [35]). *The binary unbiased black-box complexity of* OneMax *and that of any other monotone function is at most linear in the problem dimension* n.

This bound is attained by the algorithm that keeps in the memory two strings x and y that agree in those positions for which the optimal entry has been identified already, and which differ in all other positions. In every iteration, the algorithm flips a fair random coin and, depending on the outcome of this coin flip, flips exactly one bit in x or one bit in y. The bit to be flipped is chosen uniformly at random from those bits in which x and y disagree. The offspring so created replaces its parent if and only if its function value is larger. In this case, the Hamming distance between x and y reduces by one. Since the probability of choosing the right parent equals $1/2$, it is not difficult to show that, with high probability, for all constants $\varepsilon > 0$, this algorithm optimizes ONEMAX within at most $(2+\varepsilon)n$ iterations. Together with Lemma 3.2.8, this proves Theorem 3.6.8.

A drawback of this algorithm is that it is very problem-specific, and it has been an open question whether or not a "natural" binary evolutionary algorithm can achieve an $o(n \log n)$ (or better) expected running time on ONEMAX. This question was answered affirmatively in [25] and [21], as we shall discuss in Section 3.9.

Whether or not the linear bound in Theorem 3.6.8 is tight remains an open problem. In general, proving lower bounds for unbiased black-box models of arities larger than one remains one of the biggest challenges in black-box complexity theory. Owing to the greatly enlarged computational power of black-box algorithms using higher-arity operators, proving lower bounds in these models seems to be significantly harder than in the unary unbiased model. As a matter of fact, the best lower bound that we have for the binary unbiased black-box complexity of ONEMAX is the $\Omega(n/\log n)$ one stated in Theorem 3.3.3, and not even constant-factor improvements of this bound exist.

The k-Ary Unbiased Black-Box Complexity of ONEMAX

In [35], a general bound for the k-ary unbiased black-box complexity of ONEMAX of order $n/\log k$ was presented (see Theorem 9 in [35]). This bound has been improved in [45].

Theorem 3.6.9 (Theorem 3 in [45]). *For every $2 \leq k \leq \log n$, the k-ary unbiased black-box complexity of* ONEMAX *is of order at most n/k. For $k \geq \log n$, it is $\Theta(n/\log n)$.*

Note that for $k \geq \log n$, the lower bound in Theorem 3.6.9 follows from the $\Omega(n/\log n)$ unrestricted black-box complexity of ONEMAX discussed in Theorem 3.3.3.

The main idea used to achieve the results of Theorem 3.6.9 can be easily described. For a given k, the bit string is split into blocks of size $k-2$. This has to be done in an unbiased way, so that the "blocks" are not consecutive bit positions, but some random $k-2$ positions not previously optimized. Similarly to the binary case, two reference strings x and y are used to encode which $k-2$ bit positions are currently under investigation, namely the $k-2$

bits in which x and y disagree. Using the same encoding, two other strings x' and y' store which bits have been optimized already, and which ones have not been investigated so far. To optimize the $k-2$ bits in which x and y differ, the derandomized version of the result of Erdős and Rényi (Theorem 3.3.4) is used. Applied in our context, this result states that there exists a sequence of $\Theta(k/\log k)$ queries which uniquely determines the entries in the $k-2$ positions. Since $\Theta(n/k)$ such blocks need to be optimized, the claimed total expected optimization time of $\Theta(n/\log k)$ follows. Some technical difficulties need to be overcome to implement this strategy in an unbiased way. To this end, in [45] a generally applicable *encoding strategy* was presented that with k-ary unbiased variation operators simulates a memory of 2^{k-2} bits that can be accessed in an unrestricted fashion.

3.6.3.3 LeadingOnes

The Unary Unbiased Black-Box Complexity of LeadingOnes

Since LeadingOnes is a classic benchmark problem, unsurprisingly, Lehre and Witt had already presented in [72] a first bound for the unbiased black-box complexity of this function.

Theorem 3.6.10 (Theorem 2 in [72]). *The unary unbiased black-box complexity of* LeadingOnes *is* $\Theta(n^2)$.

Theorem 3.6.10 can be proven by drift analysis. To this end, in [72] a potential function was defined that maps the state of the search process at time t (i.e., the sequence $\{(x^{(1)}, f(x^{(1)})), \ldots, (x^{(t)}, f(x^{(t)}))\}$ of the pairs of search points evaluated so far and their respective function values) to the largest number of initial ones and initial zeros in any of the $t+1$ strings $x^{(1)}, \ldots, x^{(t)}$. It was then shown that a given potential k cannot increase in one iteration by more than an additive term $4/(k+1)$, in expectation, provided that k is at least $n/2$. Since with probability at least $1-e^{-\Omega(n)}$ any unary unbiased black-box algorithm reaches a state in which the potential is between $n/2$ and $3n/4$, and since from this state a total potential of at least $n/4$ must be gained, the claimed $\Omega(n^2)$ bound follows from a variant of the additive drift theorem. More precisely, using these bounds, the additive drift theorem shows that the total optimization time of any unary unbiased black-box algorithm is at least $(n/4)/\big(4/(n/2)\big) = \Omega(n^2)$.

The Binary Unbiased Black-Box Complexity of LeadingOnes

Similarly to the case of OneMax, the binary unbiased black-box complexity of LeadingOnes is much smaller than its unary counterpart.

Theorem 3.6.11 (Theorem 14 in [35]). *The binary unbiased black-box complexity of* LEADINGONES *is* $O(n \log n)$.

The algorithm achieving this bound borrows its main idea from the binary unbiased algorithm used to optimize ONEMAX in linear time, which we described after Theorem 3.6.8. We recall that the key strategy was to use two strings to encode those bits that have been optimized already. In the $O(n \log n)$ algorithm for LEADINGONES, this approach is combined with a *binary search* for the (unique) bit position that needs to be flipped next. Such a binary search step requires $O(\log n)$ steps in expectation. Iterating it n times gives the claimed $O(n \log n)$ bound.

As in the case of ONEMAX, it is not known whether or not the bound in Theorem 3.6.11 is tight. The best known lower bound is the $\Omega(n \log\log n)$ one for the unrestricted black-box model discussed in Theorem 3.3.12.

The Complexity of LEADINGONES in Unbiased Black-Box Models of Higher Arity

The $O(n \log n)$ bound presented in Theorem 3.6.11 reduces further to at most $O(n \log(n)/\log\log n)$ in the ternary unbiased black-box model.

Theorem 3.6.12 (Theorems 2 and 3 in [41]). *For every $k \geq 3$, the k-ary unbiased black-box complexity of* LEADINGONES *is of order at most $n \log(n)/\log\log n$. This bound also holds in the combined k-ary unbiased ranking-based black-box model, in which instead of absolute function values the algorithm can make use only of the ranking of the search points induced by the optimization problem instance f.*

The algorithm that certifies the upper bound in Theorem 3.6.12 uses the additional power gained through the larger arity to *parallelize* the binary search of the binary unbiased algorithm described after Theorem 3.6.11. More precisely, the optimization process is split into phases. In each phase, the algorithm identifies the entries of up to $k := O(\sqrt{\log n})$ positions. It can be shown that each phase takes $O(k^3/\log k^2)$ steps in expectation. Since there are n/k phases, a total expected optimization time of $O(nk^2/\log k^2) = O(n \log(n)/\log\log n)$ follows.

The idea of parallelizing the search for several indices was later taken up and further developed in [1], where an iterative procedure with *overlapping* phases was used to derive the asymptotically optimal $\Theta(n \log\log n)$ unrestricted black-box algorithm that proves Theorem 3.3.12.

It seems plausible that higher arities allow a larger degree of parallelization, but no formal proof of this intuition exists. In the context of LEADINGONES, it would be interesting to derive a lower bound on the smallest value of k such that an asymptotically optimal k-ary unbiased $\Theta(n \log\log n)$ black-box algorithm for LEADINGONES exists. As a first step towards answering this question, the encoding and sampling strategies sketched above could be applied to the algorithm presented in [1], to understand the smallest arity needed to implement this algorithm in an unbiased way.

3.6.3.4 Jump Functions

Jump functions are benchmark functions, which are observed to be difficult for evolutionary approaches because of their large plateaus of constant and low fitness around the global optimum. One would expect that this would be reflected in the unbiased black-box complexity, at least in the unary model. Surprisingly, this is not the case. In [28], it was shown that even extreme jump functions that reveal only the three different fitness values 0, $n/2$, and n have a small polynomial unary unbiased black-box complexity. That is, they can be optimized quite efficiently by unary unbiased approaches. This result indicates that efficient optimization is not necessarily restricted to problems for which the function values reveal a lot of information about the instance at hand.

As discussed in Section 3.3.7, the literature is not unanimous with respect to how to generalize the jump function defined in [53] to a problem class. The results stated in the following apply to the jump function defined in (3.3.1). In the unbiased black-box model, we can assume without loss of generality that the underlying target string is the all-ones string $(1,\ldots,1)$. That is, to simplify our notation, we drop the subscript z and assume that for every $\ell < n/2$ we consider the function that assigns to every $x \in \{0,1\}^n$ the function value

$$\textsc{Jump}_\ell(x) := \begin{cases} n & \text{if } |x|_1 = n, \\ |x|_1 & \text{if } \ell < |x|_1 < n-\ell, \\ 0 & \text{otherwise.} \end{cases}$$

The results in [28] cover a broad range of different combinations of jump sizes ℓ and arities k.

Arity	Short jump $\ell = O(n^{1/2-\varepsilon})$	Long jump $\ell = (1/2-\varepsilon)n$	Extreme jump $\ell = n/2-1$
$k=1$	$\Theta(n\log n)$	$O(n^2)$	$O(n^{9/2})$
$k=2$	$O(n)$	$O(n\log n)$	$O(n\log n)$
$3 \le k \le \log n$	$O(n/k)$	$O(n/k)$	$\Theta(n)$

Table 3.1 Known bounds for the unbiased black-box complexity of $\textsc{Jump}_\ell$

Theorem 3.6.13 ([28]). *Table 3.1 summarizes the known bounds for the unbiased black-box complexity of* $\textsc{Jump}_\ell$ *in the different models.*

To discuss the bounds in Theorem 3.6.13, we proceed by problem type. Almost all proofs are rather involved, and so we sketch here only the main ideas.

Short Jumps, i.e., $\ell = O(n^{1/2-\varepsilon})$

A comparison with the bounds discussed in Section 3.6.3.2 shows that the bounds for the k-ary unbiased black-box complexities of short jump functions stated above are of the same order as those for OneMax (which can be seen as a jump function with parameter $\ell = 0$). In fact, it was shown in [28, Lemma 3] that a black-box algorithm having access to a jump function with $\ell = O(n^{1/2-\varepsilon})$ can retrieve (with high probability) the true OneMax value of a search point using only a constant number of queries. The other direction is of course also true, since from the OneMax value we can compute the Jump_ℓ value without further queries. This implies that the black-box complexities of short jump functions are of the same asymptotic order as those of OneMax. Any improved bound for the k-ary unbiased black-box complexity OneMax therefore immediately carries over to short jump functions.

The fact that the $\Theta(n \log n)$ bound for the unary unbiased black-box complexity carries over to so-called Plateau functions, which assign to all sub-optimal solutions of Hamming distance at most ℓ the same function value $n - \ell - 1$ and are identical to Jump otherwise, has been discussed in [3].

Long Jumps, i.e., $\ell = (1/2 - \varepsilon)n$

Despite the fact that the above-mentioned Lemma 3 in [28] can probably not be directly extended to long jump functions, the bounds for arities $k \geq 3$ nevertheless coincide with those for OneMax. In fact, it was shown in [28, Theorem 6] that for all $\ell < (1/2 - \varepsilon)n$ and for all $k \geq 3$ the k-ary unbiased black-box complexity of Jump_ℓ is at most of the same asymptotic order as the $(k-2)$-ary unbiased black-box complexity of OneMax. For $k > 3$, this proves the bounds stated in Theorem 3.6.13. The linear bound for $k = 3$ follows from the case of extreme jumps.

A key ingredient for the bound on the *unary* unbiased black-box complexity of long jump functions is a procedure that samples a number of neighbors at some fixed distance d and studies the empirical expected function values of these neighbors to decide upon the direction in which the search for the global optimum is continued. More precisely, it uses the samples to estimate the OneMax value of the currently investigated search point. Strong concentration bounds are used to bound the probability that this approach gives an accurate estimation of the correct OneMax values.

The $O(n \log n)$ bound for the *binary* unbiased black-box complexity of long jump functions follows from its extreme analogue.

Extreme Jump, i.e., $\ell = n/2 - 1$

The work [28] first considered the *ternary* unbiased black-box complexity of the extreme jump function. A strategy that allowed individual bits to be tested was derived. Testing each bit individually in an efficient way (using the encoding strategies originally developed in [35] and described in Section 3.6.3.2 above) gives the linear bound.

In the *binary* case, the bits cannot be tested as efficiently anymore. The main idea is nevertheless to flip individual bits and to test whether the flip was in a "good" or a "bad" direction. This test is done by estimating the distance to a reference string with $n/2$ ones. Implementing this strategy in $O(n \log n)$ queries requires one to overcome a few technical difficulties imposed by the restriction of sampling only from binary unbiased distributions, resulting in a rather complex bookkeeping procedure, and a rather technical proof 4.5 pages long.

Finally, the polynomial unary unbiased black-box complexity of the extreme jump function can be proven as follows. Similarly to the cases discussed above, individual bits are flipped in a current "best" solution candidate x. A sampling routine is used to estimate whether the bit flip was in a "good" or a "bad" direction, i.e., whether it created a string that was closer to the global optimum or to its bitwise complement than the previous string. The sampling strategy works as follows. Depending on the estimated parity of $|x|_1$, exactly $n/2$ or $n/2 - 1$ bits are flipped in x. The fraction of offspring so created with function value $n/2$ (the only value that is "visible" apart from that of the global optimum) is recorded. This fraction depends on the distance from x to the global optimum $(1,\dots,1)$ or its complement $(0,\dots,0)$ and is slightly different for different distances. A key step in the analysis of the unary unbiased black-box complexity of the extreme jump function is therefore a proof that shows that a polynomial number of such samples are sufficient to determine the OneMax value of x with sufficiently large probability.

Comments on the Upper Bounds in Theorem 3.6.13

Note that even for long jump functions, the search points having a function value of 0 form plateaus around the optimum $(1,\dots,1)$ and its complement $(0,\dots,0)$ of exponential size. For the extreme jump function, all but a $\Theta(n^{-1/2})$ fraction of the search points form one single fitness plateau. Problem-unspecific black-box optimization techniques will therefore typically not find the optimum of long and extreme jump functions in subexponential time.

Lower Bound

The $\Omega(n \log n)$ lower bound in Theorem 3.6.13 follows from the more general result discussed in Theorem 3.2.5 and the $\Omega(n \log n)$ bound for ONEMAX in the unary unbiased black-box model, which we discussed in Section 3.6.3.2. Note also that Theorem 3.2.5, together with the $\Omega(n/\log n)$ unrestricted black-box complexity of ONEMAX, implies a lower bound for JUMP_ℓ of the same asymptotic order (for all values of ℓ). The linear lower bound for the extreme jump function can be easily proven by the information-theoretic arguments presented in Theorem 3.2.4. Intuitively, the algorithm needs to learn a linear number of bits, while it receives only a constant number per function evaluation.

Insights from These Bounds and Open Questions

The proof sketches provided above highlight the fact that one of the key novelties presented in [28] are the sampling strategies that are used to estimate the ONEMAX values of a current string of interest. The idea of accumulating some statistical information about the fitness landscape could be an interesting concept for the design of novel heuristics, in particular for optimization in the presence of noisy function evaluations or for dynamic problems, which change over time.

3.6.3.5 Number Partition

Number partition is one of the best-known NP-hard problems. Given a set $S \subset \mathbb{N}^n$ of n positive integers, this partition problem asks us to decide whether or not it is possible to split S into two disjoint subsets such that the sums of the integers in these two subsets are identical, i.e., whether or not two disjoint subsets S_1 and S_2 of S with $S_1 \cup S_2 = S$ and $\sum_{s \in S_1} s = \sum_{s \in S_2} s$ exist. The optimization version of partition asks us to split S into two disjoint subsets such that the absolute discrepancy $\left|\sum_{s \in S_1} s - \sum_{s \in S_2} s\right|$ is as small as possible.

In [26], a subclass of partition was studied in which the integers in S are pairwise different. The problem remains NP-hard under this assumption. It is thus unlikely that it can be solved efficiently. For two different formulations of this problem (using a signed and an unsigned function assigning to each partition S_1, S_2 of S the discrepancy $\sum_{s \in S_1} s - \sum_{s \in S_2} s$ or the absolute value of this expression, respectively), it was shown that the unary unbiased black-box complexity of this subclass is nevertheless of small polynomial order. More precisely, it was shown that there are unary unbiased black-box algorithms that need only $O(n \log n)$ function evaluations to optimize any $\text{Partition}_{\neq}$ instance. The proof techniques are very similar to the ones presented in Section 3.2.4: the algorithm achieving the $O(n\log n)$ expected optimization time first uses $O(n\log n)$ steps to learn the problem instance at hand. After some (possibly – and probably – nonpolynomial-time) offline computation of an optimal solution for this instance, this optimum is then created via an additional $O(n\log n)$ function evaluations, needed to move the integers of the partition instance to the right subset. Learning and moving the bits can be done in linear time in the unrestricted model. The factor $\log n$ stems from the fact that here, in this unary unbiased model, in every iteration a random bit is moved, so that a coupon collector process results in a logarithmic overhead.

This result and those for the different versions of jump functions described in Section 3.6.3.4 show that the unary unbiased black-box complexity can be much smaller than the typical performance of mutation-only black-box heuristics. This indicates that the unary unbiased black-box model does not always give a good estimation of the difficulty of a problem when optimized by mutation-based algorithms. As we shall discuss in Section 3.7, a possible direction to obtain more meaningful results could be to restrict the class of algorithms even further, for example through bounds on the memory size or the selection operators.

3.6.3.6 Minimum Spanning Trees

Having a formulation over the search space $\{0,1\}^m$, the minimum spanning tree problem considered in Section 3.3.8.1 can be directly studied in the unbiased black-box model proposed by Lehre and Witt. The following theorem summarizes the bounds proven in [37] for this problem. We see here that [37] also studied the black-box complexity of a model that combines the restrictions imposed by the ranking-based and the unbiased black-box models. We will discuss this model in Section 3.7 but, for the sake of brevity, will state the bounds now for this combined model.

Theorem 3.6.14 (Theorem 10 in [37]). *The* unary unbiased *black-box complexity of the MST problem is $O(mn\log(m/n))$ if there are no duplicate weights, and $O(mn\log n)$ if there are. The* ranking-based unary unbiased *black-box complexity of the MST problem is $O(mn\log n)$. Its* ranking-based binary unbiased *black box-complexity is $O(m\log n)$ and its* ranking-based 3-ary unbiased *black-box complexity is $O(m)$.*

For every k, the k-ary unbiased black-box complexity of MST for m edges is at least as large as the k-ary unbiased black-box complexity of OneMax$_m$.

As in the unrestricted case of Theorem 3.3.14, the upper bounds in Theorem 3.6.14 are obtained by modifying Kruskal's algorithm to fit the black-box setting at hand. For the lower bound, the path P on $m+1$ vertices with unit edge weights shows that OneMax$_m$ is a subproblem of the MST problem. More precisely, for all bit strings $x \in \{0,1\}^m$, the function value

$f(x) = (\textsc{OneMax}_m(x), m+1-\textsc{OneMax}_m(x))$ of the associated MST fitness function reveals the OneMax value of x.

3.6.3.7 Other Results

Motivated to introduce a class of functions for which the unary unbiased black-box complexity is $\Theta(2^m)$ for some parameter m that can be scaled between 1 and n, Lehre and Witt introduced in [72] the following function:

$$\textsc{OM-Needle}: \{0,1\}^n \to [0..n], x \mapsto \sum_{i=1}^{n-m} x_i + \prod_{i=1}^{n} x_i.$$

It is easily seen that this function has its unique global optimum in the all-ones string $(1,\ldots,1)$. All other search points whose first $n-m$ entries are equal to one are located on a plateau of function value $n-m$. In the unbiased model, this part is thus similar to the Needle functions discussed in Section 3.3.1. Lehre and Witt showed that for $0 \le m \le n$ the unary unbiased black-box complexity of this function is at least 2^{m-2} [72, Theorem 3]. Note that this function is similar in flavor to the version of the jump function proposed in [62] (see Section 3.3.7.3).

3.6.4 Beyond Pseudo-Boolean Optimization: Unbiased Black-Box Models for Other Search Spaces

In this section we discuss an extension of the pseudo-Boolean unbiased black-box model of Lehre and Witt [72] to more general search spaces. To this end, we first recall from Definition 3.6.1 that the unbiased model is defined through a set of invariances that must be satisfied by the probability distributions from which unbiased algorithms sample their solution candidates. It is therefore quite natural to first generalize the notion of an unbiased operator in the following way.

Definition 3.6.15 (Definition 1 in [37]). Let $k \in \mathbb{N}$, let S be some arbitrary set, and let $\mathcal{G}$ be a set of bijections on S that forms a group, i.e., a set of one-to-one maps $g: S \to S$ that is closed under composition $(\cdot \circ \cdot)$ and under inversion $(\cdot)^{-1}$. We call $\mathcal{G}$ the *set of invariances*.

A *k-ary $\mathcal{G}$-unbiased distribution* is a family of probability distributions $\left(D(\cdot \mid y^1,\ldots,y^k)\right)_{y^1,\ldots,y^k \in S}$ over S such that for all *inputs* $y^1,\ldots,y^k \in S$ the condition

$$\forall x \in S \forall g \in \mathcal{G}: D(x \mid y^1,\ldots,y^k) = D(g(x) \mid g(y^1),\ldots,g(y^k))$$

holds. An operator sampling from a k-ary $\mathcal{G}$-unbiased distribution is called a *k-ary $\mathcal{G}$-unbiased variation operator.*

For $S := \{0,1\}^n$ and when $\mathcal{G}$ is the set of Hamming automorphisms, it is not difficult to verify that Definition 3.6.15 extends Definition 3.6.1. A k-ary $\mathcal{G}$-unbiased black-box algorithm is one that samples all search points from k-ary $\mathcal{G}$-unbiased variation operators.

In [81], Rowe and Vose gave the following very general, but rather indirect, definition of unbiased distributions.

Definition 3.6.16 (Definition 2 in [81]). Let $\mathcal{F}$ be a class of functions from a search space S to some set Y. We say that a one-to-one map $\alpha : S \to S$ *preserves* $\mathcal{F}$ if, for all $f \in \mathcal{F}$, it holds that $f \circ \alpha \in \mathcal{F}$. Let $\Pi(\mathcal{F})$ be the class of all such $\mathcal{F}$-preserving bijections α.

A *k-ary generalized unbiased distribution (for $\mathcal{F}$)* is a k-ary $\Pi(\mathcal{F})$-unbiased distribution.

It was argued in [81] that $\Pi(\mathcal{F})$ in fact forms a group, so that Definition 3.6.16 satisfies the requirements of Definition 3.6.15.

To apply the framework of Definition 3.6.16, one has to make precise the set of invariances covered by the class $\Pi(\mathcal{F})$. This can be quite straightforward in some cases [81] but may require some more effort in others [37]. In particular, it is often inconvenient to define the whole family of unbiased distributions from which a given variation operator originates. Luckily, in many cases this effort can be considerably reduced, to proving only the unbiasedness of the variation operator itself. The following theorem demonstrates this for the case $S = [n]^{n-1}$, which is used, for example, in the single-source shortest-path problem considered in the next subsection. In this case, condition (ii) in the theorem states that it suffices to show the k-ary $\mathcal{G}$-unbiasedness of the distribution $D_{\mathbf{z}}$, without making precise the whole family of distributions associated to it.

Theorem 3.6.17. *Let $\mathcal{G}$ be a set of invariances, i.e., a set of permutations of the search space $S = [n]^{n-1}$ that form a group. Let $k \in \mathbb{N}$, and let $\mathbf{z} = (z^1, \ldots, z^k) \in S^k$ be a k-tuple of search points. Let*

$$\mathcal{G}_0 := \{g \in \mathcal{G} \mid g(z^i) = z^i \text{ for all } i \in [k]\}$$

be the set of all invariances that leave $z^1, \ldots, z^k$ fixed.

Then, for any probability distribution $D_{\mathbf{z}}$ on $[n]^{n-1}$, the following two statements are equivalent.

(i) There exists a k-ary $\mathcal{G}$-unbiased distribution $(D(\cdot \mid \mathbf{y}))_{\mathbf{y} \in S^k}$ on S such that $D_{\mathbf{z}} = D(\cdot \mid \mathbf{z})$.
(ii) For every $g \in \mathcal{G}_0$ and for all $x \in S$, it holds that $D_{\mathbf{z}}(x) = D_{\mathbf{z}}(g(x))$.

3.6.4.1 Alternative Extensions of the Unbiased Black-Box Model for the SSSP problem

As discussed in Section 3.3.8.2, several formulations of the single-source shortest-path problem (SSSP) coexist. In the unbiased black-box setting, the multicriteria formulation is not very meaningful, as the function values explicitly distinguish between the vertices, so that treating them in an unbiased fashion seems unreasonable. For this reason, in [37] only the single-objective formulation was investigated in the unbiased black-box model. Note that for this formulation, the unbiased black-box model for pseudo-Boolean functions needs to be extended to the search space $S_{[2..n]}$. The work [37] discussed three different extensions:

(a) a *structure-preserving* unbiased model in which, intuitively speaking, the operators do not consider the *labels* of different nodes, but only their local structure (e.g., the size of their neighborhoods);
(b) the *generalized* unbiased model proposed in [81] (this model follows the approach presented in Section 3.6.4 above); and
(c) a *redirecting* unbiased black-box model in which, intuitively, a node may choose to change its predecessor in the shortest-path tree but, if it decides to do so, then all possible predecessors must be equally likely to be chosen.

Whereas all three notions a priori seem to capture different aspects of what unbiasedness in the SSSP problem could mean, two of them were shown to be too powerful. More precisely, it was shown that even the *unary* structure-preserving unbiased black-box complexity of SSSP and the *unary* generalized unbiased black-box complexity are almost identical to the unrestricted black-box complexity. The three models were proven to differ by at most one query in [37, Theorem 25 and Corollary 32].

It was then shown that the redirecting unbiased black-box model yields more meaningful black-box complexities.

Theorem 3.6.18 (Corollary 28, Theorem 29, and Theorem 30 in [37]). *The unary ranking-based redirecting unbiased black-box complexity of SSSP is $O(n^3)$. Its binary ranking-based redirecting unbiased black-box complexity is $O(n^2 \log n)$. For all $k \in \mathbb{N}$, the k-ary redirecting unbiased black-box complexity of SSSP is $\Omega(n^2)$.*

The unary bound is obtained by a variant of RLS which, in every step, redirects one randomly chosen node to a random predecessor. For the binary bound, the problem instance is learned in a two-phase step. An optimal solution is then created by an imitation of Dijkstra's algorithm. For the lower bound, drift analysis can be used to prove that every redirecting unbiased algorithm needs $\Omega(n^2)$ function evaluations to reconstruct a given path on n vertices.

3.7 Combined Black-Box Complexity Models

The black-box models discussed in the previous sections either study the complexity of a problem with respect to *all* black-box algorithms (in the unrestricted model) or restrict the class of algorithms with respect to *one* particular feature of common optimization heuristics, such as the size of their memory, their selection behavior, or their sampling strategies. As we have seen, many classical black-box optimization algorithms are members of several of these classes. At the same time, a nonnegligible number of the upper bounds stated in the previous sections can, to date, be certified only by algorithms that satisfy an individual restriction, but clearly violate other requirements that are not controlled by the respective model. In the unbiased black-box model, for example, several of the upper bounds are obtained by algorithms that make use of a rather large memory size. It is therefore natural to ask if and how the black-box complexity of a problem increases if two or more of the different restrictions proposed in the previous sections are combined in a new black-box model. This is the focus of this section, which surveys results obtained in such combined black-box complexity models.

3.7.1 Unbiased Ranking-Based Black-Box Complexity

Even some of the very early publications on the unbiased black-box model considered a combination with the ranking-based model. In fact, the binary unbiased algorithm in [35], which solves OneMax with $\Theta(n)$ queries on average, uses only comparisons, and does not make use of knowledge of absolute fitness values. It was shown in [44] that, also, the other upper bounds for the k-ary black-box complexity of OneMax proven in [35] hold also in the ranking-based version of the k-ary unbiased black-box model.

Theorem 3.7.1 (Theorem 6 and Lemma 7 in [44]). *The unary unbiased ranking-based black-box complexity of* OneMax_n *is* $\Theta(n \log n)$. *For constant* k, *the* k*-ary unbiased ranking-based black-box complexity of* OneMax_n *and that of every strictly monotone function is at most* $4n-5$. *For* $2 \leq k \leq n$, *the* k*-ary unbiased ranking-based black-box complexity of* OneMax_n *is* $O(n/\log k)$.

In light of Theorem 3.6.9, it seems plausible that the upper bounds for the case $2 \leq k \leq n$ can be reduced to $O(n/k)$, but we are not aware of any result proving such a claim.

Also, the binary unbiased algorithm achieving an expected $O(n \log n)$ optimization time on LeadingOnes uses only comparisons.

Theorem 3.7.2 (follows from the proof of Theorem 14 in [35]; see Theorem 3.6.11). *The binary unbiased ranking-based black-box complexity of* LeadingOnes *is* $O(n \log n)$.

For the ternary black-box complexity we have mentioned already in Theorem 3.6.12 that the $O(n\log(n)/\log\log n)$ bound also holds in the ranking-based version of the ternary unbiased black-box model.

For the two combinatorial problems MST and SSSP, it has already been mentioned in Theorems 3.6.14 and 3.6.18 that the bounds hold also in models in which we require the algorithms to base all decisions on only the ranking of previously evaluated search points, and not on absolute function values.

3.7.2 Parallel Black-Box Complexity

The (unary) unbiased black-box model was also the starting point for the authors of [4], who introduced a black-box model to investigate the effects of a parallel optimization. Their model can be seen as a unary unbiased $(\infty+\lambda)$ memory-restricted black-box model. More precisely, their model covers all algorithms following the scheme of Algorithm 3.6.

The model covers the $(\mu+\lambda)$ EA and the (μ,λ) EA, cellular evolutionary algorithms, and unary unbiased evolutionary algorithms working in the island model. The restriction to unary unbiased variation operators can of course be relaxed to obtain general models for λ-parallel k-ary unbiased black-box algorithms.

We see that Algorithm 3.6 forces the algorithm to query λ new solution candidates in every iteration. Thus, intuitively, for every two positive integers k and ℓ with $k/\ell \in \mathbb{N}$ and for all problem classes $\mathcal{F}$, the ℓ-parallel unary unbiased black-box complexity of $\mathcal{F}$ is at most as large as its k-parallel unary unbiased black-box complexity.

Algorithm 3.6: A blueprint for λ-parallel unary unbiased black-box algorithms for the optimization of an unknown function $f: S \to \mathbb{R}$

---

1 **Initialization:**
2   **for** $i = 1,\ldots,\mu$ **do** Sample $x^{(i,0)}$ uniformly at random from S and query $f\left(x^{(i,0)}\right)$;
3     $\mathcal{I} \leftarrow \{f\left(x^{(1,0)}\right),\ldots,f\left(x^{(\lambda,0)}\right)\}$;
4 **Optimization: for** $t = 1,2,3,\ldots$ **do**
5   **for** $i = 1,\ldots,\lambda$ **do**
6   Depending only on the multiset $\mathcal{I}$ choose a pair of indices $(j,\ell) \in [\lambda] \times [0..t-1]$;
7   Depending only on the multiset $\mathcal{I}$ choose a unary unbiased probability distribution $D^{(i,t)}(\cdot)$ on S, sample $x^{(i,t)} \leftarrow D^{(i,t)}(x^{(j,\ell)})$ and query $f\left(x^{(i,t)}\right)$;
8     $\mathcal{I} \leftarrow \mathcal{I} \cup \{f\left(x^{(1,t)}\right),\ldots,f\left(x^{(\lambda,t)}\right)\}$;

The following bounds for the λ-parallel unary unbiased black-box complexity are known.

Theorem 3.7.3 (Theorems 1, 3 and 4 in [4]). *The λ-parallel unary unbiased black-box complexity of* LEADINGONES *is $\Omega\big(\frac{\lambda n}{\max\{1,\log(\lambda/n)\}}+n^2\big)$. It is of order at most $\lambda n+n^2$.*

For any $\lambda \leq e^{\sqrt{n}}$, the λ-parallel unary unbiased black-box complexity of any function having a unique global optimum is $\Omega\big(\frac{\lambda n}{\log(\lambda)}+n\log n\big)$. This bound is asymptotically tight for ONEMAX.

For LEADINGONES, the upper bound is attained by a $(1+\lambda)$ EA investigated in [70]. The lower bound was shown by means of drift analysis, building upon the arguments used in [72] to prove Theorem 3.6.10.

For ONEMAX, a $(1+\lambda)$ EA with fitness-dependent mutation rates was shown to achieve an $O\big(\frac{\lambda n}{\log(\lambda)}+n\log n\big)$ expected optimization time in [4, Theorem 4]; see Chapter 5.6 for details.

The lower bound for the λ-parallel unary unbiased black-box complexity of functions having a unique global optimum uses additive drift analysis. The proof is similar to the proof of Theorem 3.6.6 in [72], but requires a very precise tail bound for hypergeometric variables (Lemma 2 in [4]).

3.7.3 Distributed Black-Box Complexity

To study the effects of the migration topology on the efficiency of distributed evolutionary algorithms, the λ-parallel unary unbiased black-box model was extended in [5] to a distributed version, in which islands exchange their accumulated information along a given graph topology. Commonly employed topologies are the complete graph (in which all nodes exchange information with each other), the ring topology, the grid of equal side lengths, and the torus. [5] presents an unrestricted and a unary unbiased version of the distributed black-box model. In this context, it is interesting to study how the black-box complexity of a problem increases with sparser migration topologies or with the infrequency of migration. The model of [5] allows all nodes to share all the information that they have accumulated so far. Another interesting extension of the distributed model would be to study the effects of bounding the amount of information that can be shared in any migration phase. We will not present the model, nor all results obtained in [5], in detail. The main result which is interpretable and comparable to the others presented in this chapter is summarized by the following theorem.

Theorem 3.7.4 (Table 1 in [5]). *The λ-distributed unary unbiased black-box complexity of the class of all unimodal functions with $\Theta(n)$ different function values satisfies the bounds stated in Table 3.2. The lower bound for*

	Ring Topology	Grid/Torus	Complete Topology
Upper Bound	$O(\lambda n^{3/2}+n^2)$	$O(\lambda n^{4/3}+n^2)$	$\Theta(\lambda n+n^2)$
Lower Bound	$\Omega(\lambda n+\lambda^{2/3}n^{5/3}+n^2)$	$\Omega(\lambda n+\lambda^{3/4}n^{3/2}+n^2)$	

Table 3.2 The λ-distributed unary unbiased black-box complexity of the class of all unimodal functions with $\Theta(n)$ different function values

the grid applies to arbitrary side lengths, while the upper bound holds for the grid with $\sqrt{\lambda}$ islands in each of the two dimensions.

The upper bounds in Theorem 3.7.4 are achieved by a parallel (1+1) EA, in which every node migrates its complete information after every round. The lower bounds were shown to hold even for a subproblem called the "random short path," which is a collection of problems which all have a global optimum in some point with exactly $n/2$ ones. A short path of Hamming-1 neighbors leads to this optimum. The paths start at the all-ones string. Search points that do not lie on the path lead the optimization process towards the all-ones string; their objective values equal the number of ones in the string.

3.7.4 Elitist Black-Box Complexity

One of the most relevant questions in black-box optimization is how to avoid getting stuck in local optima. Essentially, two strategies have been developed.

- **Nonelitist selection.** The first idea is to allow the heuristics to direct their search towards search points that are, a priori, less favorable than the current best solutions in the memory. This can be achieved, for example, by accepting into the memory ("population") search points with function values that are smaller than the current best solutions. We refer to such selection procedures as *nonelitist selection.* Nonelitist selection is used, for example, in the Metropolis algorithm [75], Simulated Annealing [66], and, more recently, the biology-inspired "Strong Selection, Weak Mutation" framework [80].
- **Global sampling.** A different strategy to overcome local optima is *global sampling.* This approach is used, most notably, by evolutionary and genetic algorithms, but also by swarm optimizers such as ant colony optimization techniques [52] and estimation-of-distribution algorithms (EDAs, see Chapter 7 in this book). The underlying idea of global sampling is to select new solution candidates not only locally in some predefined neighborhood of the current population, but also to reserve some positive probability to sample far away from these solutions. Very often, a truly global sampling operation is used, in which *every* point $x \in S$ has a positive probability of being sampled. This probability typically decreases with increasing dis-

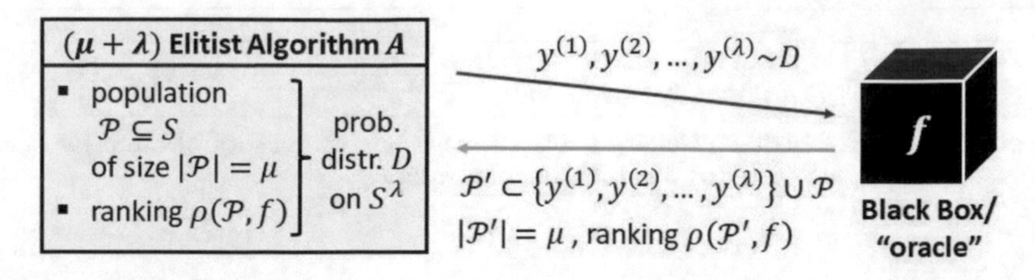

Fig. 3.7 A $(\mu+\lambda)$ elitist black-box algorithm stores the μ previously evaluated search points of largest function value (ties broken arbitrarily or according to some specified rule) and the ranking of these points induced by f. Based on this information, it decides upon a strategy according to which the next λ search points are sampled. From the $\mu+\lambda$ parent and offspring solutions, those μ search points that have the largest function values form the population for the next iteration.

tance to the current best solutions. Standard bit mutation with bit flip probabilities $p < 1/2$ is such a global sampling strategy.

Global sampling and nonelitist selection can certainly be combined, and several attempts in this direction have been made. The predominant selection strategy used in combination with global sampling, however, is *truncation selection*. Truncation selection is a natural implementation of Darwin's "survival of the fittest" paradigm in an optimization context: given a collection $\mathcal{P}$ of search points and a population size μ, truncation selection chooses from $\mathcal{P}$ the μ search points of largest function values and discards the others, breaking ties arbitrarily or according to some rule such as favoring offspring over parents or favoring genotypic or phenotypic diversity.

To understand the influence that this *elitist selection* behavior has on the performance of black-box heuristics, the *elitist black-box model* was introduced in [46] (a journal version has appeared as [49]). The elitist black-box model combines features of the memory-restricted and the ranking-based black-box models with an enforced truncation selection. More precisely, the $(\mu+\lambda)$ *elitist black-box model* covers all algorithms that follow the pseudocode in Algorithm 3.7. We use here an adaptive initialization phase. A nonadaptive version, as in the $(\mu+\lambda)$ memory-restricted black-box model, can also be considered. This and other subtleties such as the tie-breaking rules for search points of equal function values can result in different black-box complexities. It is therefore important to make very precise the model with respect to which a bound is claimed or shown to hold.

The elitist black-box model covers, in particular, all $(\mu+\lambda)$ EAs, RLS, and other hill climbers. It does not cover algorithms using nonelitist selection rules such as Boltzmann selection, tournament selection, or fitness-proportional selection. Figure 3.7 illustrates the $(\mu+\lambda)$ elitist black-box model. As a seemingly subtle but possibly influential difference from the parallel black-box complexities introduced in Section 3.7.2, note that in the elitist black-box

Algorithm 3.7: The $(\mu+\lambda)$ elitist black-box algorithm for maximizing an unknown function $f: S \to \mathbb{R}$

1 **Initialization:**
2 $X \leftarrow \emptyset$;
3 **for** $i = 1, \ldots, \mu$ **do**
4 Depending only on the multiset X and the ranking $\rho(X, f)$ of X induced by f, choose a probability distribution $D^{(i)}$ over S and sample from it $x^{(i)}$;
5 Set $X \leftarrow X \cup \{x^{(i)}\}$ and query the ranking $\rho(X, f)$ of X induced by f;
6 **Optimization: for** $t = 1, 2, 3, \ldots$ **do**
7 Depending only on the multiset X and the ranking $\rho(X, f)$ of X induced by f choose a probability distribution $D^{(t)}$ on S^λ and sample from it $y^{(1)}, \ldots, y^{(\lambda)} \in S$;
8 Set $X \leftarrow X \cup \{y^{(1)}, \ldots, y^{(\lambda)}\}$ and query the ranking $\rho(X, f)$ of X induced by f;
9 **for** $i = 1, \ldots, \lambda$ **do** Select $x \in \arg\min X$ and update $X \leftarrow X \setminus \{x\}$;

model the offspring sampled in the optimization phase do not need to be independent of each other. If, for example, an offspring x is created by crossover, in the $(\mu+\lambda)$ elitist black-box model with $\lambda \geq 2$ we allow another offspring y to be created from the same parents, whose entries y_i in those positions i in which the parents do not agree equal $1 - x_i$. These two offspring are obviously not independent of each other. It is nevertheless required in the $(\mu+\lambda)$ elitist black-box model that the λ offspring are created *before* any evaluation of the offspring happens. That is, the k-th offspring may *not* depend on the ranking or fitness of the first $k-1$ offspring.

In addition to combining several features of previous black-box models, the elitist black-box model can be further restricted to cover only those elitist black-box algorithms that sample from unbiased distributions. For this *unbiased elitist black-box model*, we require that the distribution $p^{(t)}$ in line 7 of Algorithm 3.7 is unbiased (in the sense of Section 3.6). Some of the results mentioned below also hold for this more restrictive class.

3.7.4.1 Nonapplicability of Yao's Minimax Principle

An important difficulty in the analysis of elitist black-box complexities is the fact that Yao's minimax principle (Theorem 3.2.3) cannot be directly applied to the elitist black-box model, since in this model the previously exploited fact that randomized algorithms are convex combinations of deterministic ones does not apply; see [49, Section 2.2] for an illustrated discussion. As discussed in the previous sections, Yao's minimax principle is *the* most important tool for proving lower bounds in the black-box complexity context, and we can hardly do without it. A natural workaround that allows us to nevertheless employ this technique is to extend the collection $\mathcal{A}$ of elitist black-box algorithms to some superset $\mathcal{A}'$ in which every randomized algorithm *can* be expressed as a probability distribution over deterministic ones. A lower bound shown for this broader class $\mathcal{A}'$ applies trivially to all elitist black-box algorithms. Finding extensions $\mathcal{A}'$ that do not decrease the lower bounds by too much is the main difficulty to be overcome in this strategy.

3.7.4.2 Exponential Gaps to Previous Models

In [49, Section 3], it was shown that even for quite simple function classes there can be an exponential gap between the efficiency of elitist and nonelitist black-box algorithms; and this applies even in the very restrictive (1+1) unary unbiased elitist black-box complexity model. This shows that heuristics can sometimes benefit quite crucially from eventually giving preference to search points of fitness inferior to that of the current best search points. The underlying intuition for these results is that elitist algorithms do not work very well if there are several local optima that the algorithm needs to explore in order to determine the best one of them.

3.7.4.3 The Elitist Black-Box Complexity of OneMax

As we have discussed in Sections 3.4 and 3.5, respectively, the (1+1) memory-restricted and the ranking-based black-box complexity of OneMax are only of order $n/\log n$. In contrast, it is easy to see that the combined (1+1) memory-restricted ranking-based black-box model does not allow algorithms that are faster than linear in n, as can easily be seen by standard information-theoretic considerations. In [47] (a journal version has appeared as [50]) it was shown that this linear bound is tight. Whether or not it applies to the (1+1) elitist model remains unsolved, but it was shown in [50] that the expected time needed to optimize OneMax with probability at least $1-\varepsilon$ is linear for every constant $\varepsilon > 0$. This is the so-called Monte Carlo black-box complexity, which we shall briefly discuss in Section 3.11. The following theorem summarizes the bounds presented in [50]. Without detailing this further, we note that [50, Section 9] also introduced and studied a comma-variant of the elitist black-box model.

Theorem 3.7.5 ([50]). *The (1+1) memory-restricted ranking-based black-box complexity of* OneMax *is* $\Theta(n)$.

For $1 < \lambda < 2^{n^{1-\varepsilon}}$, $\varepsilon > 0$ *being an arbitrary constant, the* $(1+\lambda)$ *memory-restricted ranking-based black-box complexity of* OneMax *is* $\Theta(n/\log \lambda)$ *(in terms of generations), while for* $\mu = \omega(\log^2(n)/\log\log n)$ *its* $(\mu+1)$ *memory-restricted ranking-based black-box complexity is* $\Theta(n/\log \mu)$.

For every constant $0<\varepsilon<1$, there exists a (1+1) elitist black-box algorithm that finds the optimum of any OneMax *instance in time $O(n)$ with probability at least $1-\varepsilon$, and this running time is asymptotically optimal.*

For constant μ, the $(\mu+1)$ elitist black-box complexity of OneMax *is at most $n+1$.*

For $\delta>0$, $C>0$, $2\le\lambda<2^{n^{1-\delta}}$, and a suitably chosen $\varepsilon = O(\log^2(n)\log\log(n)\log(\lambda)/n)$, there exists a $(1+\lambda)$ elitist black-box algorithm that needs at most $O(n/\log\lambda)$ generations to optimize OneMax *with probability at least $1-\varepsilon$.*

For $\mu=\omega(\log^2(n)/\log\log n)\cap O(n/\log n)$ and every constant $\varepsilon>0$, there is a $(\mu+1)$ elitist black-box algorithm optimizing OneMax *in time $\Theta(n/\log\mu)$ with probability at least $1-\varepsilon$.*

There exists a constant $C>1$ such that for $\mu\ge Cn/\log n$, the $(\mu+1)$ elitist black-box complexity is $\Theta(n/\log n)$.

3.7.4.4 The Elitist Black-Box Complexity of LeadingOnes

The (1+1) elitist black-box complexity of LeadingOnes was studied in [48] (a journal version has appeared as [51]). Using the approach sketched in Section 3.7.4.1, the following result was derived.

Theorem 3.7.6 (Theorem 1 in [51]). *The (1+1) elitist black-box complexity of* LeadingOnes *is $\Theta(n^2)$. This bound holds also in the case where the algorithms have access to (and can make use of) the absolute fitness values of the search points in the population, and not only their rankings, i.e., in the (1+1) memory-restricted black-box model with enforced truncation selection.*

The $(1+1)$ elitist black-box complexity of LeadingOnes is thus considerably larger than its unrestricted black-box complexity, which is known to be of order $n\log\log n$, as discussed in Theorem 3.3.12.

It is well known that the quadratic bound in Theorem 3.7.6 is matched by classical $(1+1)$-type algorithms such as the (1+1) EA, RLS, and others.

3.7.4.5 The Unbiased Elitist Black-Box Complexity of Jump Functions

Some shortcomings of previous models can be eliminated when they are combined with an elitist selection requirement. This was shown in [49] for the Jump_k function already discussed.

Theorem 3.7.7 (Theorem 9 in [49]). *For $k=0$, the unary unbiased (1+1) elitist black-box complexity of Jump_k is $\Theta(n\log n)$. For $1\le k\le \frac{n}{2}-1$, it is of order $\binom{n}{k+1}$.*

Model	Lower bound	Upper bound
Unrestricted	$\Theta(n/\log n)$	
Unbiased, arity 1	$\Theta(n \log n)$	
Unbiased, arity $2 \leq k \leq \log n$	$\Omega(n/\log n)$	$O(n/k)$
Ranking-based (unrestricted)	$\Theta(n/\log n)$	
Ranking-based unbiased, arity 1	$\Theta(n \log n)$	
Ranking-based unbiased, arity $2 \leq k \leq n$	$\Omega(n/\log n)$	$O(n/\log k)$
(1+1) comparison-based	$\Theta(n)$	
(1+1) memory-restricted	$\Theta(n/\log n)$	
λ-parallel unbiased, arity 1	$\Theta\left(\frac{\lambda n}{\log(\lambda)} + n \log n\right)$	
(1+1) elitist Las Vegas	$\Omega(n)$	$O(n \log n)$
(1+1) elitist $\log n/n$-Monte Carlo	$\Theta(n)$	
(2+1) elitist Monte Carlo/Las Vegas	$\Theta(n)$	
(1+λ) elitist Monte Carlo (# generations)	$\Theta(n/\log \lambda)$	
(μ+1) elitist Monte Carlo	$\Theta(n/\log \mu)$	
$(1,\lambda)$ elitist Monte Carlo/Las Vegas (# generations)	$\Theta(n/\log \lambda)$	

Table 3.3 Summary of known black-box complexities of ONEMAX_n in the different black-box complexity models

The bound in Theorem 3.7.7 is nonpolynomial for $k = \omega(1)$. This is in contrast to the unary unbiased black-box complexity of JUMP_k, which, according to Theorem 3.6.13, is polynomial even for extreme values of k.

3.8 Summary of Known Black-Box Complexities for OneMax and LeadingOnes

For better identification of open problems concerning the black-box complexity of the two benchmark functions ONEMAX and LEADINGONES, we summarize the bounds that have been presented in previous sections.

Table 3.3 summarizes the known black-box complexities of ONEMAX_n in the different models. The bound for the λ-parallel black-box model assumes $\lambda \leq e^{\sqrt{n}}$. The bounds for the $(1+\lambda)$ and the $(1,\lambda)$ elitist model assume $1 < \lambda < 2^{n^{1-\varepsilon}}$ for some $\varepsilon > 0$. Finally, the bound for the $(\mu+1)$ model assumes that $\mu = \omega(\log^2 n/\log\log n)$ and $\mu \leq n$.

Table 3.4 summarizes the known black-box complexities of LEADINGONES_n. The upper bounds for the unbiased black-box models also hold in the ranking-based variants.

Model	Lower bound	Upper bound
unrestricted	$\Theta(n \log\log n)$	
unbiased, arity 1	$\Theta(n^2)$	
unbiased, arity 2	$\Omega(n \log\log n)$	$O(n \log n)$
unbiased, arity ≥ 3	$\Omega(n \log\log n)$	$O(n \log(n)/\log\log n)$
λ-parallel unbiased, arity 1	$\Omega\left(\frac{\lambda n}{\log(\lambda/n)} + n^2\right)$	$O(\lambda n + n^2)$
(1+1) elitist	$\Omega(n^2)$	$O(n^2)$

Table 3.4 Summary of known black-box complexities of LEADINGONES$_n$ in the different black-box complexity models

3.9 From Black-Box Complexity to Algorithm Design

In the previous sections, the focus of our attention has been on computing performance limits for black-box optimization heuristics. In some cases, for example for ONEMAX in the unary unbiased black-box model and for LEADINGONES in the (1+1) elitist black-box model, we have obtained lower bounds that are matched by the performance of well-known standard heuristics such as RLS or the (1+1) EA. For several other models and problems, however, we have obtained black-box complexities that are much smaller than the expected running times of typical black-box optimization techniques. As discussed in the introduction, two possible reasons for this discrepancy exist. Either the respective black-box models do not capture very well the complexity of the problems for heuristic approaches, or there are ways to improve classical heuristics by novel design principles.

In the case of the restrictive models discussed in Sections 3.4-3.7, we have seen that there is some truth in the first possibility. For several optimization problems, we have seen that their complexity increases if the class of black-box algorithms is restricted to subclasses of heuristics that all share some properties that are commonly found in state-of-the-art optimization heuristics. Here, in this section, we shall demonstrate that this is nevertheless not the end of the story. We discuss two examples where a discrepancy between black-box complexity and the performance of classical heuristics can be observed, and we show how the analysis of typically rather artificial problem-tailored algorithms can inspire the design of new heuristics.

3.9.1 The $(1+(\lambda,\lambda))$ Genetic Algorithm

Our first example is a binary unbiased algorithm, which optimizes ONEMAX more efficiently than any classical unbiased heuristic, and provably faster than any unary unbiased black-box optimizer.

We recall from Theorems 3.6.7 and 3.6.8 that the unary unbiased black-box complexity of $\textsc{OneMax}_n$ is $\Theta(n \log n)$, while its binary unbiased black-box complexity is only $O(n)$. The linear-time algorithm flips one bit at a time, and uses a simple but clever encoding to store which bits have been flipped already. In this way, it is a rather problem-specific algorithm, since it "knows" that a bit that has been tested already does not need to be tested again. The algorithm is therefore not very suitable for nonseparable problems, where the influence of an individual bit depends on the value of several or all other bits.

Until recently, all existing running-time results have indicated that general-purpose unbiased heuristics need $\Omega(n \log n)$ function evaluations to optimize OneMax, so that the question of whether the binary unbiased black-box model is too "generous" arose. In [21, 25] this question was answered negatively, through the presentation of a novel binary unbiased black-box heuristic that optimizes OneMax in expected linear time. This algorithm is the $(1+(\lambda,\lambda))$ GA. Since the algorithm itself will be discussed in more detail in Chapter 5.6, we present here only the main ideas behind it.

Disregarding some technical subtleties, one observation that we can make when considering the linear-time binary unbiased algorithm for OneMax is that when it test the value of a bit, the amount of information that it obtains is the same whether or not the offspring has a better function value. In other words, the algorithm benefits equally from offspring that are better or worse than the previously best. A similar observation applies to all of the $O(n/\log n)$ algorithms for OneMax discussed in Sections 3.3-3.7. These algorithms do not strive to sample search points of large objective value, but rather aim at maximizing the amount of information that they can learn about the problem instance at hand. This way, they benefit substantially also from those search points that are (much) worse than other ones already evaluated.

Most classical black-box heuristics are different. They store only the best solutions so far, or use inferior search points only to create diversity in the population. Thus, in general, they are not very efficient in learning from "bad" samples (where we consider a search point to be "bad" if it has a small function value). When a heuristic is close to a local or a global optimum (in the sense that it has identified search points that are not far from these optima), it samples, in expectation, a fairly large number of search points that are wore than the current best solutions. Not learning from these offspring results in a significant number of "wasted" iterations, from which the heuristic does not benefit. This observation was the starting point for the development of the $(1+(\lambda,\lambda))$ GA.

Since the unary unbiased black-box complexity of OneMax is $\Omega(n \log n)$, it was clear in the development of the $(1+(\lambda,\lambda))$ GA that an $o(n \log n)$ unbiased algorithm must be at least binary. This led to the question of how recombination can be used to learn from inferior search points. The following idea emerged. For illustration purposes, assume that we have identified a search point x of function value $\mathrm{OM}_z(x) = n-1$. From the function value, we know that there exists exactly one bit that we need to flip in order to obtain the global optimum z. Since we want to be unbiased, the best mutation seems to be a random 1-bit flip. This has a probability $1/n$ of returning z. If we were to do this until we identified z, the expected number of samples would be n, and even if we stored which bits had been flipped already, we would need $n/2$ samples on average.

Assume now that, in the same situation, we flip $\ell > 1$ bits of x. Then, with a probability that depends on ℓ, we may have only flipped already optimized bits (i.e., bits in positions i for which $x_i = z_i$) to $1 - z_i$, thus resulting in an offspring of function value $n - 1 - \ell$. However, the probability that the position j in which x and z differ is among the ℓ positions is ℓ/n. If we repeat this experiment some λ times, independently of each other and always starting with x as the "parent," then the probability that j has been flipped in at least one of the offspring is $1-(1-\ell/n)^{\lambda}$. For moderately large ℓ and λ, this probability is sufficiently large for us to assume that among the λ offspring there is at least one in which j has been flipped. Such an offspring is distinguished from the others by a function value of $n-1-(\ell-1)+1 = n-\ell+1$ instead of $n-\ell-1$. Assume that there is one such offspring x' among the λ independent samples created from x. When we compare x' with x, they differ in exactly ℓ positions. In $\ell-1$ of these, the entry of x equals that of z. Only in the j-th position is the situation reversed: $x'_j = z_j \neq x_j$. We would therefore like to identify this position j, and to incorporate the bit value x'_j into x.

So far, we have used only mutation, which is a unary unbiased operation. At this point, we want to compare and merge two search points, which is one of the driving motivations behind *crossover*. Since x clearly has more "good" bits than x', a uniform crossover, which takes for each position i its entry uniformly at random from either of its two parents, does not seem to be a good choice. We would like to add some bias to the decision-making process, in favor of choosing the entries of x. This yields a biased crossover, which takes for each position i its entry y_i from x' with some probability $p < 1/2$, and from x otherwise. The hope is to choose p in such a way that in the end only good bits are chosen. Where x and x' are identical, there is nothing to worry about, as these positions are correct already (and, in general, we have no indication to flip the entry in this position). So, we only need to look at those ℓ positions in which x and x' differ. The probability of making only good choices, i.e., of selecting $\ell-1$ times the entry from x and, only for the j-th position, the entry from x' equals $p(1-p)^{\ell-1}$. This probability may not be very large, but when we do λ independent trials again, the probability of having created z in at least one of the trials equals $1-(1-p(1-p)^{\ell-1})^{\lambda}$. As we shall see, for suitable values of the parameters p, λ, and ℓ, this expression is sufficiently large to gain over the $O(n)$ strategies discussed above. Since we want to sample exactly one out of the ℓ bits in which x and x' differ, it seems intuitive to set $p = 1/\ell$; see the discussion in [25, Section 2.1].

Before we summarize the main findings, let us briefly reflect on the structure of the algorithm. In the *mutation step*, we have created λ offspring from x, by a mutation operator that flips ℓ random bits in x. This is a unary unbiased operation. From these λ offspring, we have selected one offspring x' with the largest function value among all offspring (with ties broken at random). In the *crossover phase*, we have then created λ offspring again, by recombining x and x' using a biased crossover. This biased crossover is a binary unbiased variation operator. The algorithm now chooses from these λ recombined offspring one that has the largest function value (for OneMax, ties can again be broken at random, but for other problems it can be better to favor individuals that are different from x; see [25, Section 4.3]). This selected offspring y replaces x if it is at least as good as x, i.e., if $f(y) \geq f(x)$.

We see that we have employed only unbiased operations, and that the largest arity in use is two. Both of the variation operators, mutation and biased crossover, are standard operators in the evolutionary computation literature. What is novel is that crossover is used as a *repair mechanism*, and after the mutation step.

We also see that this algorithm is ranking-based, and even comparison-based in the sense that it can be implemented in a way in which, instead of querying absolute function values, only a comparison of the function values of two search points is asked for. Using information-theoretic arguments as described in Section 3.2.2, it is then not difficult to show that for any (adaptive or nonadaptive) parameter setting the best expected performance of the $(1+(\lambda,\lambda))$ GA on OneMax_n is at least linear in the problem dimension n.

The following theorem summarizes some of the results on the expected running time of the $(1+(\lambda,\lambda))$ GA on OneMax. An exhaustive discussion of these results can be found in [24]. The fitness-dependent and self-adjusting choice of the parameters will also be discussed in Section 6.5.2.1 in this book.

Theorem 3.9.1 (from [19, 21, 22, 25]). *The $(1+(\lambda,\lambda))$ GA is a binary unbiased black-box algorithm. For a mutation strength ℓ sampled from the binomial distribution $Bin(n,k/n)$ and a crossover bias $p=1/k$, the following holds:*

- *For $k=\lambda=\Theta(\sqrt{\log(n)\log\log(n)/\log\log\log(n)})$ the expected optimization time of the $(1+(\lambda,\lambda))$ GA on OneMax_n is $O(n\sqrt{\log(n)\log\log\log(n)/\log\log(n)})$.*
- *No static parameter choice of $\lambda \in [n]$, $k \in [0..n]$, and $p \in [0,1]$ can give a better expected running time.*
- *There exists a fitness-dependent choice of λ and $k=\lambda$ such that the $(1+(\lambda,\lambda))$ GA has a linear expected running time on OneMax.*
- *A linear expected running time can also be achieved by a self-adjusting choice of $k=\lambda$.*

Note that these results answer one of the most prominent long-standing open problems in evolutionary computation: the usefulness of crossover in an optimization context. Previous and other recent examples exist where crossover has been shown to be beneficial [16, 18, 33, 36, 40, 56, 64, 83, 84], but in all of these publications, either nonstandard problems or operators were considered or the results hold only for uncommon parameter settings, or substantial additional mechanisms such as diversity-preserving selection schemes are needed to make crossover really work. To our knowledge, Theorem 3.9.1 is thus the first example that proves advantages of crossover in a natural algorithmic setting for a simple hill-climbing problem.

Without going into detail, we mention that the $(1+(\lambda,\lambda))$ GA has also been analyzed on a number of other benchmark problems, both by theoretical [8] and by empirical [25, 58, 76] means. These results indicate that the concept of using crossover as a repair mechanism can be useful far beyond OneMax.

3.9.2 Randomized Local Search with Self-Adjusting Mutation Strengths

Another example highlighting the impact that black-box complexity studies can have on the design of heuristic optimization techniques was presented in [30]. This work built on [31], where the tight bound for the unary unbiased black-box complexity of OneMax stated in Theorem 3.6.7 was presented. This bound is attained, up to an additive difference that is sublinear in n, by a variant of RLS that in each iteration chooses a value r that depends on the function value $\mathrm{OM}_z(x)$ of the current best search point x and then uses the flip_r variation operator introduced in Definition 3.6.3 to create an offspring y. The offspring y replaces x if and only if $\mathrm{OM}_z(y) \geq \mathrm{OM}_z(x)$. The dependence of r on the function value OM_z is rather complex and difficult to compute directly; see the discussion in [31]. Quite surprisingly, a self-adjusting choice of r is capable of identifying the optimal mutation strengths r in all but a small fraction of the iterations. This way, RLS with this self-adjusting parameter choice achieves an expected running time on OneMax that is only worse by an additive $o(n)$ term than that of the theoretically optimal unary unbiased black-box algorithm.

The algorithm in [30] will be discussed in Section 5.6 of this book. In the context of black-box optimization, it is interesting to note that the idea of taking a closer look at self-adjusting parameter choices, as well as our ability to investigate the optimality of such nonstatic parameter choices, is deeply rooted in the study of black-box complexities.

3.10 From Black-Box Complexity to Mastermind

In [29], the black-box complexity studies for OneMax were extended to the following generalization of OneMax to functions over an alphabet of size k.

For a given string $z \in [0..k-1]^n$, the *Mastermind* function f_z assigns to each search point $x \in [0..k-1]^n$ the number of positions in which x and z agree. Thus, formally,

$$f_z : [0..k-1]^n \to \mathbb{R}, x \mapsto |\{i \in [n] \mid x_i = z_i\}|.$$

The collection $\{f_z \mid z \in [0..k-1]^n\}$ of all such Mastermind functions forms the Mastermind problem of n *positions* and k *colors*.

The Mastermind problem models the homonymous board game, which was very popular in North America and in the western parts of Europe in the 1970s and 1980s. More precisely, it models a variant of this game, as in the original Mastermind game information is provided also about *colors* x_i that appear in z but not in the same position i; see [29] for details and results about this Mastermind variant using *black and white pegs*.

Mastermind and similar *guessing games* were studied in the computer science literature long before the release of Mastermind as a commercial board game. As we have discussed in Section 3.3, the case of $k = 2$ colors (this is the OneMax problem) had already been considered by Erdős and Rényi and several other authors in the early 1960s. These authors were mostly interested in the complexity- and information-theoretic aspects of this problem, and/or its cryptographic nature. The playful character of the problem, in turn, was the motivation of Knuth [67], who computed an optimal strategy that solves any Mastermind instance with $n = 4$ positions and $k = 6$ colors in at most five guesses.

The first to study the general case with arbitrary values of k was Chvátal [14].

Theorem 3.10.1 (Theorem 1 in [14]). *For every $k \geq 2$ the unrestricted black-box complexity of the Mastermind game with n positions and k colors is $\Omega(n \log k / \log n)$. For $\varepsilon > 0$ and $k \leq n^{1-\varepsilon}$, it is at most $(2+\varepsilon)n(1+2\log k)/\log(n/k)$.*

Note that for $k \leq n^{1-\varepsilon}$, $\varepsilon > 0$ being a constant, Theorem 3.10.1 gives an asymptotically tight bound of $\Theta(n \log k / \log n)$ for the k-color, n-position Mastermind game. Similarly to the random guessing strategy of Erdős and Rényi, it is sufficient to perform this many *random* queries, chosen independently and uniformly at random from $[0..k-1]^n$. That is, no adaptation is needed for such combinations of n and k to *learn* the secret target vector z.

The situation changes for the regime around $k = n$, which was the focus of several subsequent publications [13, 59, 60]. These publications all showed bounds of order $n \log n$ for the $k = n$ Mastermind problem. Originally motivated by the study of black-box complexities for randomized black-box heuristics, these bounds were improved to $O(n \log\log n)$ in [29].

Theorem 3.10.2 (Theorem 2.1 in [29]). *For Mastermind with n positions and $k = \Omega(n)$ colors, the unrestricted black-box complexity of the n-*

position, k-color Mastermind game is $O(n\log\log n + k)$*. For* $k = o(n)$*, it is* $O\left(n\log\left(\frac{\log n}{\log(n/k)}\right)\right)$.

Like the $O(n/\log n)$ bound for the case $k = 2$, the bounds in Theorem 3.10.2 can be achieved by *deterministic* black-box algorithms [29, Theorem 2.3]. On the other hand, and unlike the situation considered in Theorem 3.10.1, it can be shown that any (deterministic or randomized) $o(n\log n)$ algorithm for the Mastermind game with $k = \Theta(n)$ colors has to be *adaptive,* showing that in this regime adaptive strategies are indeed more powerful than nonadaptive ones.

Theorem 3.10.3 (Theorem 4.1 and Lemma 4.2 in [29]). *The nonadaptive unrestricted black-box complexity of the Mastermind problem with n positions and k colors is* $\Omega\left(\frac{n\log k}{\max\{\log(n/k),1\}}\right)$*. For* $k = n$*, this bound is tight, i.e., the nonadaptive unrestricted black-box complexity of the Mastermind problem with n positions and n colors is* $\Theta(n\log n)$*.*

Whether or not the $O(n\log\log n)$ upper bound in Theorem 3.10.2 can be further improved remains a – seemingly quite challenging – open problem. To date, the best known lower bound is the linear one reported in [14]. Some numerical results for different values of $k = n$ can be found in [7], but extending these numbers to asymptotic results may require a substantially new idea or technique for proving lower bounds in the unrestricted black-box complexity model.

3.11 Conclusion and Selected Open Problems

In this chapter we have surveyed theory-driven approaches that shed light on the performance limits of black-box optimization techniques such as local search strategies, nature-inspired heuristics, and pure random search. We have presented a detailed discussion of existing results for these black-box complexity measures. We now highlight a few avenues for future work in this young research discipline.

3.11.1 Extension to Other Optimization Problems

In line with the existing literature, our focus has been on classes of classical benchmark problems such as the OneMax, LeadingOnes, Jump, MST, and SSSP problems, since for these problems we can compare the black-box complexity results with known running-time results for well-understood heuristics. As with running-time analysis, it would be highly desirable to extend these results to other problem classes.

3.11.2 Systematic Investigation of Combined Black-Box Models

In the years before around 2013, most research on black-box complexity was centered around the question of how individual characteristics of state-of-the-art heuristics influence their performance. With this aim in mind, various black-box models have been developed that each restrict the algorithms with respect to some specific property, for example their memory size, or the properties of their variation operators or of the selection mechanisms in use. Since 2013 we have observed an increasing interest in combining two or more such restrictions to obtain a better picture of what is needed to design algorithms that significantly excel over existing approaches. A systematic investigation of such combined black-box models constitutes one of the most promising avenues for future research.

3.11.3 Tools to Derive Lower Bounds

To date, the most powerful technique to prove lower bounds on the black-box complexity of a problem is the information-theoretic approach, most notably in the form of Yao's minimax principle, and the simple information-theoretic lower bound presented in Theorem 3.2.4. Refined variants of this theorem have been designed to capture the situation in which the number of possible function values depends on the state of the optimization process, or where the probabilities for different objective values are nonhomogeneous. Unfortunately, either the verification that the conditions under which these theorems apply or the computation of a closed expression that summarizes the resulting bounds is often very tedious, making these extensions rather difficult to apply. Alternative tools for the derivation of lower bounds in black-box complexity contexts form another of the most desirable directions for future work.

In particular, for the k-ary unbiased black-box complexity with arities $k \geq 2$, we do not have any model-specific lower bounds. We therefore do not know, for example, if the linear bound on the binary unbiased black-box complexity of $\textsc{OneMax}_n$ or the $O(n \log n)$ bound on the binary unbiased black-box complexity of $\textsc{LeadingOnes}_n$ is tight, or whether the power of recombination is even larger than what these bounds, in comparison with the unary unbiased black-box complexities, indicate.

Another specifically interesting problem is raised by the $\Omega(n^2)$ lower bound on the (1+1) elitist black-box complexity of $\textsc{LeadingOnes}_n$ presented in Theorem 3.7.6. It has been conjectured in [51, Section 4] that this bound holds even for the (1+1) memory-restricted setting. A more systematic investigation of lower bounds for memory-restricted black-box models would help us to understand better the role of large populations in evolutionary computation, a question that is not very well understood to date.

3.11.4 Beyond Worst-Case Expected Optimization Time as Unique Performance Measure

Black-box complexity, as introduced in this chapter, takes the *worst-case expected optimization time* as the performance measure. This measure reduces the whole optimization procedure to one single number. This, naturally, has several disadvantages. The same critique applies to running-time analysis in general, which is very much centered around this single performance measure. Complementary performance indicators such as *fixed-budget* (see [65]) and *fixed-target* (see [11]) results have been proposed in the literature, but unfortunately have not yet attracted significant attention. Since these measure give a better picture of the *anytime behavior* of black-box optimization techniques, we believe that an extension of existing black-box complexity results to such anytime statements would make it easier to communicate and to discuss the results with practitioners, for whom the anytime performance is often at least as important as the expected optimization time.

In the same context, one may ask if the *expected* optimization time should be the only measure considered. Clearly, when the optimization time $T(A, f)$ of an algorithm A on a function f is highly concentrated, its expectation is often very similar to its median, and is in particular of the same or similar asymptotic order. Such concentration can be observed for the running time of classical heuristics on most of the benchmark problems considered in this chapter. At the same time, it is also not very difficult to construct problems for which such a concentration provably does not hold. In particular, for multimodal problems, in which two or more local optima exist, the running time is often not concentrated. In [49, Section 3] examples were presented for which the probability of finding a solution within a small polynomial given bound is rather large, but where – owing to excessive running times in the remaining cases – the expected optimization time is very large. This motivated the authors of [49] to introduce the concept of *p-Monte Carlo black-box complexity*. The *p-Monte Carlo black-box complexity* of a class $\mathcal{F}$ of functions measures the time it takes to optimize any problem $f \in \mathcal{F}$ with failure probability at most p. It was shown that even for small p, the p-Monte Carlo black-box complexity of a function class $\mathcal{F}$ can be smaller by an exponential factor than its traditional (expected) black-box complexity, which is referred to as the *Las Vegas black-box complexity* in [49].

Acknowledgements This work was supported by a public grant as part of the Investissement d'avenir project, reference ANR-11-LABX-0056-LMH, LabEx LMH, in a joint call with the Gaspard Monge Program for optimization, operations research, and their interactions with data sciences.

References

[1] Afshani, P., Agrawal, M., Doerr, B., Doerr, C., Larsen, K.G., Mehlhorn, K.: The query complexity of finding a hidden permutation. In: Space-Efficient Data Structures, Streams, and Algorithms - Papers in Honor of J. Ian Munro on the Occasion of His 66th Birthday, *Lecture Notes in Computer Science*, vol. 8066, pp. 1–11. Springer (2013)
[2] Anil, G., Wiegand, R.P.: Black-box search by elimination of fitness functions. In: Proc. of Foundations of Genetic Algorithms (FOGA'09), pp. 67–78. ACM (2009)
[3] Antipov, D., Doerr, B.: Precise runtime analysis for plateaus. In: Proc. of Parallel Problem Solving from Nature (PPSN'18), *Lecture Notes in Computer Science*, vol. 11102, pp. 117–128. Springer (2018)
[4] Badkobeh, G., Lehre, P.K., Sudholt, D.: Unbiased black-box complexity of parallel search. In: Proc. of Parallel Problem Solving from Nature (PPSN'14), *Lecture Notes in Computer Science*, vol. 8672, pp. 892–901. Springer (2014)
[5] Badkobeh, G., Lehre, P.K., Sudholt, D.: Black-box complexity of parallel search with distributed populations. In: Proc. of Foundations of Genetic Algorithms (FOGA'15), pp. 3–15. ACM (2015)
[6] Bshouty, N.H.: Optimal algorithms for the coin weighing problem with a spring scale. In: Proc. of the 22nd Conference on Learning Theory (COLT'09). Omnipress (2009)
[7] Buzdalov, M.: An algorithm for computing lower bounds for unrestricted black-box complexities. In: Companion Material for Proc. of Genetic and Evolutionary Computation Conference (GECCO'16), pp. 147–148. ACM (2016)
[8] Buzdalov, M., Doerr, B.: Runtime analysis of the $(1+(\lambda,\lambda))$ Genetic Algorithm on random satisfiable 3-CNF formulas. In: Proc. of Genetic and Evolutionary Computation Conference (GECCO'17), pp. 1343–1350. ACM (2017)
[9] Buzdalov, M., Doerr, B., Kever, M.: The unrestricted black-box complexity of jump functions. Evolutionary Computation **24**(4), 719–744 (2016). DOI 10.1162/EVCO_a_00185. URL `https://doi.org/10.1162/EVCO_a_00185`
[10] Cantor, D.G., Mills, W.H.: Determining a subset from certain combinatorial properties. Canadian Journal of Mathematics **18**, 42–48 (1966)
[11] Carvalho Pinto, E., Doerr, C.: Discussion of a more practice-aware runtime analysis for evolutionary algorithms. In: Proc. of Artificial Evo-

lution (EA'17), pp. 298–305 (2017). URL https://ea2017.inria.fr//EA2017_Proceedings_web_ISBN_978-2-9539267-7-4.pdf
[12] Cathabard, S., Lehre, P.K., Yao, X.: Non-uniform mutation rates for problems with unknown solution lengths. In: Proc. of Foundations of Genetic Algorithms (FOGA'11), pp. 173–180. ACM (2011)
[13] Chen, Z., Cunha, C., Homer, S.: Finding a hidden code by asking questions. In: Proc. of the 2nd Annual International Conference on Computing and Combinatorics (COCOON'96), pp. 50–55. Springer (1996)
[14] Chvátal, V.: Mastermind. Combinatorica **3**, 325–329 (1983)
[15] Corus, D., He, J., Jansen, T., Oliveto, P.S., Sudholt, D., Zarges, C.: On easiest functions for mutation operators in bio-inspired optimisation. Algorithmica **78**, 714–740 (2017). DOI 10.1007/s00453-016-0201-4. URL https://doi.org/10.1007/s00453-016-0201-4
[16] Dang, D., Friedrich, T., Kötzing, T., Krejca, M.S., Lehre, P.K., Oliveto, P.S., Sudholt, D., Sutton, A.M.: Escaping local optima with diversity mechanisms and crossover. In: Proc. of Genetic and Evolutionary Computation Conference (GECCO'16), pp. 645–652. ACM (2016)
[17] Dang, D., Lehre, P.K.: Runtime analysis of non-elitist populations: From classical optimisation to partial information. Algorithmica **75**, 428–461 (2016). DOI 10.1007/s00453-015-0103-x. URL https://doi.org/10.1007/s00453-015-0103-x
[18] Dang, D.C., Friedrich, T., Kötzing, T., Krejca, M.S., Lehre, P.K., Oliveto, P.S., Sudholt, D., Sutton, A.M.: Escaping local optima using crossover with emergent diversity. IEEE Transactions on Evolutionary Computation **22**(3), 484–497 (2018)
[19] Doerr, B.: Optimal parameter settings for the $(1+(\lambda,\lambda))$ genetic algorithm. In: Proc. of Genetic and Evolutionary Computation Conference (GECCO'16), pp. 1107–1114. ACM (2016)
[20] Doerr, B., Doerr, C.: Black-box complexity: from complexity theory to playing Mastermind. In: Companion Material for Proc. of Genetic and Evolutionary Computation Conference (GECCO'14), pp. 623–646. ACM (2014). URL http://doi.acm.org/10.1145/2598394.2605352
[21] Doerr, B., Doerr, C.: Optimal parameter choices through self-adjustment: Applying the 1/5-th rule in discrete settings. In: Proc. of Genetic and Evolutionary Computation Conference (GECCO'15), pp. 1335–1342. ACM (2015)
[22] Doerr, B., Doerr, C.: A tight runtime analysis of the $(1+(\lambda,\lambda))$ genetic algorithm on OneMax. In: Proc. of Genetic and Evolutionary Computation Conference (GECCO'15), pp. 1423–1430. ACM (2015)
[23] Doerr, B., Doerr, C.: The impact of random initialization on the runtime of randomized search heuristics. Algorithmica **75**, 529–553 (2016). URL https://doi.org/10.1007/s00453-015-0019-5
[24] Doerr, B., Doerr, C.: Optimal static and self-adjusting parameter choices for the $(1+(\lambda,\lambda))$ genetic algorithm. Algorithmica **80**, 1658–1709 (2018)

[25] Doerr, B., Doerr, C., Ebel, F.: From black-box complexity to designing new genetic algorithms. Theoretical Computer Science **567**, 87–104 (2015)

[26] Doerr, B., Doerr, C., Kötzing, T.: The unbiased black-box complexity of partition is polynomial. Artificial Intelligence **216**, 275–286 (2014). URL https://doi.org/10.1016/j.artint.2014.07.009

[27] Doerr, B., Doerr, C., Kötzing, T.: Solving problems with unknown solution length at (almost) no extra cost. In: Proc. of Genetic and Evolutionary Computation Conference (GECCO'15), pp. 831–838. ACM (2015). URL http://doi.acm.org/10.1145/2739480.2754681

[28] Doerr, B., Doerr, C., Kötzing, T.: Unbiased black-box complexities of jump functions. Evolutionary Computation **23**, 641–670 (2015). URL https://doi.org/10.1162/EVCO_a_00158

[29] Doerr, B., Doerr, C., Spöhel, R., Thomas, H.: Playing Mastermind with many colors. Journal of the ACM **63**, 42:1–42:23 (2016). URL http://dl.acm.org/citation.cfm?id=2987372

[30] Doerr, B., Doerr, C., Yang, J.: k-bit mutation with self-adjusting k outperforms standard bit mutation. In: Proc. of Parallel Problem Solving from Nature (PPSN'16), *Lecture Notes in Computer Science*, vol. 9921, pp. 824–834. Springer (2016)

[31] Doerr, B., Doerr, C., Yang, J.: Optimal parameter choices via precise black-box analysis. In: Proc. of Genetic and Evolutionary Computation Conference (GECCO'16), pp. 1123–1130. ACM (2016)

[32] Doerr, B., Fouz, M., Witt, C.: Quasirandom evolutionary algorithms. In: Proc. of Genetic and Evolutionary Computation Conference (GECCO'10), pp. 1457–1464. ACM (2010). DOI 10.1145/1830483.1830749. URL http://doi.acm.org/10.1145/1830483.1830749

[33] Doerr, B., Happ, E., Klein, C.: Crossover can provably be useful in evolutionary computation. Theoretical Computer Science **425**, 17–33 (2012)

[34] Doerr, B., Jansen, T., Witt, C., Zarges, C.: A method to derive fixed budget results from expected optimisation times. In: Proc. of Genetic and Evolutionary Computation Conference (GECCO'13), pp. 1581–1588. ACM (2013). DOI 10.1145/2463372.2463565. URL http://doi.acm.org/10.1145/2463372.2463565

[35] Doerr, B., Johannsen, D., Kötzing, T., Lehre, P.K., Wagner, M., Winzen, C.: Faster black-box algorithms through higher arity operators. In: Proc. of Foundations of Genetic Algorithms (FOGA'11), pp. 163–172. ACM (2011)

[36] Doerr, B., Johannsen, D., Kötzing, T., Neumann, F., Theile, M.: More effective crossover operators for the all-pairs shortest path problem. Theoretical Computer Science **471**, 12–26 (2013)

[37] Doerr, B., Kötzing, T., Lengler, J., Winzen, C.: Black-box complexities of combinatorial problems. Theoretical Computer Science **471**, 84–106 (2013)

[38] Doerr, B., Kötzing, T., Winzen, C.: Too fast unbiased black-box algorithms. In: Proc. of Genetic and Evolutionary Computation Conference (GECCO'11), pp. 2043–2050. ACM (2011)
[39] Doerr, B., Le, H.P., Makhmara, R., Nguyen, T.D.: Fast genetic algorithms. In: Proc. of Genetic and Evolutionary Computation Conference (GECCO'17), pp. 777–784. ACM (2017). DOI 10.1145/3071178.3071301. URL http://doi.acm.org/10.1145/3071178.3071301
[40] Doerr, B., Theile, M.: Improved analysis methods for crossover-based algorithms. In: Proc. of Genetic and Evolutionary Computation Conference (GECCO'09), pp. 247–254. ACM (2009)
[41] Doerr, B., Winzen, C.: Black-box complexity: Breaking the $O(n \log n)$ barrier of LeadingOnes. In: Artificial Evolution (EA'11), Revised Selected Papers, *Lecture Notes in Computer Science*, vol. 7401, pp. 205–216. Springer (2012)
[42] Doerr, B., Winzen, C.: Memory-restricted black-box complexity of OneMax. Information Processing Letters **112**(1-2), 32–34 (2012). URL https://doi.org/10.1016/j.ipl.2011.10.004
[43] Doerr, B., Winzen, C.: Playing Mastermind with constant-size memory. Theory of Computing Systems **55**, 658–684 (2014). URL https://doi.org/10.1007/s00224-012-9438-8
[44] Doerr, B., Winzen, C.: Ranking-based black-box complexity. Algorithmica **68**, 571–609 (2014). URL https://doi.org/10.1007/s00453-012-9684-9
[45] Doerr, B., Winzen, C.: Reducing the arity in unbiased black-box complexity. Theoretical Computer Science **545**, 108–121 (2014). URL https://doi.org/10.1016/j.tcs.2013.05.004
[46] Doerr, C., Lengler, J.: Elitist black-box models: Analyzing the impact of elitist selection on the performance of evolutionary algorithms. In: Proc. of Genetic and Evolutionary Computation Conference (GECCO'15), pp. 839–846. ACM (2015). URL http://doi.acm.org/10.1145/2739480.2754654
[47] Doerr, C., Lengler, J.: OneMax in black-box models with several restrictions. In: Proc. of Genetic and Evolutionary Computation Conference (GECCO'15), pp. 1431–1438. ACM (2015)
[48] Doerr, C., Lengler, J.: The (1+1) elitist black-box complexity of LeadingOnes. In: Proc. of Genetic and Evolutionary Computation Conference (GECCO'16), pp. 1131–1138. ACM (2016). URL http://doi.acm.org/10.1145/2908812.2908922
[49] Doerr, C., Lengler, J.: Introducing elitist black-box models: When does elitist behavior weaken the performance of evolutionary algorithms? Evolutionary Computation **25** (2017). DOI 10.1162/evco_a_00195. URL https://doi.org/10.1162/evco_a_00195
[50] Doerr, C., Lengler, J.: OneMax in black-box models with several restrictions. Algorithmica **78**, 610–640 (2017). URL https://doi.org/10.1007/s00453-016-0168-1

[51] Doerr, C., Lengler, J.: The (1+1) elitist black-box complexity of LeadingOnes. Algorithmica **80**, 1579–1603 (2018). DOI 10.1007/s00453-017-0304-6. URL `https://doi.org/10.1007/s00453-017-0304-6`

[52] Dorigo, M., Stützle, T.: Ant colony optimization. MIT Press (2004)

[53] Droste, S., Jansen, T., Wegener, I.: On the analysis of the (1+1) evolutionary algorithm. Theoretical Computer Science **276**, 51–81 (2002)

[54] Droste, S., Jansen, T., Wegener, I.: Upper and lower bounds for randomized search heuristics in black-box optimization. Theory of Computing Systems **39**, 525–544 (2006)

[55] Erdős, P., Rényi, A.: On two problems of information theory. Magyar Tudományos Akadémia Matematikai Kutató Intézet Közleményei **8**, 229–243 (1963)

[56] Fischer, S., Wegener, I.: The Ising model on the ring: Mutation versus recombination. In: Proc. of Genetic and Evolutionary Computation Conference (GECCO'04), *Lecture Notes in Computer Science*, vol. 3102, pp. 1113–1124. Springer (2004)

[57] Fournier, H., Teytaud, O.: Lower bounds for comparison based evolution strategies using VC-dimension and sign patterns. Algorithmica **59**, 387–408 (2011)

[58] Goldman, B.W., Punch, W.F.: Fast and efficient black box optimization using the parameter-less population pyramid. Evolutionary Computation **23**, 451–479 (2015)

[59] Goodrich, M.T.: On the algorithmic complexity of the Mastermind game with black-peg results. Information Processing Letters **109**, 675–678 (2009)

[60] Jäger, G., Peczarski, M.: The number of pessimistic guesses in generalized black-peg Mastermind. Information Processing Letters **111**, 933–940 (2011)

[61] Jansen, T.: Black-box complexity for bounding the performance of randomized search heuristics. In: Y. Borenstein, A. Moraglio (eds.) Theory and Principled Methods for the Design of Metaheuristics, Natural Computing Series, pp. 85–110. Springer (2014). DOI 10.1007/978-3-642-33206-7_5. URL `https://doi.org/10.1007/978-3-642-33206-7_5`

[62] Jansen, T.: On the black-box complexity of example functions: The real jump function. In: Proc. of Foundations of Genetic Algorithms (FOGA'15), pp. 16–24. ACM (2015)

[63] Jansen, T., Sudholt, D.: Analysis of an asymmetric mutation operator. Evolutionary Computation **18**, 1–26 (2010). DOI 10.1162/evco.2010.18.1.18101. URL `https://doi.org/10.1162/evco.2010.18.1.18101`

[64] Jansen, T., Wegener, I.: The analysis of evolutionary algorithms - a proof that crossover really can help. Algorithmica **34**, 47–66 (2002)

[65] Jansen, T., Zarges, C.: Performance analysis of randomised search heuristics operating with a fixed budget. Theoretical Computer Science **545**, 39–58 (2014). URL https://doi.org/10.1016/j.tcs.2013.06.007

[66] Kirkpatrick, S., Gelatt, C.D., Vecchi, M.P.: Optimization by simulated annealing. Science **220**(4598), 671–680 (1983)

[67] Knuth, D.E.: The computer as Master Mind. Journal of Recreational Mathematics **9**, 1–5 (1977)

[68] Kötzing, T., Neumann, F., Sudholt, D., Wagner, M.: Simple max-min ant systems and the optimization of linear pseudo-boolean functions. In: Proc. of Foundations of Genetic Algorithms (FOGA'11), pp. 209–218. ACM (2011). DOI 10.1145/1967654.1967673. URL http://doi.acm.org/10.1145/1967654.1967673

[69] de Perthuis de Laillevault, A., Doerr, B., Doerr, C.: Money for nothing: Speeding up evolutionary algorithms through better initialization. In: Proc. of Genetic and Evolutionary Computation Conference (GECCO'15), pp. 815–822. ACM (2015). URL http://doi.acm.org/10.1145/2739480.2754760

[70] Lässig, J., Sudholt, D.: Analysis of speedups in parallel evolutionary algorithms and (1+λ) EAs for combinatorial optimization. Theoretical Computer Science **551**, 66–83 (2014). DOI 10.1016/j.tcs.2014.06.037. URL https://doi.org/10.1016/j.tcs.2014.06.037

[71] Lehre, P.K., Witt, C.: Black-box search by unbiased variation. In: Proc. of Genetic and Evolutionary Computation Conference (GECCO'10), pp. 1441–1448. ACM (2010)

[72] Lehre, P.K., Witt, C.: Black-box search by unbiased variation. Algorithmica **64**, 623–642 (2012)

[73] Lindström, B.: On a combinatory detection problem i. Mathematical Institute of the Hungarian Academy of Science **9**, 195–207 (1964)

[74] Lindström, B.: On a combinatorial problem in number theory. Canadian Mathematical Bulletin **8**, 477–490 (1965)

[75] Metropolis, N., Rosenbluth, A.W., Rosenbluth, M.N., Teller, A.H., Teller, E.: Equation of state calculations by fast computing machines. The Journal of Chemical Physics **21**, 1087–1092 (1953)

[76] Mironovich, V., Buzdalov, M.: Hard test generation for maximum flow algorithms with the fast crossover-based evolutionary algorithm. In: Companion Material for Proc. of Genetic and Evolutionary Computation Conference (GECCO'15), pp. 1229–1232. ACM (2015)

[77] Motwani, R., Raghavan, P.: Randomized Algorithms. Cambridge University Press (1995)

[78] Neumann, F., Witt, C.: Runtime analysis of a simple ant colony optimization algorithm. Algorithmica **54**, 243–255 (2009)

[79] Neumann, F., Witt, C.: Bioinspired Computation in Combinatorial Optimization – Algorithms and Their Computational Complexity. Springer (2010)

[80] Paixão, T., Pérez Heredia, J., Sudholt, D., Trubenová, B.: Towards a runtime comparison of natural and artificial evolution. Algorithmica **78**, 681–713 (2017)

[81] Rowe, J., Vose, M.: Unbiased black box search algorithms. In: Proc. of Genetic and Evolutionary Computation Conference (GECCO'11), pp. 2035–2042. ACM (2011)

[82] Storch, T.: Black-box complexity: Advantages of memory usage. Information Processing Letters **116**(6), 428–432 (2016). DOI 10.1016/j.ipl.2016.01.009. URL `https://doi.org/10.1016/j.ipl.2016.01.009`

[83] Sudholt, D.: Crossover is provably essential for the Ising model on trees. In: Proc. of Genetic and Evolutionary Computation Conference (GECCO'05), pp. 1161–1167. ACM Press (2005)

[84] Sudholt, D.: Crossover speeds up building-block assembly. In: Proc. of Genetic and Evolutionary Computation Conference (GECCO'12), pp. 689–702. ACM (2012)

[85] Sudholt, D.: A new method for lower bounds on the running time of evolutionary algorithms. IEEE Transactions on Evolutionary Computation **17**, 418–435 (2013)

[86] Teytaud, O., Gelly, S.: General lower bounds for evolutionary algorithms. In: Proc. of Parallel Problem Solving from Nature (PPSN 2006), *Lecture Notes in Computer Science*, vol. 4193, pp. 21–31. Springer (2006)

[87] Witt, C.: Tight bounds on the optimization time of a randomized search heuristic on linear functions. Combinatorics, Probability & Computing **22**, 294–318 (2013)

[88] Yao, A.C.C.: Probabilistic computations: Toward a unified measure of complexity. In: Proc. of Foundations of Computer Science (FOCS'77), pp. 222–227. IEEE (1977)

Chapter 4
Parameterized Complexity Analysis of Randomized Search Heuristics

Frank Neumann and Andrew M. Sutton

Abstract This chapter compiles a number of results that apply the theory of parameterized algorithmics to the running-time analysis of randomized search heuristics such as evolutionary algorithms. The parameterized approach articulates the running time of algorithms solving combinatorial problems in finer detail than traditional approaches from classical complexity theory. We outline the main results and proof techniques for a collection of randomized search heuristics tasked to solve NP-hard combinatorial optimization problems such as finding a minimum vertex cover in a graph, finding a maximum leaf spanning tree in a graph, and the traveling salesperson problem.

4.1 Introduction

Randomized search heuristics (RSHs) are a class of general-purpose algorithms that are often deployed to tackle hard combinatorial optimization problems that arise in practice. Instances of practical, real-world problems are usually structured or restricted in some way, and it is typically assumed that RSH techniques are successful when the underlying strategy is able to exploit the structural properties of the resulting search space.

The mathematical analysis of the running time of randomized search heuristics on discrete optimization problems has advanced in the last decade. For a wide array of these techniques, rigorous and precise asymptotic bounds on the performance as a function of problem size are now available. However,

Frank Neumann
Optimisation and Logistics Group, School of Computer Science, The University of Adelaide, Adelaide, Australia e-mail: `frank.neumann@adelaide.edu.au`

Andrew M. Sutton
Department of Computer Science, University of Minnesota Duluth, Duluth, MN, USA
e-mail: `amsutton@d.umn.edu`

B. Doerr, F. Neumann (eds.), *Theory of Evolutionary Computation*,
Natural Computing Series, https://doi.org/10.1007/978-3-030-29414-4_4

many of these kinds of results are restricted only to toy problems. While such analyses are useful for gaining an understanding of the general working principles underlying RSH techniques, it is often not clear how they might be interpreted in the context of classically hard problems in computer science.

Unless $\mathsf{P} = \mathsf{NP}$, the worst-case runtime of an NP-hard problem cannot be bounded from above by a polynomial in the input size. This is a rather restrictive view, and it often tells us nothing about the typical behavior of algorithms on problems that are likely to be encountered in practice. For example, many experimental studies confirm that randomized search heuristics such as evolutionary algorithms (EAs), ant colony optimization, simulated annealing, and simple hill-climbing perform well on practical instances of NP-hard problems. An important research question for RSH techniques applied to combinatorial optimization is: which features of a given instance determine its hardness, and how do such parameters influence the runtime?

The field of *parameterized complexity* offers a refinement of classical time complexity by analyzing the running time of an algorithm not just as a function of problem size, but also as a function of further parameters of the input, for example, solution size, structural restrictions, or quality of approximation [12, 15]. The idea is to capture the essence of what makes a problem instance hard, and try to isolate this hardness to some structural feature of the instance or its solution. The inevitable combinatorial explosion in the runtime is confined to a function of this parameter, with only polynomial dependence on the input size. Even large instances may exhibit a very restricted structure and can be easier to solve, independent of size. Parameterized complexity is therefore an obvious candidate for systematically studying what features of a particular problem are hard for RSH techniques. It can also offer advice on what types of problem might be soluble or insoluble by such approaches, and guide algorithm design. It should be noted that parameterized analysis can also be applied to study the efficiency of modules of an evolutionary algorithm. A good example is the hypervolume indicator, which has been widely applied in the area of evolutionary multiobjective optimization. Computing the optimal hypervolume is hard when the dimension grows, and the computation of the hypervolume has been investigated in [5] from a parameterized and average-case perspective.

Many hard problems have "easy parts" that can be efficiently solved in order to effectively shrink a problem to its computationally hard core structure. This can be done by efficiently reducing the problem instance to a smaller instance (kernelization), or constraining the search tree to a manageable size that is still guaranteed to contain a solution (bounded search tree method). A slower exact algorithm (even brute-force search) can then be run on the resulting smaller instance or search space. With little to no hope of a polynomial-time solution, one instead seeks algorithms that can solve a problem in time that grows polynomially with the problem size, although perhaps superpolynomially with respect to some instance parameter. In other words, if the parameter is fixed to be small, the problem class is tractable, even as its instances grow large. Such a problem class (and corresponding algorithm) is called *fixed-parameter tractable* (FPT). A slightly less desirable situation is an algorithm that runs in so-called *slicewise polynomial time* (XP). Here the runtime is a polynomial in the problem size, but a polynomial whose degree depends on the parameter.

This kind of demarcation into hard and easy components can also be useful for the analysis of RSH techniques. At the extreme end of the spectrum are functions such as NEEDLE, whose black-box complexity establishes that no RSH could even beat simple random sampling in expectation. At the other extreme are problems from the ONEMAX class that are solved efficiently by even very simple approaches. Likely, practical optimization problems lie somewhere between these two extremes, containing some mixture of components that can be efficiently exploited by randomized search heuristics and components that essentially require random sampling. If the hard core component that demands random sampling is guaranteed to be small by the nature of the problem class, then RSH techniques can be a reasonable approach. The theory of parameterized complexity is therefore useful for isolating the structural features that can be efficiently exploited by RSH techniques from the hard "core" of a problem, on which an approach must resort to some kind of stochastic brute-force search behavior such as random walks, lucky jumps, or explicit restarts.

It should therefore not come as a surprise that analyzing randomized search heuristics from the perspective of parameterized complexity can lead to useful theoretical insights into algorithm design. For example, it has been shown that the specific choice of search operator can directly influence the fixed-parameter tractability of an algorithm on certain problems, for example, tree-preserving mutation on the maximum-leaf spanning tree problem [24] or standard uniform crossover on the closest-string problem [39].

The aim of this chapter is to discuss a number of results in the field of parameterized complexity applied to RSH techniques. We begin in Section 4.2 by introducing some background and technical details. In Section 4.3, we consider the maximum-leaf spanning tree problem and show that the use of a mutation operator commonly used for spanning trees reduces the XP runtime to FPT runtime when compared with standard bit mutations. In Section 4.4, we discuss multiobjective evolutionary algorithms that quickly focus their search on a kernel of minimum vertex cover instances, and subsequently perform random sampling on that kernel, resulting in FPT runtime. Decomposing the runtime analysis of an algorithm into a set of instance parameters is useful in its own right to better understand the components of a problem that influence the behavior of search heuristics. In Section 4.5, we present results on the maximization of submodular functions under different constraints. These results derive the expected time that simple evolutionary algorithms need to produce approximations as a function of both the problem size and additional parameters of the input. In Section 4.6, we describe the analysis of a standard evolutionary algorithm (EA) applied to the Euclidean traveling salesperson problem (TSP), which bounds the running time in the context of a well-known TSP parameterization (the number of points interior to the convex hull). In this case, it is possible to prove that the performance of the algorithm is bounded by the number of interior points, although this is not enough to obtain the desired fixed-parameter tractable runtime. On the other hand, if the EA is allowed to use some problem-specific information (namely, the cyclic order of points as they appear on the convex hull), it can explicitly focus its search on a small subset of states. This dramatic search space reduction yields fixed-parameter tractable runtimes for algorithms on parameterized TSP instances. We summarize the chapter in Section 4.7 and briefly discuss some open research problems.

4.2 Parameterized Complexity Analysis

Extending traditional runtime analysis by parameterization requires conducting a rigorous runtime analysis of an algorithm on a *parameterization* of a problem class. A parameterization of a problem class is a mapping of problem instances into the set of natural numbers. The running time of the algorithm is then expressed in terms of both the problem size and this extra parameter.

Let L be a language over a finite alphabet Σ. A *parameterization* of L is a mapping $\kappa : \Sigma^* \to \mathbb{N}$. The corresponding *parameterized problem* is the pair (L, κ). For a string $x \in \Sigma^*$, let $k = \kappa(x)$ and $n = |x|$. An algorithm deciding $x \in L$ in time bounded by $n^{g(k)}$ is called a *slicewise polynomial-time* algorithm (or XP algorithm). Here, $g : \mathbb{N} \to \mathbb{N}$ is an arbitrary but computable function. An algorithm deciding $x \in L$ in time bounded by $g(k) \cdot n^{O(1)}$ is called a *fixed-parameter tractable* (or FPT) algorithm for the parameterization κ. Both kinds of algorithms run in polynomial time for fixed k, but an XP algorithm allows the degree of the polynomial to depend on the parameter, while the degree of the polynomial for the running time is independent of both n and k for an FPT algorithm.

Randomized search heuristics are typically stochastic processes that are allowed to run for a certain number of iterations, after which the best-so-far result is collected and returned. In each iteration, the process keeps a set of one or more candidate solutions, and evaluates their quality via a *fitness* or *objective* function. The candidate solutions for the next iteration are then computed using a number of transformation operations.

To analyze this class of algorithm, we consider a random variable T that measures the number of basic iterations (usually measured in calls to the objective function) until a solution is first discovered. Here, a solution may be, depending on the context, an element that maximizes or minimizes the objective function. This allows us to treat optimization problems in the same manner as one would treat decision problems. Specifically, given a class of instances of an optimization problem, for each N one can construct a decision problem $L \subseteq \Sigma^*$ as the set of all instances on which the maximum (or, minimum) objective function value is at least (or, at most) a particular value.

The quantity $E[T]$ is the *expected optimization time*, and is the most commonly used performance measure in the rigorous runtime analysis of randomized search heuristics. We say an algorithm is a *Monte Carlo FPT algorithm* for a parameterized problem (L, κ) if it accepts $x \in L$ with probability at least $1/2$ in time $g(\kappa(x)) \cdot |x|^{O(1)}$ and accepts $x \notin L$ with probability zero. Thus, any randomized search heuristic with a bound $E[T] \leq g(\kappa(x)) \cdot |x|^{O(1)}$ on L can be trivially transformed into a Monte Carlo FPT algorithm by stopping its execution after $2g(\kappa(x)) \cdot |x|^{O(1)}$ iterations.

Note that the parameter is allowed to depend on the input in more or less an arbitrary way. The selection of a *meaningful* parameterization depends strongly on what a "typical" problem instance looks like. In most cases, one hopes to choose a parameter that is assumed to be small over the set of problems one wishes to solve. Ideally, the parameter should somehow capture the source of exponential complexity for the problem [15].

The goal of applying parameterized complexity analysis to the field of randomized search heuristics is thus to somehow understand how much information from the fitness function can be exploited in more detail. At the worst extreme, there is no exploitable information in the fitness of solutions at all (i.e., the fitness of a solution tells us nothing about its relationship to a global optimum), and we are in a blind NEEDLE-like case. Any RSH technique that employs such a fitness function must then rely entirely on getting lucky enough to stumble on an optimal solution. However, as previously mentioned, for most realistic problems we conjecture that there exists some structure in the fitness function that can be implicitly used by the RSH technique. Parameterized analysis can be seen as a technique that allows us to inspect the fitness function to assist in bounding how much "luck" is required to solve the problem.

4.3 Maximum-Leaf Spanning Trees

The classical minimum spanning tree problem, which can be solved in polynomial time by well-known deterministic algorithms such as those of Kruskal and Prim, has gained significant attention in the evolutionary computation literature [11, 32]. This includes the investigations of Witt [43], who considered an additional structural parameter of the given graph. He gave an upper bound on the runtime of simple evolutionary algorithms for the minimum spanning tree problem that depends on the circumference of the given graph. We will not present the details here, as the focus of this chapter is on NP-hard problems. We instead refer the interested reader to the original articles.

We start our investigations by considering an NP-hard variant of a spanning tree problem where the choice of mutation operator affects the parameterized runtime. Specifically, the commonly used standard bit mutation operation results in XP runtime, whereas a mutation operator that creates feasible solutions produces FPT runtime.

The problem we consider is the maximum-leaf spanning tree problem, and we summarize the results given in [24]. Given an undirected, connected graph $G=(V,E)$, the goal is to find a spanning tree T^* of G such that the number of leaves is maximum.

The authors of [24] considered two simple evolutionary algorithms that differ in the choice of the mutation operator. The first algorithm uses a general mutation operator carrying out standard bit mutations, and the second is specific to spanning tree problems. Both algorithms start with an arbitrary spanning tree T of G. We denote by m the number of edges in G, and by $\ell(T)$ the number of leaves of the spanning tree T. A new solution is accepted only if it is a spanning tree whose number of leaves is at least as high as the number of leaves in the current solution. The algorithm called the Generic (1+1) EA is given in Algorithm 4.1.

Algorithm 4.1: Generic (1+1) EA

1 Choose a spanning tree of T uniformly at random;
2 **repeat** *forever*
3 Produce T' by swapping each edge of T independently with probability $1/m$;
4 **if** T' *is a tree* **and** $\ell(T') \geq \ell(T)$ **then** $T \leftarrow T'$;

Swapping an edge in the mutation step of the Generic (1+1) EA means that if an edge is present in T then it is not contained in T' with probability $1/m$. On the other hand, if an edge is not present in T then it is contained in T' with probability $1/m$. An edge does not change from T to T' with probability $1-1/m$ in each mutation step, independently of the other edges.

The mutation operator of Algorithm 4.1 does not necessarily create an offspring that is a tree. If the offspring is not a tree, then this individual is discarded, as it represents an infeasible solution.

The second algorithm we consider is called the Tree-Based (1+1) EA and is illustrated in Algorithm 4.2. This approach uses a problem-specific mutation operator that ensures valid solutions, i.e., spanning trees. It is well known that, given a spanning tree T, a new spanning tree T' can be created by introducing an edge $e \in E \setminus T$ and removing an edge from the resulting cycle. Mutation operators based on this idea are commonly used when applying evolutionary algorithms to NP-hard spanning tree problems.

Our goal is to point out the differences between the two algorithms. To do this, we compare the expected optimization time $E[T]$ of the two algorithms. This shows that the problem-specific mutation operator of Algorithm 4.2 makes the difference between a fixed-parameter evolutionary algorithm and

Algorithm 4.2: Tree-Based (1+1) EA

1 Choose an arbitrary spanning tree T of G;
2 **repeat** *forever*
3 Choose S according to a Poisson distribution with parameter $\lambda = 1$ and perform sequentially S random edge-exchange operations to obtain a spanning tree T'. A random exchange operation applied to a spanning tree $\tilde{T}$ chooses an edge $e \in E \setminus \tilde{T}$ uniformly at random. The edge e is inserted and one randomly chosen edge of the cycle in $\tilde{T} \cup \{e\}$ is deleted;
4 **if** $\ell(T') \geq \ell(T)$ **then** $T \leftarrow T'$;

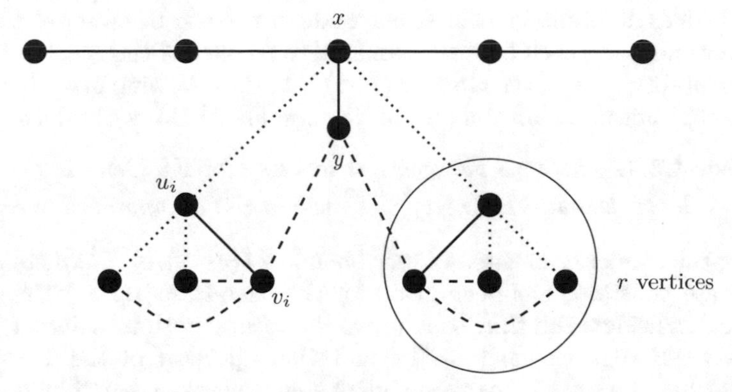

Fig. 4.1 Local optimum, shown with dashed edges, and global optimum, shown with dotted edges; shared edges are drawn solid.

an evolutionary algorithm that cannot compute an optimal solution in expected FPT time.

For the Generic (1+1) EA, the authors of [24] gave a lower bound which showed that the algorithm cannot solve the problem in FPT time. They considered the graph given in Fig. 4.1. The instance contains a local optimum, which has a distance to the global optimum in terms of the number of edges that have to be exchanged. The number of these edge exchanges depends on the number of nodes, r, the magnitude of which can be chosen to make it hard or easy to escape from the local optimum.

Formally, our graph, called G_{loc} (see Fig. 4.1) contains two components consisting of r vertices each. In component i, $1 \leq i \leq 2$, two vertices u_i and v_i are connected to all the other vertices in that component. The vertex u_i is connected to vertex x, which lies outside the component. Similarly, vertex v_i is connected to vertex y. In addition, x and y share an edge. The graph is completed by attaching a path of $n - 2r - 2$ vertices to the vertex x. A tree has to contain all the edges of the path attached to x. In addition, at least one of the edges $\{u_i, x\}$ and $\{v_i, y\}$ has to be chosen for each i. For a given component, the maximum number of possible leaves is at most $r - 1$. This can be obtained by attaching all nodes of the component either to u_i or to v_i.

The graph contains a local optimum T_{lopt} which consists of all edges attached to the vertices v_i, $1 \leq i \leq 2$, the edge $\{x, y\}$, and all path edges. The global optimum T_{opt} consists of all edges attached to the vertices u_i, $1 \leq i \leq 2$, the edge $\{x, y\}$, and all path edges. Compared with T_{lopt}, T_{opt} has an extra leaf, namely the vertex y. However, T_{lopt} and T_{opt} differ by $4(r-1)$, edges which make it hard for the algorithms under consideration to obtain T_{opt} if T_{lopt} has been produced before.

T_{lopt} can only by improved by swapping at least $2(r-2)$ edges, as all nonsolid edges adjacent to at least one node v_i need to be swapped to reach an improvement. As each bit corresponding to an edge of the graph is flipped with probability $1/m$ in the Generic (1+1) EA, the following lower bound on the expected optimization time of the Generic (1+1) EA is obtained.

Theorem 4.3.1. *The expected optimization time of the Generic (1+1) EA on G_{loc} is lower bounded by $(m/c)^{2(r-2)}$ where c is an appropriate constant.*

Using the same arguments, a lower bound of $((r-2)/c)^{r-2}$ where c is an appropriate constant, has been given for the Tree-Based (1+1) EA. Again the bound considers the time to improve the locally optimal solution, which requires $r-2$ edge exchanges. The mutation operator of the Tree-Based (1+1) EA has the benefit that a spanning tree is always created by introducing an edge and removing an edge from the resulting cycle, which results in a lower bound that is smaller than the one obtained for the Generic (1+1) EA. In terms of upper bounds, the Tree-Based (1+1) EA runs in FPT time when the value of an optimal solution k is the parameter.

The proof of the main result builds on the following lemma, which upper bounds the number of edges and the number of nodes of degree at least three as a function of k.

Lemma 4.3.2. *Any connected graph G on n nodes and with a maximum number of k leaves in any spanning tree has at most $n+5k^2-7k$ edges and at most $10k-14$ nodes of degree at least three.*

Each spanning tree has $n-1$ edges, which implies that the number of edge exchanges to obtain a maximum-leaf spanning tree from any spanning tree is $n+5k^2-7k-(n-1) \leq 5k^2$. Furthermore, a nonoptimal spanning tree can be improved by removing an edge of degree two from the cycle. The number of nodes of degree at least 3 is at most $10k-14$, which gives a lower bound of $1/20k$ on the probability of removing an edge of degree two from the cycle.

The upper bound for the Tree-Based (1+1) EA is given in the following theorem, and the proof uses the arguments stated above.

Theorem 4.3.3. *If the maximum number of leaf nodes in any spanning tree of G is k, then the Tree-Based (1+1) EA finds an optimal solution in expected time $O(2^{15k^2 \log k})$.*

4.4 Minimum Vertex Cover

The minimum vertex cover problem is an important classical NP-hard combinatorial optimization problem. Given an undirected connected graph $G = (V,E)$, the task is to find a minimum set of vertices $V' \subseteq V$ such that each edge $e \in E$ is covered by one of the chosen nodes, i.e., $e \cap V' \neq \emptyset$ holds for each $e \in E$. A set of vertices V' covering each edge $e \in E$ is called a vertex cover.

Using a binary variable x_i for each vertex $v_i \in V$, the minimum vertex cover problem can be formulated as the following integer linear program (ILP):

$$\begin{aligned} \text{minimize} \quad & \sum_{i=1}^{n} x_i \\ \text{subject to} \quad & x_i + x_j \geq 1, \quad \forall \{i,j\} \in E, \\ & x_i \in \{0,1\}, \qquad 1 \leq i \leq n. \end{aligned}$$

The linear program (LP) relaxation is obtained by relaxing the requirement $x_i \in \{0,1\}$ to $x_i \in [0,1], 1 \leq i \leq n$

The vertex cover problem is the most prominent problem in the area of parameterized complexity. As stated before, this area usually deals with decision problems. In the case of the vertex cover problem, one asks whether a given graph G has a vertex cover of at most k nodes.

Earlier studies [16, 33] on the performance of the $(1+1)$ EA have shown that this algorithm may get stuck in the smaller component of a complete bipartite graph when the two partitions have different sizes. Escaping this local optimum requires the algorithm to flip all bits belonging to the global optimum at once, and therefore has a waiting time of $\Omega(n^{\mathsf{OPT}})$, where OPT is the value of an optimal solution. Furthermore, if the two partitions V_1 and V_2 of the bipartite graph are extremely unbalanced, say $|V_1| = n^{\varepsilon}$ and $|V_2| = n^{1-\varepsilon}$, where $\varepsilon > 0$ is an arbitrary small constant, then the approximation ratio achieved by getting stuck in a local optimum is only $n^{1-\varepsilon}/n^{\varepsilon} = n^{1-2\varepsilon}$ and can therefore be made very close to the trivial approximation achieved by selecting all vertices of the given graph.

4.4.1 Global SEMO

We consider the search space $\{0,1\}^n$, where each bit x_i of a search point x corresponds to a vertex v_i of the given graph G. The vertex v_i is chosen in the solution x iff $x_i = 1$. The task is to find a solution x with a minimum number of vertices that covers all edges. This motivates us to introduce a fitness function based on the number of edges left uncovered by x.

We denote by $E(x)$ the set of edges covered by the cover x, i.e., $E(x) := \{e \mid e \cap V_x \neq \emptyset\}$, where $V_x := \{v_i \mid x_i = 1, 1 \leq i \leq n\}$ is the subset of vertices chosen by x.

Kratsch and Neumann [25] considered two fitness functions for minimum vertex cover. The first fitness function was

$$f_1(x) = (|x|_1, u(x)),$$

where $|x|_1 = |\{i : x_i = 1\}|$ corresponds to the number of chosen vertices and $u(x) := |E \setminus E(x)|$ is the number of edges left uncovered by x. Note that $u(x)$ is useful for directing the search process towards a feasible solution, i.e., a solution x for which $u(x) = 0$ holds. This function had already been considered in [16] in the context of approximations.

In addition, the authors of [25] examined a second fitness function that uses additional information obtained from a linear program. Let $G(x) = (V, E \setminus E(x))$ be the graph obtained from G by removing all edges covered by nodes in x. We also consider the fitness function

$$f_2(x) = (|x|_1, LP(x)),$$

where $LP(x)$ denotes the optimum value of the relaxed vertex cover ILP for $G(x)$, i.e., the cost of an optimal fractional vertex cover of $G(x)$.

Algorithm 4.3: Global SEMO

```
1  Choose an initial solution x ∈ {0,1}^n uniformly at random;
2  Determine f(x) and initialize P ← {x};
3  repeat forever
4      Choose x ∈ P randomly;
5      Create x' by flipping each bit of x independently with probability 1/n;
6      Determine f(x');
7      if ∃x'' ∈ P, f(x'') ≤ f(x') and f(x'') ≠ f(x') then
8          P is unchanged
9      else
10         exclude all x'' where f(x') ≤ f(x'') from P and add x' to P
```

The multiobjective approach uses the Global SEMO algorithm (see Algorithm 4.3). The algorithm starts with a bit string chosen uniformly at random. In each iteration, one individual x of the current population P is selected uniformly at random and undergoes standard bit mutation to produce an offspring x'. The offspring x' is added to the population iff it is not strictly dominated by any other individual in P. In this case, all individuals in P that are (weakly) dominated by x' are removed from P. We will examine Global SEMO for the minimum vertex cover problem in this section and for maximization in several different types of problem involving submodular functions in the next section.

When minimizing the number of uncovered edges and the number of chosen vertices at the same time, Global SEMO achieves an approximation to within a factor of $O(\log n)$ for the minimum vertex cover problem. These results may be generalized to the wider class of set cover problems. Kratsch and Neumann [25] have used a modification of Global SEMO (called Global SEMO$_{alt}$) and shown that their approach computes an optimal solution in FPT time.

Algorithm 4.4: Alternative mutation operator in Global SEMO$_{alt}$

1 Let $U(x) \subseteq E$ denote the set of edges that are not covered by x;
2 Let $S(x) \subseteq \{1,\ldots,n\}$ denote the vertices that are incident on the edges in $U(x)$;
3 Choose $b \in \{0,1\}$ uniform at random;
4 **if** $b=0$ **or** $S(x)=\emptyset$ **then**
5 flip each bit of x independently with probability $1/n$
6 **else**
7 flip each bit of $S(x)$ independently with probability $1/2$;
8 flip each bit of $\{1,\ldots,n\} \setminus S(x)$ independently with probability $1/n$

The results presented rely on an alternative mutation operator (see Algorithm 4.4) that has the ability to perform bit flips with a high probability if the corresponding node is adjacent to at least one uncovered edge (line 7 of Algorithm 4.4). This allows the algorithm to perform random sampling on the subgraph consisting of the uncovered edges. If this subgraph constitutes a kernel of the problem, the random sampling process is similar to a brute-force search on the kernel. We will summarize those results in the following.

We outline the results for the algorithms introduced in this section, but should also mention that the vertex cover problem has been subject to further parameterized analyses in the context of randomized search heuristics. For example, the investigations of the vertex cover problem that we present in this section have been extended to the weighted vertex cover problem [35]. Gao et al. [18] have studied random initialization heuristics as well as local search algorithms in terms of parameterized complexity and approximation. Furthermore, the vertex cover problem has been analyzed in dynamic settings where edges can be removed from or added to the graph [34].

4.4.2 Parameterized Analysis

The first parameterized result in the context of optimal vertex covers considers Global SEMO$_{alt}$ together with the objective function f_1, which uses the number of uncovered edges as the second objective. The population size of the algorithm is upper bounded by $n+1$, as the main objective (number of chosen nodes) can only take on that many different values. The same upper bound on the population size is applied when using f_2.

The first analysis relies on the following basic insight. Let OPT be the value of an optimal solution; then an optimal solution has to include all nodes of degree at least $\mathsf{OPT}+1$. This is based on the simple observation that if a node v of degree $\mathsf{OPT}+1$ is not selected, all neighbors of v have to be selected, resulting in a nonoptimal solution.

Theorem 4.4.1. *The expected optimization time of Global SEMO$_{alt}$ for the minimum vertex cover problem using the fitness function f_1 is upper bounded by $O(\mathsf{OPT}\cdot n^4+n\cdot 2^{\mathsf{OPT}+\mathsf{OPT}^2})$.*

The proof of the theorem proceeds in several different phases. First, the expected time until the search point 0^n is included in the population is analyzed. The proof for this part focuses on selecting the individual with the smallest number of 1-bits, which happens with probability at least $1/(n+1)$, as the number of different values for $|x|_1$ is at most $n+1$. Producing a solution with a smaller number of 1-bits is always accepted, and the problem can be seen as maximizing the number of 0-bits, slowed down by a population of size at most $n+1$. Hence, after an expected number of $O(n^2 \log n)$ steps of Global SEMO or Global SEMO$_{alt}$ using f_1 or f_2, the search point 0^n is included in the population.

We now consider f_1 and assume that the search point 0^n is already included in the population. Subsequently, the expected number of steps where the population does not contain a solution x for f_1 that is a kernel for the problem is upper bounded by $O(\mathsf{OPT}\cdot n^4)$. For f_1, x is a kernel iff the vertices chosen by x constitute a subset of an optimal solution and the maximum degree of $G(x)$ is at most OPT. In order to upper bound the number of steps where the population does not contain a solution x that is a kernel, a potential function with $O(n^2\mathsf{OPT})$ different values is taken into account that measures the population with respect to the number of uncovered edges that its individuals have. It can be shown that the potential can always be improved with probability at least $\Omega(1/n^2)$ if no kernel is contained in the population. As the potential cannot increase, the expected number of steps where the population does not contain a kernel is $O(n^4\cdot\mathsf{OPT})$

Denoting by $\hat{x}$ the resulting vertex cover, the kernel instance $G(\hat{x})$ has at most $\mathsf{OPT}^2+\mathsf{OPT}$ nonisolated nodes. In this case, the alternative mutation operator is able to produce the optimal solution from $\hat{x}$ in expected time $O(n\cdot 2^{\mathsf{OPT}+\mathsf{OPT}^2})$. In this upper bound, the factor n accounts for selecting the individual $\hat{x}$ with probability at least $1/(n+1)$ and the term $O(2^{\mathsf{OPT}+\mathsf{OPT}^2})$ accounts for mutating this individual into an optimal solution. The exponential component of the runtime arises from the waiting time to make a lucky random jump, but this jump is now required only on a reasonably small kernel instance.

The runtime bound can be improved if the value of an optimal linear program $LP(x)$ for the graph $G(x)$ consisting only of the uncovered edges is used as the second criterion, leading to the fitness function f_2. The goal is to minimize the penalty $LP(x)$, and we have $LP(x) = 0$ iff x is a vertex cover.

The analysis is based on the following result of Nemhauser and Trotter [31], who proved a very strong relation between optimal fractional vertex covers and minimum vertex covers.

Theorem 4.4.2. *Let x^* be an optimal fractional vertex cover and let $P_0, P_1 \subseteq V$ be the vertices whose corresponding components of x^* are 0 or 1, respectively. Then there exists a minimum vertex cover that contains P_1 and no vertex of P_0.*

Theorem 4.4.2 implies that one can take all vertices set to 1 in an optimal fractional vertex cover and reduce the size of the problem in this way. Furthermore, it is well known that every basic feasible solution x of the vertex cover LP relaxation is half-integral, i.e., we have $x \in \{0, 1/2, 1\}^n$ [4]. Using these properties, the following result has been shown.

Theorem 4.4.3. *The expected optimization time of Global SEMO$_{alt}$ for the minimum vertex cover problem using the fitness function f_2 is upper bounded by $O(n^2 \cdot \log n + \mathsf{OPT} \cdot n^2 + n \cdot 4^{\mathsf{OPT}})$.*

We now explain the key ideas of the proof. We already know that the population contains the search point 0^n after an expected number of $O(n^2 \log n)$ steps. After 0^n has been included in the population, the number of steps where the population does not contain a kernel is investigated. For f_2, a solution x is a kernel iff $LP(x) = LP(0^n) - |x|_1$ and each optimal fractional vertex cover assigns $1/2$ to each nonisolated vertex of $G(x)$. The number of steps where P does not contain such a kernel x after 0^n has been included in the population can be bounded by $O(\mathsf{OPT} \cdot n^2)$ using the following arguments. Solutions with objective value $(r, LP(0^n) - r)$ are Pareto optimal. The proof proceeds by considering the solution x with objective vector $(r, LP(0^n) - r)$ and the largest value of r in the population. If x is not a kernel, that x can be chosen for mutation with a probability of at least $1/(n+1)$ and one specific bit can be flipped with a probability of at least $1/(en)$ to produce a Pareto-optimal offspring x' with objective vector $(r+1, LP(0^n) - r - 1)$. As the value of the LP is upper bounded by OPT, at most OPT of such steps can happen. This upper bounds the number of additional steps (after 0^n has been included in the population) by $O(n^2 \cdot \mathsf{OPT})$.

Let $\hat{x}$ be the kernel with objective vector $(r, LP(0^n) - r)$, where r is the maximum such that all nonisolated vertices of $G(x)$ obtain a value of $1/2$ in $LP(\hat{x})$. $G(\hat{x})$ has at most $2(\mathsf{OPT} - |\hat{x}|_1) \leq 2 \cdot \mathsf{OPT}$ nonisolated vertices, as the vertices that are chosen belong to an optimal solution and every nonisolated vertex contributes $1/2$ to the LP value. The expected time to produce an optimal solution after a kernel $\hat{x}$ has been included in the population is $O(n \cdot$

$2^{2\cdot\mathsf{OPT}}) = O(n\cdot 4^{\mathsf{OPT}})$, as the optimal solution can be obtained by choosing $\hat{x}$ for mutation and flipping exactly the bits corresponding to the nonisolated nodes of an optimal solution while not flipping the remaining bits.

Kratsch and Neumann have also given the following trade-off results with respect to runtime and approximation. These results show the previous FPT time bound ($\varepsilon = 0$), as well as that Global SEMO$_{alt}$ achieves a 2-approximation ($\varepsilon = 1$) in expected polynomial time.

Theorem 4.4.4. *Using the fitness function f_2, the expected number of iterations of Global SEMO$_{alt}$ until it has generated a $(1+\varepsilon)$-approximate vertex cover, i.e., a solution of fitness $(r,0)$ with $r \leq (1+\varepsilon)\cdot\mathsf{OPT}$, is $O(n^2\cdot\log n + \mathsf{OPT}\cdot n^2 + n\cdot 4^{(1-\varepsilon)\cdot\mathsf{OPT}})$.*

The proof of Theorem 4.4.4 uses the same kernelization arguments as the proof of Theorem 4.4.3. Once a solution $\hat{x}$ that is a kernel of the problem has been produced, it is shown that if $\hat{x}$ is selected for mutation then it will mutate with probability $\Omega((1/4)^{(1-\varepsilon)\cdot\mathsf{OPT}'})$ into a solution x' for which

$$|x'|_1 + 2\cdot LP(x') \leq (1+\varepsilon)\cdot\mathsf{OPT}$$

holds. Such a solution x' can be turned into a vertex cover by single mutation steps that reduce $LP(x)$ by at least $1/2$ while increasing the size of the vertex cover by one, leading to a vertex cover of size at most $(1+\varepsilon)\cdot\mathsf{OPT}$.

4.5 Submodular Functions with Constraints

Submodular functions constitute a broad class of interesting problems. A function $f\colon 2^X \to \mathbb{R}$ is submodular iff $f(A\cup B)+f(A\cap B) \leq f(A)+f(B)$ for all $A,B \subseteq X$. In the context of optimizing a submodular function f, we will often consider the incremental value of adding a single element, leading to an equivalent definition. We denote by $F_i(A) = f(A\cup\{i\}) - f(A)$ the marginal value of i with respect to A. A function f is submodular iff $F_i(A) \geq F_i(B)$ for all $A \subseteq B \subseteq X$ and $i \in X\setminus B$.

We consider the problem of maximizing a given submodular function f. The problem is NP-hard, as it generalizes many NP-hard combinatorial optimization problems, such as maximum cut [14, 19] and several others [1, 7, 14, 21], The class of submodular functions also includes the class of linear functions that have been well studied in the area of theory of evolutionary computation. Friedrich and Neumann [17] have analyzed the maximization of submodular functions with different constraints and carried out runtime analyses depending on the parameters of the given constraint. We will summarize the results in this section.

Friedrich and Neumann considered the maximization of a given submodular function f under a given set of matroid constraints. A matroid is a pair

$(X,\mathcal{I})$ composed of a ground set X and a nonempty collection $\mathcal{I}$ of subsets of X satisfying (1) if $A \in \mathcal{I}$ and $B \subseteq A$ then $B \in \mathcal{I}$ and, (2) if $A, B \in \mathcal{I}$ and $|A| > |B|$ then $B + x \in \mathcal{I}$ for some $x \in A \setminus B$. The sets in $\mathcal{I}$ are called *independent*, and the *rank* of a matroid is the size of any maximal independent set. We will consider several different classes of submodular functions together with different types of matroid constraints.

Friedrich and Neumann analyzed the (1+1) EA and Global SEMO as baseline algorithms. For the (1+1) EA, the fitness function $h(x) = (v(x), f(x))$ was considered. Here, $v(x)$ measures the constraint violation of x. Generalizing the fitness function used by Reichel and Skutella [37] for the intersection of two matroids, they considered problems with k matroid constraints $M_1, \ldots, M_k$,

$$v(x) = k \cdot |x|_1 - \sum_{j=1}^{k} r_j(x),$$

where $r_j(x)$ denotes the rank of x in matroid M_j, i.e.,

$$r_j(X) = \max\{|Y| \colon Y \subseteq X, Y \in I_j\}$$

for the set X given by x.

We have $v(x) = 0$ iff x is a feasible solution and $v(x) > 0$ otherwise. The function $h(x)$ is optimized in lexicographic order, i.e.,

$$h(y) \geq h(x) \text{ holds iff } (v(y) < v(x)) \vee (v(y) = v(x) \wedge f(y) \geq f(x)).$$

We denote by F the set of feasible solutions. For Global SEMO, Friedrich and Neumann set $z(x) = f(x)$ iff $x \in F$ and $z(x) = -1$ iff $x \notin F$ and considered the multiobjective problem $g(x) := (z(x), |x|_0)$, where $|x|_0 = \sum_{i=1}^{n}(1 - x_i)$ denotes the number of 0-bits in the given bit string x. Adding the number of 0-bits as the second objective to be maximized forces the empty set to be Pareto optimal, and allows the algorithm to construct solutions greedily.

4.5.1 Monotone Functions with Uniform Constraints

We now summarize the results for the special class of monotone submodular functions under one uniform matroid constraint. A function f is monotone iff $f(A) \leq f(B)$ for all $A \subseteq B$. A uniform matroid constraint of size r means that a set is feasible iff it consists of at most r elements, i.e., $\mathcal{I} = \{A \subseteq X \colon |A| \leq r\}$.

A key property of Global SEMO that is often employed in theoretical analysis is that it constructs solutions in a manner similar to a greedy algorithm. Furthermore, the population size can be bounded by $n + 1$, as the number of different objective values for the second objective is $n + 1$. This implies that one particular individual that is needed for the analysis is selected with probability $\Omega(1/n)$. The algorithm removes elements in order to maximize the number of zeros. Using the number of zeros as the second objective implies that the algorithm maintains a population where the solution with the smallest number of elements is never removed. Furthermore, each solution that has a smaller number of selected elements than the solutions previously found is included in the population. Eventually, this leads to a population which includes the solution consisting of the empty set. In terms of the first objective (the overall goal function), the algorithm tries to maximize its objective value in a greedy manner. It does so by adding elements that provide the largest benefit to a current solution. Putting these arguments together, the following approximation result can be obtained for Global SEMO and the maximization of monotone submodular functions with a uniform constraint.

Theorem 4.5.1. *The expected time until Global SEMO has obtained a* $(1-1/e)$*-approximation for a monotone submodular function* f *under a uniform constraint of size* r *is* $O(n^2(\log n + r))$.

The proof of the theorem uses the fact that the population size is always bounded by $n+1$ and therefore one particular individual is selected with probability at least $1/(n+1)$ in each step. The first phase of the proof shows that the empty set, represented by the bit string 0^n, is included in the population in expected time $O(n^2 \log n)$. Similarly to the analysis for vertex cover in the previous section, this bound is obtained by considering the factor $O(n)$ for the population size and bounds on a coupon collector process for maximizing the number of 0-bits. The $O(n^2 r)$ term accounts for the greedy process where the correct individual in the population is selected with probability $\Omega(1/n)$ and the appropriate greedy step is applied to this individual with probability $\Omega(1/n)$. Finally, there are at most r of these steps, as no more than r elements can be inserted owing to the given constraint. The approximation ratio follows from the greedy process.

4.5.2 Monotone Submodular Functions under Matroid Constraints

Now we take a look at more complex problems. Again we consider monotone submodular functions but with k matroid constraints. The algorithm that we consider is the (1+1) EA. The number of these matroid constraints is the important parameter that we consider and it determines the approximation ratio that is achieved, as well as the exponent of the runtime. Furthermore, there is a parameter $p \geq 1$ that allows for a fixed value of k to trade off the approximation quality and runtime of the algorithm.

Theorem 4.5.2. *For any integers* $k \geq 2$, $p \geq 1$ *and a real value* $\varepsilon > 0$, *the expected time until the (1+1) EA has obtained a* $(1/(k+1/p+\varepsilon))$-

approximation for any monotone submodular function f under k matroid constraints is $O\Big(\frac{1}{\varepsilon}\cdot n^{2p(k+1)+1}\cdot k\cdot \log n\Big)$.

We summarize the main ideas of the proof here. The first part of the proof consists of showing that the algorithm reaches a feasible solution x with $f(x)\geq \mathsf{OPT}/n$. The expected time until the $(1+1)$ EA has obtained such a solution can be upper bounded by $O(n^{k+1})$. To attain this bound, the proof first argues that the $(1+1)$ EA obtains a feasible solution in expected time $O(kn(\log k+\log n))$ by using the fitness level method applied to the value of the penalty $v(x)$. Afterwards, it is shown that, from any feasible solution x, a feasible solution y with $f(x)\geq \mathsf{OPT}/n$ can be obtained by flipping $k+1$ specific bits. The expected waiting time for this event is $O(n^{k+1})$.

A p-exchange operation applied to the current solution x introduces at most $2p$ new elements and deletes at most $2kp$ elements of x. A solution y that can be obtained from x by a p-exchange operation is called a p-exchange neighbor of x. According to [27], every solution x for which there exists no p-exchange neighbor y with $f(y)\geq (1+\frac{\varepsilon}{n(k+1)})\cdot f(x)$ is a $(1/(k+1/p+\varepsilon))$-approximation for any monotone submodular function. So, the proof works by analyzing the time until a feasible solution has been obtained. Afterwards, it uses the fact that there is still a p-exchange neighbor unless the desired approximation quality has already been obtained.

4.5.3 Symmetric Submodular Functions under Matroid Constraints

We now summarize the main result for Global SEMO for the optimization of symmetric submodular functions under k matroid constraints. The following theorem makes use of the greedy and local search ability that the algorithm Global SEMO has.

Theorem 4.5.3. *The expected number of iterations until Global SEMO attains a* $\left(\frac{1}{(k+2)(1+\varepsilon)}\right)$*-approximation for any symmetric submodular function under k matroid constraints is* $O\big(\frac{1}{\varepsilon}n^{k+6}\log n\big)$, *for any constant* $\varepsilon>0$.

The analysis makes use of the following result in [26], which shows that there are always locally improving steps as long as the desired approximation quality has not been obtained.

Lemma 4.5.4. *Let x be a solution such that no solution with fitness at least* $\left(1+\frac{\varepsilon}{n^4}\right)\cdot f(x)$ *can be achieved by deleting one element or by inserting one element and deleting at most k elements. Then x is a* $\left(\frac{1}{(k+2)(1+\varepsilon)}\right)$*-approximation.*

The proof of Theorem 4.5.3 uses this lemma together with the fact that Global SEMO introduces the search point 0^n into the population after an expected number of $O(n^2 \log n)$ steps. As the search point 0^n is Pareto optimal, it stays in the population once it has been introduced. Selecting 0^n for mutation and inserting the element that leads to the largest increase in the f-value produces a solution y with $f(y) \geq \mathsf{OPT}/n$. The reason for this is that the number of elements is limited by n and that f is submodular. Global SEMO will also always have a solution with the largest f-value obtained so far in the population. Selecting this solution x for mutation and flipping at most $k+1$ specific bits according to Lemma 4.5.4 produces a solution y with $f(y) \geq \left(1+\frac{\varepsilon}{n^4}\right) \cdot f(x)$ as long as x does not yet have the desired approximation quality. The expected waiting time for this event is $O(n^{k+2})$, as at most $k+1$ specific bits of x have to be flipped and the population size is at most $n+1$.

The number of steps that improve the solution with the largest f-value needed in order to achieve the desired $\left(\frac{1}{(k+2)(1+\varepsilon)}\right)$-approximation is upper bounded by

$$\log_{\left(1+\frac{\varepsilon}{n^4}\right)} \frac{\mathsf{OPT}}{\mathsf{OPT}/n} = O\left(\frac{1}{\varepsilon} n^4 \log n\right)$$

which implies that the expected time to achieve a $\left(\frac{1}{(k+2)(1+\varepsilon)}\right)$-approximation is $O\left(\frac{1}{\varepsilon} n^{k+6} \log n\right)$.

4.6 Euclidean TSP

Given a set of n points $V = \{v_1, v_2, \ldots, v_n\}$ in the plane, the objective of the Euclidean TSP is to find a permutation $\pi \colon V \to V$ that minimizes the cost function

$$c(\pi) = \sum_{i=1}^{n} d(v_{\pi(i)}, v_{\pi(i+1)}), \tag{4.6.1}$$

where $d(v_i, v_j)$ denotes the Euclidean distance separating the points v_i and v_j and arithmetic is taken to be modulo n. The Euclidean TSP is NP-hard, but can be approximated to within a factor $(1+\varepsilon)$ for every fixed ε in polynomial time [2].

It is convenient to consider the complete undirected graph $G = (V, E)$ and define the Hamiltonian cycle $C(\pi) \subseteq E$ induced by the edges followed by a given permutation π:

$$C(\pi) = \{\{v_{\pi(1)}, v_{\pi(2)}\}, \{v_{\pi(2)}, v_{\pi(3)}\}, \ldots, \{v_{\pi(n-1)}, v_{\pi(n)}\}, \{v_{\pi(n)}, v_{\pi(1)}\}\}.$$

We will refer to the cycle $C(\pi)$ as a *tour*.

Iterative improvement methods rely on the iterated exchange of a small number of edges and are powerful approaches for solving large-scale TSP instances in practice. These heuristics move through the space of candidate solutions by repeatedly applying move or mutation operators to pivot between tours. For the TSP, this is typically some variant of the powerful k-opt operation. The k-opt move considers some candidate tour $C(\pi)$, and deletes k mutually disjoint edges and reassembles the remaining fragments into a new valid tour $C(\pi')$. The operation induces a neighborhood structure on the search space of tours, and thus serves as a strong and easy-to-implement local search operator. However, instances exist where this approach is provably inefficient. For example, local search algorithms employing a k-opt neighborhood operator can take exponential time even to find a locally optimal solution [6]. This even holds for the Euclidean case [13].

The convex hull of V is the smallest convex set containing V. A point $v \in V$ is called an *inner point* if v lies in the interior of the convex hull of V. We denote by $\mathsf{Inn}(V) \subset V$ the set of inner points of V, and define $\mathsf{Out}(V) := V \setminus \mathsf{Inn}(V)$. The TSP parameterized by $k = \mathsf{Inn}(V)$ is in FPT. Specifically, Deĭneko et al. [9] showed that if a Euclidean TSP instance with n vertices has k vertices interior to the convex hull, there is a dynamic programming FPT algorithm. Other parameterizations are not as propitious; for example, finding a local optimum in the k-opt neighborhood for the metric TSP is hard for $\mathsf{W}[1]$ [28]. $\mathsf{FPT} \subseteq \mathsf{W}[1]$, but the containment is conjectured to be proper [15], in which case no such FPT algorithm can exist.

Parameterized results for evolutionary algorithms for the Euclidean TSP have been developed in a series of papers [29, 30, 40, 41] in the context of the inner-point parameterization of Deĭneko et al. [9]. We also would like to mention that the generalized traveling salesperson problem has been investigated in the context of parameterized complexity. In this problem, the cities belong to different clusters and the goal is to compute a shortest tour that visits each cluster exactly once. We refer the interested reader for details of the generalized TSP to Corus et al. [8].

The remainder of this section sketches these results, starting with the setting in which the algorithm is oblivious to problem-specific information (other than the cost of a tour) and ending with algorithms that exploit problem-specific structure.

4.6.1 Black-Box Algorithms

In the black-box setting, heuristics are not allowed any access to domain-specific knowledge about the instance other than the cost of a tour. For Euclidean TSP instances with $k = \mathsf{Inn}(V)$ inner points, it is possible to show that the $(\mu+\lambda)$ EA generates an optimal solution in slicewise polynomial time (that is, in time $n^{g(k)}$, where g depends only on k). Later, in Section 4.6.2, we will discuss how it is possible to improve this to FPT time when domain knowledge is incorporated into the design of the algorithm.

The 2-opt operator mentioned above corresponds to segment reversal in the linear form of the corresponding tour permutation. We refer to the 2-opt operation as the *inversion* operation and illustrate it in Fig. 4.2. We consider random local search (RLS), defined in Algorithm 4.5, and the $(\mu+\lambda)$ EA, defined in Algorithm 4.6. Note that RLS maintains a population of size one, and performs exactly one inversion operation in each iteration. On the other hand, the $(\mu+\lambda)$ EA maintains a population of μ permutations and produces λ offspring in each generation by applying Poisson mutation (see Function `mutate`).

Definition 4.6.1. The *inversion* operation σ^{I}_{ij} transforms permutations into one another by segment reversal in their linear forms.

A permutation x is transformed into a permutation $\sigma^{\mathrm{I}}_{ij}[x]$ by inverting the subsequence of the linear form of x from position i to position j, where $1 \le i < j \le n$:

$$x = (x(1),\dots,x(i-1),x(i),x(i+1),\dots,x(j-1),x(j),x(j+1),\dots,x(n)),$$
$$\sigma^{\mathrm{I}}_{ij}[x] = (x(1),\dots,x(i-1),x(j),x(j-1),\dots,x(i+1),x(i),x(j+1),\dots,x(n)).$$

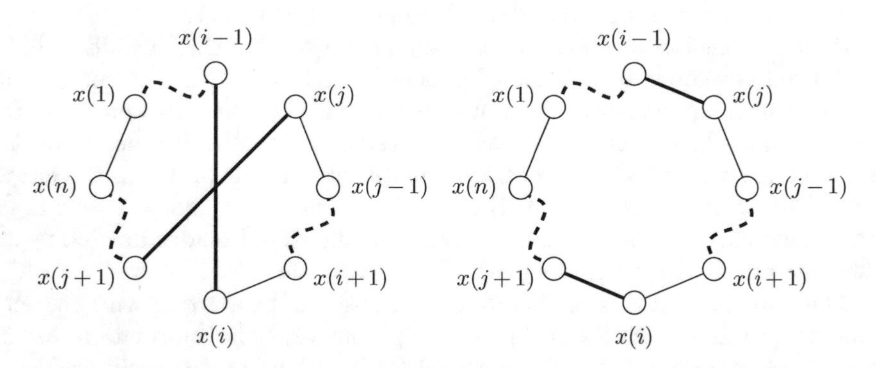

Fig. 4.2 The effect of the inversion operation σ^{I}_{ij} on a tour. Inverting a subsequence in the permutation representation corresponds to a 2-opt move in which a pair of edges in the current tour is replaced by a pair of edges not in the tour.

We also consider the permutation *jump* operator studied by Scharnow, Tinnefeld, and Wegener [38] in the context of sorting problems.

Definition 4.6.2. The *jump* operation σ^{J}_{ij} transforms permutations into one another by position shifts in their linear form. A permutation x is transformed into a permutation $\sigma^{\mathrm{J}}_{ij}[x]$ by moving the element in position i in the linear form of x into position j in the linear form of $\sigma^{\mathrm{J}}_{ij}[x]$ while the other elements between position i and position j are shifted in the appropriate direction. Without loss of generality, suppose $i < j$. Then,

$$x = (x(1),\dots,x(i-1),x(i),x(i+1),\dots,x(j-1),x(j),x(j+1),\dots,x(n)),$$
$$\sigma^{\mathrm{J}}_{ij}[x] = (x(1),\dots,x(i-1),x(i+1),\dots,x(j-1),x(j),x(i),x(j+1),\dots,x(n)).$$

Algorithm 4.5: Randomized local search (RLS)

1 Choose a random permutation x on V;
2 **repeat** *forever*
3   choose a random distinct pair of elements (i,j) from $[n]$;
4     $y \leftarrow \sigma^{\mathrm{I}}_{ij}[x]$;
5   **if** $f(y) \leq f(x)$ **then** $x \leftarrow y$;

Function `mutate(`x`)`

1 $y \leftarrow x$;
2 draw s from a Poisson distribution with unit expectation;
3 perform $s+1$ random inversion operations on y;
4 **return** y;

Algorithm 4.6: The $(\mu+\lambda)$ EA

1 Choose a multiset P of μ random permutations on V;
2 **repeat** *forever*
3     $P' \leftarrow \{\}$;
4   **repeat** λ *times*
5   choose x uniformly at random from P;
6         $y \leftarrow$ `mutate(`x`)`;
7         $P' \leftarrow P' \uplus \{y\}$;
8     $P \leftarrow$ `select(`$P \uplus P'$`)` ;

Every tour $C(\pi)$, for all permutations π on V, corresponds to a set of edges that describe a closed polygon in the plane. If V is noncollinear (no three points are collinear), the vertices on the boundary of the convex hull of V appear in their cyclic order in a minimum-cost tour, and no edge is intersecting [36]. When a tour contains a pair of edges that intersect at a point p, those edges form the diagonals of a convex quadrilateral. The interior edges of this figure describe nondegenerate triangles in the Euclidean plane. Thus, as long as no three points are collinear, removing these edges and replacing them with the corresponding nonintersecting edges results in a strictly shorter tour. This is illustrated in Fig. 4.3.

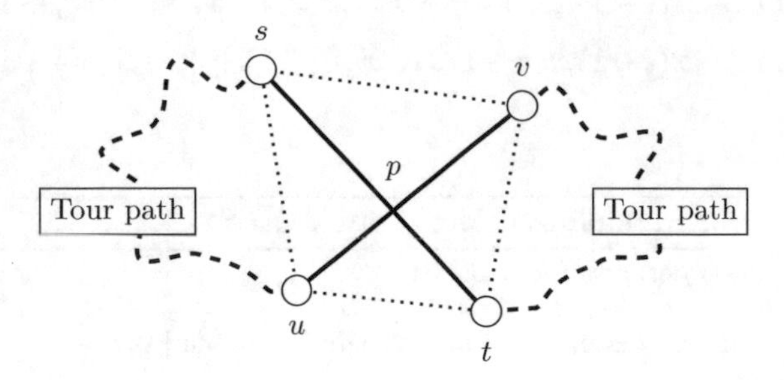

Fig. 4.3 Removing the intersecting edges (s,t) and (u,v) and reconnecting the two disconnected tour path segments with edges (s,v) and (u,t) results in a strictly shorter tour.

4.6.1.1 Avoiding Arbitrarily Small Improvements

Worst-case proofs for 2-opt on the TSP exploit the fact that when points are allowed in arbitrary positions, the smallest change in fitness between neighboring solutions can be made arbitrarily small [13]. This allows the possibility of exponential-length paths between a candidate solution and a reachable local optimum. Sutton and Neumann [40] circumvented this is by imposing bounds on the angles between points. A set of points V is *angle-bounded by* ε for some $0 < \varepsilon < \pi/2$ if, for any three points $u, v, w \in V$, $0 < \varepsilon < \theta < \pi - \varepsilon$, where θ denotes the angle formed by the line from u to v and the line from v to w. Under this condition, the runtime bound depends on the angle bound ε, and so we may consider it as an additional parameterization of the instance. This is also applicable to the class of TSP instances whose points are embedded in an $m \times m$ grid (with the further restriction that no three points are collinear). This kind of quantization can result when the coordinates of each point are rounded to the nearest value in a set of m equidistant values. In these cases, the changes in cost between neighboring solutions can be bounded from below, avoiding exponentially long improvement chains to reach a local optimum.

Definition 4.6.3. Let V be a set of points angle-bounded by ε. We define

$$A(\varepsilon) = \left(\frac{d_{\max}}{d_{\min}} - 1\right)\left(\frac{\cos(\varepsilon)}{1 - \cos(\varepsilon)}\right)$$

where $d_{\max}$ and $d_{\min}$ denote the maximum and minimum Euclidean distances, respectively, between points in V.

Quantized instances yield a more meaningful interpretation of $A(\varepsilon)$, as is captured by the following proposition.

Proposition 4.6.4. *Let V be a set of points embedded in an $m \times m$ grid with no three points collinear. Then V is angle-bounded by ε such that*

$$A(\varepsilon) = m^5.$$

Proposition 4.6.4 follows from Definition 4.6.3 and the fact that V is angle-bounded by $\arctan\left(1/(2(m-2)^2)\right)$ and $d_{\max} = O(m)$.

4.6.1.2 Instances in Convex Position

A set of points V are in convex position when $\mathsf{Inn}(V) = \emptyset$. In this case, we must wait only for the process to remove all intersecting edges. Upper bounds on the time until RLS and the $(\mu+\lambda)$ EA have removed all such edges (and thus produced an optimal tour) can be expressed as a function of the angle-bounding function A. More conveniently, when an instance is embedded in an $m \times m$ grid, both processes can solve the instance in time polynomial in both n and m.

Theorem 4.6.5. *Let V be a set of planar points in convex position angle-bounded by ε. The expected time for RLS to solve the TSP on V is $O(n^3 A(\varepsilon))$, where A is as defined in Definition 4.6.3.*

The proof of Theorem 4.6.5 relies on the fact that any 2-opt move that replaces a pair of intersecting edges with a pair of nonintersecting edges in an angle-bounded instance results in an improvement of the tour by at least

$$2d_{\min}\left(1-\cos(\varepsilon)\right)/\left(\cos(\varepsilon)\right). \tag{4.6.2}$$

Any pair of intersecting edges can be removed with a particular 2-opt operation (each of which occurs with probability $\Omega(n^{-2})$), and thus we can derive a straightforward bound on the waiting time until all such intersections have been removed.

Theorem 4.6.6. *Let V be a set of planar points in convex position angle-bounded by ε. The expected number of fitness evaluations needed by the $(\mu+\lambda)$ EA using 2-opt mutation to solve the TSP on V is bounded from above by $O\left(n \cdot A(\varepsilon) \cdot \max\left\{\mu n^2, \lambda\right\}\right)$, where A is as defined in Definition 4.6.3.*

The proof of Theorem 4.6.6 is similar to the proof of Theorem 4.6.5, except that we must account for any slowdown incurred by selecting from a population. Specifically, the probability that at least one of the λ offspring improves on the current best-so-far point is at least $1-\left(1-\frac{1}{\mu e n(n-1)/2}\right)^{\lambda}$. When $\lambda \geq \mu n(n-1)/2$, an intersection is removed with constant probability in each generation and we must wait only $O(nA(\varepsilon))$ generations to find an intersection-free tour (owing to the improvement guarantee from (4.6.2)). On the other hand, when $\lambda < \mu en(n-1)/2$, the improvement probability can be as low as $\lambda/(\mu en^2)$. The runtime bound follows by accounting for this and the extra $\mu+\lambda$ fitness evaluations that need to occur in each generation.

4.6.1.3 Bounded Number of Inner Points

The polynomial-time results on angle-bounded instances in convex position raise the question of what kind of influence the number of inner points can have on the running time of the above-mentioned algorithms. In this section, we discuss how the Euclidean TSP parameterized by the number of inner points can be solved in *slicewise polynomial time* in the black-box setting.

Theorem 4.6.7. *Let V be a set of points angle-bounded by ε such that $|\mathsf{Inn}(V)| = k$. The expected number of fitness evaluations needed for the $(\mu+\lambda)$ EA using 2-opt mutation to solve the TSP on V is bounded from above by*

$$O\Big(n \cdot A(\varepsilon) \cdot \max\left\{\mu n^2, \lambda\right\} + \mu n^{4k}(2k-1)!\Big),$$

and the expected optimization time for the $(1+1)$ *EA is*

$$O\Big(n^3 \cdot A(\varepsilon) + n^{4k}(2k-1)!\Big).$$

Theorem 4.6.7 can be proved by partitioning the amount of time the $(\mu+\lambda)$ EA spends on tours that contain intersections and tours that do not contain intersections. In particular, let $x^{(t)}$ be the best-so-far tour found by generation t of the $(\mu+\lambda)$ EA. If $C(x^{(t)})$ contains a pair of intersecting edges, the probability of the EA creating a strictly improving tour via a 2-opt mutation on $x^{(t)}$ is bounded from below. Moreover, the angle-boundedness of the instance guarantees an additional lower bound on the amount of actual fitness improvement when such a mutation occurs. Hence, the total expected time that the process spends on tours with intersecting edges is bounded as in Theorem 4.6.6.

In the case where $x^{(t)}$ contains no intersecting edges, the vertices on the boundary of the convex hull must appear in $x^{(t)}$ in their correct cyclic order for a minimum-cost tour [36]. An optimal tour can then be produced from $x^{(t)}$ by rearranging the points in $\mathsf{Inn}(V)$ to the correct positions. Poisson mutation (see Function `mutate`) is capable of performing this rearrangement by selecting at most $2|\mathsf{Inn}(V)| = 2k$ specific inversion operations. This occurs with probability at least

$$\frac{1}{en^{4k}(2k-1)!},$$

which yields a simple upper bound on the waiting time to jump from an intersection-free tour to an optimal solution. The claim then follows by carefully accounting for the correct parent selection probabilities and summing the bounds on the expected time spent on tours with intersections and nonoptimal intersection-free tours.

4.6.1.4 Mixed-Mutation Strategies

The proofs of the theorems in the preceding sections rely on the inversion operator to construct an intersection-free tour, but then rely on the inversion operator to simulate a jump operation in order to transform the intersection-free tour into an optimal solution. The analysis can be improved by relying on a mixed-mutation strategy (see Function `mixed-mutation`) that performs a mixture of both inversion and jump operations, each with constant probability. This improves the upper bound on the running time by a factor of $\Omega\big(n^{2k}(2k-1)!/(k-1)!\big)$.

Function `mixed-mutation`(x)

1 $y \leftarrow x$;
2 draw r from a uniform distribution on the interval $[0,1]$;
3 draw s from a Poisson distribution with unit expectation;
4 **if** $r < 1/2$ **then** perform $s+1$ random inversion operations on y;
5 **else** perform $s+1$ random jump operations on y;
6 **return** y;

Theorem 4.6.8. *Let V be a set of points angle-bounded by ε such that $|\mathsf{Inn}(V)| = k$. The expected number of fitness evaluations needed for the $(\mu+\lambda)$ EA using mixed mutation to solve the TSP on V is bounded from above by*

$$O\Big(n \cdot A(\varepsilon) \cdot \max\left\{\mu n^2, \lambda\right\} + \mu n^{2k}(k-1)!\Big),$$

and the expected optimization time for the $(1+1)$ EA is bounded from above by

$$O\Big(n^3 \cdot A(\varepsilon) + n^{2k}(k-1)!\Big).$$

The proof is similar to the proof of Theorem 4.6.7. With mixed mutation, a 2-opt operation still occurs with constant probability, so the likelihood of a sufficient improvement is asymptotically equivalent to the case of Theorem 4.6.7. A jump operation occurs also with constant probability, but the probability that such an operation jumps to an optimal solution (by correctly rearranging the positions of the points in $\mathsf{Inn}(V)$) is bounded from below by

$$\Omega\left(\frac{1}{n^{2k}(k-1)!}\right).$$

4.6.2 FPT Evolutionary Algorithms

In the case where search heuristics have access to problem-specific information, FPT results are also available. Specifically, we consider heuristics that have access to both fitness values and the cyclic ordering of the points on the convex hull. This ordering can be precomputed in polynomial time [20] and stored so that it is available to the heuristic at any time.

4.6.2.1 A Population-Based Approach

Building on a previous study of Theile [42], Sutton et al. [41] constructed a population-based evolutionary algorithm that efficiently solves the Euclidean TSP when the number of inner points is not too large. They showed that a small modification to Theile's $(\mu+1)$ EA that carefully maintains the invariant that the points in $\mathsf{Out}(V)$ remain in correct convex-hull order for each individual results in an FPT evolutionary algorithm for the inner-point parameterization of the Euclidean TSP.

The EA maintains a large population of permutations on subtours in the graph $G=(V,E)$ (a *subtour* is a Hamiltonian cycle on a subset of V). In each generation, a new offspring is created via a specialized mutation operator that extends the subtour by incorporating an additional randomly chosen vertex, and a modified truncation selection is applied that chooses the best individual for a subtour. The EA can be seen as an evolutionary approach to dynamic programming, the framework for which was presented in [10].

For a set of n points V in the plane with $|\mathsf{Inn}(V)|=k$, we denote by $\gamma:=(p_1,p_2,\ldots,p_{n-k})$ a linear order on the points of $\mathsf{Out}(V)$ such that for all $i\in\{1,\ldots,n-k\}$, p_i and p_{i+1} are adjacent on the boundary of the convex hull of V. For any subset $U\subseteq V$, a permutation on U is a bijection $x\colon U\to U$. We say that a permutation x on $U\subseteq V$ is γ-respecting if and only if, for all $p_i,p_j\in U$, $x^{-1}(p_i)<x^{-1}(p_j) \implies i<j$. We call U the *ground set* of the permutation x on U. We refer to the first element $x(1)$ in the linear order of such a permutation as the *head vertex* and the last element $x(|U|)$ as the *tail vertex*.

The $(\mu+\lambda)$ EA maintains a population P of γ-respecting permutations on subsets of V. For each subset $S\subseteq\mathsf{Inn}(V)$ and each $i\in[n-k]$, the population P contains permutations on the ground set $S\cup\{p_1,p_2,\ldots,p_i\}$. There are $(|S|+i)!$ possible permutation on this ground set. If we were to allow all of them in the population, $|P|$ would be exponential in n. Hence, the key to the FPT running time of the EA is the realization that in an optimal solution, the points in $\mathsf{Out}(V)$ must always appear in their order around the hull. Therefore it is wasteful to consider permutations that are not γ-respecting.

To exploit this, for each possible ground set $S \cup \{p_1, p_2, \ldots, p_i\}$, the population contains exactly $|S|+1$ γ-respecting permutations on that ground set, one for each possible unique tail vertex from the ground set. Specifically, for every $S \subseteq \mathsf{Inn}(V)$ and every $i \in [n-k]$ there is a permutation x for every $r \in S \cup \{p_i\}$ such that

(a) the head vertex of x is $x(1) = p_1$,
(b) the tail vertex of x is $x(|S|+i) = r$, and
(c) x is γ-respecting.

We denote a permutation over the ground set $S \cup \{p_1, p_2, \ldots, p_i\}$ with tail vertex r by $x_{(i,S,r)}$. The corresponding subtour of a $x_{(i,S,r)}$ is a cycle $(x(1) = p_1, v_{x(2)}, \ldots, v_{x(|S|+i-1)}, r, p_1)$ that starts at p_1 and runs through each point of the ground set U exactly once (the i points of $\mathsf{Out}(V)$ are visited in the order in which they appear in γ). Finally, the cycle visits r before returning to p_1. An illustration of a subtour for an example permutation $x_{(i,S,r)}$ on a small ground set is depicted in Fig. 4.4. The fitness function utilized by the $(\mu+\lambda)$ EA is simply the cost of the subtour of an individual:

$$f(x_{(i,S,r)}) = \sum_{j=1}^{|S|+i} d(v_{x(j)}, v_{x(j+1)}), \tag{4.6.3}$$

where the summation indices are taken to be modulo $|S|+i$.

For any given $S \subseteq \mathsf{Inn}(V)$, there are $n-k$ ways to construct a ground set (by choosing i) and $|S|+1$ ways to choose the tail vertex from $S \cup \{p_i\}$. The total number of individuals in the population is thus

$$\mu = |P| = (n-k) \sum_{s=0}^{k} \binom{k}{s} (s+1) = O(2^k kn).$$

The specially designed mutation operator extends a permutation $x = x_{(i,S,r)}$ by adding exactly one new point to its ground set, preserving the validity constraints. In particular, a vertex v is chosen uniformly at random from the remaining vertices in $(\mathsf{Inn}(V) \setminus S) \cup \{p_{i+1}\}$.[1] A new permutation x' is constructed from x by concatenating v with the linear order described by x; that is, for $j \in \{1, \ldots, |S|+i+1\}$,

$$x'(j) = \begin{cases} v & \text{if } j = |S|+i+1, \\ x(j) & \text{otherwise.} \end{cases}$$

Thus x' is a permutation over the ground set $S \cup r$ and uses v as the new tail vertex:

[1] We have abused notation slightly by taking $\{p_{|\mathsf{Out}(V)|+1}\}$ to mean $\emptyset$.

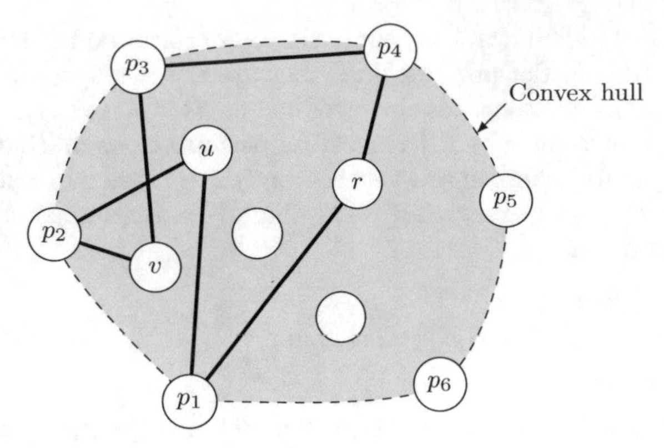

Fig. 4.4 The subtour defined by the permutation $x_{(i,S,r)} = (p_1, u, p_2, v, p_3, p_4, r)$ where $S = \{u, v, r\}$ and $i = 4$. The positions of the points $p_i \in \mathsf{Out}(V)$ in the linear order of the permutation respect their cyclic order around the convex hull.

$$x' = \begin{cases} x'_{(i,S\cup\{v\},v)} & \text{if } v \in \mathsf{Inn}(V), \\ x'_{(i+1,S,v)} & \text{if } v = p_{i+1}. \end{cases}$$

When $i = n-k$ and $S = \mathsf{Inn}(V)$, the mutation operator has no effect, since the ground set cannot be extended for such an individual.

In each generation of the $(\mu+\lambda)$ EA, λ individuals are selected uniformly at random from P. For each selected individual x, an offspring is generated by composing the mutation operator described above $s+1$ times, where s is drawn from a Poisson distribution with unit expectation. Survival selection proceeds by ensuring that each mutated offspring may replace only the individual in the parent population with the same ground set and tail vertex, and this replacement occurs only when the fitness of the offspring is at least as good as the fitness of the corresponding parent. In this way, the surviving population maintains the invariant that each valid combination of ground set and tail vertex is represented exactly once.

Theorem 4.6.9. *Let V be a set of n points in the Euclidean plane with $|\mathsf{Inn}(V)| = k$. After $O(\max\{2^k k^2 n^2 \lambda^{-1}, n\})$ generations, the $(\mu+\lambda)$ EA solves the TSP on V to optimality in expectation and with probability $1 - e^{-\Omega(n)}$.*

Note that this bound translates to $O(\max\{2^k k^2 n^2, \lambda n\})$ fitness evaluations in expectation, by taking the random numbers counting fitness evaluations and generations to be T_f and T_g, respectively, and noting that for Algorithm 4.7, $E[T_f] = \mu + \lambda E[T_g]$. The proof of Theorem 4.6.9 proceeds by bounding the time it takes to increase the set of optimal subtours in the population. In particular, we say that a population is *solved to order m* when it contains an individual permutation on a ground set of size m that corresponds to an optimal subtour on that ground set. Obviously, such subtours are never lost (since they cannot be replaced by a suboptimal subtour), and

Algorithm 4.7: $(\mu+\lambda)$ EA

```
1  P ← ∅;
2  foreach i ∈ {1,...,n−k} do
3      foreach S ⊆ Inn(V) do
4          foreach r ∈ S ∪ p_i do
5              x ← a permutation on the ground set S ∪ {p_1, p_2, ..., p_i} such that
                 x(|S|+i) = r and x respects γ;
6              P ← P ∪ x;
7  repeat forever
8      P' ← {};
9      repeat λ times
10         Select an individual z ← x_(i,S,r) ∈ P uniformly at random;
11         Draw s from a Poisson distribution with unit expectation;
12         Generate z' ← x_(i',S',r') by applying the mutation operator s+1 times;
13         Let f(z') be the cost of TSP tour generated by z';
14         P' ← P' ∪ z';
       /* truncation selection based on the same ground set */
15     foreach offspring z' in P' do
16         Let z'' ← x_(i',S'r') ∈ P be an individual defined on the same ground set
             as z' having the same end vertex if such an individual exists in the
             population;
17         if f(z') ≤ f(z'') then P ← P ∪ z' \ z'';
```

the initial population is solved to order 1 since it contains the individual $x_{(p_1,\emptyset,p_1)}$. The claim follows by bounding the probability of a transformation from a population solved to order m to one solved to order $m+1$, and subsequently taking the waiting time to get a population solved to order n.

4.6.2.2 Inner-Point Permutations

As we saw in Section 4.6.2.1, incorporating domain knowledge into the design of an EA can allow us to create a randomized FPT algorithm for a particular parameterization of the Euclidean TSP. Algorithm 4.7, however, potentially needs a large population, specifically $\mu = O(2^k kn)$. Another approach is to keep a small population and use an EA to search for the optimal ordering on the inner points. Specifically, we let $\gamma = (p_1, p_2, \ldots, p_{n-k})$ be the fixed order of points in $\mathsf{Out}(V)$ as they appear on the convex hull. For any permutation $x\colon \mathsf{Inn}(V) \to \mathsf{Inn}(V)$, it is straightforward to compute the value of the optimal tour through $\mathsf{Inn}(V)$ and $\mathsf{Out}(V)$ respecting the order of both γ and x. The naive approach is to try all $O(n^k)$ possible ways of merging the linear orders of the permutations γ and x. This would violate our FPT requirement, since the parameter appears in the power of the polynomial. Instead, to preserve our FPT conditions, we can directly use a dynamic programming approach to compute the fitness of the permutation x on $\mathsf{Inn}(V)$.

We define two $(n-k)\times(k+1)$ matrices F^{Out} and F^{Inn}, where $F^{\mathrm{Out}}[i,j]$ (or $F^{\mathrm{Inn}}[i,j]$) stores the value of the minimum-weight subtour of all tours through points $p_1,p_2,\ldots,p_i$ and $x(1),x(2),\ldots,x(j)$ such that they respect the orders of both γ and x, and they end on an outer point (or inner point, respectively). Then the optimal tour given the permutations γ and x is

$$Dyn(x)=\min\{F^{\mathrm{Out}}[n-k,k]+d(p_{n-k},p_1),F^{\mathrm{Inn}}[n-k,k]+d(x(k),p_1)\}.$$

Taking the boundary case as $F^{\mathrm{Out}}[1,0]=0$ (the subtour consisting only of p_1), we can compute

$$F^{\mathrm{Inn}}[i,j]=\min\{F^{\mathrm{Out}}[i,j-1]+d(p_i,x(j)),F^{\mathrm{Inn}}[i,j-1]+d(x(j-1),x(j))\}$$

for $i\in\{1,2,\ldots,n-k\}$ and $j\in\{1,\ldots,k\}$, and

$$F^{\mathrm{Out}}[i,j]=\min\{F^{\mathrm{Out}}[i-1,j]+d(p_{i-1},p_i),F^{\mathrm{Inn}}[i-1,j]+d(x(j),p_i)\}$$

for $i\in\{2,3,\ldots,n-k\}$ and $j\in\{0,\ldots,k\}$. Entries that do not correspond to valid subtours, namely $F^{\mathrm{Out}}[1,j]$ for $j\geq 1$ (since the tour cannot end on p_1 and then return to p_1) and $F^{\mathrm{Inn}}[i,0]$ for $i\geq 1$ (since a subtour cannot end on an inner point when the inner-point set is empty), are set to ∞.

The two F matrices can be computed in $O(nk)$ time using dynamic programming. Thus, the time complexity of the fitness evaluation of $Dyn(x)$ is $O(nk)$.

Algorithm 4.8: $(\mu+\lambda)$ EA^k

```
1  Choose a multiset P of μ random permutations on V;
2  repeat forever
3      P' ← {};
4      repeat λ times
5          Choose x uniformly at random from P;
6          Draw s from a Poisson distribution with unit expectation;
7          Construct x' from x by applying s+1 random basic operations;
8          Let f(x') be Dyn(x');
9          P' ← P' ∪ x';
10     P ← select(P ⊎ P');
```

Theorem 4.6.10. *Let V be a set of n points in the Euclidean plane with $|\mathsf{Inn}(V)|=k$. Assuming $\lambda=O(\mu)$, the $(\mu+\lambda)$ EA^k solves the TSP on V using at most $O(\mu+(k-1)!k^{2k})$ fitness evaluations with the jump operation as the basic mutation operation. This bound can be improved to $O(\mu+(k-$*

2)!k^{2k-2}) by using 2-opt mutation. Moreover, each fitness evaluation has time complexity $O(nk)$.

Note that we state the theorem slightly differently than in [41], in which the expected number of *generations* was proved to be $O(\max\{(k-1)!k^{2k}\lambda^{-1},1\})$ for jumps and $O(\max\{(k-2)!k^{2k-2}\lambda^{-1},1\})$ for 2-opt mutation. The bounds stated in Theorem 4.6.10 follow by noting that the number of fitness evaluations in T_g generations of Algorithm 4.8 is $\mu+\lambda T_g$, and the added assumption about λ. The proof of Theorem 4.6.10 relies again on the probability that a given mutation correctly arranges the inner points. Since the mutation operation performs $s+1$ random basic operations, where s is Poisson distributed, the probability that it performs ℓ basic operations is $e^{-1}/(\ell-1)!$. On a permutation of length k, a distinct jump (or 2-opt) move is chosen uniformly at random with probability at least k^{-2}, so the probability that a *specific sequence* of ℓ basic operations occurs is at least

$$p(k,\ell) = \frac{1}{e(\ell-1)!k^{2\ell}}.$$

Therefore, the waiting time to create a globally optimal offspring is bounded by the diameter of the search space induced by the mutation operator. For 2-opt, this bound is at most $k-1$ [3], and for the jump operation, the bound is k. In the case of jump, the probability that at least one of the λ offspring created in any generation is optimal is at least $1-(1-p(k,k))^{\lambda} \geq \min\{\lambda p(k,k), 1-e^{-1}\}$. The claim follows from a standard waiting-time argument. We improve the bound for 2-opt by substituting $p(k,k-1)$ in the above transformation probability.

4.7 Conclusion

In this chapter, we have presented an outline of recent results on the parameterized complexity analysis of randomized search heuristics. This approach of incorporating additional salient parameters into running-time analysis allows a finer-grained understanding of the influence of problem structure on the behavior of these general-purpose optimization techniques.

We have seen that a parameterized analysis can illuminate the inherent efficiency of particular search operators, as well as reveal the difficult components that might arise in the search space of a problem instance. This is the case for the maximum-leaf spanning tree problem. On graphs where k is the maximum number of leaves in a spanning tree, a tree-preserving mutation operator guarantees that the (1+1) EA can find such a tree in fixed-parameter tractable time $O(2^{15k^2 \log k})$. This is in contrast to standard mutation, for which there exist graphs with m edges requiring $(m/c)^{\Omega(k)}$ steps.

We have also observed that the concept of kernelization from the theory of parameterized complexity can be useful. Multiobjective algorithms using a specialized mutation operator can focus the search on a problem kernel of the vertex cover problem, leading to an FPT running time. We have explored how parameterized analysis can help to strengthen an understanding of the components of very general problem classes on simple evolutionary algorithms. This is the case, for example, with the maximization of submodular functions under different constraints.

For the Euclidean TSP, the inner-point parameterization of Deĭneko et al. [9] illuminates the difficulty for RSH techniques arising from the number of points that lie inside the convex hull of the instance. This informs the design of FPT problem-specific evolutionary algorithms, but so far the best known black-box analysis for this parameterization remains in XP time. An open problem is therefore either to prove that this is a lower bound for the parameterization, or to improve the upper bound to FPT time.

Traditional running-time analyses of randomized search heuristics on some artificial benchmark functions have already implicitly used a parameterized perspective. One clear example is for the JUMP function, the running time analysis of which is typically parameterized by the jump-gap size (k) and the string length (n). Indeed, the running-time dichotomy between mutation-only evolutionary algorithms ($\Omega(n^k)$ [22]) and recombinant evolutionary algorithms ($O(4^k \operatorname{poly}(n))$ [22, 23]) already exhibits an "FPT-like" flavor. The application of parameterized analysis to running-time analysis of randomized search heuristics on combinatorial optimization problems with well-established parameterizations from the classical community is therefore a very natural research direction.

Perhaps the most significant research requirement is the need for good problem parameterizations. This requires theoreticians to work closely with practitioners in order to understand what problem components are the most meaningful and relevant in the real world, i.e., what features are most likely to be manifested (or be restricted) in practice, and what problem characteristics might be exploitable by different techniques. This emphasizes the importance of a strong and vibrant relationship between theory and practice.

References

[1] Ageev, A.A., Sviridenko, M.: An 0.828-approximation algorithm for the uncapacitated facility location problem. Discrete Applied Mathematics **93**(2-3), 149–156 (1999)

[2] Arora, S.: Polynomial time approximation schemes for Euclidean traveling salesman and other geometric problems. J. ACM **45**(5), 753–782 (1998). DOI 10.1145/290179.290180. URL `http://doi.acm.org/10.1145/290179.290180`

[3] Bafna, V., Pevzner, P.A.: Genome rearrangements and sorting by reversals. SIAM Journal of Computing **25**(2), 272–289 (1996)

[4] Balinski, M.L.: On maximum matching, minimum covering and their connections. In: Proceedings of the Princeton Symposium on Mathematical Programming, pp. 434–445 (1970)

[5] Bringmann, K., Friedrich, T.: Parameterized average-case complexity of the hypervolume indicator. In: C. Blum, E. Alba (eds.) Proceedings of the Genetic and Evolutionary Computation Conference, pp. 575–582. ACM (2013). DOI 10.1145/2463372.2463450. URL `http://doi.acm.org/10.1145/2463372.2463450`

[6] Chandra, B., Karloff, H., Tovey, C.: New results on the old k-opt algorithm for the traveling salesman problem. SIAM Journal on Computing **28**(6), 1998–2029 (1999)

[7] Cornuejols, G., Fisher, M., Nemhauser, G.L.: On the uncapacitated location problem. In: Studies in Integer Programming, *Annals of Discrete Mathematics*, vol. 1, pp. 163 – 177. Elsevier (1977)

[8] Corus, D., Lehre, P.K., Neumann, F., Pourhassan, M.: A parameterised complexity analysis of bi-level optimisation with evolutionary algorithms. Evolutionary Computation **24**(1), 183–203 (2016). DOI 10.1162/EVCO_a_00147. URL `https://doi.org/10.1162/EVCO_a_00147`

[9] Deĭneko, V.G., Hoffman, M., Okamoto, Y., Woeginger, G.J.: The traveling salesman problem with few inner points. Operations Research Letters **34**, 106–110 (2006)

[10] Doerr, B., Eremeev, A.V., Neumann, F., Theile, M., Thyssen, C.: Evolutionary algorithms and dynamic programming. Theoretical Computer Science **412**(43), 6020–6035 (2011)

[11] Doerr, B., Johannsen, D., Winzen, C.: Multiplicative drift analysis. Algorithmica **64**(4), 673–697 (2012)

[12] Downey, R.G., Fellows, M.R.: Parameterized Complexity. Springer (1999)

[13] Englert, M., Röglin, H., Vöcking, B.: Worst case and probabilistic analysis of the 2-opt algorithm for the TSP. In: Proceedings of the Eighteenth Annual ACM-SIAM Symposium on Discrete Algorithms, pp. 1295–1304. Society for Industrial and Applied Mathematics (2007)

[14] Feige, U., Goemans, M.X.: Approximating the value of two power proof systems, with applications to MAX 2SAT and MAX DICUT. In: Third Israel Symposium on Theory and Computing Systems (ISTCS), pp. 182–189 (1995)

[15] Flum, J., Grohe, M.: Parameterized complexity theory. Springer-Verlag (2006)

[16] Friedrich, T., He, J., Hebbinghaus, N., Neumann, F., Witt, C.: Approximating covering problems by randomized search heuristics using multi-objective models. Evolutionary Computation **18**(4), 617–633 (2010)

[17] Friedrich, T., Neumann, F.: Maximizing submodular functions under matroid constraints by evolutionary algorithms. Evolutionary Computation **23**(4), 543–558 (2015). DOI 10.1162/EVCO_a_00159. URL https://doi.org/10.1162/EVCO_a_00159

[18] Gao, W., Friedrich, T., Neumann, F.: Fixed-parameter single objective search heuristics for minimum vertex cover. In: J. Handl, E. Hart, P.R. Lewis, M. López-Ibáñez, G. Ochoa, B. Paechter (eds.) Proceedings of the Fourteenth International Conference on Parallel Problem Solving from Nature, *Lecture Notes in Computer Science*, vol. 9921, pp. 740–750. Springer (2016). DOI 10.1007/978-3-319-45823-6_69. URL https://doi.org/10.1007/978-3-319-45823-6_69

[19] Goemans, M.X., Williamson, D.P.: Improved approximation algorithms for maximum cut and satisfiability problems using semidefinite programming. Journal of the ACM **42**(6), 1115–1145 (1995)

[20] Graham, R.L.: An efficient algorithm for determining the convex hull of a finite planar set. Information Processesing Letters **1**(4), 132–133 (1972). DOI 10.1016/0020-0190(72)90045-2. URL https://doi.org/10.1016/0020-0190(72)90045-2

[21] Håstad, J.: Some optimal inapproximability results. Journal of the ACM **48**(4), 798–859 (2001)

[22] Jansen, T., Wegener, I.: The analysis of evolutionary algorithms: A proof that crossover really can help. Algorithmica **34**(1), 47–66 (2002)

[23] Kötzing, T., Sudholt, D., Theile, M.: How crossover helps in pseudo-Boolean optimization. In: Proceedings of the Genetic and Evolutionary Computation Conference, pp. 989–996 (2011)

[24] Kratsch, S., Lehre, P.K., Neumann, F., Oliveto, P.S.: Fixed parameter evolutionary algorithms and maximum leaf spanning trees: A matter of mutation. In: R. Schaefer, C. Cotta, J. Kolodziej, G. Rudolph (eds.) Proceedings of the Eleventh Conference on Parallel Problem Solving from Nature, *Lecture Notes in Computer Science*, vol. 6238, pp. 204–213. Springer-Verlag (2010)

[25] Kratsch, S., Neumann, F.: Fixed-parameter evolutionary algorithms and the vertex cover problem. Algorithmica **65**(4), 754–771 (2013). DOI 10.1007/s00453-012-9660-4. URL https://doi.org/10.1007/s00453-012-9660-4

[26] Lee, J., Mirrokni, V.S., Nagarajan, V., Sviridenko, M.: Non-monotone submodular maximization under matroid and knapsack constraints. In: Proceedings of the Forty-First Annual ACM Symposium on Theory of Computing, pp. 323–332 (2009)

[27] Lee, J., Sviridenko, M., Vondrák, J.: Submodular maximization over multiple matroids via generalized exchange properties. Mathematics of Operations Research **35**(4), 795–806 (2010)

[28] Marx, D.: Searching the k-change neighborhood for TSP is W[1]-hard. Operations Research Letters **36**(1), 31–36 (2008). DOI 10.1016/

j.orl.2007.02.008. URL http://www.sciencedirect.com/science/article/pii/S0167637707000302

[29] Nallaperuma, S., Sutton, A.M., Neumann, F.: Fixed-parameter evolutionary algorithms for the Euclidean traveling salesperson problem. In: IEEE Congress on Evolutionary Computation (CEC'13), pp. 2037–2044. IEEE (2013)

[30] Nallaperuma, S., Sutton, A.M., Neumann, F.: Parameterized complexity analysis and more effective construction methods for ACO algorithms and the Euclidean traveling salesperson problem. In: Proceedings of the IEEE Congress on Evolutionary Computation, pp. 2045–2052. IEEE (2013). DOI 10.1109/CEC.2013.6557810. URL http://dx.doi.org/10.1109/CEC.2013.6557810

[31] Nemhauser, G.L., Trotter, L.E.: Vertex packings: Structural properties and algorithms. Mathematical Programming **8**, 232–248 (1975)

[32] Neumann, F., Wegener, I.: Randomized local search, evolutionary algorithms, and the minimum spanning tree problem. Theoretical Computer Science **378**(1), 32–40 (2007)

[33] Oliveto, P.S., He, J., Yao, X.: Analysis of the (1+1) EA for finding approximate solutions to vertex cover problems. IEEE Trans. Evolutionary Computation **13**(5), 1006–1029 (2009). DOI 10.1109/TEVC.2009.2014362. URL https://doi.org/10.1109/TEVC.2009.2014362

[34] Pourhassan, M., Gao, W., Neumann, F.: Maintaining 2-approximations for the dynamic vertex cover problem using evolutionary algorithms. In: Proceedings of the Conference on Genetic and Evolutionary Computation, GECCO '15, pp. 903–910. ACM, New York, NY, USA (2015). DOI 10.1145/2739480.2754700. URL http://doi.acm.org/10.1145/2739480.2754700

[35] Pourhassan, M., Shi, F., Neumann, F.: Parameterized analysis of multiobjective evolutionary algorithms and the weighted vertex cover problem. In: Proceedings of the Fourteenth International Conference of Parallel Problem Solving from Nature, pp. 729–739. Springer International Publishing (2016). DOI 10.1007/978-3-319-45823-6_68. URL https://doi.org/10.1007/978-3-319-45823-6_68

[36] Quintas, L.V., Supnick, F.: On some properties of shortest Hamiltonian circuits. The American Mathematical Monthly **72**(9), 977–980 (1965)

[37] Reichel, J., Skutella, M.: Evolutionary algorithms and matroid optimization problems. Algorithmica **57**(1), 187–206 (2010)

[38] Scharnow, J., Tinnefeld, K., Wegener, I.: The analysis of evolutionary algorithms on sorting and shortest paths problems. Journal of Mathematical Modelling and Algorithms **3**(4), 349–366 (2004)

[39] Sutton, A.M.: Crossover can simulate bounded tree search on a fixed-parameter tractable optimization problem. In: H.E. Aguirre, K. Takadama (eds.) Proceedings of the Genetic and Evolutionary Computation Conference, pp. 1531–1538. ACM (2018). DOI 10.

1145/3205455.3205598. URL http://doi.acm.org/10.1145/3205455.3205598

[40] Sutton, A.M., Neumann, F.: A parameterized runtime analysis of evolutionary algorithms for the Euclidean traveling salesperson problem. In: Proceedings of the Twenty-Sixth Conference on Artificial Intelligence (AAAI'12), pp. 1105–1111. AAAI Press (2012)

[41] Sutton, A.M., Neumann, F., Nallaperuma, S.: Parameterized runtime analyses of evolutionary algorithms for the planar Euclidean traveling salesperson problem. Evolutionary Computation **22**(4), 595–628 (2014). DOI 10.1162/EVCO_a_00119. URL https://doi.org/10.1162/EVCO_a_00119

[42] Theile, M.: Exact solutions to the traveling salesperson problem by a population-based evolutionary algorithm. In: C. Cotta, P. Cowling (eds.) Evolutionary Computation in Combinatorial Optimization, *Lecture Notes in Computer Science*, vol. 5482, pp. 145–155. Springer-Verlag (2009). DOI 10.1007/978-3-642-01009-5_13. URL http://dx.doi.org/10.1007/978-3-642-01009-5_13

[43] Witt, C.: Revised analysis of the (1+1) EA for the minimum spanning tree problem. In: D.V. Arnold (ed.) Genetic and Evolutionary Computation Conference, GECCO '14, Vancouver, BC, Canada, July 12-16, 2014, pp. 509–516. ACM (2014). DOI 10.1145/2576768.2598237. URL http://doi.acm.org/10.1145/2576768.2598237

Chapter 5
Analysing Stochastic Search Heuristics Operating on a Fixed Budget

Thomas Jansen

Abstract When stochastic search heuristics are used for optimisation they are often stopped after some time has passed and the best search point they have found at this point is used as the solution. Fixed-budget analysis is an analytical perspective that delivers results about the expected quality of the solution in this situation. It allows the comparison of different stochastic search heuristics when only a fixed computational budget is available and it offers a very different perspective from runtime analysis. This chapter introduces and motivates this approach to the theoretical analysis of stochastic search heuristics. It provides basic results, describes a general technique to derive such results from runtime results, covers analytical methods that have been applied and describes a range of different results that have been obtained so far.

5.1 Introduction

Evolutionary algorithms [25], ant colony optimisation [11], artificial immune systems [38], simulated annealing [17] and random local search [29] are just five examples of different stochastic search heuristics. Each heuristic implements a specific idea of how the search for an optimal solution should be conducted in general, often borrowing this idea from a natural example. Evolutionary algorithms mimic the process of natural evolution, ant colony optimisation is based on the foraging behaviour of ants, artificial immune systems are modelled on the immune system of vertebrates, simulated annealing takes inspiration from annealing in metallurgy, and local search can be described as a greedy search that always looks for the next improving move. A theoretical

---

Thomas Jansen
Department of Computer Science, Aberystwyth University, Aberystwyth, UK. e-mail: t.jansen@aber.ac.uk

B. Doerr, F. Neumann (eds.), *Theory of Evolutionary Computation*,
Natural Computing Series, https://doi.org/10.1007/978-3-030-29414-4_5

understanding of these heuristics – their potential and their limitations – can help in selecting the right heuristic for a given problem and help in applying it in a way that makes it more efficient on the given problem. This motivates the theoretical analysis of this kind of general heuristic.

The past 25 years have been mostly dominated by a theoretical approach that can be summarised as runtime analysis. It can be argued that when Mühlenbein wanted to find out 'how genetic algorithms really work' [31] he obtained the first runtime result, namely that the $(1+1)$ EA has an expected runtime of $O(n \log n)$ on OneMax. We now have a large number of results, textbooks covering this topic [18, 33] and a large number of analytical tools that are of invaluable help in improving our understanding of how these heuristics work ([1] provides a good overview). The starting point of runtime analysis is that all these stochastic search heuristics are really stochastic search algorithms and, consequently, they should be analysed as algorithms. This makes correctness and efficiency the two most important aspects of a theoretical analysis. Correctness translates to 'Will the algorithm find an optimum eventually almost surely?' and can often be answered positively with little difficulty (see [39] for a more in-depth study of this topic). Efficiency, measured as runtime for classical algorithms, translates to 'How long will it take on average until the algorithm finds an optimum?' and the study of this question is known as runtime analysis. While it appears natural to ask how long an algorithm takes to accomplish its goal, this perspective is at odds with the way stochastic search heuristics are actually applied in practice. They are used when a problem is not well understood and no good problem-specific algorithm is known. They are run for a limited time in the hope of finding a good solution, not necessarily an optimal one or even one that approximates an optimal solution with a preset quality. Even if an optimal solution is found, it might not be possible to recognise this and be sure. In this situation, runtime analysis does not offer much insight. It has been argued [20] that practitioners are more interested in other kinds of questions that runtime analysis cannot answer: 'What is the expected quality of a solution if I run my heuristic for this long? What better quality can I expect if I double that time?' These kinds of question are answered by fixed-budget analysis.

In the next section we will provide formal definitions and introduce an overview of the goals of fixed-budget analysis. We will formally introduce random local search and the $(1+1)$ EA as a toy example to see how fixed-budget results can be derived. In Section 5.3 we present and discuss a method to transfer runtime results in a systematic way into fixed-budget results. This is an important step because not only is the number of runtime results large but there are also a number of powerful analytical tools that can be utilised to develop results for fixed-budget analysis in this way. We consider different analytical methods in Section 5.4. We provide a brief overview of the kind of results and insights that have been obtained using the fixed-budget perspective in Section 5.5. We summarise the chapter in Section 5.6 and mention possible directions of future research.

5.2 Analytical Perspective and Basic Results

Runtime analysis considers the number of function evaluations $T_{A,f}$ a given stochastic search heuristic A makes on a set objective function f until an optimum is found for the first time. Usually, one analyses the expectation of this random variable $\mathrm{e}[T_{A,f}]$, the expected optimisation time. We could say that we fix the solution quality (we care only about finding an optimum and discard any worse solution) and care about the time it takes.

In fixed-budget analysis we reverse this. We fix the time (by deciding on the computational budget we want to spend) and care about the solution quality we are able to reach. Let x_t denote the best search point after A has evaluated t search points when running on f. For a given computational budget, $b \in \mathbb{N}$, we analyse the expected function value $\mathrm{e}[f(x_t)]$ for all $t \leq b$. Similarly to runtime analysis, there are situations where it might be more useful to consider not (only) the expectation of the random variable in question ($T_{A,f}$ for runtime analysis and $f(x_t)$ for fixed-budget analysis) but (also) other aspects of its distribution.

Concentrating on the best search point found after t search points is a natural thing for elitist algorithms and also makes sense for non-elitist algorithms. If our aim is optimisation, we will output the best search point found even if the population of a non-elitist search heuristic does not contain it any more. It is possible, of course, to also consider the function values of other search points of interest (e.g. the best search point in the current population for a population-based search heuristic). However, we will stick to the best search point here. This choice aligns fixed-budget analysis with the classical best-so-far curves known from experimental studies of search heuristics, where one plots the currently best function value found against the number of steps.

It is important to note that we are interested not only in $f(x_b)$ but also in all $f(x_t)$ with $t \leq b$. Thus, we can visualise our results by plotting $\mathrm{e}[f(x_t)]$ against t (or the bounds we have for it). This yields graphs that look exactly like best-so-far curves (see Fig. 5.1 on page 254 for an example). Thus, fixed-budget results are a direct theoretical equivalent of the empirical best-so-far curves and therefore are a more accessible type of theoretical result than runtime results for practitioners.

We use the well-known $(1+1)$ EA as a simple example and define it here formally for the sake of completeness and clarity. In the form given it is used to maximise a function $f\colon \{0,1\}^n \to \mathbb{R}$ that maps bit strings of fixed length n to real numbers.

We leave the choice of the initial search point x_0 open. In most cases it will be selected uniformly at random. However, for the sake of the discussion here, it is sometimes more convenient to consider other initial search points.

To introduce fixed-budget analysis, it makes sense to consider an even simpler algorithm, a variant of local search. We also formulate it for maximi-

(1+1) evolutionary algorithm ((1+1) EA)

```
1  t := 0. Select x_t ∈ {0,1}^n. Evaluate f(x_t).
2  While t+1 < b do
3      t := t+1. y := x_{t−1}.
4      For each i ∈ {1,2,...,n}: With probability 1/n flip i-th bit in y.
5      Evaluate f(y).
6      If f(y) ≥ f(x_{t−1}) then x_t := y else x_t := x_{t−1}.
```

sation of $f\colon \{0,1\}^n \to \mathbb{R}$. Again, and for the same reason, we choose not to define the choice of the initial search point x_0 here.

Random local search (RLS)

```
1  t := 0. Select x_t ∈ {0,1}^n. Evaluate f(x_t).
2  While t+1 < b do
3      t := t+1. y := x_{t−1}.
4      Select i ∈ {1,2,...,n} uniformly at random. Flip i-th bit in y.
5      Evaluate f(y).
6      If f(y) ≥ f(x_{t−1}) then x_t := y else x_t := x_{t−1}.
```

For a given algorithm A and an objective function f, it is usually the case that $f(x_t)$ approaches $\max\limits_{x\in\{0,1\}^n} f(x)$ when t approaches $\mathrm{e}[T_{A,f}]$. For a computational budget $b > \mathrm{e}[T_{A,f}]$, the most useful question probably is what is the probability that $f(x_b) < \max\limits_{x\in\{0,1\}^n} f(x)$ still holds. In most cases, the expectation $\mathrm{e}[f(x_b)]$ will be so close to $\max\limits_{x\in\{0,1\}^n} f(x)$ that it will be not very informative. This is also illustrated in the example in Fig. 5.1 on page 254. Therefore, we restrict our interest to budgets b with $b \leq \mathrm{e}[T_{A,f}]$. Note that, ideally, we would like to get results up to $b = \mathrm{e}[T_{A,f}]$. However, obtaining results for $b = o(\mathrm{e}[T_{A,f}])$ might be easier and can still be very informative.

We have mentioned in the introduction that one kind of question that practitioners would like to see answered is what they can expect to gain in the quality of the solution if they double the computation time, for example. In fixed-budget analysis, it is therefore very desirable to obtain exact results that reveal not only the order of growth but also the leading multiplicative constants. While less precise results can still be valuable it should be kept in mind that exact results are more desirable and important in the fixed-budget perspective than they are in general runtime results.

One other notable difference from runtime analysis is that in fixed-budget analysis, in general, we are more interested in lower bounds than in upper bounds. The reasons are exactly the same as in runtime analysis, and the difference simply stems from the change of perspective. In practice, guarantees about the performance are usually more useful (or at least reassuring) than negative results. Thus, while in runtime analysis upper bounds on the expected runtime are most useful, in fixed-budget analysis lower bounds on the expected function value are most useful (if we consider maximisation; in the case of minimisation, upper bounds are more useful).

He [14] has pointed out that fixed-budget analysis has some similarity to the study of the convergence rate. However, there are significant differences that make studying the performance of stochastic search heuristics in the fixed-budget perspective an essentially different endeavour.

We introduce fixed-budget analysis using a very well-known example from classical probability theory: the coupon collector's problem ([30] is one of many textbooks that contain a good description). The collector wants to get a complete collection of n different coupons. The coupons are purchased one by one and each time each coupon type is equally likely. It is well known that the expected number of coupons the collector needs to buy to get a complete collection is $n \ln n + O(n)$ and the reason is simply that when the current collection contains i different coupons the probability of getting a new one equals $(n-i)/n$. This implies that, in expectation, one needs to buy $n/(n-i)$ coupons to increase the collection from i to $i+1$ different coupons, and the result follows from the linearity of expectation.

It is not difficult to see that the result of the classical coupon collector's problem is a runtime result. The random process is exactly the same as the one defined by starting random local search in the all 0-bit string 0^n on the classical ONEMAX problem, $\text{ONEMAX}(x) = \sum_{i=1}^{n} x[i]$, when we read the i-th bit $x[i]$ as an indicator variable that has value 1 if the collection contains a coupon of type i and 0 otherwise. Thus, $\text{e}[T_{\text{RLS,ONEMAX}}] = n \ln n + O(n)$ holds.

In the fixed-budget scenario, we ask a different question. We have enough money to buy b coupons; how many different coupons can we expect to find in our collection after buying b? We actually want an answer to this question for any number $t \leq b$ of coupons bought. Note that this is exactly the same as asking for the expected function value of RLS on ONEMAX when starting with 0^n. It is rather elementary to observe that the probability of having bought a coupon of type i after buying t coupons equals $1-(1-1/n)^t$ and that this is the expected value of the i-th bit in x_t. By linearity of expectation, it follows that $\text{e}[\text{ONEMAX}(x_t)] = n \cdot \left(1-(1-1/n)^t\right)$ holds. We have plotted this result in Fig. 5.1 as an example and to demonstrate the similarity to best-so-far curves.

Using the more common random initialisation in RLS, the process is not much different. We expect $\text{ONEMAX}(x_0) = n/2$ in this case (because each bit is set to 1 with probability 1/2). For the remaining $n/2$ bits the random process is unchanged, so $\text{e}[\text{ONEMAX}(x_t)] = (n/2) + (n/2) \cdot \left(1-(1-1/n)^t\right)$ follows for random initialisation (see [21] for a more detailed proof).

Another example function that is useful for theoretical study is LEADINGONES. It yields as the function value the number of consecutive 1-bits counting from left to right (expressed in a formula as $\text{LEADINGONES}(x) =$

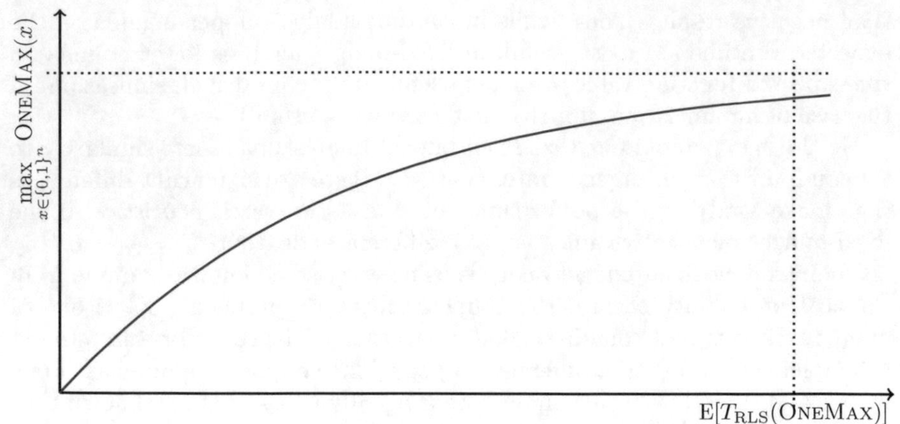

Fig. 5.1 Expected number of different coupons after buying t coupons, and also e[OneMax(x_t)] for RLS on OneMax with initialisation with the all 0-bit string 0^n. The x-axis denotes t; the expected value is on the y-axis.

$\sum_{i=1}^{n}\prod_{j=1}^{i} x[j]$). It is appealing because the probability of increasing the function value by means of mutation does not change a lot during a run, as it is always necessary and sufficient to flip exactly the leftmost 0-bit. Thus, the probability is always exactly $1/n$ for RLS and between $1/n$ and $(1/n)(1-1/n)^{n-1} \approx 1/(en)$ for the (1+1) EA. This is very different from OneMax, where the probability is $\Theta(1)$ as long as there are still $\Theta(n)$ 0-bits but drops to $\Theta(1/n)$ when the number of remaining 0-bits has dropped to $O(1)$.

It is not difficult to see that the bits after the leftmost 0-bit are always uniformly randomly distributed because they never have played a role in selection and the mutation operators are unbiased with respect to the role of 0-bits and 1-bits. This makes it intuitive that the expected increase in function value is $2/n$ for RLS. And, indeed, it can be shown that for RLS with random initialisation e[LeadingOnes(x_t)] $= 1 + (2t/n) - 2^{-\Omega(n)}$ holds (but the proof is rather long and technical, see [21]).

For RLS and simple functions, it is not too difficult to get very precise fixed-budget results, as we have seen for OneMax and LeadingOnes. The same is true for other well-known example functions. There are, however, a number of less precise results for other functions (see e.g. [21] for results on Jump$_k$ and Ridge; see [18] or [21] for definitions and discussion).

Making the seemingly small step from RLS to the (1+1) EA changes this significantly. The reason is the increased degree of variability that the standard bit mutations used in the (1+1) EA have in comparison with the local mutations that are used in RLS.

Considering the (1+1) EA with random initialisation on LeadingOnes, it is still possible to prove e[LeadingOnes(x_t)] $= 1 + (2t/n) - o(t/n)$ [21] if the budget is much smaller than $n^2/2$, i.e. $b = o(n^2)$, because in this case we are still far from reaching the optimum and the probability of having reached it is so small that it does not contribute significantly to the expected function value. For larger budgets $b = cn^2$ with a constant $0 < c < 1/2$, it can be shown that

$$n \cdot ce^{-c}\left(1+e^{-ce^{-ce^{-c}}}\right) \le \textsc{LeadingOnes}(x_b) \le n \cdot ce^{-c}\left(1+e^{-ce^{-c}}\right)$$

holds asymptotically almost surely (i.e. with probability $1-o(1)$) [21]. One can improve the analysis numerically to some extent (see [21] for a detailed analysis) but this is still some way from a precise result for larger budgets.

Surprisingly, results for the $(1+1)$ EA on OneMax are even more difficult to obtain. Direct, elementary analysis seems to be too complicated to deal with this still very simple case. This motivates us to look for more powerful methods to derive fixed-budget computation results, something we begin in the next section.

5.3 Reusing Known Runtime Results

Whereas runtime analysis has been around for at least 25 years, fixed-budget analysis was introduced much more recently [20]. This implies that the number of methods available to perform fixed-budget analysis is much smaller than for runtime analysis and, also, that there are many more runtime results available than fixed-budget results. It is therefore significant that there is a method that allows runtime results to be transferred to the perspective of fixed-budget analysis.

We know that one is usually interested in $T_{A,f}$, the time a heuristic A needs to find a global optimum of a function f for the first time. We can generalise this notion slightly and say that $T_{A,f}(v)$ is the time A needs to find a search point with a function value of at least v for the first time. This way, the runtime $T_{A,f}$ becomes $T_{A,f}\left(\max\limits_{x\in\{0,1\}^n} f(x)\right)$. This slightly more general notion presents us with $T_{A,f}$ as a function that maps function values to runtimes. Clearly, its inverse function $T_{A,f}^{-1}$ maps times to function values. So, for a deterministic algorithm A, it suffices to have $T_{A,f}$. We can then compute the inverse function $T_{A,f}^{-1}$ and get, for each computational budget b, the function value $T_{A,f}^{-1}(b)$ that can be reached in b steps. Unfortunately, for randomised algorithms A, things are not quite so simple. We can still compute the inverse function of $\mathrm{e}[T_{A,f}(v)]$, but this does not necessarily yield the expected function value after a given number of steps owing to the variability of the random process. But if we have not only the expectation but also some concentration bounds, we can use this directly to transform results about the runtime into fixed-budget results about the function value.

For the sake of clarity, we summarise the four steps that are involved in the process before we discuss an example. We start with a known result on the expected runtime $\mathrm{e}[T_{A,f}]$ of a heuristic A on a function f. In the first step, we generalise this to a result on $\mathrm{e}[T_{A,f}(v)]$, the expected time A needs on f to find a function value of at least v for the first time. This is usually not too hard. In the second step we need deviation bounds, i. e., statements of the form $\Pr[T_{A,f}(v) \leq \mathrm{e}[T_{A,f}(v)] - d_l] \leq p_l$ and $\Pr[T_{A,f}(v) \geq \mathrm{e}[T_{A,f}(v)] + d_u] \leq p_u$. We can still obtain a result if we have only either the upper or the lower bound. The third step is, given these bounds, to compute the inverse functions $\left(\mathrm{e}[T_{A,f} - d_l]\right)^{-1}$ and $\left(\mathrm{e}[T_{A,f} + d_u]\right)^{-1}$. While this might be a little tedious, it is certainly not difficult. Now, in the fourth and final step, this yields the fixed-budget results

$$\Pr\left[f(x_t) \leq \left\lceil \left(\mathrm{e}[T_{A,f} + d_u]\right)^{-1}(t) \right\rceil\right] \leq p_u$$

and

$$\Pr\left[f(x_t) \geq \left\lceil \left(\mathrm{e}[T_{A,f} - d_l]\right)^{-1}(t) \right\rceil\right] \leq p_l$$

as a direct consequence.

The method was presented together with an application to the $(1+1)$ EA on LeadingOnes [10] because there are very precise results for the runtime available for this setting. There are very exact results for the runtime [2] that can easily be extended to an exact result $\mathrm{e}\left[T_{(1+1)\ \mathrm{EA},\textsc{LeadingOnes}}(a)\right] = \left(n^2 - n\right)/2 \cdot \left(\left(1 + 1/(n-1)\right)^a - 1\right)$ for any function value a. One still needs to add concentration bounds to the result for the expected runtime, and there are several ways of doing that. One approach is to rely on the Chebyshev inequality (see [4] for an exposition that is tailored towards the kind of analysis we are performing). This yields

$$\Pr\left[\textsc{LeadingOnes}(x_b) = n \ln(1 + 2b/n^2) \pm \Theta(d/n)\right] = 1 - o(1)$$

for any $d = \omega(n^{3/2})$ and budget $b \leq \mathrm{e}[T_{(1+1)\ \mathrm{EA},\textsc{LeadingOnes}}]$ (see [10] for a complete proof). While the bounds are not too strong (because the Chebyshev inequality is a very general result that consequently does not yield the strongest bounds), they are easy to obtain and it is easy to derive precise numerical statements (and not only asymptotic ones). For concrete applications, this is appealing.

If one wants stronger bounds, the application of Chernoff bounds for the derivation of concentration results is possible. It is technically more involved (again, see [10] for a complete proof), but we obtain a much stronger result. This way, one can prove that

$$\Pr\left[\textsc{LeadingOnes}(x_b) = n \ln(1 + 2b/n^2) \pm \Theta(d/n)\right] = 1 - e^{-\Omega(d^2/n^3)}$$

holds for any $d \leq 2n^2$ and budget $b \leq \mathrm{e}[T_{(1+1)\ \mathrm{EA},\textsc{LeadingOnes}}]$. While this is much tighter asymptotically, it does not lend itself to obtaining numerical results for small n, which might be an issue, depending on the application.

Using results that derive sharp concentration bounds for variable drift analysis [7], one can also obtain bounds by drift analysis, even sharper but technically even more involved. This yields

$$\Pr\left[\frac{(n^2-n)\,/\,(n^2-n+2b-2d-o(d))}{\ln(1-1/n)} \leq \textsc{LeadingOnes}(x_b) \leq \frac{(n^2-n)\,/\,(n^2-n+2b+2d-o(d))}{\ln(1-1/n)}\right] = 1-2^{-\Omega(n^{\varepsilon})}$$

for any $d = \Omega(n^{(3/2)+\varepsilon})$ ($\varepsilon > 0$ constant) and budget $b \leq \mathrm{e}[T_{(1+1)\ \mathrm{EA},\textsc{LeadingOnes}}]$. These are also of asymptotical nature, and it is difficult to derive good numerical bounds for small n in this way.

We see that the method is general and has the potential to derive good bounds. It depends on the runtime result, as well as the concentration bounds used, how tight and useful the results are. Depending on the application, some kinds of result might be more applicable than others.

5.4 Advanced Methods

Direct analysis based on first principles (such as that for RLS on OneMax seen in Section 5.2) is not likely to be feasible for more than the simplest scenarios. The method of transferring results about the expected runtime to fixed-budget results with the help of concentration results that we discussed in the previous section has the potential to yield helpful results, but it is limited to situations where runtime results are available and where sufficiently good concentration bounds can be found. To obtain results for other settings, different analytical approaches may be needed.

It can be argued that the most versatile and successful collection of analytical tools for runtime analysis is what is known as drift analysis. Drift analysis covers a range of different results that consider the expected change in one step of a random process over a state space. It clearly pre-dates runtime analysis of evolutionary algorithms [13]. It was introduced into the runtime analysis of evolutionary algorithms, among other things, to simplify the analysis of the (1+1) EA on the class of linear functions [15], i.e. OneMax-like functions where each bit does not necessarily contribute 1 to the overall function value but has a fixed weight w_i that it contributes, $f(x) = \sum_{i=1}^{n} w_i \cdot x[i]$. In addition to additive drift [15] we now have tools for considering multiplicative drift [8], negative drift (for proving lower bounds) [35] and variable drift [9] ([36] provides an overview).

Since drift analysis is based on the expected change in one step of a fairly general and abstract random process and is not tied to runtime analysis, it should not come as a surprise that it has been successfully applied to fixed-budget analysis. Lengler and Spooner [28] considered the $(1+1)$ EA on the class of linear functions, changing the algorithm so that it performed minimisation instead of maximisation. This change in perspective has the advantage that the roles of upper and lower bounds on the expected function value play the same role as upper and lower bounds on the runtime (while, with maximisation, the roles are reversed as we have discussed earlier). They introduced a drift theorem that they called fixed-budget multiplicative drift and that provides an upper- as well as a lower-bound result.

Theorem 5.4.1 (Theorem 1 in [28]). *Let $(X_t)_{t\geq 0}$ be a stochastic process, let $b \in \mathbb{N}$, and $0<\delta<1$.*

Upper bound. *If for all $t \leq b$ we have $\mathrm{e}[X_t - X_{t+1} \mid X_t = x] \geq \delta x$ then $\mathrm{e}[X_b \mid X_0] \leq X_0 \cdot (1-b)^b \leq X_0 \cdot e^{-\delta b}$.*

Lower bound. *If for all $t \leq b$ we have $\mathrm{e}[X_t - X_{t+1} \mid X_t = x] \leq \delta x$ then $\mathrm{e}[X_b \mid X_0] \geq X_0 \cdot (1-b)^b \geq X_0 \cdot e^{-2\delta b}$ for $\delta \leq 0.797$.*

If we are minimising OneMax with the $(1+1)$ EA, it is not too difficult to see that the expected decrease in function value is at least $\text{OneMax}(x)/(en)$: with probability $(1/n)(1-1/n)^{n-1} \geq 1/(en)$ we flip exactly one bit and there are $\text{OneMax}(x)$ bits that all decrease the number of remaining 1-bits by 1. This yields $\mathrm{e}[\text{OneMax}(x_t)] \leq (n/2) \cdot e^{-t/(en)}$ as a direct consequence of Theorem 5.4.1.

When we are considering arbitrary linear functions and the expected function value, it is clear that the concrete value depends on the concrete weights in a very significant way. In order to obtain statements that are independent of the concrete function values (in particular, to allow the derivation of statements that allow a direct comparison with OneMax), we can restrict our interest to linear functions with only positive weights that are sorted with respect to size and sum up to n, i.e. $w_1 \geq w_2 \geq \cdots \geq w_n > 0$ and $\sum_{i=1}^{n} w_i = n$. Considering only positive weights is of no consequence for the $(1+1)$ EA because it is completely symmetric with respect to the roles of 0-bits and 1-bits (i.e. we can replace w_i by $-w_i$ and swap the roles of 0-bits and 1-bits at position i without significantly changing the situation for the $(1+1)$ EA). Assuming that the weights are sorted is of no consequence for the $(1+1)$ EA, because it is completely symmetric with respect to bit positions. Finally, assuming that the weights now add up to n is of no consequence for the $(1+1)$ EA, because it is sensitive only to the ordering of the function values, not their absolute values. Thus, changing all weights from w_i to some $s \cdot w_i$ with the same s does not change anything significant for the $(1+1)$ EA.

Using this setting and following the same ideas as for the $(1+1)$ EA on OneMax, one can prove that

$$\frac{n}{2} \cdot e^{-t/n} - O\left(\frac{t}{n}\right) \leq \mathrm{e}[f(x_t)] \leq \frac{n}{2} \cdot e^{-t/(en)}$$

holds [28] for any such linear function. In the upper bound, we recognise the result for OneMax, which is provably the easiest linear function for the $(1+1)$ EA (as measured by runtime analysis [42]).

We also see that the bound is only tight for rather small budgets $b = o(n)$ (remember that b is the total computational budget we are allowed to spend, and thus $t \leq b$ in the formula above), much smaller than the expected runtime $\Theta(n \log n)$. We see here again that it is often much easier to obtain tight results for earlier stages of a run. The difficulty is that later in the run we have a non-negligible probability that the optimum has already been found. If that is the case, no further improvements in the function value are possible, so that this case significantly reduces the expected contribution of any steps at later stages. Lengler and Spooner [28] suggested considering conditional results that they called 'a posteriori results': results that are conditioned on the assumption that, at the point of time we consider, the run has not yet finished. For the $(1+1)$ EA on OneMax, they were able to prove the following result of this flavour.

Theorem 5.4.2 (Theorem 19 in [28]). *Let γ, ρ be two constants with $0 < \gamma < \rho < 1$. Consider the $(1+1)$ EA on OneMax for t steps and assume that $\text{OneMax}(x_t) = o(n/\log n)$ and $\text{OneMax}(x_t) = \Omega(n^\rho)$. Then*

$$\Pr\left[\text{OneMax}(x_{t+\beta}) = \text{OneMax}(x_t) \cdot e^{-\beta/(en)} \cdot (1 \pm o(1))\right] = 1 - o(1)$$

for all $\beta \leq (\rho - \gamma) \cdot en \ln n$.

We see that even for OneMax we do not have an unconditional tight result for the expected function value that the $(1+1)$ EA can deliver after t steps for budgets that can get arbitrarily close to the expected runtime. For general linear functions, the situation is even more open.

A fundamentally different approach is based on stochastic differential equations [16]. It is very similar in spirit to much earlier attempts to improve our understanding of evolutionary algorithms by employing methods to analyse dynamic systems inspired by statistical mechanics [37]: one introduces some assumptions that facilitate the analysis and are reasonable but not proven to hold in all cases. Based on these assumptions, it becomes possible to derive results that otherwise elude the state of the art of analytical methods. It is important to note that the assumptions made are explicit and the subsequent analysis is completely rigorous. The only non-rigorous step is the definition of the model that is considered instead of the actual random process as defined by the algorithm and the fitness function. While this is not the first time that stochastic differential equations have been used in the analysis of evolutionary algorithms (see [40] for a different example), it is the first time that analytical results have been obtained that are clearly beyond the previous body of knowledge.

This approach is based on the analysis of stochastic processes that can be described as a one-dimensional Itô process [34], a process $(X_t)_{t\geq 0}$ on $(\Omega, \mathcal{F}, P)$ of the form $X_t = X_0 + \int_0^t b(s, X_s)\ ds + \int_0^t \sigma(s, X_s)\ dB_s$ where dB_t denotes a Brownian motion process, b is absolutely integrable in $[0,t]$ and σ is square integrable in $[0,t]$. Considering the probability density $p(x,t)$ of the random process X_t, we approximate the density by considering it as a time-continuous process (instead of the time-discrete process it actually is) and applying a Taylor expansion to the Chapman-Kolmogorov equation $p(x, t+\Delta t) = \int \Delta(\delta \mid x) \cdot p(x-\delta, t) \cdot d\delta$ describing the probability density of X_t. This leads to the diffusion equation (also known as the Fokker-Planck equation)

$$\Delta t \cdot \frac{\partial p(x,t)}{\partial t} \approx -\frac{\partial}{\partial x}(p(x,t) \cdot \mathrm{e}[\Delta]) + \frac{1}{2} \cdot \frac{\partial}{\partial x^2}(p(x,t) \cdot \mathrm{e}[\Delta^2])$$

and one considers the stochastic differential equation associated with it,

$$dX_t \approx \mathrm{e}[\Delta]\ dt + \sqrt{\mathrm{e}[\Delta^2]}\ dB_t,$$

which turns out to be an Itô process [34]. Based on this, the main assumption made is the following (Hypothesis 1 in [16]). The dynamics of RLS and the (1+1) EA can be approximated by the Itô process

$$\begin{aligned} dX_t &= b(X_t, t)\ dt + \rho(X_t, t)\ dB_t, \\ b(X_t, t) &= \mathrm{e}[X_{t+1} - X_t \mid X_t], \\ \sigma^2(X_t, t) &= \mathrm{e}\left[(X_{t+1} - X_t)^2 \mid X_t\right], \end{aligned}$$

where B_t is a Brownian motion process.

Approximation errors are introduced by considering a time-continuous random process instead of the actual algorithm that operates in discrete time steps, and by the Gaussian approximation that depends on the operators the algorithm applies and the fitness function. It is worth mentioning that the method is not tied to RLS and the (1+1) EA; in particular, it can cover a much wider range of selection operators, and a number of results are already available (see [16] for details).

Using this framework, it is not difficult to derive a number of results, we consider fixed-budget results for LeadingOnes as an example. We start with RLS and remember that $\mathrm{e}[\textsc{LeadingOnes}(x_t)] = 1 + (2t/n) - 2^{-\Omega(n)}$ holds (with the long and technical proof to be found in [21]). The approach here yields $1 + (t/n) - 2^{-n} \leq \mathrm{e}[\textsc{LeadingOnes}(x_t)] \leq 1 + (2t/n) - 2^{-n}$ (and we see that the upper bound is actually tight and the lower bound off by a factor of 2), but also a result for the variance of the process, namely $2 + (t/n) - 2^{-\Omega(n)} \leq \mathrm{Var}[\textsc{LeadingOnes}(x_t)] \leq 2 + (4t/n) - 2^{-\Omega(n)}$.

For the (1+1) EA, the rigorous analysis becomes much more involved and we have the tight result $\mathrm{e}[\textsc{LeadingOnes}(x_t)] = 1 + (2t/n) - o(t/n)$ only for not too large budgets $b = o(n^2)$ [21]. Using the approximation approach, we obtain $1 + (t/(en)) - 2^{-\Omega(n)} \leq \mathrm{e}[\textsc{LeadingOnes}(x_t)] \leq 1 + (2t/n) - 2^{-\Omega(n)}$ and also $2 + (t/(en)) - 2^{-\Omega(n)} \leq \mathrm{Var}[\textsc{LeadingOnes}(x_t)] \leq 2 + (4t/n) - 2^{-\Omega(n)}$ without a restriction on the budget. The result, however, is again not tight. The deviation seems to indicate that the results are rather pessimistic, resulting in lower bounds that tend to be too small. This trend is confirmed by other results [16]. It is also possible to use the same approach to derive fixed-budget results for non-elitist trajectory-based algorithms, i.e. for algorithms that follow the same general algorithmic framework as RLS and the (1+1) EA (Algorithms 5.2 and 5.2, respectively) but make use of a selection that allows one to stay with an inferior search point under some circumstances (usually purely stochastic with some fixed probability that might depend on the difference in function value, the Metropolis algorithm and simulated annealing are examples of this kind of algorithm [17]). We point the interested reader to the original paper [16] and will not expand on the brief description of the method presented so far.

5.5 Results Obtained by Using the Fixed-Budget Perspective

When introducing the fixed-budget perspective, it has been argued that compared to runtime analysis, fixed-budget results have greater potential to answer questions that practitioners actually care about [20]. This might lead to the belief that results obtained in the fixed-budget framework should give answers to questions for which runtime analysis remains silent. While this is actually the case, there is more to fixed-budget results. They might actually lead to answers that are to some extent contradictory to the insights that runtime analysis has provided. We will consider a range of different results that have been obtained using the fixed-budget perspective, and the kind of insights that have been obtained with the help of it.

Artificial immune systems [38] are a class of stochastic search heuristics that are algorithmically very similar to evolutionary algorithms while being very different in their motivation and quite different in the concrete implementation of operators. Artificial immune systems take inspiration from different theories about the functioning of the immune system of vertebrates. One aspect of artificial immune systems that is, algorithmically speaking, very different from evolutionary algorithms is the way mutation works. In evolutionary algorithms, mutation operators create small random choices and allow a search that concentrates mostly on search points that are rather close to the current population [18]. The standard bit mutation used in the (1+1) EA (Algorithm 5.2) is a very typical example. Artificial immune systems make use of so-called hypermutations, which have a much higher probability of producing new search points that have a huge Hamming distance from the original search point. One can compare the performance of such hypermutations with the performance of other mutation operators such as standard bit mutations by taking the $(1+1)$ EA and replacing only the mutation operator (see [19] for an example of this approach). When one analyses the performance on common simple example functions such as OneMax and LeadingOnes, the usual finding is that standard bit mutations perform much better than hypermutations (or, expressed a little more drastically and less precisely, evolutionary algorithms perform much better than artificial immune systems). However, when the fixed-budget perspective is used to perform the same analysis, a quite different and more varied picture emerges. We discuss the findings here in some more detail (taking the setting and results from [22]).

One example of a hypermutation operator is the somatic contiguous hypermutations that are used in the B-cell algorithm [26]. In these somatic contiguous hypermutations one picks a starting position p and a mutation length l uniformly at random and flips a contiguous region of l bits starting at position p (wrapping around and continuing at the beginning of the bit string if $l+p>n$). Another example is the inversely fitness-proportional mutations that have been used in CLONALG [3] and other artificial immune systems. Like standard bit mutations, these flip each bit with some probability, but instead of using a fixed probability $1/n$ they use $e^{-\rho f}$, where $\rho \in R^+$ is a parameter and f is a normalised fitness that is in the interval $[0,1]$. It is not difficult to show that both mutation operators perform poorly on OneMax in comparison with the $(1+1)$ EA and RLS in terms of expected runtime. However, when a very small computational budget b (e.g. $b=o(n^{1/4})$) is considered, it can be shown that both somatic contiguous hypermutations and inversely fitness-proportional mutations perform much better than the $(1+1)$ EA with standard bit mutations and RLS in terms of expected fitness values [22]. The reason is that both of these hypermutation operators have a very good chance of making large improving steps (large in the sense of $\Omega(n^{1/4})$), while RLS and the $(1+1)$ EA are restricted to improving the function value in each generation in expectation by not more than a rather small constant. Thus, during these first steps, hypermutations perform much better and it depends on the context of the concrete application which kind of mutation operator should be preferred.

One example where one might be much more interested in the short-term performance of a heuristic than in its ability to locate an optimal solution in a much longer time frame is dynamic optimisation, where we are confronted with an objective function that changes over time. In this context, it can be much more important to reliably track the global optimum and be able to keep the distance to the local optimum limited than to locate it occasionally. With this specific measure of performance in mind, one can for example compare evolutionary algorithms and artificial immune systems when confronted with a dynamic problem. We consider one such example, where an evolutionary algorithm using standard bit mutations is compared with the B-cell algorithm on a dynamic bistable example problem [24]. In this example problem, the fitness $f(x)$ is given as the number of bits in which x agrees with the unique global optimum (this is precisely the same as OneMax when the global optimum is the all-1-bits string 1^n). The global optimum oscillates between two search points that are the bitwise complement of each other. The transition from one to the other is gradual, flipping one bit after the other in a consecutive fashion. Once the global optimum is equal to one of the two special search points, it remains unchanged for a period of fixed length. The performance of the different algorithms depends crucially on the length of this stable period, as well as on the speed of the execution platform (which defines how many function evaluations an algorithm can perform in the time during which the dynamic fitness function does not change). While the evolutionary algorithm is generally better in locating the global optimum in the stable phases (given that the stable phase is long enough and the execution platform fast enough), during the transition phases the artificial immune systems are better able to keep up with the moving optimum. Such results are easier to describe and analyse precisely in the fixed-budget perspective. We point the interested reader to [24] for a complete picture of different performance results in this specific context.

The results on OneMax establish that RLS and the (1+1) EA are much more efficient hill-climbers than the B-cell algorithm. One example function that was introduced in the context of the role that crossover plays is H-Iff [41]. To compute the function value H-Iff(x), we consider x as a concatenation of $n/2^l$ bit strings of equal length 2^l for each $l \in \{0,1,\ldots,\log n\}$ (and assume for the sake of simplicity that n is a power of 2). One such block contributes 2^l to the overall function value if it is either an all 0-bits or an all 1-bits block. Otherwise, it does not make any contribution to the function value. We see that a bit string and its bitwise complement have the same function value and that the all-1-bits string and the all-0-bits string are the two global optima. Random local search and the (1+1) EA both get stuck (either forever in the case of RLS or for very long times in the case of the (1+1) EA) in local optima which are easy to find. This might lead to the speculation that in the fixed-budget perspective considering very small budgets RLS and the (1+1) EA outperform the B-cell algorithm on H-Iff, while in the runtime perspective the B-cell algorithm is much more efficient: it can be shown using simple fitness-level arguments that its expected optimisation time is $O(n^3 \log n)$ [23]. We remark that the same problem and algorithm have been subject to an intense analysis in a bigger study that, however, concentrated only on results from the expected runtime perspective [43]. Considering the fixed-budget perspective [23], the somewhat surprising finding is that RLS and the (1+1) EA do not have a clear advantage on H-Iff even for rather small computational budgets.

Analysing the performance of stochastic search heuristics from the fixed-budget perspective is not restricted to artificial example problems. The first example of a combinatorial optimisation problem where fixed-budget results have been obtained for simple stochastic search heuristics (namely, RLS and (a variant of) the $(1+1)$ EA) is the traveling salesperson problem (TSP) [32]. While there are no provably tight results, there are lower bounds on the expected gain in function value in one generation that can be used to obtain performance guarantees for a stochastic search heuristic. We have argued above that results of this type are most relevant in practical settings, and it is important that [32] provides an example of results of this kind for a practically relevant, hard combinatorial optimisation problem.

In the TSP, we are looking for an optimal order of all nodes of the given input graph so that visiting the nodes in this order is a tour of minimum length. While it is possible to encode this problem in a binary search space, it is more natural to have the algorithms operate in the natural search space: the set of all permutations. This implies that it becomes necessary to define a mutation operator for this search space. Nallaperuma et al. [32] used a 2-opt step as a local mutation for RLS. In a 2-opt step, one selects from a tour two edges $\{u_1,u_2\}$, $\{v_1,v_2\}$ that do not share any node, and appear in the tour in that order. They are removed and replaced by the edges $\{u_1,v_1\}$ and $\{u_2,v_2\}$. One can in some sense 'simulate' standard bit mutations (concentrating on the fact that a mutation flipping k bits can be viewed as executing k local mutations and that this happens with probability $\approx e^{-1}/k!$) by selecting a number k from a Poisson distribution with parameter $\lambda = 1$ and performing $k+1$ 2-opt steps. Note that performing $k+1$ 2-opt steps (instead of k) excludes the case where no change is made (something that happens for the $(1+1)$ EA with probability $\approx e^{-1}$).

The analysis considers random instances that are generated by placing n points in a d-dimensional unit hypercube, each point chosen independently with its own probability density function $f_i\colon [0,1]^d \to [0,\phi]$, where $\phi > 1$ is a parameter. The probability densities are chosen by an adversary who has the aim of creating difficult random instances. The power of the adversary increases with increasing ϕ (where $\phi = 1$ would equal a powerless adversary because there is only one valid choice and the instances are drawn with respect to the uniform distribution).

For a fixed-budget analysis results, on the expected improvement for a 2-opt step are very helpful. For two variants of the TSP, the TSP where distances are given by the Manhattan metric and the Euclidean TSP, Englert et al. [12] showed that the probability that the smallest possible improvement for a 2-opt step is less than ε is bounded from above by $576\varepsilon n^4\phi$ for the Manhattan metric and by $n^2\varepsilon\log(1/\varepsilon)\phi^3$ for the Euclidean metric. Using this it can be shown for the Manhattan metric that RLS and the $(1+1)$ EA both achieve an expected fitness gain of $\Omega(t/(n^6\phi))$ or reach a local optimum with expected approximation ratio $O(\sqrt{\phi})$ in t steps (see [32] for the proof). By considering not just single 2-opt steps but sequences of 2-opt steps, this bound can be improved to $\Omega(t/(n^5\phi))$ for RLS if the number of steps t is sufficiently large, namely $t > (3/2)n^3$. For the $(1+1)$ EA, the analysis is more complicated and the proof in [32] works only when one makes the assumption that the (1+1) EA never removes an edge in a 2-opt step that is later reinserted in the sequence of steps considered. It can be hoped that this assumption does not improve the performance and the result also holds for the (1+1) EA, because excluding such steps only rejects steps that would lead to an improvement. However, proving this is an open problem. Under this assumption the same improved bound of $\Omega(t/(n^5\phi))$ can be proved if the number of steps t is sufficiently large, namely $t > (3/2)e^2n^3$ for the (1+1) EA. For the Euclidean metric, the bound obtained by analysing a single step is $\Omega(t\log(n\phi)/(n^6\phi^3))$ for RLS and the (1+1) EA (and, as above, both algorithms might alternatively find a local optimum with an expected approximation ratio of $O(\sqrt{\phi})$). Again, for both algorithms this can be improved by considering sequences of steps (under the condition that the number of steps t is large enough, namely $t > (3/2)n^3$ for RLS and $t > (3/2)e^2n^3$ for the (1+1) EA) to $\Omega(t\sqrt{\log(n\phi)}/(n^5\phi^{5/2}))$. Again, for the (1+1) EA the additional assumption about not accepting 2-opt steps that remove an edge that is later re-inserted is required for the proof. It is possible to generalise the analysis to cover variants of RLS and the (1+1) EA that do not produce one mutated search point in each round but instead produce λ mutated search points in each round (independently and identically distributed) and select a best one. The detailed results and how the expected improvement in t steps can be translated into a result about the expected approximation ratio can be found in [32].

One other example that demonstrates the usefulness of the fixed-budget perspective stems from the area of the design of stochastic search heuristics. One motivation for theoretical analysis of general stochastic search heuristics is that it leads to a better understanding that should enable us to design better heuristics. Doerr et al. [6] presented one example of this kind of research. They considered the generalised OneMax problem: the function value $f(x)$ is given by the number of bits where x agrees with the unknown target bit string. If the target bit string is the all-1-bits string 1^n, the function is OneMax. For this problem, they considered heuristics that are unbiased in the sense of unbiased black-box complexity [27]: intuitively speaking, the search heuristic has to be completely symmetric with respect to the roles of 0-bits and 1-bits as well as bit positions. Moreover, selection can depend only on the function values and not the search points themselves. In this context, Doerr et al. considered a variant of RLS with two modifications. The first modification is that if $f(x) < n/2$ (i.e. more than half of the bits in the current search point are wrong), the current search point x is replaced by its bitwise complement $\overline{x}$. Since $f(x) = n - f(\overline{x})$ for all generalised OneMax functions f, this is guaranteed to be an improvement. The second motivation is that, instead of flipping precisely one bit, the algorithm flips a number of bits that depends on the function value. To determine how many bits should be flipped Doerr et al. performed a very precise analysis of the expected improvement in function value, the drift, and based on this defined a function that yields the number of bits to be flipped for each function value. This optimal number is 1 if the number of incorrect bits is less than $n/3$ and increases monotonically with the number of incorrectly set bits after that. Moreover, it is always an odd number. Using the method of variable drift analysis [7], they derived precise results for the expected runtime that show the expected runtime to be $n\ln(n) - cn \pm o(n)$ for a constant c with $0.2539 < c < 0.2665$. This is at most εn ($\varepsilon > 0$ constant) worse than the black-box complexity, the lower bound for any unbiased algorithm. However, the actual advantage appears to be small compared with unmodified RLS which has an expected runtime $n\ln(n) + \gamma n \pm o(n)$. Thus, the new algorithm has an advantage of at most $0.151n$ while the expected runtime is $n\ln(n) \pm \Theta(n)$, a not very impressive advantage. However, by performing an analysis using the fixed-budget perspective one can show that in comparison with RLS the expected function value is about 13% larger for the new algorithm, a clear and tangible advantage. The main difficulty in deriving the fixed-budget result is in proving a concentration result for an aspect of the expected runtime (something that is not surprising remembering the results in Section 5.3). Proof details can be found in [6]. A similar result can be proven for a variant that self-adjusts the choice of the number of bits to be flipped [5].

5.6 Summary

The fixed-budget perspective is an alternative to the 'classical' analytical perspective that runtime analysis offers. Instead of asking how long we expect a stochastic search heuristic to run, we ask what solution quality we can expect when we run the heuristic for a fixed number of steps. Since stochastic search heuristics are often applied to hard problems that are not well understood, it is often the case that they do not find an optimal solution, and even in the case where they do the user is not able to recognise that this has happened. Runtime analysis is not a good match to this way of applying stochastic search heuristics. The fixed-budget perspective is more useful in these situations because it provides information about the expected performance in a set time frame.

Considering RLS on OneMax, we have seen that in such very simple cases it is possible to obtain very precise results. However, in even slightly more realistic cases the analysis becomes much harder. This motivates the search for analytical tools and methods to derive fixed-budget results.

So far, there is only one method that is tailored towards derivation of fixed-budget results. It allows one to transfer runtime results if one is able to generalise the runtime result to a runtime result for all possible function values and if one is able to provide concentration bounds for this generalised runtime result. The main difficulty is obtaining a strong concentration result.

The strength of the concentration bound is directly reflected in the quality of the fixed-budget result.

Results for the $(1+1)$ EA on the class of linear functions show that existing tools such as drift analysis can be applied to obtain fixed-budget results. There is certainly a lot of room to extend this direction of research and see how other methods can be used and adapted to yield fixed-budget results.

Our lack of knowledge when it comes to the $(1+1)$ EA on the most studied example function, ONEMAX, and its cousins, the class of linear functions (see Section 5.4), demonstrates that fixed-budget analysis is still in an early stage and that there are many fundamental problems still unsolved that have already been answered for runtime analysis some time ago. Answering these kinds of questions has led to very significant research efforts in the runtime analysis community with a large number of results, deep insights and, most importantly, the development of a number of powerful and very useful analytical tools. It can be hoped that open problems that are so fundamental, easy to state and yet very difficult to answer will lead to a similar effort and success in the area of fixed-budget computations and their analysis.

It is currently unclear if the approximative approach via stochastic differential equations that is based on an approximation and in this sense yields only non-rigorous results is a productive step forward. This approach has just been published [16], and the results that come with it are currently restricted to weaker versions of known results and results for less well-known algorithms. It remains to be seen if it will deliver significant new insights in the future.

There are a number of results that show that fixed-budget analysis can yield relevant and important insights. When one is designing novel stochastic search heuristics (in the example we considered, based on insights that stem from theoretical analysis), the fixed-budget perspective has the potential to show that an advantage that might appear to be very small (or even insignificant) in the runtime perspective can be relevant and meaningful from the fixed-budget point of view. Considering artificial immune systems and revisiting runtime results that demonstrated that evolutionary algorithms are faster in optimising some example functions has revealed that while the artificial immune systems took much longer to get to an optimum, they made much quicker progress at the beginning of a run, implying that they would be the preferred choice if the available computational budget was small. The same effect can be exploited in the context of dynamic optimisation, where it can be most important to make some progress quickly in situations where the landscape is changing rapidly.

The travelling salesperson problem is, to the best of our knowledge, the first and so far only combinatorial optimisation problem where fixed-budget results are available, even for a range of algorithms. The way the results were obtained demonstrates that insights that were gained in a different context for different purposes can sometimes be reused to provide significant new fixed-budget results. It would be useful to find more examples of this kind to increase our knowledge about the performance of stochastic search heuristics in combinatorial optimisation from the perspective of fixed-budget analysis.

References

[1] Anne Auger and Benjamin Doerr, editors. *Theory of Randomized Search Heuristics.* World Scientific, 2011.

[2] Süntje Böttcher, Benjamin Doerr, and Frank Neumann. Optimal fixed and adaptive mutation rates for the LeadingOnes problem. In *Proceedings of the 11th International Conference on Parallel Problem Solving From Nature (PPSN 2010)*, pages 1–10, 2010.

[3] Leandro N. de Castro and Fernando J. Von Zuben. Learning and optimization using the clonal selection principle. *IEEE Transactions on Evolutionary Computation*, 6(3):239–251, 2002.

[4] Benjamin Doerr. Analyzing randomized search heuristics: Tools from probability theory. In Auger and Doerr [1], pages 1–20.

[5] Benjamin Doerr, Carola Doerr, and Jing Yang. k-bit mutation with self-adjusting k outperforms standard bit mutation. In *Proceedings of the 14th International Conference on Parallel Problem Solving From Nature (PPSN 2016)*, pages 824–834, 2016.

[6] Benjamin Doerr, Carola Doerr, and Jing Yang. Optimal parameter choices via precise black-box analysis. In *Proceedings of the Genetic and Evolutionary Computation Conference (GECCO 2016)*, pages 1123–1130, 2016.

[7] Benjamin Doerr, Mahmoud Fouz, and Carsten Witt. Sharp bounds by probability-generating functions and variable drift. In *Proceedings of the Genetic and Evolutionary Computation Conference (GECCO 2011)*, pages 2083–2090, 2011.

[8] Benjamin Doerr and Leslie Goldberg. Drift analysis with tail bounds. In *Proceedings of the 11th International Conference on Parallel Problem Solving From Nature (PPSN 2010)*, pages 174–183, 2010.

[9] Benjamin Doerr and Leslie Goldberg. Adaptive drift analysis. *Algorithmica*, 65(1):224–250, 2013.

[10] Benjamin Doerr, Thomas Jansen, Carsten Witt, and Christine Zarges. A method to derive fixed budget results from exptected optimisation times. In *Proceedings of the Genetic and Evolutionary Computation Conference (GECCO 2013)*, pages 1581–1588, 2013.

[11] Marco Dorigo and Thomas Stützle. *Ant Colony Optimisation.* Springer, 2004.

[12] Mathias Englert, Heiko Röglin, and Berthold Vöcking. Worst case and probabilistic analysis of the 2-Opt algorithm for the TSP. *Algorithmica*, 68(1):190–264, 2014.

[13] Bruce Hajek. Hitting-time and occupation-time bounds implied by drift analysis with applications. *Advances in Applied Probability*, 13(3):502–505, 1982.

[14] Jun He. An analytic expression of relative approximation error for a class of evolutionary algorithms. In *Proceedings of the IEEE World Congress on Computational Intelligence (CEC 2016)*, pages 4366–4373, 2016.

[15] Jun He and Xin Yao. Drift analysis and average time complexity of evolutionary algorithms. *Artificial Intelligence*, 127(1):57–85, 2001.

[16] Jorge Pérez Heredia. Modelling evolutionary algorithms with stochastic differential equations. *Evolutionary Computation*, 26(4):657–686, 2018.

[17] Thomas Jansen. Simulated annealing. In Auger and Doerr [1], pages 171–196.

[18] Thomas Jansen. *Analyzing Evolutionary Algorithms: The Computer Science Perspective.* Springer, 2013.

[19] Thomas Jansen and Christine Zarges. Analyzing different variants of immune inspired somatic contiguous hypermutations. *Theoretical Computer Science*, 412(6):517–533, 2011.

[20] Thomas Jansen and Christine Zarges. Fixed budget computations: a different perspective on run time analysis. In *Proceedings of the Genetic and Evolutionary Computation Conference (GECCO 2012)*, pages 1325–1332, 2012.

[21] Thomas Jansen and Christine Zarges. Performance analysis of randomised search heuristics operating with a fixed budget. *Theoretical Computer Science*, 545:39–58, 2014.

[22] Thomas Jansen and Christine Zarges. Reevaluating immune-inspired hypermutations using the fixed budget perspective. *IEEE Transactions on Evolutionary Computation*, 18(5):674–688, 2014.

[23] Thomas Jansen and Christine Zarges. Understanding randomised search heuristics. Lessons from the evolution of theory: a case study. In *Proceedings of the 20th International Conference on Soft Computing (MENDEL 2014)*, pages 293–298, 2014.

[24] Thomas Jansen and Christine Zarges. Analysis of randomised search heuristics for dynamic optimisation. *Evolutionary Algorithms*, 23(4):513–541, 2015.

[25] Kenneth A. De Jong. *Evolutionary Computation: A Unified Approach.* MIT Press, 2006.

[26] Johnny Kelsey and Jon Timmis. Immune inspired somatic contiguous hypermutation for function optimisation. In *Proceedings of the Genetic and Evolutionary Computation Conference (GECCO 2003)*, pages 207–218, 2003.

[27] Per Kristian Lehre and Carsten Witt. Black-box search by unbiased variation. *Algorithmica*, 64(4):623–642, 2012.

[28] Johannes Lengler and Nick Spooner. Fixed budget performance of the (1+1) EA on linear functions. In *Proceedings of the 2015 ACM Confer-*

ence on Foundations of Genetic Algorithms XIII (FOGA 2015), pages 52–61, 2015.

[29] Wil Michiels, Emile Aarts, and Jan Korst. *Theoretical Aspects of Local Search.* Springer, 2007.

[30] Rajeev Motwani and Prabhakar Raghavan. *Randomized Algorithms.* Cambridge University Press, 1995.

[31] Heinz Mühlenbein. How genetic algorithms really work: Mutation and hillclimbing. In *Proceedings of the 2nd International Conference on Parallel Problem Solving from Nature (PPSN II)*, pages 15–26, 2002.

[32] Samadhi Nallaperuma, Frank Neumann, and Dirk Sudholt. Expected fitness gains of randomized search heuristics for the traveling salesperson problem. *Evolutionary Computation*, 25(4): 673–705, 2017.

[33] Frank Neumann and Carsten Witt. *Bioinspired Computation in Combinatorial Optimization: Algorithms and Their Computational Complexity.* Springer, 2010.

[34] Bernt Øksendal. *Stochastic Differential Equations: An Introduction with Applications.* University of Michigan Press, 2003.

[35] Pietro Simone Oliveto and Carsten Witt. Simplified drift analysis for proving lower bounds in evolutionary computation. *Algorithmica*, 59(3):369–386, 2011.

[36] Pietro Simone Oliveto and Xin Yao. Runtime analysis of evolutionary algorithms for discrete optimization. In Auger and Doerr [1], pages 21–52.

[37] Adam Prügel-Bennett. Modelling evolving populations. *Journal of Theoretical Biology*, 185(1):81–95, 1997.

[38] Mark Read, Paul S. Andrews, and Jon Timmis. An introduction to artificial immune systems. In *Handbook of Natural Computing*, pages 1575–1597. Springer, 2012.

[39] Günter Rudolph. Stochastic convergence. In *Handbook of Natural Computing*, pages 847–869. Springer, 2012.

[40] Tom Schaul. Natural evolution strategies converge on sphere functions. In *Proceedings of the Genetic and Evolutionary Computation Conference (GECCO 2012)*, pages 329–336, 2012.

[41] Richard A. Watson, Gregory S. Hornby, and Jordan B. Pollack. Modeling building-block interdependency. In *Proceedings of the Fifth International Conference on Parallel Problem Solving From Nature (PPSN 1998)*, pages 97–108, 1998.

[42] Carsten Witt. Tight bounds on the optimization time of a randomized search heuristic on linear functions. *Combinatorics, Probability and Computing*, 22(2):294–318, 2013.

[43] Xiaoyun Xia and Yuren Zhou. On the effectiveness of immune inspired mutation operators in some discrete optimization problems. *Information Sciences*, 426:87–100, 2018.

Chapter 6
Theory of Parameter Control for Discrete Black-Box Optimization: Provable Performance Gains Through Dynamic Parameter Choices

Benjamin Doerr and Carola Doerr

Abstract Parameter control is aimed at realizing performance gains through a dynamic choice of the parameters which determine the behavior of the underlying optimization algorithm. In the context of evolutionary algorithms, this research line has for a long time been dominated by empirical approaches. With the significant advances in running-time analysis achieved in the last ten years, the parameter control question has become accessible to theoretical investigations. A number of running-time results for a broad range of different parameter control mechanisms have been obtained in recent years. This chapter surveys these results, and puts them into context by proposing an updated classification scheme for parameter control.

6.1 Introduction

Evolutionary algorithms and many other iterative black-box optimization heuristics are parameterized algorithms, i.e., their search behavior depends (to a large extent) on a set of parameters which the user needs to specify, or which are set by the algorithm designer to some default values. It is today well understood that the parameter choice can have a very decisive influence on the performance of a heuristic [71]. Understanding how to best choose the parameters is therefore an important task. It is referred to as the *parameter selection problem.*

The parameter-setting problem is difficult for several reasons.

Benjamin Doerr
École Polytechnique, CNRS, Laboratoire d'Informatique (LIX), Palaiseau, France

Carola Doerr
Sorbonne Université, CNRS, LIP6, Paris, France

B. Doerr, F. Neumann (eds.), *Theory of Evolutionary Computation*,
Natural Computing Series, https://doi.org/10.1007/978-3-030-29414-4_6

- *Complexity of performance prediction.* Despite significant research efforts devoted to this problem, predicting how the performance of an algorithm depends on the chosen parameter values remains a very challenging problem – both with empirical and with theoretical methods. In fact, determining optimal parameter values can be very complex even for a single parameter. Many black-box optimization heuristics, however, rely on two or more parameters. Rigorously analyzing the interdependency between these parameters is often infeasible by state-of-the-art technology.
- *Problem and instance dependence.* It is well known that no globally good parameter values exist, but that suitable parameter values can differ substantially between different optimization problems, and even between different instances of the same problem.
- *State dependence.* It is, furthermore, widely acknowledged that the best parameter values can change during the optimization process. For example, it is often beneficial to use larger mutation rates at the beginning of an optimization process, to allow faster exploration, and to shrink the search radius over time, to allow better exploitation in the later stages, see Section 6.2 for a detailed example.

To overcome these difficulties, a large number of different parameter-setting techniques have been developed. Following standard terminology in evolutionary computation, they can be classified into two main approaches.

- *Static parameter settings: parameter tuning.* Parameter tuning is aimed at identifying parameter values that are, for a given algorithm on a given problem (instance), globally suitable throughout the whole optimization process. The parameters are initialized with these values and do not change during the optimization process. Parameter tuning thus addresses the above-mentioned problem and instance dependence of optimal parameter choices, but not their state dependence.
  In *empirical studies*, parameter tuning often requires an initial set of experiments that support an informed decision. Automated tools that help the user to identify reasonable static parameter values are available, and have been shown to provide significant performance gains over a manual tuning process. [3, 11, 51, 52, 72] are examples of automated parameter-tuning approaches that have been used in (and to some extend specifically designed for) evolutionary optimization contexts.
  In *theoretical studies,* parameter tuning requires running-time bounds that depend on the parameters under investigation. The minimization of these performance bounds then suggests suitable parameter values. A prime example of such a mathematical approach to parameter tuning is the precise running-time bound for the $(1+1)$ EA with mutation rate $p=c/n$ on linear functions. Witt [86] has shown that this expected optimization time is $(1 \pm o(1))\frac{e^c}{c} n \ln(n)$. This bound, together with larger running-time bounds for mutation rates $p \neq c/n$, proves that the often recommended choice $p=1/n$ is indeed optimal for the $(1+1)$ EA on this problem. Such

precise upper and lower bounds, however, are rare. Even worse, only few running-time bounds that depend on two or more parameters exist; see Section 6.4.3.

- **Dynamic parameter setting: parameter control.** Parameter control, in contrast, is aimed at benefiting from a non-static choice of the parameters, with the underlying idea that the flexibility in the behavior can be used to adjust the algorithms' behavior to the current state of the optimization process. Put differently, parameter control is aimed not only at *identifying* parameter values that are a good compromise for the whole optimization process, but also at *tracking* the evolution of the best parameter values. Even when the optimal parameter values are rather stable, the role of parameter control is to identify these values *on the fly*, without a dedicated tuning step that precedes the actual optimization process.

This chapter focuses on non-static parameter choices, and thus on parameter control mechanisms. We survey existing theoretical studies of parameter control in the context of evolutionary algorithms and other standard black-box optimization heuristics. We also summarize a few standard techniques used in the empirical research literature.[1] We structure our presentation with a new classification scheme for parameter control mechanisms. This taxonomy builds on the well-known classification by Eiben, Hinterding, and Michalewicz [40], but modifies it to better reflect the developments that parameter control has witnessed in the last 20 years.

This chapter is structured as follows. We indicate the motivation for the use of non-static parameter choices in Section 6.2 by demonstrating a simple example where adaptive parameter selection is provably beneficial. We then introduce our revised classification scheme in Section 6.3. In the subsequent Sections 6.4-6.8, we survey existing theoretical results. In Section 6.9, we conclude this chapter with a discussion of promising avenues for future work. A summary of selected theoretical running-time results covered in this chapter can be found in Table 6.2.

6.2 A Motivating Example: (1+1) EA and RLS on LeadingOnes

We start this section with an example that demonstrates the potential advantages of parameter control mechanisms. To this end, we study the well-known LeadingOnes benchmark, the problem of minimizing an unknown function of the type

[1] Readers interested in empirical studies of parameter control are referred to [61] for an exhaustive survey. Additional pointers can be found in the systematic literature review [2], the book chapter [41] (and other chapters in the same collection), and the seminal paper [40].

$$\mathrm{LO}_{z,\sigma} : \{0,1\}^n \to [0..n] := \{0,1,\dots,n\},$$
$$x \mapsto \max\{i \in [0..n] \mid \forall j \in [1..i] : x_{\sigma(j)} = z_{\sigma(j)}\},$$

where $z \in \{0,1\}^n$ and σ is a permutation (one-to-one map) of the set $[1..n]$. Optimizing $\mathrm{LO}_{z,\sigma}$ corresponds to identifying z, the unique optimum of $\mathrm{LO}_{z,\sigma}$. Note that for every x the function value $\mathrm{LO}_{z,\sigma}(x)$ is the length of the longest prefix that x and z have in common, when comparing the strings in the order prescribed by σ.

It has been shown in [12] that the $(1+1)$ EA with a static mutation rate $0<p<1$ needs $\frac{1}{2p^2}[(1-p)^{1-n}-(1-p)]$ iterations, on average, to optimize a LEADINGONES instance. This term is minimized for $p \approx 1.59/n$, which yields an expected optimization time of around $0.77n^2$. It was observed in [12] that a fitness-dependent choice of the mutation rate gives a better optimization time. More precisely, when x denotes the current best individual, and we choose in the next iteration $p = 1/(\mathrm{LO}(x)+1)$ as the mutation rate, the expected optimization time decreases to around $(e/4)n^2 \approx 0.68n^2$. This is almost 21% better than the expected optimization time of the $(1+1)$ EA with standard mutation rate $p=1/n$ and about 11.7% better than the $0.77n^2$ expected running time mentioned above which the best static mutation rate, $p \approx 1.59/n$, achieves.

Also Randomized Local Search (RLS), the algorithm which in each iteration flips one uniformly selected bit and selects the better of the two offspring as the parent individual for the next iteration, can profit from a non-static choice of the *step size*, i.e., the number of bits that it flips in every iteration. It is well known that RLS needs $n^2/2$ iterations, in expectation, to optimize an n-dimensional LEADINGONES instance. In Fig. 6.1 we take a closer look at the optimization process, and plot the expected number of iterations (y-axis) needed by RLS to identify, on the 1000-dimensional LEADINGONES problem, a solution of fitness value at least $\mathrm{LO}(x)$ (x-axis). This is the blue straight line. We also illustrate in the same figure the corresponding expected *fixed-target running times* of the RLS variant which in each iteration flips exactly two and three pairwise different bits, respectively. These are the yellow and gray curves, respectively. The lowermost, black line illustrates the expected performance of the RLS variant which chooses in each iteration the best of these three parameter values. We observe that this adaptive variant has an expected optimization time that is around 20% smaller than that of standard 1-bit-flip RLS. We also see that for LO values smaller than $n/2$, it is advisable to flip more than one bit per iteration, while 1-bit flips are optimal once a solution of LO value $\geq n/2$ has been identified. This can be best seen by comparing the slopes of the curves in this plot of fixed-target running times. The ultimate goal of parameter control is the design of mechanisms that detect such transitions and suggest the best possible parameter values for the different stages in an automated way.

We note that in the example discussed in this section, "only" constant factors could be gained through the dynamic parameter choice, but that

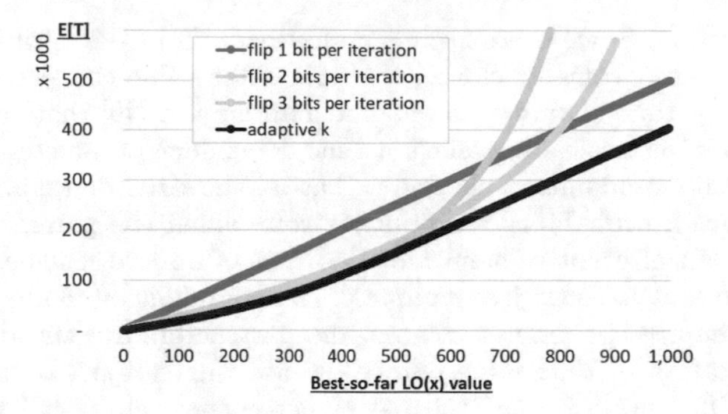

Fig. 6.1 Expected fixed-target running times of RLS variants flipping in each iteration exactly one, two, three, or an adaptive number of bits on the 1,000-dimensional LeadingOnes function. The adaptive variant, which chooses the best among the three parameter values, has a total expected optimization time that is about 20% better than RLS, which always flips one bit per iteration.

in general asymptotic performance gains can also be expected. An example where this has been rigorously proven will be discussed in Section 6.4.3.

6.3 Classification of Parameter Control Mechanisms

A considerable obstacle to be overcome when searching for previous work on non-static parameter choices is the lack of a commonly agreed-upon terminology. This has led to a situation in which similar techniques have significantly different names, and, conversely, the same term has been used for two fundamentally different concepts. Since 1999 a widely accepted classification scheme for parameter setting has been the taxonomy proposed by Eiben, Hinterding, and Michalewicz in [40]. We present this classification in Section 6.3.1, and modify it to cope with the developments in parameter control in the last twenty years in Section 6.3.2.

6.3.1 The Classification Scheme of Eiben, Hinterding, and Michalewicz

Eiben, Hinterding, and Michalewicz [40] distinguished three different types of parameter control, namely deterministic, self-adaptive, and adaptive parameter settings.

- A dynamic parameter choice is called *deterministic* if the choice of the parameter value does not depend on the fitness landscape encountered by

the algorithm. Since there is thus no feedback from the optimization process into the parameter choice, the parameter value can depend only on iteration or time counters. It was noted already in [40] that the term "deterministic" is misleading, since a time-dependent parameter choice may still contain randomized elements, that is, the time or iteration counter determines a probability distribution from which the parameter value is sampled. As alternative names for this class of update schemes, the terms *scheduled* and *feedback-free parameter control* might be more appropriate.
- In *self-adaptive* parameter choices, the parameters are encoded into the representation of the search points and are thus subject to variation operators. The hope is that the better parameter values will yield better offspring and will thus be more likely to survive the evolutionary process. By this, implicitly, the choice of the parameters depends on the optimization process and thus, in particular, on the fitness function.
- *Adaptive* parameter choices are dynamic parameter settings in which there is an explicit dependence of the parameters on the optimization process. This large category includes structurally simple success-based update rules such as those resembling the one-fifth success rule in evolution strategies, and also learning-inspired techniques which choose the parameter values depending on statistics from the optimization process so far.

6.3.2 A Revised Classification Scheme

At the time of writing of [40], the three different types of parameter control discussed in Section 6.3.1 were of similar importance. In the last almost twenty years, however, we have observed an increasing interest (and massive progress) in the subcategory of adaptive parameter control schemes, which also play a predominant role in theoretical studies. In particular, in recent years it has become quite clear that the substantial differences between, say, a simple deterministic fitness-dependent choice of a parameter value and a parameter choice via reinforcement-learning approaches are a reason for us not to have both in the same category. We therefore present in the next subsection an alternative classification scheme, which takes into account this development.

- *State-dependent parameter control.* We classify as *state-dependent parameter control* those mechanisms that depend only on the current state of the search process, for example the current population, its fitness values, and its diversity, but also those that depend on a time or iteration counter. Hence this subsumes the previous "deterministic" category (containing time-dependent parameter choices) and all other parameter-setting mechanisms which determine the current parameter values via a prespecified function mapping algorithm states to parameter values, possibly in a randomized manner. All these mechanisms require the user to precisely specify how the parameter value depends on the current state and, as such, need a substantial understanding of the problem to be solved.

- *Success-based parameter control.* To overcome the usability challenges and the inflexibility of state-dependent parameter control mechanisms, several approaches to setting the parameters in a *self-adjusting* manner have been proposed. As one important type of self-adjusting parameter control mechanism, we classify as *success-based* parameter settings all those mechanisms that change the parameters from one iteration to the next. In other words, the parameter value to be used in the current iteration is determined (possibly in a randomized manner) by the parameter value used in the previous iteration and by an evaluation of how successful the previous iteration was. The success measure can be simple binary information such as whether a solution with superior fitness was found, but it could also take into account quantitative information such as the fitness gain or loss in this iteration. Depending on the parameter to be set, also other quantities than the fitness can be taken into account, for example the evolution of the diversity of the population.
  The most common forms of success-based rules are multiplicative updates of parameters, which increase or decrease the parameter value by suitable factors depending on whether the previous iteration was classified as a success or not. Success-based rules other than multiplicative updates have been designed as well. For example, in [33] the offspring were generated with two different parameter values and the information about which parameter value led to the best offspring determined the parameters of the next iteration; see Section 6.5.2.3 for a detailed discussion.
- *Learning-inspired parameter control.* As the second main type of self-adjusting parameter control mechanism, we classify as *learning-inspired parameter control* mechanisms all those schemes which are aimed at exploiting a longer search history than just one iteration. To allow such learning mechanisms also to adapt quickly to changing environments, older information is taken into account to a lesser extent than more recent information. This can be achieved by considering only information from (static or sliding) *time windows* or by discounting the importance of older information via weights that decrease (usually exponentially) with the age of the data.
  Most learning-inspired parameter control mechanisms that have been experimented with in the evolutionary computation context borrow tools from machine learning, where a similar problem known as the *multiarmed bandit problem* has been studied; see Section 6.6.1.
- *Endogenous parameter control (self-adaptation).* This category corresponds to the self-adaptive parameter control mechanisms in the taxonomy of [40]. We prefer the name "endogenous parameter control" as it best emphasizes the structural difference of these mechanisms, which is that they encode the parameters in the genome and let them evolve via the usual variation and selection mechanisms of the evolutionary system.

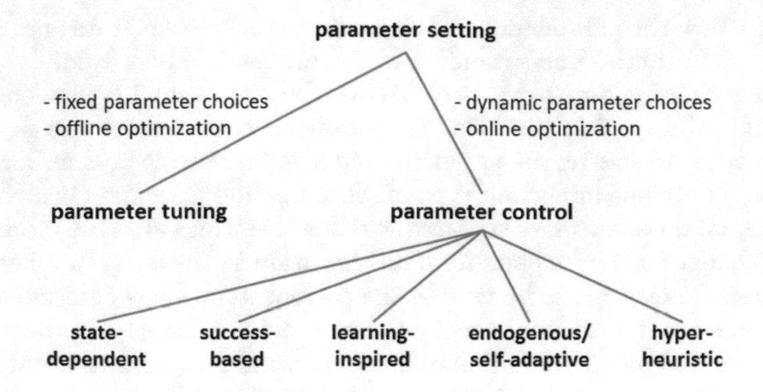

Fig. 6.2 Classification of parameter control mechanisms. We also call success-based and learning-inspired mechanisms *self-adjusting*.

- *Hyper-heuristics.* Hyper-heuristics are algorithms that operate on a set of low-level heuristics, select from it an algorithm, and run it for some time, before reevaluating which of the low-level heuristics to use next. The main hope is that hyper-heuristics will automate the algorithm selection and configuration process, in a way that allows maximizing the profit from different algorithmic ideas in the different stages of the optimization process. Similarly to the motivation behind endogenous parameter control, the use of a high-level hyper-heuristic is guided by the belief that the high complexity of the parameter control problem calls for efficient heuristic approaches.

Figure 6.2 summarizes our classification scheme. Existing theoretical results are summarized in the following sections, which are structured according to this taxonomy.

We emphasize that our classification is partly driven by the historical development of the field. For example, it would be more logical not to have hyper-heuristics (as long as they essentially optimize parameters) as a separate category, but rather to classify them as success-based or learning-inspired parameter control schemes. Since, historically, the area of hyper-heuristics developed relatively independently (partially owing to the fact that there are many hyper-heuristics that cannot be seen as parameter control mechanisms), we prefer to maintain a separate category for hyper-heuristics.

6.4 State-Dependent Parameter Control

We recall from the previous section that state-dependent parameter selection schemes are those mechanisms which choose the parameter values based only on the current state of the algorithm, without making use of the search history. One of the best known examples of state-dependent parameter control is the so-called *cooling schedule* used in Simulated Annealing. The idea of this cooling schedule is to start the heuristic with a rather generous acceptance behavior and to increase the selective pressure during the optimization process; see Section 6.4.1 for a more detailed description. The cooling schedule, as the name suggests, is a time-dependent selection mechanism, which maps the iteration counter to a temperature value that defines the selective pressure.

As we shall see in this section, time-dependent parameter selection schemes have also been experimented with in the context of evolutionary computation. In addition, other state-dependent parameter settings, such as rank- and fitness-based mutation rates and diversity-based parameter choices, have been analyzed empirically, but have received considerably less attention in the theory-of-evolutionary-algorithms community.

6.4.1 Time-Dependent Parameter Choices

Simulated Annealing is typically not regarded as an evolutionary algorithm, since it draws inspiration from the physical phenomenon of an annealing process. We nevertheless decided to discuss it in this chapter, as it is structurally very similar to Randomized Local Search, and certainly falls into the class of iterative randomized black-box optimization heuristics.

Simulated Annealing [63] is a (1+1)-type search heuristic that uses a Boltzmann selection rule to decide whether or not to replace the previous parent individual x by a new solution y. More precisely, the algorithm keeps in its memory only one previously evaluated solution x, and modifies it by a local variation. In the case of pseudo-Boolean maximization this local move is identical to that of RLS, i.e., the offspring y is created from x by flipping exactly one bit, the position of which is chosen uniformly at random. The new solution y always replaces x if it is better, and it replaces x with probability $\exp\big((f(y)-f(x))/T\big)$ otherwise. That is, the better y is, the larger the probability that it survives the selection procedure. The novelty of Simulated Annealing over its predecessor, the Metropolis algorithm [74], is an adaptive choice of the "temperature" T in the Boltzmann selection rule: while the Metropolis algorithm uses the same T throughout the whole optimization process, the value of T is decreased over time in Simulated Annealing, either with each iteration or, more commonly, after a fixed number τ of iterations. The adaptive selective pressure results in a more generous acceptance behavior at the beginning of the optimization process (to allow faster exploration), and a more and more elitist selection towards the end ("exploitation"). Algorithm 6.1 summarizes this algorithm. For constant $T_t = T$, we obtain from Algorithm 6.1 the pseudocode for the Metropolis algorithm. Numerous successful applications and more than 43,000 citations[2] of [63] witness that this idea of controlling the selective pressure during the optimization process can have an impressive impact on performance.

Algorithm 6.1: Simulated Annealing for the maximization of a pseudo-Boolean function $f:\{0,1\}^n \to \mathbb{R}$

1 **Initialization:** Choose $x \in \{0,1\}^n$ uniformly at random;
2 **Optimization: for** $t = 1,2,3,\ldots$ **do**
3 Create from x a new solution candidate y by flipping exactly one bit in x;
4 **if** $f(y) \geq f(x)$ **then**
5 $x \leftarrow y$
6 **else**
7 $x \leftarrow y$ with probability $\exp((f(y) - f(x))/T_t)$

A number of theoretical results analyzing the performance of Simulated Annealing exist. Most of these prove convergence to a global optimum for suitably chosen parameter settings; see the book chapter [50] for a summary of selected theoretical and empirical results. In addition to the results mentioned there, a plethora of running-time results exist for combinatorial optimization problems on graphs, including most notably matching [81] and graph bisection problems [15, 53, 59]. Selected theoretical studies that concentrate on the advantages of dynamic parameter choices are summarized below.

Answering an open problem posed in [58], Wegener presented in [85] a problem class for which Simulated Annealing outperforms its static counterpart, the Metropolis algorithm, regardless of how the temperature value is chosen in the latter. More precisely, Wegener proved that Simulated Annealing with a multiplicative temperature decay $T(t) = \alpha T(1)$ ($\alpha < 1$ being a constant and the initial value $T(1)$ being ignorant of the instance, but possibly depending on the number of edges m and the maximum weight $w_{\max}$) has a better expected optimization time on some subclasses of the minimum spanning tree (MST) problem than the Metropolis algorithm with any fixed temperature. Previous examples of this phenomenon had been presented in [83] and [39], but were of a rather artificial nature. The novelty of [85] was thus to prove this statement for a natural combinatorial optimization problem. A particular instance of the MST problem for which Wegener proved the superiority of Simulated Annealing is a graph that has the form of connected triangles. Wegener also showed a provable advantage for ε-separated graphs, in which nonequal weights are apart from each other by a constant factor of at least $1+\varepsilon$ [85, Section 5].

One of the first studies analyzing a classic evolutionary algorithm with a dynamic parameter setting was presented by Droste, Jansen, and Wegener in the above-mentioned publication [39]. Besides a *time-dependent selection*

[2] This number of citations is according to Google Scholar as of April 12, 2018.

strategy, these authors also analyzed the (1+1) EA with a *time-dependent mutation rate* $p \in \{2^k/n \mid k = 0,1,2,\ldots,\lceil \log_2(n) \rceil - 2\}$. In this algorithm, the mutation rate is initialized to $1/n$ and doubled in every iteration until p exceeds $1/2$, in which case it is reset to $1/n$. An example function, PATHTOJUMP, was presented for which the (1+1) EA with the time-dependent mutation rate needs only $O(n^2 \log n)$ steps, on average, to locate the optimum, while the (1+1) EA with the static mutation rate $p = 1/n$ does not optimize PATHTOJUMP in expected polynomial time. The authors of [39] also showed a converse result in which the dynamic (1+1) EA is much slower than the classical static one. It is not difficult to see that the dynamic (1+1) EA performs worse than the static (1+1) EA on most classical benchmark functions such as ONEMAX and LEADINGONES; see [39, Section 4]. This work was later extended and simplified by Jansen and Wegener in [55].

In [56] a comparison was made between the (1+1) EA with static and with time-dependent mutation rates on the one hand, and Simulated Annealing and the Metropolis algorithm on the other hand, but the focus of this work was not on the advantages of adaptive parameter choices, but rather on a comparison of the different selection schemes.

6.4.2 Rank-Dependent Parameter Control

Motivated by empirical work reported in [16], Oliveto, Lehre, and Neumann analyzed in [77] a $(\mu+1)$ EA with *rank-based mutation rates*. In this algorithm, the individuals of the parent population are ranked according to their fitness values, and the mutation rate applied in an iteration t depends on the rank of the (uniformly selected) individual undergoing mutation. The intuition behind these rank-based mutation rates is that individuals with larger ranks (i.e., worse fitness) should be modified more aggressively (suggesting large mutation rates), while the best individuals of the population should be modified with caution, suggesting small mutation rates.

To be more precise, the algorithm proposed in [16] uses standard bit mutation with mutation rate p_i, where for the i-th ranked search point the value of p_i is set to $p_{\min} + (p_{\max} - p_{\min})(i-1)/m$ (linear interpolation ensuring a minimum mutation rate of $p_{\min} > 0$ and a maximum mutation rate $p_{\max}$). The variant studied in [77] uses $p_{\min} = 1/n$, $p_{\max} = 1$, and $m = \mu$. Theorem 6.4.1 below gives a general upper bound for the rank-based $(\mu+1)$ EA, which is better than the $\Theta(n^n)$ expected running time of the (1+1) EA on functions such as NEEDLE or TRAP.

Theorem 6.4.1 (Theorems 1 and 2 in [77]). *For $\mu \geq 2$ and $\mu = poly(n)$, the expected optimization time of the $(\mu+1)$ EA with rank-based mutation* *rates is at most*[3] $7 \cdot 3^n$ *for any pseudo-Boolean function* $f : \{0,1\}^n \to \mathbb{R}$, *and it is* $O(\mu n \log n)$ *for* ONEMAX.[4]

In addition to these results, examples were constructed for which the $(\mu+1)$ EA with rank-based mutation rates performs significantly worse [77, Section V] and significantly better [77, Section VI] than the classical $(\mu+1)$ EA with standard bit mutation rate $p = 1/n$.

6.4.3 Fitness-Dependent Parameter Control

While rank-based parameter selection was originally introduced with the hope of finding a generally well-functioning control scheme, fitness-based parameter selection schemes are often highly problem-tailored, and cannot be assumed to work particularly well when applied to different objective functions. The theoretical results stated below should therefore not be considered as a suggestion for generally applicable parameter control mechanisms, but rather as a point of comparison for more plausible, general-purpose parameter update techniques, i.e., we should use these results only as a lower bound for the performance of a best possible parameter update scheme. This way, the results form a baseline that helps us understand and judge the limits of parameter control.

6.4.3.1 Fitness-Dependent Mutation Rates for the (1+1) EA on LeadingOnes

The first study showing a significant advantage of a fitness-dependent choice of the mutation rate was presented in [12], where the following result was shown.[5]

Theorem 6.4.2 (Theorems 3-6 in [12]). *On* LEADINGONES, *the expected number of iterations needed by the* $(1+1)$ *EA with a* static *mutation rate* $p \in (0,1)$ *to identify the optimal solution is* $\frac{1}{2p^2}[(1-p)^{1-n} - (1-p)]$. *This expression is minimized for* $p \approx 1.59/n$, *which gives an expected optimization time of around* $0.77n^2$.

[3] This bound is mistakenly stated as $O(2^n)$ in [77, Theorem 1], but the proof clearly shows the upper bound stated here.

[4] We recall that ONEMAX is the function that assigns to each $x \in \{0,1\}^n$ the number of ones in it, i.e., $\mathrm{OM}(x) = \sum_{i=1}^n x_i$. All running-time bounds that we state in this chapter for the optimization of ONEMAX also apply to the optimization of the functions $\mathrm{OM}_z : \{0,1\}^n \to \mathbb{R}, x \mapsto |\{i \in [n] \mid x_i = z_i\}|$, whose fitness landscape is isomorphic to that of $\mathrm{OM} = \mathrm{OM}_{(1,\dots,1)}$.

[5] Prior to [12], fitness-dependent mutation rates had also been analyzed in *immune algorithms* [87, 88], but no advantage of the parameter choices analyzed could be shown.

For the (1+1) EA variant that chooses in every iteration the fitness-dependent mutation rate $p=1/(\mathrm{LO}(x)+1)$*, where* x *denotes the solution that undergoes modification, the expected optimization time decreases to around* $(e/4)n^2 \approx 0.68n^2$*. No other fitness-dependent mutation rate can achieve a better expected optimization time.*

In this result the expected optimization time of the fitness-dependent (1+1) EA is almost 21% better than the expected optimization time of the (1+1) EA with the standard mutation rate $p=1/n$ and about 11.7% better than the $0.77n^2$ expected running time which the best static mutation rate, $p \approx 1.59/n$, achieves.

6.4.3.2 Fitness-Dependent Mutation Rates for the $(1+\lambda)$ EA on OneMax

Interestingly, the question of how to best control the mutation rate during the optimization process gained relevance with the establishment of *black-box complexity* as a measure for the best possible running time that any randomized search heuristic of a certain type can achieve (see Chapter 3 of this book for a survey of publications on this complexity notion). By comparing existing algorithms with the theoretically best possible performance, one can judge how well suited a given approach is. Unsurprisingly, the best-possible algorithms take into account the state of the optimization process, and adjust their parameters accordingly.

In this context and, more precisely, in the context of analyzing lower bounds for the performance of unbiased parallel evolutionary algorithms (see Section 3.7.2 for a more detailed description of the motivation) Badkobeh, Lehre, and Sudholt analyzed in [10] the optimal fitness-dependent mutation rate for the $(1+\lambda)$ EA on OneMax. The main result is summarized by the following theorem.

Theorem 6.4.3 (Theorems 3 and 4 in [10]). *For* $\lambda \leq e^{\sqrt{n}}$ *the* $(1+\lambda)$ *EA that uses in each iteration the mutation rate* $p(x) := \max\left\{1/n, \frac{\ln\lambda}{n\ln(en/(n-\mathrm{OM}(x)))}\right\}$ *(where* x *denotes the parent individual held in the memory at the beginning of the iteration) has an expected optimization time on OneMax equal to* $\Theta\left(n\log n + \frac{\lambda n}{\log\lambda}\right)$.

This performance is the best possible among all unary unbiased black-box algorithms that create λ *offspring in parallel.*

The performance of this fitness-dependent $(1+\lambda)$ EA for many values of λ is superior to the performance of the $(1+\lambda)$ EA with the static mutation rates considered so far, which is $\Theta(n\log n + \frac{\lambda n\log\log n}{\log n})$ for a mutation rate $p=c/n$, where c is a constant [34, 46], and $\Theta\left(\sqrt{\lambda}n\log n + \frac{\lambda n}{\log\lambda}\right)$ for $p=\ln(\lambda)/(2n)$ and $\lambda \in \omega(1) \cap n^{O(1)}$ [33, Lemma 1.2].

In Section 6.5.2.3 we will see an example of a purely success-based adaptation scheme which achieves the same expected performance as the $(1+\lambda)$ EA with a fitness-dependent mutation rate. More recently, a self-adaptive $(1,\lambda)$ EA has been designed, which also achieves the same bound. This algorithm will be discussed in Section 6.7.

6.4.3.3 Fitness-Dependent Mutation Strengths for RLS on OneMax

While the result in Section 6.4.3.2 is of asymptotic order only, one might hope to get more precise results for selected values of λ. Unfortunately, the precise relationship between function values and optimal mutation rates is not known even in the very special case $\lambda = 1$. What is known, however, is the following.

In [30] it was shown that the best possible running time on ONEMAX that any unary unbiased black-box algorithm can achieve is $n\ln(n) - cn \pm o(n)$ for a constant c between 0.2539 and 0.2665 (see Section 3.6 for a discussion of unbiased black-box algorithms). It cannot be better by more than an additive $o(n)$ term than the expected optimization time attained by the RLS variant that in every iteration chooses the mutation strength (i.e., the number of bits to be flipped) in a way that maximizes the expected progress. By the symmetry of the ONEMAX function, this *drift-maximizing mutation strength* depends only on the fitness of the current best solution, and not on the structure of this search point. More precisely, when ℓ different bits of the search point x are flipped to create y, the expected progress $E[\max\{\mathrm{OM}(y) - \mathrm{OM}(x), 0\}]$ equals

$$\sum_{i=\lceil \ell/2 \rceil}^{\ell} \frac{\binom{n-\mathrm{OM}(x)}{i}\binom{\mathrm{OM}(x)}{\ell-i}(2i-\ell)}{\binom{n}{\ell}}. \tag{6.4.1}$$

The drift-maximizing mutation strength $r_{\mathrm{opt}}(x)$ is the value of ℓ that maximizes this expression.[6]

Theorem 6.4.4 (Theorem 9 in [30]). *The expected optimization time $E[T]$ of the drift-maximizing algorithm with fitness-dependent mutation strengths $r_{\mathrm{opt}}(x)$ is $n\ln(n) - cn \pm o(n)$ for a constant c between 0.2539 and 0.2665. The unary unbiased black-box complexity is smaller than $E[T]$ by an additive term of at most $o(n)$.*

Compared with RLS or the RLS variant using an optimized initialization phase presented and analyzed in [65], the bound in Theorem 6.4.4 is smaller

---

[6] No easy-to-interpret algebraic relationship between x and $r_{\mathrm{opt}}(x)$ could be established in [30], and an approximation of $r_{\mathrm{opt}}(x)$ was therefore used in that publication. It was shown, however, that this affects the overall performance by at most $o(n)$ iterations.

by an additive term between $0.138n \pm o(n)$ and $0.151n \pm o(n)$. For problem dimensions $\leq 10,000$, the advantage of the drift-maximizing algorithm over classic RLS is around 2%.

In the language of fixed-budget computation as introduced by Jansen and Zarges in [57], the drift-maximizing algorithm with a budget of at least $0.2675n$ iterations computes a solution with an expected fitness distance to the optimum roughly 13% smaller than the output produced by RLS [30, Section 6].

6.4.3.4 Fitness-Dependent Offspring Population Sizes in the $(1+(\lambda,\lambda))$ Genetic Algorithm

All the results above concern the control of the mutation rate. A fitness-dependent choice of the *offspring population sizes* was considered in [26] for the $(1+(\lambda,\lambda))$ GA on OneMax. Since this algorithm later gave rise to a growing interest in parameter control (note that the conference version [25] appeared before most of the other results mentioned in this section), we describe this algorithm in more detail. Note in particular that, in contrast to the purely mutation-based algorithms mentioned above, the $(1+(\lambda,\lambda))$ GA also uses crossover.

The $(1+(\lambda,\lambda))$ GA (Algorithm 6.2) works with a parent population of size one. This population $\{x\}$ is initialized with a search point chosen from $\{0,1\}^n$ uniformly at random. The $(1+(\lambda,\lambda))$ GA then proceeds in iterations, each consisting of a mutation phase, a crossover phase, and a final elitist selection step determining the new parent population.

In the *mutation phase*, a step size ℓ is chosen at random from the binomial distribution $\mathrm{Bin}(n,p)$, where the parameter p is called the mutation rate of the algorithm. Then, independently, λ offspring are created by flipping exactly (i.e., pairwise different) ℓ random bits in x. In an intermediate selection step, one best mutation offspring x' is selected as the mutation winner. In the *crossover phase*, again λ offspring are created, this time via a biased uniform crossover between x and x', taking each entry from x' with probability c only and taking the entry from x otherwise. Again, an intermediate selection chooses one of the best crossover offspring y as the crossover winner. In the final *selection step*, this y replaces x if its *fitness* is at least as large as the fitness of x, i.e., if and only if $f(y) \geq f(x)$ holds.

The $(1+(\lambda,\lambda))$ GA thus has three parameters that need to be set prior to any execution: the offspring population size λ, the mutation rate p, and the crossover bias c. Using intuitive considerations, it was suggested in [26] that $p = \lambda/n$ and $c = 1/\lambda$ should be used. With these choices, the three-dimensional parameter space is reduced to a one-dimensional one, and only λ needs to be set. In [26] it was shown that choosing $\lambda = \Theta(\sqrt{\log n})$ yields an expected running time of $O\left(\max\left\{\frac{n\log(n)}{\lambda}, \lambda n\right\}\right)$

Algorithm 6.2: The $(1+(\lambda,\lambda))$ GA maximizing a given function $f:\{0,1\}^n \to \mathbb{R}$ with offspring population size λ, mutation rate p, and crossover bias c. The mutation operator σ_ℓ generates an offspring from one parent by flipping exactly ℓ random bits (without replacement). The crossover operator cross_c performs a biased uniform crossover, taking bits independently with probability c from the second argument.

1 **Initialization:** Choose $x \in \{0,1\}^n$ uniformly at random (u.a.r.);
2 **Optimization: for** $t = 1,2,3,\ldots$ **do**
3 **Mutation phase:**
4 Sample ℓ from $\text{Bin}(n,p)$;
5 **for** $i = 1,\ldots,\lambda$ **do** $x^{(i)} \leftarrow \sigma_\ell(x)$;
6 Choose $x' \in \{x^{(1)},\ldots,x^{(\lambda)}\}$ with $f(x') = \max\{f(x^{(1)}),\ldots,f(x^{(\lambda)})\}$ u.a.r.;
7 **Crossover phase:**
8 **for** $i = 1,\ldots,\lambda$ **do** $y^{(i)} \leftarrow \text{cross}_c(x,x')$;
9 Choose $y \in \{y^{(1)},\ldots,y^{(\lambda)}\}$ with $f(y) = \max\{f(y^{(1)}),\ldots,f(y^{(\lambda)})\}$ u.a.r.;
10 **Selection step: if** $f(y) \geq f(x)$ **then** $x \leftarrow y$;

for the $(1+(\lambda,\lambda))$ GA on the OneMax problem. This bound was later improved to $F^* = \Theta(n\sqrt{\log(n)\log\log\log(n)/\log\log(n)})$ in [23]; this expected running time is attained for a slightly larger value of λ, namely $\lambda^* = \Theta(\sqrt{\log(n)\log\log(n)/\log\log\log(n)})$. Finally, [21] showed that the suggested dependencies $p = \lambda/n$ and $c = 1/\lambda$ are asymptotically optimal in the sense that any static parameter combination (p,c,λ) that gives an expected running time of $O(F^*)$ needs to satisfy $p = \Omega(\lambda^*/n)$, $p = (1/n)\exp(O(\sqrt{\log(n)\log\log\log(n)/\log\log(n)}))$, $c = \Theta(1/(pn))$, and $\lambda = \Theta(\lambda^*)$. No parameter combination can achieve an asymptotically better running time than $\Theta(F^*)$.

The results mentioned above all concern static parameter values. In terms of dynamic parameters, it had already been observed in [26] that a better expected running time, namely a linear one, can be achieved by the $(1+(\lambda,\lambda))$ GA on OneMax if we allow the parameters to depend on the function values. This linear expected performance was later shown to be asymptotically optimal.

Theorem 6.4.5 (Theorem 8 in [26] and Sections 5 and 6.5 in [24]). *The expected optimization time of the* $(1+(\lambda,\lambda))$ *GA with* $p = \lambda/n$, $c = 1/\lambda$, *and* $\lambda = \sqrt{n/(n-f(x))}$ *on* OneMax *is* $\Theta(n)$, *and this is asymptotically the best possible among all dynamic parameter choices. For any static parameter values* (p,c,λ), *the expected running time of the* $(1+(\lambda,\lambda))$ *GA on* OneMax *is of order at least* $n\sqrt{\log(n)\log\log\log(n)/\log\log(n)}$, *and thus strictly larger than linear.*

In Section 6.5 we will discuss a success-based parameter control mechanism that identifies and tracks good values for λ in an automated way.

6.5 Success-Based Parameter Control

We have classified as success-based parameter control mechanisms all those which change the parameters from one iteration to the next, based on the outcome of the iteration. This includes, in particular, multiplicative update rules which change parameters by constant factors depending on whether the iteration was considered a success or not.

6.5.1 The One-Fifth Success Rule and Other Multiplicative Success-Based Updates

Even the very early studies of evolution strategies used a simple, yet powerful technique to adapt the parameters online. The so-called *one-fifth success rule*, which was independently discovered in [19, 79, 82], suggests that one should set the step size of an evolution strategy in such a manner that 1/5 of the iterations lead to a fitness improvement. The idea behind this is that when the success rate is higher, then most likely the step size is too small and time is wasted on minor improvements; however, when the success rate is lower, then time is wasted by waiting too long for an improvement. The value 1/5 was derived from some theoretical considerations about the performance of the (1+1) evolution strategy on the *sphere* problem $f : \mathbb{R}^n \to \mathbb{R}, x \mapsto \sum_{i=1}^n x_i^2$. Rechenberg showed that a success rate of about 20% yields an optimal expected gain for this problem (and also on another problem with a so-called inclined ridge; see [79] for details).

The first implementations of this one-fifth success rule were not success-based in our language, but rather observed the success rate over several iterations and then adjusted the step size if a discrepancy from the target success rate of 1/5 was detected. In [62], a simpler success-based implementation was proposed. Here, the step size is multiplied by some number $F > 1$ in the case of success and divided by $F^{1/4}$ in the case of no success. The hyper-parameter F is called the *update strength* of the adaptation rule.

We next present two examples of success-based parameter control suggested in the literature.

- *Example 1: the one-fifth success rule applied to the* $(1+(\lambda,\lambda))$ *GA.* It may be surprising that a simple multiplicative success-based rule can work. We therefore present an illustrated example, the self-adjusting $(1+(\lambda,\lambda))$ GA, which was originally proposed in [26] and later formally analyzed on the OneMax problem in [22]. We will describe this algorithm in more detail in Section 6.5.2.1, but note here only that when the recommended dependencies $p = \lambda/n$ and $c = 1/\lambda$ are used, the self-adjusting $(1+(\lambda,\lambda))$ GA requires one to set the offspring population size λ as the only parameter. The value of λ is adapted based on the success of a full iteration, using the implementation of the one-fifth success rule suggested in [62]. Figure 6.3 shows how well the optimal fitness-dependent value of the offspring population size λ suggested by Theorem 6.4.5 (smooth black curve) is approximated by this multiplicative success-based update rule (irredular red curve). The uppermost (blue) curve shows the evolution of the current-best fitness value, from which the optimal fitness-dependent mutation rate is computed. Note that in this figure we show the optimal mutation rates *per iteration*, each of which costs 2λ function evaluations. The update strength F in this illustration was set to 1.5.

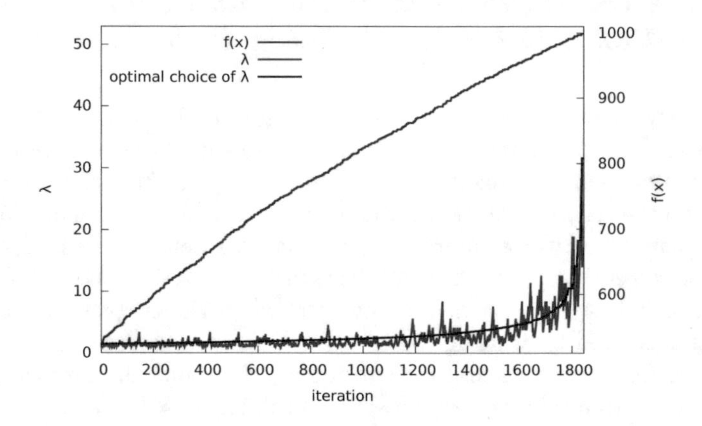

Fig. 6.3 The application of the one-fifth success rule to the offspring population size λ in the $(1+(\lambda,\lambda))$ GA on OneMax shows a very good fit to the optimal fitness-dependent population size.

- *Example 2: the* $(1,\lambda)$ *EA with success-based offspring population size* λ. A different success-based parameter control was suggested in [48] for the control of the offspring population size λ in a non-elitist $(1,\lambda)$ evolution strategy (ES). Motivated by a theoretical result that proves that in the $(1,\lambda)$ ES the so-called local serial progress is maximized when the expected progress of the second best offspring created in one iteration is zero (this result applies to any function $f:\mathbb{R}^n \to \mathbb{R}$), the following multiplicative success-based update rule for the offspring population size λ was suggested. Denoting by $x^{(t)}$ the parent individual of the t-th iteration, by $\lambda(t)$ the selected offspring population size, and by $x^{(t),1},\dots,x^{(t),\lambda(t)}$ the offspring created in the t-th iteration, sorted by decreasing function values, the offspring population size for the next iteration is set to

$$\lambda^{(t+1)} := \max\left\{2, \lambda^{(t)} \exp\left(\frac{-\beta(f(x^{(t),2}) - f(x^{(t)}))}{\sqrt{\sum_{i=1}^{\lambda(t)} (f(x^{(t),)i}) - f(x^{(t)}))^2/(\lambda-1)}}\right)\right\}, \tag{6.5.1}$$

where $\beta \in (0,1)$ is a hyper-parameter that controls the speed of the adaptation. While this update mechanism, to the best of our knowledge, has not been formally analyzed, it was shown in [48] to give good performance on the hyperplane and the hypersphere problems.

6.5.2 Theoretical Results for Success-Based Parameter Control

In this section, we describe the theoretical results known for success-based based parameter control mechanisms. We note that some studies of hyper-heuristics resemble closely a success-based parameter control. The reader can find these in Section 6.8.3.

6.5.2.1 The Self-Adjusting $(1+(\lambda,\lambda))$ GA on OneMax and on MaxSAT

We have seen in Theorem 6.4.5 that the $(1+(\lambda,\lambda))$ GA with mutation rate $p=\lambda/n$, crossover bias $c=1/\lambda$, and fitness-dependent population size $\lambda=\sqrt{n/(n-f(x))}$ takes an expected number of $\Theta(n)$ function evaluations to optimize a OneMax instance of problem dimension n. This is the asymptotically best running time among all static and dynamic parameter choices. A substantial drawback of this result is the rather complex dependence of λ on the current best function value $f(x)$. The question of whether this relationship can be detected by a parameter control mechanism in an automated way suggests itself. In fact, in [26] a success-based choice of λ had already been suggested, and was shown to achieve a very similar empirical performance to the fitness-dependent choice, across all tested problem dimensions $n \leq 5{,}000$. In [24], the efficiency of this success-based variant of the $(1+(\lambda,\lambda))$ GA, which we will describe in more detail below, could be formally proven.

Theorem 6.5.1 (Theorem 9 in [24]). *The expected optimization time of the self-adjusting $(1+(\lambda,\lambda))$ GA (Algorithm 6.3) with mutation rate $p=\lambda/n$, crossover bias $c=1/\lambda$, and a sufficiently small update strength $F>1$ on* OneMax *is $\Theta(n)$.*

The success-based choice of the parameter λ uses the above-mentioned implementation of the one-fifth success rule considered in [62]. That is, after an iteration that leads to an increase in the best observed function value ("success"), λ is reduced by a constant factor $F>1$. If an iteration is not successful, λ is increased by a multiplicative factor $F^{1/4}$. Consequently, after a series of iterations with an average success rate of $1/5$, this mechanism ends up with the initial value of λ.

Algorithm 6.3: The self-adjusting $(1+(\lambda,\lambda))$ GA with mutation probability p, crossover bias c, and update strength F

1 **Initialization:** Sample $x \in \{0,1\}^n$ uniformly at random (u.a.r.);
2 Initialize $\lambda \leftarrow 1$;
3 **Optimization: for** $t = 1,2,3,\ldots$ **do**
4   **Mutation phase:**
5   Sample ℓ from $\mathrm{Bin}(n,p)$;
6   **for** $i = 1,\ldots,\lambda$ **do** $x^{(i)} \leftarrow \sigma_\ell(x)$;
7   Choose $x' \in \{x^{(1)},\ldots,x^{(\lambda)}\}$ with $f(x') = \max\{f(x^{(1)}),\ldots,f(x^{(\lambda)})\}$ u.a.r.;
8   **Crossover phase:**
9   **for** $i = 1,\ldots,\lambda$ **do** $y^{(i)} \leftarrow \mathrm{cross}_c(x,x')$;
10   Choose $y \in \{y^{(1)},\ldots,y^{(\lambda)}\}$ with $f(y) = \max\{f(y^{(1)}),\ldots,f(y^{(\lambda)})\}$ u.a.r.;
11   **Selection and update step:**
12   **if** $f(y) > f(x)$ **then** $x \leftarrow y$; $\lambda \leftarrow \max\{\lambda/F, 1\}$;
13   **if** $f(y) = f(x)$ **then** $x \leftarrow y$; $\lambda \leftarrow \min\{\lambda F^{1/4}, n\}$;
14   **if** $f(y) < f(x)$ **then** $\lambda \leftarrow \min\{\lambda F^{1/4}, n\}$;

Since $p = \lambda/n$, the value of λ is capped at n. Likewise, it is capped from below at 1. The value of λ is allowed to be non-integral. Where an integer is required (i.e., in lines 6, 7, 9, and 10 of Algorithm 6.3), λ is rounded to the closest integer. That is, in these four lines, instead of λ we consider $\lfloor \lambda \rfloor = \lambda - \{\lambda\}$ if the fractional part $\{\lambda\}$ of λ is less than 1/2, and we consider $\lceil \lambda \rceil = \lfloor \lambda \rfloor + 1$ otherwise.

In the experiments conducted in [25] (see in particular Figure 4 there), all update strengths $F \in [1,2]$ worked well. While this indicates some robustness of the result in Theorem 6.5.1 with respect to the F-value, it has been argued in [24, Section 6.4] that update strengths F greater than 2.25 may lead to an exponential expected optimization time on OneMax. A commonly used value for F, also used in Auger's implementation [7], is $F = 1.5$. This is also the value with which Fig. 6.3 was created.

One may wonder, further, how important the relationship between the two multiplicative updates, that is, the exponent 1/4 is. It was argued in [24, Section 6.4] that a similar result to that in Theorem 6.5.1 is likely to hold for a range of other exponents as long as the exponent is not too large. Hence, in discrete optimization, there is no particular reason for a one-fifth rule (that is, the exponent 1/4). This has also been observed in a recent study of image composition, where a success-based one-kth success rule was used to adjust the length of a random walk that was part of the mutation operator [75]. In a set of initial experiments $k = 9$ seemed to be a suitable value, and was used for empirical evaluations.

Being the first algorithm which can provably reduce the expected optimization time by applying a success-based parameter control mechanism, the self-adjusting $(1+(\lambda,\lambda))$ GA has also been analyzed on other functions, by empirical and by theoretical means. In [26, Section 4], a promising empirical performance for linear functions $f:\{0,1\}^n \to \mathbb{R}, x \mapsto \sum_{i=1}^n w_i x_i$ with random weights $w_i \in [1,2]$ and for so-called royal road functions with block size 5 had already been reported. In [47], the self-adjusting $(1+(\lambda,\lambda))$ GA was tested on a number of combinatorial problems. In particular, for the maximum satisfiability problem, the self-adjusting $(1+(\lambda,\lambda))$ GA showed very good performance, beaten only by the parameterless population pyramid proposed in the same publication. Inspired by this result, a mathematical running-time analysis of the $(1+(\lambda,\lambda))$ GA on random satisfiability instances was conducted in [14]. This confirmed that the $(1+(\lambda,\lambda))$ GA has better performance than solely mutation-based algorithms (see, e.g., [37]). This work, however, also showed that weaker fitness–distance correlation of the satisfiability instances can lead to the effect that when offspring are created with a high mutation rate, the algorithm has problems in determining the structurally better ones. This difficulty can be overcome by imposing an upper limit on the population size λ, which determines the mutation rate $p=\lambda/n$.

6.5.2.2 The $(1+\lambda)$ EA with Success-Based Offspring Population Size λ

For the $(1+\lambda)$ EA, the following success-based adaptation of the offspring population size was suggested in [54, Section 5]. The offspring population size λ is initialized to one. After each iteration, the number s of offspring having a function value that is at least as large as that of the parent fitness is determined. When $s=0$ (i.e., if the iteration has been unsuccessful), the offspring population size λ is doubled, otherwise it is replaced by $\lfloor \lambda/s \rfloor$. The intuition for this adaptive choice of the offspring population is to have a value of λ that is inversely proportional to the probability of creating an offspring that replaces its parent. This algorithm, which we call the $(1+\{2\lambda,\lambda/s\})$ EA, was not analyzed by mathematical means in [54], but showed encouraging empirical performance on OneMax, LeadingOnes, and a benchmark function called SufSamp.

The idea of a success-based offspring population size was taken up in [66], where a theoretical analysis of two similar success-based update schemes was performed. The first update scheme, the $(1+\{2\lambda,1\})$ EA, doubles λ if no strictly better search point can be identified and sets λ to one otherwise. The second $(1+\lambda)$ EA variant, the $(1+\{2\lambda,\lambda/2\})$ EA, also doubles λ if no solution of quality better than the parent is found, and reduces λ to $\max\{1,\lfloor \lambda/2 \rfloor\}$ otherwise. While these schemes do not result in an improved overall running time in terms of function evaluations, they are both able to achieve a significant reduction of the *parallel* optimization time on selected benchmark problems. That is, the average number of *generations* needed before an optimal solution is evaluated for the first time is smaller than that for classical sequential evolutionary algorithms, which do not perform any evaluations in parallel. The precise results are as follows.

Function	Algorithm	$E[T^{\text{seq}}]$	$E[T^{\text{par}}]$
OneMax	$(1+\{2\lambda,1\})$ EA	$\Theta(n\log n)$	$O(n)$ (*)
	$(1+\{2\lambda,\lambda/2\})$ EA	$\Theta(n\log n)$	$O(n)$
LeadingOnes	$(1+\{2\lambda,1\})$ EA	$\Theta(n^2)$	$\Theta(n\log n)$
	$(1+\{2\lambda,\lambda/2\})$ EA	$\Theta(n^2)$	$O(n)$
unimodal with d different function values	$(1+\{2\lambda,1\})$ EA	$O(dn)$	$O(d\log n)$
	$(1+\{2\lambda,\lambda/2\})$ EA	$O(dn)$	$O(d+\log n)$
Jump_k, $k\geq 2$	$(1+\{2\lambda,1\})$ EA	$O(n^k)$	$O(n+k\log n)$ (*)
	$(1+\{2\lambda,\lambda/2\})$ EA	$O(n^k)$	$O(n+k\log n)$

Table 6.1 Expected sequential and parallel running times of the $(1+\{2\lambda,\lambda/2\})$ EA and the $(1+\{2\lambda,1\})$ EA on selected benchmark problems [66]. For the two bounds marked (*), we have slightly improved the original bound of $O(n\log n)$ via an elementary argument; see the proof below Theorem 6.5.2

Theorem 6.5.2 (Theorem 7 in [66] and the proof below for the results marked (*) in Table 6.1). *The sequential and parallel expected running times of the $(1+\{2\lambda,\lambda/2\})$ EA and the $(1+\{2\lambda,1\})$ EA satisfy the bounds given in Table 6.1.*

Proof. Using the classic fitness level method, the expected parallel running time of the $(1+\{2\lambda,1\})$ EA on OneMax was bounded from above by $2\sum_{i=1}^{n-1}\log(\frac{2en}{n-i})$ in [66]. This expression was bounded further by $2n\log(2en)=O(n\log n)$. However, a closer look reveals that, with Stirling's formula, we easily obtain

$$2\sum_{i=1}^{n-1}\log\left(\frac{2en}{n-i}\right)\leq 2\log\left(\frac{(2en)^n}{n!}\right)\leq 2\log\left(\frac{(2en)^n}{(n/e)^n}\right)=2\log((2e^2)^n)=O(n).$$

This improved bound immediately carries over to the bound for Jump_k, $k\geq 2$, where the expected parallel running time of the $(1+\{2\lambda,1\})$ EA is bounded by the expected parallel running time on OneMax plus the time needed to "jump" from the local optimum to the global one, which is of order at most $k\log n$. □

6.5.2.3 The Two-Rate $(1+\lambda)$ EA with Success-Based Mutation Rates

In the previous examples, we have studied different ways to control the offspring population size. We now turn our attention to a success-based adaptation of the mutation rates in a $(1+\lambda)$ EA with a fixed offspring population size λ, which was introduced and analyzed in [33]. The $(1+\lambda)$ EA$_{(2r,r/2)}$ stores a parameter r that controls the mutation rate. This parameter is adjusted after each iteration by the following mechanism. In each iteration, the $(1+\lambda)$ EA$_{(2r,r/2)}$ creates $\lambda/2$ offspring by standard bit mutation with mutation rate $r/(2n)$, and it creates $\lambda/2$ offspring with mutation rate $2r/n$. At the end of the iteration a random coin is flipped. With probability $1/2$, the value of r is replaced randomly by either $r/2$ or $2r$ and with the remaining probability of $1/2$, it is set to the value that the winning individual in the last iteration has been created with. Finally, the value of r is capped at 2 if it is smaller, and at $n/4$ if it exceeds this value. Algorithm 6.4 summarizes this two-rate $(1+\lambda)$ EA variant.

Algorithm 6.4: The 2-rate $(1+\lambda)$ EA$_{(2r,r/2)}$ with adaptive mutation probabilities and static population size for the maximization of a pseudo-Boolean function $f:\{0,1\}^n \to \mathbb{R}$

```
1  Initialization: Sample x ∈ {0,1}^n uniformly at random (u.a.r.);
2  Initialize r ← r^init;
3  Optimization: for t = 1,2,3,... do
4      for i = 1,...,λ/2 do
5          Create y^(i) by flipping each bit in x independently with probability
           r/(2n);
6      for i = λ/2+1,...,λ do
7          Create y^(i) by flipping each bit in x independently with probability 2r/n;
8      x* ← argmax{f(y^(1)),...,f(y^(λ))} (ties broken u.a.r.);
9      if f(x*) ≥ f(x) then x ← x*;
10     With prob. 1/4 replace r by max{r/2,2}, with prob. 1/4 by min{2r,n/4}, and
         with the remaining prob. by the rate with which x* has been created
         (capped again at 2 and n/4);
```

Theorem 6.5.3 (Theorem 1.1 in [33]). *Let* $\lambda = \omega(1)$ *and* $\lambda = n^{O(1)}$. *The expected optimization time of the* $(1+\lambda)$ *EA*$_{(2r,r/2)}$ *on* OneMax *is* $\Theta(n\log n + n\lambda/\log\lambda)$.

By the result presented in Theorem 6.4.3 above, the $\Theta(n\log n + n\lambda/\log\lambda)$ expected running time achieved by the $(1+\lambda)$ EA$_{(2r,r/2)}$ is the best possible among all λ-parallel black-box algorithms.

6.5.2.4 Success-Based Mutation Strengths for the Multivariate OneMax Problem

In [27], a success-based choice of the mutation strength was proven to be very efficient for a multivariate generalization of the OneMax problem.

Concretely, the authors of [27] studied three different classes of generalized OneMax functions. Denoting the size of the alphabet by r, the first class contains, for all $z \in [0..r-1]^n$, the functions $\mathrm{OM}_z^{(1)} : [0..r-1]^n \to [0..n]; x \mapsto |\{i \in [1..n] \mid x_i = z_i\}|$, and the second all functions $\mathrm{OM}_z^{(2)} : [0..r-1]^n \to [0..n(r-1)]; x \mapsto \sum_{i=1}^n |x_i - z_i|$, while the third class subsumes all r^n functions $\mathrm{OM}_z^{(3)} : [0..r-1]^n \to [0..n(r-1)]; x \mapsto \min\{x_i-(z_i-r), |x_i-z_i|, (z_i+r)-x_i\}$. Unlike all other settings considered in this chapter, [28] studied the *minimization* of these OneMax generalizations. In our description below, we stick to this optimization target, to ease comparison with the original publication.

The self-adjusting algorithm studied in [27] was an RLS variant, which flips one coordinate in every iteration. For each coordinate i, a *velocity* v_i is stored, which denotes the mutation strength at this coordinate. When, in iteration t, coordinate i is chosen for modification, the entry x_i of the current best solution x is replaced by $x_i - \lfloor v_i \rfloor$ with probability $1/2$ and by $x_i + \lfloor v_i \rfloor$ otherwise. The entries in positions $j \neq i$ are not subject to mutation. The resulting string y replaces x if its fitness is at least as good as that of x, i.e., if $f(y) \leq f(x)$ holds (we recall that we are aiming at minimizing f). If the offspring y is strictly better than its parent x, i.e., if $f(y) < f(x)$ holds, the velocity v_i in the i-th component is increased by multiplying it by a fixed constant $A > 1$ and v_i is decreased to bv_i otherwise, where $b < 1$ is again some fixed constant. If the value of v_i drops below 1 or exceeds $\lfloor r/4 \rfloor$, it is capped at these values.

Theorem 6.5.4 (Theorem 17 in [28]). *For constants A, b satisfying $1 < A \leq 2$, $1/2 < b \leq 0.9$, $2Ab - b - A > 0$, $A + b > 2$, and $A^2 b > 1$, the expected running time of $RLS_{A,b}$ on any of the generalized r-valued* OneMax *functions $\mathrm{OM}_z^{(i)}$, $i \in \{1,2,3\}$ and $z \in [0..r-1]^n$, is $\Theta(n(\log n + \log r))$. This is asymptotically the best possible among all comparison-based variants of RLS and the $(1+1)$ EA.*

In this theorem, the update strengths can be chosen, for example, as $A \in [1.6, 2]$ and $b = (1/A)^{1/4}$, imitating the above-mentioned interpretation of the one-fifth success rule proposed in [62].

Using a result proven in [20], it was argued in [28, Section 6.1] that the $\Theta(n(\log n + \log r))$ expected running time of the self-adaptive RLS variant is better by a multiplicative factor of at least $\log r$ than that of any RLS or $(1+1)$ EA variant using *static* step sizes. The optimality of the bound follows from the simple information-theoretic $\Omega(n \log r)$ lower bound which applies to all comparison-based algorithms, while the $\Omega(n \log n)$ lower bound applies to any unary unbiased black-box algorithm (see Section 3.6 for a discussion of unbiased black-box algorithms).

6.5.2.5 Success-Based Migration Intervals for Parallel EAs in the Island Model

A multiplicative success-based adaptation scheme has also been used to adjust the migration interval in a parallel (1+1) EA with a fixed number λ of islands. Mambrini and Sudholt [73] applied the two schemes described in Section 6.5.2.2 for the control of the offspring population size of the $(1+\lambda)$ EA now to control the migration interval. In their parallel EA, every island has its own migration interval, at the end of which it broadcasts its current best solution to all of its neighbors. In the $(2\tau_i, 1)$ variant of the parallel EA (Algorithm 2 in [73]), improved solutions are always broadcast instantly, to all neighboring islands, and the migration interval τ_i of the corresponding island is set to one. It is set to one also if, during the migration interval, at least one superior solution has migrated to the island. The migration interval is doubled otherwise, i.e., if at the end of the migration period no strictly better solution has been identified or migrated from a different island.

In the $(2\tau_i, \tau_i/2)$ scheme (Algorithm 3 in [73]), the broadcast happens only at the end of the migration interval, which is again doubled if no improved solution could be identified nor migrated from another island, and halved otherwise.

The $(2\tau_i, \tau_i/2)$ scheme was analyzed for the complete graph topology, for which all migration intervals τ_i are identical. For the $(2\tau_i, 1)$ variant, [73] proved results for general graph topologies with λ islands, as well as for a few selected topologies such as a unidirectional ring, a grid, or a torus. The results comprise upper bounds on the expected communication effort needed to optimize general black-box optimization benchmarks; see Sections 4 and 5 in [73]. These bounds were then applied to the same benchmark functions as those considered in Theorem 6.5.2. In some cases, including the complete graph topology, the adaptive migration intervals were shown to outperform any static choice in terms of expected communication effort, without (significantly) increasing the expected parallel running time. Table 1 in [73] summarizes the results for the selected benchmark problems. The bounds proven in [73] are upper bounds, and the question of complementing these with meaningful lower bounds seems to remain an open problem.

6.6 Learning-Inspired Parameter Control

In contrast to the success-based control mechanisms discussed in the previous section, we call all those self-adjusting parameter control mechanisms which are based on information obtained over more than one iteration *learning-inspired*.

6.6.1 Adaptive Operator Selection

An important class of parameter control schemes takes inspiration from the machine learning literature, and in particular from the *multiarmed bandit problem.* These *adaptive operator selection* techniques[7] maintain a portfolio of k possible parameter values. At each step, they decide which of the possible parameter values to use next. To this end, they assign to each possible parameter value a *confidence value.* This confidence value is supposed to be an indicator of how suitable the corresponding value is at the given stage of the optimization process. The confidence value can be, for example, an estimator of the likelihood or of the magnitude of progress we would obtain from running the algorithm with this value. These confidence values determine or modify the *probabilities* of choosing the corresponding parameter value. We present below three ways to implement this adaptive operator selection principle.

What distinguishes the parameter control setting from the classically setting considered in machine learning is the fact that the "rewards," i.e., the gain that we can obtain with a given value, can change drastically over time, compared with the static (but random) reward typically investigated in the machine learning literature. The non-static reward distributions change the complexity of the algorithms and the theoretical analysis considerably. As far as we know, the only theoretical stdy that rigorously proved an advantage of learning-based parameter control is [29], which we shall discuss in more detail in Section 6.6.2. Despite the promising empirical performance of adaptive operator selection techniques, none of the techniques mentioned below has established itself as a standard routine. Potential reasons for this situation include the complexity of these techniques, the difficulty of finding good hyper-parameters that govern the update rules, and a lack of theoretical support.

- *Probability matching.* This technique is aimed at assigning the probabilities proportionally to the confidence values, while maintaining for each parameter value a minimum probability $p_{\min}$ of being sampled. Concretely, in round t we choose the i-th parameter value with probability

$$p_t^i := p_{\min} + (1 - kp_{\min}) \frac{c_t^i}{\sum_{j=1}^k c_t^j},$$

  where k is the total number of different parameter values from which we can choose (the size of the *portfolio*), and c_t^j is the confidence in parameter value j at time t. After one iteration has been executed with the i-th parameter value, its confidence value is updated to

[7] The term "operator" is used because the adaptive operator selection mechanisms were originally designed not only to choose between different parameter values but also between different actions, such as different variation operators.

$$c_{t+1}^i := (1-\alpha)c_t^i + \alpha r^t,$$

where r^t denotes the (normalized) reward obtained in the t-th round and $0 < \alpha < 1$ is the *hyper-parameter* that determines the speed of the adaptation. The confidence values of parameter values that have not been selected in the t-th round are not updated.

- *Adaptive pursuit.* When larger portfolios are used, the previous mechanism of choosing the operator with probability roughly proportional to the confidence value might not give enough preference to the truly best choice. To this end, a more "aggressive" update rule has been suggested: *adaptive pursuit.* This selection scheme uses the same confidence values as probability matching, but applies a much more progressive update rule for the probabilities. In adaptive pursuit, the probabilities of selection are obtained from the probabilities of the previous iteration according to a "winner takes all" policy. Concretely, the "best arm," i.e., the parameter value with the highest confidence value, is assigned a probability of $(1-\beta)p_t^{i^*} + \beta p_{\max}$, while for all other parameters the probability of being sampled is reduced to $p_{t+1}^i := (1-\beta)p_t^i + \beta p_{\min}$. Empirical comparisons of probability matching and adaptive pursuit are presented in [84]. In general, it seems that adaptive pursuit is more suitable for situations in which the quality differences between the potential parameter values are small, but persistent.
- *Upper confidence bound.* The *upper confidence bound* (UCB) algorithm, originally proposed in [6], plays an important role in machine learning, as it is one of the few strategies that can be proven to behave optimally in a classical operator selection problem. More precisely, the UCB algorithm can be proven to achieve minimum *cumulative regret* in the *multiarmed bandit problem* in which the reward of each "arm" follows a static probability distribution. Interpreting the different "arms" as the different parameter values that we want the algorithm to choose from, the UCB algorithm chooses in every step the parameter value i that maximizes the expression

$$\mathrm{ER}(i) + \sqrt{c \log\left(\frac{2\sum_{j=1}^k n_{j,t}}{n_{i,t}}\right)},$$

where $\mathrm{ER}(i)$ is an estimate of the expected reward of the i-th parameter value, $n_{j,t}$ is the number of times the j-th parameter value has been chosen in the first t iterations, and c is a hyper-parameter that determines the bias between exploiting parameter values with high expected reward and exploring parameter values that have not yet been tested very often. While it is provably optimal in static settings, the UCB algorithm is rather sedate, and thus not very well suited for environments that change gradually over time – the typical situation encountered in the optimization of rather smooth optimization problems. In the parameter control context, it therefore makes sense to replace $n_{j,t}$ by an index that counts the number of occurrences in a given time interval only, instead of considering the whole history (a *sliding window*; see [44] for a detailed discussion and experimental results on two discrete benchmark problems). In contrast, when the environment changes abruptly, a combination of the UCB algorithm with a statistical test that detects significant changes in the fitness landscape has been shown to perform very well [17, 43].

6.6.2 Theoretical Results for Learning-Inspired Parameter Control

The first, and so far only, theoretical study that rigorously analyzed a learning-inspired parameter selection scheme is [29]. The algorithm proposed there is a generalized version of RLS, which selects in every step the number of bits to be flipped according to the following rule. With probability $\varepsilon > 0$ a random one of the k possible mutation strengths $1,\dots,k$ is chosen, and with the remaining probability the algorithm greedily selects the parameter value for which the expected progress (termed *velocity* in [29]) is maximized. The expected progress is estimated by a time-discounted average of the progress observed in the learning iterations. More precisely, the velocity of mutation strength r at time t is defined via

$$v_t[r] := \frac{\sum_{s=1}^{t} \mathbf{1}_{r_s=r}(1-\delta)^{t-s}(f(x_s)-f(x_{s-1}))}{\sum_{s=1}^{t} \mathbf{1}_{r_s=r}(1-\delta)^{t-s}}, \tag{6.6.1}$$

where r_s is the parameter value used in the s-th iteration, and the hyper-parameter δ determines the speed of the adaptation process. The authors of [29] refer to δ as the *forgetting rate*, inspired by the observation that the reciprocal $1/\delta$ of the forgetting rate is (apart from constant factors) the information half-life. Note here that, compared with [29], we have changed the meanings of ε and δ, to be in line with the classical literature in machine learning, where the algorithm in [29] would be classified as an ε-*greedy* selection scheme (meaning that with probability ε a random choice is made, and otherwise a greedy choice).

The main theoretical result in [29] is a proof that, for suitably selected hyper-parameters ε and δ, this algorithm essentially always uses the best possible mutation strength when run on ONEMAX. More precisely, it was shown that in all but a fraction $o(1)$ of the iterations the selected parameter value achieves an expected progress that differs from the best possible progress by at most some lower-order term. Consequently, the algorithm has the same optimization time (apart from an $o(n)$ additive lower-order term) and the same asymptotic 13% superiority in the fixed-budget perspective as the fastest algorithm which can be obtained from these mutation strengths, which again comes arbitrarily close (by taking k large) to the performance of the hand-crafted mutation strength schedule presented in Theorem 6.4.4.

Theorem 6.6.1 (Theorems 1 and 2 in [29]). *Let $T(r_{\max})$ be the minimum expected running time that any randomized local search algorithm using a fitness-dependent mutation strength of at most $r_{\max}$ can achieve on* OneMax. *Then the expected running time T of the ε-greedy RLS variant presented in [29] with hyper-parameters $\varepsilon = n^{-0.01}$, $\delta = n^{-0.99}$, and $k = r_{\max}$ is $T(r_{\max}) + o(n)$.*

In the fixed-budget perspective, the following holds. Let $x_\varepsilon^{(t)}$ be the best solution that the ε-greedy RLS variant with this parameter setting has identified within the first t iterations. Similarly, let $x_{RLS}^{(t)}$ be the best solution that the classic RLS using 1-bit flips only has found within the first t iterations. For $t \geq 0.2675n$, the expected Hamming distances to the optimum z satisfy

$$E[H(x_\varepsilon^{(t)}, z)] \leq (1 + o(1))\, 0.872\, E[H(x_{RLS}^{(t)}, z)].$$

The hyper-parameters in this result were taken as one example where this algorithm shows superior performance. As noted in [29], the particular choice of these parameters is not overly critical. Clearly, ε has to be $o(1/\log n)$ to ensure that at most $o(n)$ iterations are performed with a suboptimal mutation strength. Likewise, δ has to be $o(n)$ to ensure that information learned $\Omega(n)$ iterations ago (and thus at a time when the velocities could be substantially different) has no significant influence on the current decision.

In addition to this theoretical result, [29] also presented empirical results for the LeadingOnes and the minimum spanning tree problem. These experimental studies suggest that, for suitably chosen hyper-parameters ε, δ, and k, the average optimization time of the ε-greedy RLS variant can be significantly smaller than that of the $(1+1)$ EA. It even outperforms, empirically, RLS on LeadingOnes, and the RLS variant that always flips either one or two random bits in the current best solution on the MST problem.

6.7 Self-Adaptation: Endogenous Parameter Control

As we have seen in the previous sections, an elegant way to overcome the difficulty of finding the right parameters of an evolutionary algorithm and to cope with the fact that the optimal parameter values may change during a run of the algorithm is to let the algorithm optimize the parameters *on the fly*. However, formally speaking, this is an even more complicated task, because we now have to design a suitable parameter-setting mechanism. While a number of natural heuristics such as the one-fifth success rule have proven to be effective in certain cases, it would be even more elegant not to add an *exogenous* parameter control mechanism onto the algorithm, but rather to integrate the parameter control mechanism into the evolutionary process, that is, to attach the parameter value to the individual (consequently, there is no global parameter value, but each individual carries its own parameter value), to modify it via (extended) variation operators, and to use the fitness-based selection mechanism of the algorithm to ensure that good parameter values become dominant in the population.

This *self-adaptation* of the parameter values has two main advantages.

- It is generic, that is, the adaptation mechanism is provided by the algorithm, and only the representation of the parameter in the individual and the extension of the variation operators has to be provided by the user.
- It allows existing algorithms and existing code to be reused.

Despite these advantages, self-adaptation is not used a lot in discrete evolutionary optimization (unlike in continuous optimization) and, consequently, there is also little theoretical work on this topic.

Self-adaptation for discrete evolutionary computation was proposed in the seminal paper [8] by Bäck, which also contains a mathematical convergence proof for the mutation rate (in the particular setting proposed there). Apart from this result, only two publications on running-time analysis for self-adapting parameter choices have appeared so far. Since these results, like the paper by Bäck, are concerned with self-adaptive mutation rates, we discuss self-adaptation only for mutation rates in the following and note that other parameters could be optimized via self-adaptation in a similar way.

6.7.1 Implementing Self-Adaptive Mutation Rates

To use self-adaptation for the mutation rate, the individuals (which are usually possible solution candidates) have to be extended to also contain "their" mutation rate. In the purest possible form, as done by Bäck [8], this is implemented via appending additional bits to the bit string which represents the solution candidate. These additional bits encode in a suitable manner the mutation rate. This pure form has the advantage that any standard variation operator can be used directly on the extended individuals. The downside of this approach is that non-binary data is artificially treated like binary decision variables.

It has been argued, for example in [28], that it can be preferable to encode non-binary data in its original form and to modify it via data-specific variation operators. In the context of self-adaptation, the mutation rate has been encoded as a floating-point number in $]0,1[$ in [9, 64], which is mutated according to a log-normal distribution. In the recent theoretical studies [18, 38], only a discrete set of possible mutation rates was allowed. In [38], mutation rates r/n, with $r \in [1..n/2]$ being a power of two, were used. As a mutation, the rate r/n was replaced by a random choice between $(r/2)/n$ and $(2r)/n$.

With either representation of the mutation rate, the extended mutation operator (acting on the extended individuals) will always be such that first the encoded mutation rate is mutated and then the core individual is mutated with this new rate. This is necessary for the subsequent selection step to see an influence of the new mutation rate and thus, hopefully, prefer individuals with a more profitable mutation rate.

Finally, when designing a self-adaptive parameter optimization scheme, one may want to prefer non-elitist algorithms. An elitist algorithm carries the risk of getting stuck with individuals that have a high fitness, but a very unprofitable mutation rate. In this situation, progress can only be made when the mutation of the mutation rate in one iteration changes the rate to a value that admits an improvement. In other words, it is not possible to change the rate in several iterations if no improvement is made.

6.7.2 Theory for Self-Adaptive Mutation Rates

In the first publication analyzing self-adaptation through the running-time analysis paradigm, Dang and Lehre [18] considered the following setting. They used a simple non-elitist algorithm which, in each iteration, generates from a population of λ individuals a new population of λ individuals. This is done by independently selecting λ times an (extended) parent individual from the current population, mutating it via the (extended) mutation operator, and adding it to the new population. For the mutation rate, Dang and Lehre assumed that there is only a finite set $\mathcal{M}$ of prespecified rates (for most results, they took $|\mathcal{M}|=2$). The extended mutation operator first, with probability p, which is a global parameter of the algorithm, replaces the current rate of the individual by a random different one, and then mutates the core individual via standard bit mutation with the new rate. For the selection operator, a wide range of choices are subsumed in this publication, since the results are phrased in terms of a parameter of the selection operator, namely the reproductive rate. A selection operator (possibly depending on a fitness function f) has a reproductive rate α if, for all populations P and each individual x of the population, the expected number of times x is chosen in λ independent applications of the selection operator is at most α. For example, always selecting a best individual from the population leads to $\alpha=\lambda$, whereas a uniform random selection gives $\alpha=1$.

For this setting, the following results were shown. If a mutation rate p_1 satisfies $p_1 \geq (\ln\alpha+\delta)/n$ for some constant δ, then an algorithm that always uses the rate p_1 (equivalent to the case where $\mathcal{M}=\{p_1\}$) and random initialization needs with high probability an at least exponential time to reach the optimum of any pseudo-Boolean function with a unique optimum (this is Theorem 2 of [18] in the special case of $|\mathcal{M}|=1$).

If two rates are used, that is, $\mathcal{M} = \{p_1, p_2\}$, and the mutation operator chooses the current rate of the individual uniformly at random, then even if only one of the rates satisfies the dangerous condition $p_i \geq (\ln\alpha + \delta)/n$, the above problem can remain: if $p_1 \geq (\ln\alpha - \ln(1+\delta_1))/n$, $p_2 \geq (\ln\alpha - \ln(1-\delta_2))/n$, and $\delta_1/(\delta_1+\delta_2) \leq \frac{1}{2} - \varepsilon$ for constants $\delta_1, \delta_2, \varepsilon > 0$, then again an at least exponential running-time results with high probability (Theorem 4 in [18]). This result again applies to any pseudo-Boolean function $f : \{0,1\}^n \to \mathbb{R}$ that has a unique optimum.

The latter of these two results shows that randomly mixing a good and a bad operator can be essentially as bad as using the bad operator alone. This is not overly surprising, but points out the contrast with the following result for a self-adaptive choice of the mutation rate. For a suitable example function f, it was proven that an algorithm with a suitably initialized population, with tournament selection with tournament size 2, with population size $\lambda \geq c\ln(n)$, and with a self-adaptive choice between the two mutation rates $p_1 \geq \ln(3)$ and $p_2 = \ln(3/2) - \varepsilon$ finds the optimum of f in a polynomial running time, whereas using either of these two rates alone or randomly mixing between them leads to an at least exponential running time with high probability.

As for almost all such examples, this one also is slightly artificial and needs quite some assumptions, for example, that all λ individuals are initialized to the unique local optimum. Nevertheless, this result demonstrates that self-adaptation can outperform static parameter choices and random mixing. The reason for this is that, as the proofs reveal, the self-adaptation is able to find in a relatively short time the mutation rate which is most profitable (as opposed to fixed parameter choices) and to remember it (as opposed to random mixing).

Very recently, a less artificial example of the use of self-adaptation was presented in [38]. There it was shown that the $(1,\lambda)$ EA with a self-adaptive choice of the mutation rate can achieve an asymptotically identical performance to the self-adjusting $(1+\lambda)$ EA presented in [33] (see also Section 6.5.2.3). In the self-adaptive setting of [38], the extended individuals store their mutation rate, which is r/n for an integer $r \in [32..n/64]$. The extended mutation operator first changes r to $r/32$ or $32r$ (uniform random choice) and then performs standard bit mutation with the new mutation rate r/n. One of the offspring with maximum fitness is selected as the new parent individual. In the case of ties, individuals with a smaller rate are preferred, which creates a small extra drift towards the usually recommended rates of order $\Theta(1/n)$. It was shown that when $\lambda \geq (\ln n)^{1+\varepsilon}$, this algorithm finds the optimum of the OneMax function in an expected number of $O(n/\log\lambda + (n\log n)/\lambda)$ iterations, which is the asymptotically best possible running time for λ-parallel algorithms (see Theorem 6.4.3, which we have cited from [10]).

6.8 Hyper-Heuristics

Hyper-heuristics are search or optimization heuristics which, during the run of an algorithm, choose in a possibly adaptive manner which low-level heuristics to use. Since in some situations hyper-heuristics can closely resemble an adaptive parameter choice, we describe in this section what is known about such hyper-heuristics.

6.8.1 A Brief Introduction to Hyper-Heuristics

Hyper-heuristics either choose from a prespecified set of low-level heuristics (*selection hyper-heuristics*) or try to generate low-level heuristics from existing components (*generation hyper-heuristics*). There is a considerable amount of applied research on generation hyper-heuristics, for example, for scheduling problems, packing problems, satisfiability, and the traveling salesman problem. However, since there appears to be no theoretical work on generation hyper-heuristics and since, naturally, generation hyper-heuristics are substantially different from parameter control mechanisms, we will not detail this subarea further, and refer, as for all other topics incompletely covered here, to the recent survey [13].

As is true in general for optimization heuristics, hyper-heuristics can also be divided into *construction hyper-heuristics* and *perturbation hyper-heuristics*. The former try to construct a solution from partial solutions. This has led to interesting results, for example, in production scheduling, educational timetabling, and vehicle routing. Since constructing a solution from partial solutions is necessarily a highly problem-specific approach, it is not surprising that general theoretical results for this subarea do not yet exist.

In contrast, perturbation hyper-heuristics work, in a similar manner to classical evolutionary algorithms, with complete solution candidates, which are randomly modified in the hope of gaining superior solutions. Perturbation selection hyper-heuristics have found numerous applications, among others, in various scheduling contexts. The most common form of perturbative selection hyper-heuristics is *single-point search*, which, in a fashion analogous to $(1+\lambda)$ EAs, repeat creating one or more offspring from a single parent and selecting the next parent from these offspring and the previous parent. For such selection hyper-heuristics, some general mechanisms for choosing the low-level heuristic that create the offspring have been proposed, see Section 6.8.3.

As said above, selection hyper-heuristics are methods that select, during the run of an algorithm, which one out of several prespecified simpler algorithmic building blocks to use. When the different prespecified choices are essentially identical apart from an internal parameter, then this selection hyper-heuristic could equally well be interpreted as a dynamic choice of the internal parameter. For example, when only the two mutation operators are available that flip exactly one or exactly two bits, then a selection hyper-heuristic choosing between them could also be interpreted as a *randomized local search* heuristic using a dynamic choice of the number of bits it flips. Conversely, some of the studies described previously could equally well be described in the language of simple selection hyper-heuristics. In this section, we follow the language used by the original authors and do not aim at drawing a line between the different fields.

We now describe the main theoretical studies that have appeared in the hyper-heuristics community, as long as they resemble dynamic parameter control mechanisms, the main topic of this chapter.

6.8.2 Random Mixing of Low-Level Heuristics

6.8.2.1 Markov Chain Analyses

The first theoretical study of selection hyper-heuristics was conducted by He, He, and Dong [49]. They considered a variant of the classic $(1+1)$ EA which in each iteration selects a mutation operator from a finite set of operators according to a fixed probability distribution. In hyper-heuristics language, this is a single-point selection heuristic using a *mixed strategy.* He et al. showed that the asymptotic convergence rate and the asymptotic hitting time resulting from any mixed strategy are not worse than those resulting from exclusively using the worst of the given operators.

Some care is necessary when interpreting this result. The asymptotic hitting time as defined in [49] is not the asymptotic order of magnitude of the classical hitting time (the number of iterations until the optimum is generated), but the spectral radius $\rho(N)$ of the fundamental matrix $N=(I-T)^{-1}$ of the Markov chain describing the parent individual in a run of this single-point heuristic, where I is the identity matrix and T is the transition matrix restricted to the non-optimal search points. This asymptotic hitting time is only loosely related to the classical hitting time. Denoting by T_x the classical hitting time of this Markov chain (usually called the optimization time of the EA) when started in the state x, then only the weak relation

$$E_{\min} := \min\{E[T_x] \mid x \in S_{\text{nonopt}}\} \le \rho(N) \le \max\{E[T_x] \mid x \in S_{\text{nonopt}}\} =: E_{\max}$$

is known, where S_{nonopt} is the set of all non-optimal search points. Consequently, the asymptotic hitting time $\rho(N)$ only provides a lower bound on the worst-case expected hitting time $E_{\max}$. Note that the best-case expected hitting time $E_{\min}$ is often very small, as witnessed by search points x that are very close to the optimum. Consequently, the lower bound for the worst-case hitting time given by $\rho(N)$ can be relatively weak. Nothing is known about how the asymptotic hitting time is related to the running time starting from a random search point, which is the usual performance measure. For these reasons, it is not clear how to translate the result of [49] into the classical running-time analysis language.

6.8.2.2 Running-Time Analysis of Mixed Strategies

The first to conduct a running-time analysis for selection hyper-heuristics in the classical methodology were Lehre and Özcan [67]. In [67, Theorem 3], it was stated that the (1+1) EA[8] using the mixed strategy of choosing in each iteration the mutation operator randomly between the 1-bit-flip operator (with probability p) and the 2-bit-flip operator (with probability $1-p$) optimizes the OneMax function in an expected time of at most

$$\min\left\{\frac{n}{p}(1+\ln(n)), \frac{n^2}{1-p}\left(1-\frac{1}{n}\right)\right\} \le \begin{cases} \frac{n}{p}(1+\ln n) & \text{if } p > \frac{1+\ln n}{n+\ln n}, \\ \frac{1}{1-p}n^2 & \text{otherwise.} \end{cases} \quad (6.8.1)$$

It appears to us that this result is not absolutely correct, since, for example, in the case $p=0$ the expected optimization time is clearly infinite: if the random initial search point has an odd Hamming distance from the optimum, then the optimum cannot be reached via 2-bit flips only. For similar reasons, the expected running time has to be larger than in (6.8.1) for very small values of p. We will therefore prove the following result.

Theorem 6.8.1. *Consider the* $(1+1)$ *EA with the mixed mutation strategy of flipping a single random bit with probability* p *and flipping two (different) random bits with probability* $1-p$. *Let* T *be the running time (number of iterations) of this algorithm on the* OneMax *benchmark function. If* $p>0$, *then*

$$E[T] \le \begin{cases} \frac{n}{p} + n^2 & \text{if } p \le \frac{1}{n} \\ \frac{n}{p}\left(\ln(np) + 1 + \frac{\ln(np)}{np-1}\right) & \text{if } p > \frac{1}{n}. \end{cases}$$

If $p=0$, *then with probability* $\frac{1}{2}$ *the algorithm never finds the optimum (and thus the expected running time* $E[T]$ *is infinite).*

Proof. For the case $p=0$, we note that with probability exactly $\frac{1}{2}$ the random initial search point has an odd Hamming distance from the optimum.[9] Since 2-bit flips change the Hamming distance by -2, 0, or $+2$, the algorithm can never reach the optimum in this case.

---

[8] We note that some authors prefer to call the algorithm used in [67] a variant of *randomized local search* rather than an evolutionary algorithm, since it only creates offspring within a bounded distance from the parent.

[9] This well-known fact follows from the beautiful argument $0 = (1-1)^n = \sum_{i=0}^{n} 1^i(-1)^{n-i}\binom{n}{i}$.

Hence, let us assume $p > 0$ for the remainder of this proof. When the current search point of the (1+1) EA has a Hamming distance of $d \geq 1$ from the optimum, then the probability p_d that one iteration ends with a better search point is

$$p_d = p\frac{d}{n} + (1-p)\frac{d(d-1)}{n(n-1)} = \frac{d((1-p)d+np-1)}{n(n-1)}.$$

Using $p_1 = \frac{p}{n}$ and $p_d \geq \frac{d(d-1)}{n(n-1)}$ for all $d \geq 2$, the classic fitness level theorem yields

$$\begin{aligned}
E[T] &\leq \sum_{d=1}^{n} p_d^{-1} \\
&\leq \frac{n}{p} + n(n-1)\sum_{d=2}^{n} \frac{1}{d(d-1)} \\
&= \frac{n}{p} + n^2\left(1-\frac{1}{n}\right)^2 \leq \frac{n}{p} + n^2.
\end{aligned}$$

Above, we have used the equation $\sum_{d=2}^{n} \frac{1}{d(d-1)} = 1 - \frac{1}{n}$, valid for all $n \in \mathbb{N}$, which can be shown easily by induction.

For $p > \frac{1}{n}$, we also have the estimate

$$\begin{aligned}
E[T] &\leq \sum_{d=1}^{n} p_d^{-1} \\
&= n(n-1)\sum_{d=1}^{n} \frac{1}{d((1-p)d+np-1)} \\
&\leq n(n-1)\left(\frac{1}{(n-1)p} + \int_1^n \frac{1}{d((1-p)d+np-1)}\mathrm{d}d\right) \\
&= \frac{n}{p} + n(n-1)\left(-\frac{1}{np-1}\ln\left(\frac{(1-p)d+np-1}{d}\right)\right)\Bigg|_1^n \\
&= \frac{n}{p} + \frac{n(n-1)}{np-1}\left(\ln((n-1)p) - \ln\left(\frac{n-1}{n}\right)\right) \\
&\leq \frac{n}{p} + n^2\frac{\ln(np)}{np-1} = \frac{n}{p}\left(1+\ln(np)\left(1+\frac{1}{np-1}\right)\right) \\
&= \frac{n}{p}\left(\ln(np) + 1 + \frac{\ln(np)}{np-1}\right).
\end{aligned}$$

Note that for all $p > \frac{1}{n}$ we have $\ln(np) < np - 1$. Hence the bound above is less than $\frac{n}{p} + n^2$ and thus stronger than the first bound. $\square$

Without giving full details, we remark that better results can be obtained by using variable drift instead of the classic fitness level method. Since a 2-bit flip giving a fitness improvement automatically improves the fitness by exactly two, we have that the expected fitness gain in one iteration starting with a search point with fitness distance d is

$$h_d = p\frac{d}{n} + 2(1-p)\frac{d(d-1)}{n(n-1)} = \frac{d(2(1-p)d+np+p-2)}{n(n-1)}.$$

Now the variable drift theorem for upper bounds on hitting times (see [60], note that for processes in $\mathbb{N}_0$ the integration can be replaced by a summation) gives $E[T] \le \sum_{d=1}^{n} h_d^{-1}$, which can be estimated in a similar fashion to the term $\sum_{d=1}^{n} p_d^{-1}$ above. What is more interesting than the slightly improved upper bound is that the variable drift theorem for lower bounds [32] gives a very similar lower bound, namely $E[T] \ge \sum_{d=3}^{n} h_d^{-1}$; note again that for integer-valued processes we can replace the integration with a summation.

The above results show that, for the classical benchmark function OneMax, mixing the 1-bit-flip and 2-bit-flip operators in a random fashion gives no improvement over exclusively using the 1-bit operator. In light of the precise analysis of the performance of k-bit-flip operators on OneMax in [30], this result is not very surprising. There, it was shown that the expected fitness gain is never maximized by flipping an even number of bits. Also, from a fitness of $(\frac{2}{3}+o(1))n$ on, the 1-bit-flip operator is the only one that maximizes the expected fitness gain.

6.8.2.3 Superiority of Mixed Strategies

To demonstrate the use of mixing operators, Lehre and Özcan [67] constructed an example function GapPath, which has the property that the (1+1) EA mixing 1-bit and 2-bit flips when initialized with $x_0 = (0,\dots,0)$ can optimize GapPath only when both the 1-bit-flip and the 2-bit-flip mutation operator are chosen with positive probability. Based on this result, several ways to alternate between a low and a high p-value were discussed, including a success-based reinforcement approach. While these ideas were shown to give improvements over certain choices of p such as $p = \frac{1}{n}$, they did not outperform natural choices such as $p = \frac{1}{2}$ or $p = 1$.

An example similar to GapPath was used to show that mixing 1-bit-flip and 2-bit-flip operators can be necessary also in multi-objective optimization [78].

We note that a more natural example of the need for mixing, without being explicitly stated there, had already been considered by Neumann and Wegener [76] (and a slightly more technical example had been given even earlier by Giel and Wegener [45]). Neumann and Wegener [76] analyzed how simple randomized search heuristics solve the minimum spanning tree problem in connected undirected graphs $G = (V, E)$ having $n := |V|$ vertices and $m := |E|$ edges with integral edge weights in $[1..w_{\max}]$. They used the natural representation that individuals are sets $S = S(x)$ of edges represented via bit strings $x \in \{0,1\}^E$. As the fitness (to be minimized) of an individual, they proposed

$$f(x) = M^2(C_x - 1) + M\left(\sum_{e \in S(x)} x_e - (n-1)\right) + \sum_{e \in S(x)} x_e w(e),$$

where $M = n^2 w_{\max}$ and C_x is the number of connected components of the graph $(V, \{e \in E \mid x_e = 1\})$. This fitness function punishes connected components as first priority, then punishes the number of edges, and only then prefers solutions with smaller total weight (we do not see that the punishment of edges is necessary, but clearly it does no harm either). Besides the $(1+1)$ EA, Neumann and Wegener analyzed the performance of (in their language) a variant of the randomized local search heuristic which, in each iteration, either (uniform random choice) flips a single random bit or flips two different random bits. In hyper-heuristics language, they thus considered the same single-point selection hyper-heuristic with random mixing between the 1-bit-flip and the 2-bit-flip operator as in [67] except that they fixed the probability p to $\frac{1}{2}$.

Neumann and Wegener showed that this algorithm computes a minimum spanning tree in an expected number of $O(m^2 \log(n w_{\max}))$ iterations. It can easily be seen, and has been shown in [80], that for this algorithm, the $w_{\max}$ term in the running-time bound can be omitted, but we shall not care about this usually small improvement in the following. Neumann and Wegener do not make this explicit, but from their proofs it is clear that any other mixing which uses both operators with constant probability would give the same result. The reason why Neumann and Wegener used both 1-bit flips and 2-bit flips is that, obviously, all spanning trees are local optima of the fitness function. Consequently, using 1-bit flips only carries the risk of getting stuck in a local optimum forever. The parity argument used in the proof of Theorem 6.8.1 shows also that when only the 2-bit-flip operator is used, the algorithm has a constant probability (of exactly $\frac{1}{2}$) of never reaching an optimum.

Theorem 6.8.2 (analogous to Theorem 11 in [76]). *Consider the $(1+1)$ EA with the mixed strategy of flipping one random bit (with probability p) and two different random bits (with probability $1-p$), solving the minimum spanning tree problem in connected undirected graphs having n vertices, m edges, and integral edge weights in $[1..w_{\max}]$.*

- *If both p and $1-p$ are $\Omega(1)$, then the expected optimization time is $O(m^2 \log(n w_{\max}))$.*
- *If $p = 0$, then with probability $\frac{1}{2}$ the algorithm never finds any spanning tree.*

- *If $p = 1$ and the input graph does not have the property that each spanning tree is a minimum spanning tree, then with positive probability the algorithm never finds a minimum spanning tree.*

Consequently, this algorithm solves the minimum spanning tree problem in polynomial expected time if and only if $p \notin \{0,1\}$, that is, if there is a true mixing of the two mutation operators.

The publications [45, 76] showed that hyper-heuristics using random mixing of mutation operators could, with equal justification, just be called evolutionary algorithms using a possibly non-standard mutation operator. With equal justification, one could declare the $(1+1)$ EA or the $(1+\lambda)$ EA using the classic standard bit mutation operator (flipping each bit independently with probability $\frac{1}{n}$) a single-point selection hyper-heuristic choosing the k-bit-flip operator with probability exactly $\binom{n}{k}(\frac{1}{n})^k(1-\frac{1}{n})^{n-k}$. The same statement (with a different probability distribution) is true when the heavy-tailed mutation operator of [35] is used instead of standard bit mutation.

We end this section with a recent result giving an example where a large number of mixings give asymptotically the same performance. In [4], the plateau function PLATEAU_k is defined by $\text{PLATEAU}_k(x) = \text{ONEMAX}(x)$ if $\text{ONEMAX}(x) \in [0..n-k] \cup \{n\}$, and $\text{PLATEAU}_k(x) = n-k$ if $\text{ONEMAX}(x) \in [n-k+1..n-1]$. This function thus agrees with the ONEMAX function except that it has a large plateau of size $N = \sum_{i=1}^{k} \binom{n}{i} = \frac{n^k}{k!} + o(n^k)$ around the optimum. Consider the $(1+1)$ EA randomly mixing the k mutation operators which flip exactly $1, 2, \ldots, k$ bits. Let $p_1, p_2, \ldots, p_k \in [0,1]$ with $\sum_{i=1}^{k} p_i = 1$ be the probabilities of selecting the corresponding operators (and let us view these numbers as constants, that is, not depending on n). Assume $p_1 > 0$ to ensure that the algorithm surely converges. Then the expected optimization time is $E[T] = (1+o(1))N$ regardless of the values of $p_1, \ldots, p_k$.

6.8.3 Beyond Mixing: Advanced Selection Mechanisms[10]

The first to conduct a theoretical analysis of more sophisticated selection hyper-heuristics were Alanazi and Lehre [1]. Besides the *simple random*

[10] Warning: all results described in this section use a different definition of the 2-bit-flip operator, namely one where, independently and uniformly at random, $i \in [1..n]$ and $j \in [1..n]$ are chosen and then first the i-th bit is flipped and then the j-th bit is flipped. Consequently, this operator, with probability $1-1/n$, flips two random different bit positions. With probability $1/n$, however, we have $i = j$ and thus the two flipping operations cancel and the offspring is identical to the parent. We do not see much reason for the use of this alternative operator. We suspect (but have not checked this rigorously) that all results presented in this section hold as well for the classic 2-bit-flip operator, which flips two randomly chosen different bit positions (in other words, returns a random search point with Hamming distance 2 from the parent).

heuristic (choosing a low-level heuristic uniformly at random each time, that is, mixing with a uniform distribution), they considered the following classical selection mechanisms.

- *Random gradient*: take a random low-level heuristic and repeat using it as long as a true fitness improvement is obtained.
- *Greedy*: in each iteration, use all low-level heuristics in parallel and continue with a best search point generated by one of them (or the parent if no offspring is at least as good as the parent).
- *Permutation*: generate initially a random cyclic order of the low-level heuristics and then use them in that order. This mechanism can be seen as a quasirandom analogue of the *simple random* heuristic (see [31] for a discussion of the use of quasirandomness in evolutionary computation). Alternatively, this hyper-heuristic can be viewed as a time-dependent parameter control method. In fact, the time-dependent choices of the mutation rate discussed in [39, 55], see Section 6.4.1, can be seen as special cases of this hyper-heuristic.

Again for the choice between 1-bit flips and 2-bit flips, Alanazi and Lehre proved upper and lower bounds on the expected optimization time on the LeadingOnes benchmark function. While the results are relatively tight (the corresponding upper and lower bounds deviate by at most a factor of $6+o(1)$), the intervals of possible running times intersect. Hence this first running-time analysis for these advanced selection mechanisms does not yet give a conclusive picture.

Given that the probabilities of finding a true improvement are very low in this discrete optimization problem, one would expect that the four selection mechanisms would all use the two operators in a very balanced manner and thus lead to very similar running times. This was indeed the first set of results in the remarkable work of Lissovoi, Oliveto, and Warwicker [69]. Building on the precise analysis method of [12] instead of the fitness level method, they showed that the expected running time for all four selection mechanisms is $\frac{1}{2}\ln(3)\,n^2+o(n^2)\approx 0.549n^2$. Consequently, the more complex heuristics do not give a measurable performance gain over a simple randomized selection of the operator, and all are worse than just using 1-bit flips, which is known to give an expected running time of precisely $0.5n^2$.

Theorem 6.8.3 (Theorem 4.2 and Corollary 4.3 in [69]). *The* $(1+1)$ *EA using one of the selection mechanisms* simple random, random gradient, greedy, *or* permutation *to choose between the* 1*-bit-flip or the* 2*-bit-flip operator optimizes the* LeadingOnes *function in an expected number of* $\frac{1}{2}\ln(3)\,n^2+o(n^2)\approx 0.549n^2$ *iterations.*

Lissovoi et al. [69] built on this strong result by proposing to use a slower adaptation (a similar idea can be found already in [1], there, however, in a very problem-specific manner and only with preliminary experimental results). For the *random gradient* method, they proposed to switch the low-level heuristic only after a phase of τ iterations. More precisely, the current low-level heuristic is used for up to τ iterations. If an improvement is found, immediately another phase with this operator starts. If a phase of τ iterations does not see a fitness improvement, then a new phase is started with a random operator.

For this *generalized random gradient* mechanism with a phase length of $\tau = cn$ for a constant c, they showed (still for the LEADINGONES problem and the 1-bit and 2-bit mutation operators) an expected running time of $g(c)n^2 + o(n^2)$, where $g(c)$ is a constant depending on c only that tends to $\frac{\ln(2)+1}{4} \approx 0.423$ when c tends to infinity. Consequently, this new hyper-heuristic outperforms the previously investigated ones when c is large enough. As c tends to infinity, its performance approaches the best possible performance that can be obtained from the two mutation operators, which is, as also shown in [69], $\frac{\ln(2)+1}{4}n^2 + o(n^2)$. The following variant of this result appeared in the preprint [70].

Theorem 6.8.4 (Theorem 7 and Corollary 15 in [70]). *Consider the* $(1+1)$ *EA using the* generalized random gradient *selection heuristic with phase length* $\tau \in \omega(n)$ *and* $\tau \le cn\ln(n)$, $c < \frac{1}{2}$, *to choose between the* 1*-bit-flip and the* 2*-bit-flip operator. Then this algorithm optimizes the* LEADINGONES *function in an expected number of* $\frac{1}{4}(\ln(2)+1)n^2 + o(n^2) \approx 0.423n^2$ *iterations. This is, apart from lower-order terms, the best running time which can be achieved with these two mutation operators.*

A similarly generalized variant of the *greedy* selection hyper-heuristic was also found to improve over the classical selection heuristics, but appears not to give the same good results as the generalized random gradient method.

The generalized random gradient heuristic was extended further in [36]. There, an operator was defined as successful (which leads to another phase using this operator) if it leads to σ improvements in a phase of at most τ iterations. Hence, in this language, the previous generalized random gradient heuristic uses $\sigma = 1$. By using a larger value of σ, the algorithm is able to take more robust decisions about what is a success. This was used in [36] to determine the phase length τ in a self-adjusting manner. While the previous work [69] does not state this explicitly, the choice of τ is crucial. A τ-value of smaller asymptotic order than $\Theta(n)$ leads to typically no improvement within a phase and thus reverts the algorithm to the simple random heuristic. A τ-value of more than $cn\ln(n)$, where c is a suitable constant, results in both operators being successful in most parts of the search space. Consequently, the algorithm sticks to the first choice for a large majority of the optimization process and thus does not profit from the availability of both operators.

Since the choice of τ is so critical, a mechanism that successfully adjusts it to the right value is desirable. In [36], it was shown that by choosing $\sigma \in \Omega(\log^4 n) \cap o(\sqrt{n/\log n})$ – note that this is a quite wide range – the value of τ can be easily adjusted on the fly via a multiplicative update rule. This gives again the asymptotically optimal running time of Theorem 6.8.4.

6.9 Conclusion and Outlook

Recent years have seen a significant increase in our understanding of parameter control. The results stemming from the theory community indicate that success-based rules can easily lead to good parameter settings. These rules are easy to find owing to their intuitive hyper-parameters: if we conduct the update by multiplying the parameter by F in the case of success and by $F^{1/(\sigma-1)}$ in the case of failure, then F controls the speed of adaptation and $1/\sigma$ is the intended rate of success (e.g., $\sigma = 5$ in the case of the classical one-fifth success rule). It is also easy to observe whether such an update rule works as desired or not: if the rate of success aimed at cannot be obtained, then imbalance in the updates leads to an exponential growth or shrinking of the parameter value. Therefore, we currently see no reason not to try such a multiplicative update rule in a situation where one expects a monotonic relation between a parameter and the success of an iteration.

The increased power of learning-based approaches (being able to gather and exploit information obtained over many iterations) or self-adaptation suggests that one should not ignore these; however, our current understanding here is more limited. Indeed, we feel that making these directions more usable is among the following **open problems** we want to mention.

- *Theory for learning-inspired parameter control mechanisms.* While there has been considerable momentum for empirical studies of learning-inspired parameter control mechanisms [17, 42, 44, 68, 84], these mechanisms still lack a solid mathematical foundation. The only result that we are aware of in this context is the (almost) optimality of the ε-greedy RLS variant presented in [29]; see Section 6.6.2 above. In addition to its intrinsic motivation, this research direction will most probably result in a better reconciliation of research activities in optimization and machine learning, where many of the empirically tested techniques stem from.
- *Understanding self-adaptation.* While self-adaptation is massively used in continuous evolutionary optimization, it only plays a marginal role in discrete optimization. The general hope that the inclusion of the adaptive process into the main evolutionary algorithm will easily automate the on-the-fly control of parameters has not yet come true. The two, very recent, theoretical studies of this topic suggest, however, that self-adaptation can work. Therefore, extending these first studies towards a more profound understanding of how to use self-adaptation in discrete evolutionary optimization seems to be both a profitable and a feasible endeavor.
- *Controlling more than one parameter.* As indicated in Section 6.1, even for static parameter settings we do not have many examples of running-time bounds that depend on two or more parameters, with the exceptions of a bound for the $(1+\lambda)$ EA with mutation rate $p=c/n$ proven in [46], a tight running-time analysis for the $(\mu+\lambda)$ EA [5], and the three-dimensional analysis of the $(1+(\lambda,\lambda))$ GA presented in [24]. For non-static parameter

choices, the complexity of the analysis increases considerably, as the parameters often interact in a manner that is difficult to analyze. We are not aware of any theoretical result addressing the control of two or more parameters. According to [61], the empirical studies also focus mostly on controlling a single parameter, while for the simultaneous adaptation of two or more parameters only a few mechanisms have been tested.

Acknowledgements We thank Franziska Huth for providing Fig. 6.3. We thank Thomas Bäck for useful feedback. This work was supported by a public grant as part of the Investissement d'avenir project, reference ANR-11-LABX-0056-LMH, LabEx LMH, in a joint call with the Gaspard Monge Program for optimization, operations research, and their interactions with data sciences.

Algorithm	Dynamic parameter	Control scheme	Function	Results	Reference	Sec./Thm.
State-dependent parameter control schemes						
(1+1) EA	p	time-dep.	PathToJump	$O(n^2 \log n)$	[39, 55]	Sec. 6.4.1
$(\mu+1)$ EA	p	rank-based	OneMax	$O(\mu n \log n)$	[77]	Thm. 6.4.1
$(\mu+1)$ EA	p	rank-based	$f:\{0,1\}^n \to \mathbb{R}$	$7 \cdot 3^n$	[77]	Thm. 6.4.1
(1+1) EA	p	fitness-dep.	LeadingOnes	$0.68n^2$	[12]	Thm. 6.4.2
$(1+\lambda)$ EA	p	fitness-dep.	OneMax	$\Theta(n \log n + n\lambda/\log\lambda)$	[10]	Thm. 6.4.3
RLS	step size ℓ	fitness-dep.	OneMax	$n\ln(n) - cn \pm o(n)$	[30]	Thm. 6.4.4
$(1+(\lambda,\lambda))$ GA	λ	fitness-dep.	OneMax	$\Theta(n)$	[26]	Thm. 6.4.5
Success-based parameter control schemes						
$(1+(\lambda,\lambda))$ GA	λ	one-fifth success rule	OneMax	$\Theta(n)$	[24]	Thm. 6.5.1
$(1+(\lambda,\lambda))$ GA	λ	one-fifth success rule	MaxSAT	$O(\max\{n, (n \log n)/(\frac{m}{n})^{4+\varepsilon}\})$	[14]	Sec. 6.5.2.1
$(1+\lambda)$ EA	λ	$\{2\lambda, \lambda/2\}$, $\{2\lambda, 1\}$	see Table 6.1		[66]	Thm. 6.5.2
$(1+\lambda)$ EA	p	2-rate $\{2r, r/2\}$	OneMax	$\Theta(n \log n + n\lambda/\log\lambda)$	[33]	Thm. 6.5.3
RLS	pos.-dep. step size ℓ_i	$A\ell_i, b\ell_i$	r-ary OneMax	$\Theta(n(\log n + \log r))$	[28]	Thm. 6.5.4
parallel (1+1) EA	migration interval τ	$\{2\tau, \tau/2\}$, $\{2\tau, 1\}$	see Table 6.1		[73]	Sec. 6.5.2.5
Learning-inspired parameter control schemes						
RLS	step size ℓ	ε-greedy	OneMax	$n\ln(n) - cn \pm o(n)$	[29]	Thm. 6.6.1
Endogeneous (self-adaptive) parameter control schemes						
(λ,λ) EA	p	self-adaptation	artificial example showing that self-a. can work		[18]	Sec. 6.7.2
$(1,\lambda)$ EA	p	self-adaptation	OneMax	$\Theta(n \log n + n\lambda/\log\lambda)$	[38]	Sec. 6.7.2
Hyper-heuristics						
RLS	step size ℓ	random mixing	OneMax	$\frac{n}{p}(\ln(np) + O(1))$ for $p > 1/n$	[67], Thm. 6.8.1	Thm. 6.8.1
RLS	step size ℓ	random mixing	MST	$O(m^2 \log(nw_{\max}))$	[76]	Thm. 6.8.2
RLS	step size ℓ	classic schemes	LeadingOnes	$\approx 0.549n^2$	[69]	Thm. 6.8.3
RLS	step size ℓ	gen. rand. grad.	LeadingOnes	$\approx 0.423n^2$	[69]	Thm. 6.8.4
RLS	step size ℓ, window τ	gen. rand. grad., $(1+o(1))$ success rule	LeadingOnes	$\approx 0.423n^2$	[36]	Sec. 6.8.3

Table 6.2 Summary of selected theoretical running-time bounds, sorted by parameter control scheme. We report the expected number of function evaluations needed to identify an optimal solution, not the number of generations. For ease of comparison, we refer to the algorithms considered in Section 6.8 as RLS variants, and not as (1+1) EAs. Abbreviations: Sec.=Section; Thm.=Theorem; fitness-dep.=fitness-dependent; pos.-dep.=position-dependent; self-a.=self-adaptation; gen. rand. grad. = generalized random gradient

References

[1] Alanazi, F., Lehre, P.K.: Runtime analysis of selection hyper-heuristics with classical learning mechanisms. In: Proc. of Congress on Evolutionary Computation (CEC'14), pp. 2515–2523. IEEE (2014)

[2] Aleti, A., Moser, I.: A systematic literature review of adaptive parameter control methods for evolutionary algorithms. ACM Computing Surveys **49**, 56:1–56:35 (2016)

[3] Ansótegui, C., Malitsky, Y., Samulowitz, H., Sellmann, M., Tierney, K.: Model-based genetic algorithms for algorithm configuration. In: Proc. of International Conference on Artificial Intelligence (IJCAI'15), pp. 733–739. AAAI Press (2015)

[4] Antipov, D., Doerr, B.: Precise runtime analysis for plateaus. In: Proc. of Parallel Problem Solving from Nature (PPSN'18), *Lecture Notes in Computer Science*, vol. 11102, pp. 117–128. Springer (2018)

[5] Antipov, D., Doerr, B., Fang, J., Hétet, T.: Runtime analysis for the $(\mu+\lambda)$ EA optimizing OneMax. In: Proc. of Genetic and Evolutionary Computation Conference (GECCO'18), pp. 1459-1466. ACM (2018).

[6] Auer, P., Cesa-Bianchi, N., Fischer, P.: Finite-time analysis of the multiarmed bandit problem. Machine Learning **47**, 235–256 (2002)

[7] Auger, A.: Benchmarking the (1+1) evolution strategy with one-fifth success rule on the BBOB-2009 function testbed. In: Companion Material for Proc. of Genetic and Evolutionary Computation Conference (GECCO'09), pp. 2447–2452. ACM (2009)

[8] Bäck, T.: The interaction of mutation rate, selection, and self-adaptation within a genetic algorithm. In: Proc. of Parallel Problem Solving from Nature (PPSN'92), pp. 87–96. Elsevier (1992)

[9] Bäck, T., Schütz, M.: Intelligent mutation rate control in canonical genetic algorithms. In: International Symposium on Foundations of Intelligent Systems (ISMIS'96), *Lecture Notes in Computer Science*, vol. 1079, pp. 158–167. Springer (1996)

[10] Badkobeh, G., Lehre, P.K., Sudholt, D.: Unbiased black-box complexity of parallel search. In: Proc. of Parallel Problem Solving from Nature (PPSN'14), *Lecture Notes in Computer Science*, vol. 8672, pp. 892–901. Springer (2014)

[11] Bartz-Beielstein, T., Flasch, O., Koch, P., Konen, W.: SPOT: A toolbox for interactive and automatic tuning in the R environment. In: Proc. of the 20. Workshop Computational Intelligence, pp. 264–273. Universitätsverlag Karlsruhe (2010)

[12] Böttcher, S., Doerr, B., Neumann, F.: Optimal fixed and adaptive mutation rates for the LeadingOnes problem. In: Proc. of Parallel Problem Solving from Nature (PPSN'10), *Lecture Notes in Computer Science*, vol. 6238, pp. 1–10. Springer (2010)

[13] Burke, E.K., Gendreau, M., Hyde, M.R., Kendall, G., Ochoa, G., Özcan, E., Qu, R.: Hyper-heuristics: a survey of the state of the art. Journal of the Operational Research Society **64**, 1695–1724 (2013)
[14] Buzdalov, M., Doerr, B.: Runtime analysis of the $(1+(\lambda,\lambda))$ Genetic Algorithm on random satisfiable 3-CNF formulas. In: Proc. of Genetic and Evolutionary Computation Conference (GECCO'17), pp. 1343–1350. ACM (2017)
[15] Carson, T., Impagliazzo, R.: Hill-climbing finds random planted bisections. In: Proc. of Annual Symposium on Discrete Algorithms (SODA'01), pp. 903–909. ACM/SIAM (2001). URL http://dl.acm.org/citation.cfm?id=365411.365805
[16] Cervantes, J., Stephens, C.R.: Limitations of existing mutation rate heuristics and how a rank GA overcomes them. IEEE Transactions on Evolutionary Computation **13**, 369–397 (2009)
[17] Costa, L.D., Fialho, Á., Schoenauer, M., Sebag, M.: Adaptive operator selection with dynamic multi-armed bandits. In: Proc. of Genetic and Evolutionary Computation Conference (GECCO'08), pp. 913–920. ACM (2008)
[18] Dang, D.C., Lehre, P.K.: Self-adaptation of mutation rates in non-elitist populations. In: Proc. of Parallel Problem Solving from Nature (PPSN'16), *Lecture Notes in Computer Science*, vol. 9921, pp. 803–813. Springer (2016)
[19] Devroye, L.: The compound random search. Ph.D. dissertation, Purdue Univ., West Lafayette, IN (1972)
[20] Dietzfelbinger, M., Rowe, J.E., Wegener, I., Woelfel, P.: Tight bounds for blind search on the integers and the reals. Combinatorics, Probability & Computing **19**, 711–728 (2010)
[21] Doerr, B.: Optimal parameter settings for the $(1+(\lambda,\lambda))$ genetic algorithm. In: Proc. of Genetic and Evolutionary Computation Conference (GECCO'16), pp. 1107–1114. ACM (2016)
[22] Doerr, B., Doerr, C.: Optimal parameter choices through self-adjustment: Applying the 1/5-th rule in discrete settings. In: Proc. of Genetic and Evolutionary Computation Conference (GECCO'15), pp. 1335–1342. ACM (2015)
[23] Doerr, B., Doerr, C.: A tight runtime analysis of the $(1+(\lambda,\lambda))$ genetic algorithm on OneMax. In: Proc. of Genetic and Evolutionary Computation Conference (GECCO'15), pp. 1423–1430. ACM (2015)
[24] Doerr, B., Doerr, C.: Optimal static and self-adjusting parameter choices for the $(1+(\lambda,\lambda))$ genetic algorithm. Algorithmica **80**, 1658–1709 (2018)
[25] Doerr, B., Doerr, C., Ebel, F.: Lessons from the black-box: Fast crossover-based genetic algorithms. In: Proc. of Genetic and Evolutionary Computation Conference (GECCO'13), pp. 781–788. ACM (2013)
[26] Doerr, B., Doerr, C., Ebel, F.: From black-box complexity to designing new genetic algorithms. Theoretical Computer Science **567**, 87–104 (2015)

[27] Doerr, B., Doerr, C., Kötzing, T.: Provably optimal self-adjusting step sizes for multi-valued decision variables. In: Proc. of Parallel Problem Solving from Nature (PPSN'16), *Lecture Notes in Computer Science*, vol. 9921, pp. 782–791. Springer (2016)

[28] Doerr, B., Doerr, C., Kötzing, T.: Static and self-adjusting mutation strengths for multi-valued decision variables. Algorithmica **80**, 1732–1768 (2018). DOI 10.1007/s00453-017-0341-1. URL `https://doi.org/10.1007/s00453-017-0341-1`

[29] Doerr, B., Doerr, C., Yang, J.: k-bit mutation with self-adjusting k outperforms standard bit mutation. In: Proc. of Parallel Problem Solving from Nature (PPSN'16), *Lecture Notes in Computer Science*, vol. 9921, pp. 824–834. Springer (2016)

[30] Doerr, B., Doerr, C., Yang, J.: Optimal parameter choices via precise black-box analysis. In: Proc. of Genetic and Evolutionary Computation Conference (GECCO'16), pp. 1123–1130. ACM (2016)

[31] Doerr, B., Fouz, M., Witt, C.: Quasirandom evolutionary algorithms. In: Proc. of Genetic and Evolutionary Computation Conference (GECCO'10), pp. 1457–1464. ACM (2010). DOI 10.1145/1830483.1830749. URL `http://doi.acm.org/10.1145/1830483.1830749`

[32] Doerr, B., Fouz, M., Witt, C.: Sharp bounds by probability-generating functions and variable drift. In: Proc. of Genetic and Evolutionary Computation Conference (GECCO'11), pp. 2083–2090. ACM (2011)

[33] Doerr, B., Gießen, C., Witt, C., Yang, J.: The $(1+\lambda)$ evolutionary algorithm with self-adjusting mutation rate. In: Proc. of Genetic and Evolutionary Computation Conference (GECCO'17), pp. 1351–1358. ACM (2017)

[34] Doerr, B., Künnemann, M.: Optimizing linear functions with the $(1+\lambda)$ evolutionary algorithm—different asymptotic runtimes for different instances. Theoretical Computer Science **561**, 3–23 (2015)

[35] Doerr, B., Le, H.P., Makhmara, R., Nguyen, T.D.: Fast genetic algorithms. In: Proc. of Genetic and Evolutionary Computation Conference (GECCO'17), pp. 777–784. ACM (2017). Full version available at `http://arxiv.org/abs/1703.03334`

[36] Doerr, B., Lissovoi, A., Oliveto, P.S., Warwicker, J.A.: On the runtime analysis of selection hyper-heuristics with adaptive learning periods. In: Proc. of Genetic and Evolutionary Computation Conference (GECCO'18), pp. 1015–1022. ACM (2018).

[37] Doerr, B., Neumann, F., Sutton, A.M.: Time complexity analysis of evolutionary algorithms on random satisfiable k-CNF formulas. Algorithmica **78**, 561–586 (2017)

[38] Doerr, B., Witt, C., Yang, J.: Runtime analysis for self-adaptive mutation rates. In: Proc. of Genetic and Evolutionary Computation Conference (GECCO'18), pp. 1475–1482. ACM (2018).

[39] Droste, S., Jansen, T., Wegener, I.: Dynamic parameter control in simple evolutionary algorithms. In: Proc. of Foundations of Genetic Al-

gorithms (FOGA'00), pp. 275–294. Morgan Kaufmann (2000). DOI 10.1016/B978-155860734-7/50098-6. URL https://doi.org/10.1016/B978-155860734-7/50098-6

[40] Eiben, A.E., Hinterding, R., Michalewicz, Z.: Parameter control in evolutionary algorithms. IEEE Transactions on Evolutionary Computation **3**, 124–141 (1999)

[41] Eiben, A.E., Michalewicz, Z., Schoenauer, M., Smith, J.E.: Parameter control in evolutionary algorithms. In: Parameter Setting in Evolutionary Algorithms, *Studies in Computational Intelligence*, vol. 54, pp. 19–46. Springer (2007)

[42] Fialho, Á., Costa, L.D., Schoenauer, M., Sebag, M.: Extreme value based adaptive operator selection. In: Proc. of Parallel Problem Solving from Nature (PPSN'08), *Lecture Notes in Computer Science*, vol. 5199, pp. 175–184. Springer (2008)

[43] Fialho, Á., Costa, L.D., Schoenauer, M., Sebag, M.: Dynamic multi-armed bandits and extreme value-based rewards for adaptive operator selection in evolutionary algorithms. In: Proc. of Learning and Intelligent Optimization (LION'09), *Lecture Notes in Computer Science*, vol. 5851, pp. 176–190. Springer (2009). URL https://doi.org/10.1007/978-3-642-11169-3_13

[44] Fialho, Á., Costa, L.D., Schoenauer, M., Sebag, M.: Analyzing bandit-based adaptive operator selection mechanisms. Annals of Mathematics and Artificial Intelligence **60**, 25–64 (2010). URL https://doi.org/10.1007/s10472-010-9213-y

[45] Giel, O., Wegener, I.: Evolutionary algorithms and the maximum matching problem. In: Proc. of Symposium on Theoretical Aspects of Computer Science (STACS'03), pp. 415–426. Springer (2003)

[46] Gießen, C., Witt, C.: The interplay of population size and mutation probability in the $(1+\lambda)$ EA on OneMax. Algorithmica **78**, 587–609 (2017)

[47] Goldman, B.W., Punch, W.F.: Parameter-less population pyramid. In: Proc. of Genetic and Evolutionary Computation Conference (GECCO'14), pp. 785–792. ACM (2014)

[48] Hansen, N., Gawelczyk, A., Ostermeier, A.: Sizing the population with respect to the local progress in $(1,\lambda)$-evolution strategies - a theoretical analysis. In: Proc. of Congress on Evolutionary Computation (CEC'95), pp. 80–85. IEEE (1995)

[49] He, J., He, F., Dong, H.: Pure strategy or mixed strategy? - An initial comparison of their asymptotic convergence rate and asymptotic hitting time. In: Proc. of Evolutionary Computation in Combinatorial Optimization (EvoCOP'12), *Lecture Notes in Computer Science*, vol. 7245, pp. 218–229. Springer (2012). DOI 10.1007/978-3-642-29124-1_19

[50] Henderson, D., Jacobson, S.H., Johnson, A.W.: The theory and practice of simulated annealing. In: F. Glover, G.A. Kochenberger (eds.) Handbook of Metaheuristics, pp. 287–319. Springer (2003)

[51] Hutter, F., Hoos, H.H., Leyton-Brown, K.: Sequential model-based optimization for general algorithm configuration. In: Proc. of Learning and Intelligent Optimization (LION'11), pp. 507–523. Springer (2011)
[52] Hutter, F., Hoos, H.H., Leyton-Brown, K., Stützle, T.: ParamILS: An automatic algorithm configuration framework. Journal of Artificial Intelligence Research **36**, 267–306 (2009)
[53] Impagliazzo, R.: Hill-climbing vs. simulated annealing for planted bisection problems. In: Proc. of the 4th International Workshop on Approximation Algorithms for Combinatorial Optimization Problems (RANDOM-APPROX'01), *Lecture Notes in Computer Science*, vol. 2129, pp. 2–5. Springer (2001). DOI 10.1007/3-540-44666-4_2. URL https://doi.org/10.1007/3-540-44666-4_2
[54] Jansen, T., De Jong, K.A., Wegener, I.: On the choice of the offspring population size in evolutionary algorithms. Evolutionary Computation **13**, 413–440 (2005)
[55] Jansen, T., Wegener, I.: On the analysis of a dynamic evolutionary algorithm. Journal of Discrete Algorithms **4**, 181–199 (2006)
[56] Jansen, T., Wegener, I.: A comparison of simulated annealing with a simple evolutionary algorithm on pseudo-Boolean functions of unitation. Theoretical Computer Science **386**, 73–93 (2007). DOI 10.1016/j.tcs.2007.06.003. URL https://doi.org/10.1016/j.tcs.2007.06.003
[57] Jansen, T., Zarges, C.: Performance analysis of randomised search heuristics operating with a fixed budget. Theoretical Computer Science **545**, 39–58 (2014). URL https://doi.org/10.1016/j.tcs.2013.06.007
[58] Jerrum, M., Sinclair, A.: The Markov chain Monte Carlo method: An approach to approximate counting and integration. In: D.S. Hochbaum (ed.) Approximation Algorithms for NP-hard Problems, pp. 482–520. PWS Publishing Co. (1997). URL http://dl.acm.org/citation.cfm?id=241938.241950
[59] Jerrum, M., Sorkin, G.B.: Simulated annealing for graph bisection. In: Proc. of Annual Symposium on Foundations of Computer Science (FOCS'93), pp. 94–103. IEEE Computer Society (1993). DOI 10.1109/SFCS.1993.366878. URL https://doi.org/10.1109/SFCS.1993.366878
[60] Johannsen, D.: Random combinatorial structures and randomized search heuristics. Ph.D. thesis, Universität des Saarlandes (2010)
[61] Karafotias, G., Hoogendoorn, M., Eiben, A.: Parameter control in evolutionary algorithms: Trends and challenges. IEEE Transactions on Evolutionary Computation **19**, 167–187 (2015)
[62] Kern, S., Müller, S.D., Hansen, N., Büche, D., Ocenasek, J., Koumoutsakos, P.: Learning probability distributions in continuous evolutionary algorithms - a comparative review. Natural Computing **3**, 77–112 (2004)
[63] Kirkpatrick, S., Gelatt, C.D., Vecchi, M.P.: Optimization by simulated annealing. Science **220**, 671–680 (1983)

[64] Kruisselbrink, J.W., Li, R., Reehuis, E., Eggermont, J., Bäck, T.: On the log-normal self-adaptation of the mutation rate in binary search spaces. In: Proc. of Genetic and Evolutionary Computation Conference (GECCO'11), pp. 893–900. ACM (2011)
[65] de Perthuis de Laillevault, A., Doerr, B., Doerr, C.: Money for nothing: Speeding up evolutionary algorithms through better initialization. In: Proc. of Genetic and Evolutionary Computation Conference (GECCO'15), pp. 815–822. ACM (2015). URL http://doi.acm.org/10.1145/2739480.2754760
[66] Lässig, J., Sudholt, D.: Adaptive population models for offspring populations and parallel evolutionary algorithms. In: Proc. of Foundations of Genetic Algorithms (FOGA'11), pp. 181–192. ACM (2011)
[67] Lehre, P.K., Özcan, E.: A runtime analysis of simple hyper-heuristics: to mix or not to mix operators. In: Proc. of Foundations of Genetic Algorithms (FOGA'13), pp. 97–104. ACM (2013). DOI 10.1145/2460239.2460249. URL http://doi.acm.org/10.1145/2460239.2460249
[68] Li, K., Fialho, Á., Kwong, S., Zhang, Q.: Adaptive operator selection with bandits for a multiobjective evolutionary algorithm based on decomposition. IEEE Transactions on Evolutionary Computation **18**, 114–130 (2014). DOI 10.1109/TEVC.2013.2239648. URL https://doi.org/10.1109/TEVC.2013.2239648
[69] Lissovoi, A., Oliveto, P.S., Warwicker, J.A.: On the runtime analysis of generalised selection hyper-heuristics for pseudo-Boolean optimisation. In: Proc. of Genetic and Evolutionary Computation Conference (GECCO'17), pp. 849–856. ACM (2017)
[70] Lissovoi, A., Oliveto, P.S., Warwicker, J.A.: Hyper-heuristics can achieve optimal performance for pseudo-boolean optimisation. Arxiv e-prints **1801.07546** (2018)
[71] Lobo, F.G., Lima, C.F., Michalewicz, Z. (eds.): Parameter Setting in Evolutionary Algorithms, *Studies in Computational Intelligence*, vol. 54. Springer (2007)
[72] López-Ibáñez, M., Dubois-Lacoste, J., Cáceres, L.P., Birattari, M., Stützle, T.: The irace package: Iterated racing for automatic algorithm configuration. Operations Research Perspectives **3**, 43–58 (2016)
[73] Mambrini, A., Sudholt, D.: Design and analysis of schemes for adapting migration intervals in parallel evolutionary algorithms. Evolutionary Computation **23**, 559–582 (2015). DOI 10.1162/EVCO_a_00153. URL https://doi.org/10.1162/EVCO_a_00153
[74] Metropolis, N., Rosenbluth, A.W., Rosenbluth, M.N., Teller, A.H., Teller, E.: Equation of state calculations by fast computing machines. The Journal of Chemical Physics **21**, 1087–1092 (1953)
[75] Neumann, A., Szpak, Z.L., Chojnacki, W., Neumann, F.: Evolutionary image composition using feature covariance matrices. In: Proc. of the Genetic and Evolutionary Computation Conference, (GECCO'17), pp.

817–824. ACM (2017). DOI 10.1145/3071178.3071260. URL http://doi.acm.org/10.1145/3071178.3071260

[76] Neumann, F., Wegener, I.: Randomized local search, evolutionary algorithms, and the minimum spanning tree problem. Theoretical Computer Science **378**, 32–40 (2007)

[77] Oliveto, P.S., Lehre, P.K., Neumann, F.: Theoretical analysis of rank-based mutation - combining exploration and exploitation. In: Proc. of Congress on Evolutionary Computation (CEC'09), pp. 1455–1462. IEEE (2009)

[78] Qian, C., Tang, K., Zhou, Z.: Selection hyper-heuristics can provably be helpful in evolutionary multi-objective optimization. In: Proc. of Parallel Problem Solving From Nature (PPSN'16), *Lecture Notes in Computer Science*, vol. 9921, pp. 835–846. Springer (2016). DOI 10.1007/978-3-319-45823-6_78

[79] Rechenberg, I.: Evolutionsstrategie. Friedrich Frommann Verlag (Günther Holzboog KG), Stuttgart (1973)

[80] Reichel, J., Skutella, M.: On the size of weights in randomized search heuristics. In: Proc. of Foundations of Genetic Algorithms (FOGA'09), pp. 21–28. ACM (2009)

[81] Sasaki, G.H., Hajek, B.E.: The time complexity of maximum matching by simulated annealing. Journal of the ACM **35**, 387–403 (1988). DOI 10.1145/42282.46160. URL http://doi.acm.org/10.1145/42282.46160

[82] Schumer, M.A., Steiglitz, K.: Adaptive step size random search. IEEE Transactions on Automatic Control **13**, 270–276 (1968)

[83] Sorkin, G.B.: Efficient simulated annealing on fractal energy landscapes. Algorithmica **6**, 367–418 (1991). DOI 10.1007/BF01759051. URL https://doi.org/10.1007/BF01759051

[84] Thierens, D.: An adaptive pursuit strategy for allocating operator probabilities. In: Proc. of Genetic and Evolutionary Computation Conference (GECCO'05), pp. 1539–1546. ACM (2005)

[85] Wegener, I.: Simulated annealing beats Metropolis in combinatorial optimization. In: Proc. of International Colloquium on Automata, Languages and Programming (ICALP'05), *Lecture Notes in Computer Science*, vol. 3580, pp. 589–601. Springer (2005)

[86] Witt, C.: Tight bounds on the optimization time of a randomized search heuristic on linear functions. Combinatorics, Probability & Computing **22**, 294–318 (2013)

[87] Zarges, C.: Rigorous runtime analysis of inversely fitness proportional mutation rates. In: Proc. of Parallel Problem Solving from Nature (PPSN'08), *Lecture Notes in Computer Science*, vol. 5199, pp. 112–122. Springer (2008)

[88] Zarges, C.: On the utility of the population size for inversely fitness proportional mutation rates. In: Proc. of Foundations of Genetic Algorithms (FOGA'09), pp. 39–46. ACM (2009)

Chapter 7
Analysis of Evolutionary Algorithms in Dynamic and Stochastic Environments

Frank Neumann, Mojgan Pourhassan and Vahid Roostapour

Abstract Many real-world optimization problems occur in environments that change dynamically or involve stochastic components. Evolutionary algorithms and other bio-inspired algorithms have been widely applied to dynamic and stochastic problems. This survey gives an overview of major theoretical developments in the area of runtime analysis for these problems. We review recent theoretical studies of evolutionary algorithms and ant colony optimization for problems where the objective functions or the constraints change over time. Furthermore, we consider stochastic problems with various noise models and point out some directions for future research.

7.1 Introduction

Real-world problems are often stochastic and may have dynamic components. Evolutionary algorithms and other bio-inspired algorithmic approaches such as ant colony optimization have been applied to a wide range of stochastic and dynamic problems. The goal of this chapter is to give an overview of recent theoretical developments in the area of evolutionary computation for stochastic and dynamic problems in the context of discrete optimization.

Stochastic problems occur frequently in real-world applications owing to unpredictable factors. A good example is the scheduling of trains. Schedules give precise timings for when trains arrive and depart. However, the actual departure and arrival times may be subject to delays due to various factors

Frank Neumann e-mail: frank.neumann@adelaide.edu.au · Mojgan Pourhassan e-mail: mojgan.pourhassan@adelaide.edu.au · Vahid Roostapour e-mail: vahid.roostapour@adelaide.edu.au

Optimisation and Logistics, School of Computer Science, The University of Adelaide, Australia

B. Doerr, F. Neumann (eds.), *Theory of Evolutionary Computation*,
Natural Computing Series, https://doi.org/10.1007/978-3-030-29414-4_7

such as weather conditions and interfering schedules of other trains. When evolutionary computation is used for the optimization of stochastic problems, the uncertainty is usually reflected through a noisy fitness function. The underlying fitness function for these problems is noisy in the sense that it produces different results for the same input. Two major noise models, namely prior noise and posterior noise, have been introduced and investigated in the literature. In the case of prior noise, the solution is changed prior to the evaluation of the given fitness function, whereas in the case of posterior noise the solution is evaluated with the given fitness function and a value according to a given noise distribution is added before returning the fitness value.

Dynamic problems constitute another important part of what occurs in real-world applications. Problems can change over time owing to various components becoming unavailable or available at a later point in time. Different parts of the problem that can be subject to a change are the objective function and possible constraints of the given problem. In terms of scheduling of trains, they might become unavailable due to mechanical failures and it might be necessary to reschedule the trains in the network in order to still serve the demands of the customers well.

The area of runtime analysis has contributed many interesting studies to the theoretical understanding of bio-inspired algorithms in this area. We start by investigating popular benchmark algorithms such as randomized local search (RLS) and the $(1+1)$ EA on different dynamic problems. This includes dynamic versions of OneMax, the classical vertex cover problem, the makespan scheduling problem, and classes of the well-known knapsack problem. Afterwards, we summarize the main results for stochastic problems. Important studies in this area have considered the OneMax problem and investigated the runtime behavior of evolutionary algorithms with respect to prior and posterior noise. Moreover, the influence of populations in evolutionary algorithms for solving stochastic problems has been analyzed in the literature, and we place particular emphasis on those studies. Furthermore, we review the performance of population-based algorithms on different posterior noise functions.

Ant colony optimization (ACO) algorithms are another important type of bio-inspired algorithm that have been used and analyzed for solving dynamic and stochastic problems. Due to their different way of constructing solutions, based on sampling from the underlying search space by performing random walks on a so-called construction graph, they have a different ability to deal with dynamic and stochastic problems. Furthermore, an important parameter in ACO algorithms is the pheromone update strength, which allows one to determine how quickly previously good solutions are forgotten by the algorithms. This parameter plays a crucial role when distinguishing ACO algorithms from classical evolutionary algorithms. At the end of this chapter, we present a summary of the results obtained on dynamic and stochastic problems in the context of ACO.

This chapter is organized as follows. In Section 7.2, we summarize the dynamic and stochastic settings that have been investigated in the literature. We present the main results obtained for evolutionary algorithms in dynamic and stochastic environments in Section 7.3 and 7.4, respectively. We highlight theoretical results on the behavior of ACO algorithms for dynamic and stochastic problems in Section 7.5. Finally, we finish with some conclusions and outline some future research directions.

7.2 Preliminaries

This section includes formal definitions of the dynamic and stochastic optimization settings that are investigated in this chapter.

In dynamically changing optimization problems, some part of the problem is subject to change over time. Usually changes to the objective function or the constraints of the given problem are considered. The different problems that have been studied from a theoretical perspective will be introduced in the forthcoming subsections. In the case of stochastic optimization problems, the optimization algorithm does not have access to the deterministic fitness value of a candidate solution. Different types of noise that change the actual fitness value have been introduced. The most important ones, prior noise and posterior noise, will be introduced in Section 7.2.5.

The theoretical analysis of evolutionary algorithms for dynamic and stochastic problems concentrates on the classical algorithms such as RLS and the $(1+1)$ EA. Furthermore, the benefit of population-based approaches has been examined. These algorithms will be introduced in Section 7.2.6.

7.2.1 Dynamic OneMax *Problem*

Investigations started by considering a generalization of the classical OneMax problem. In the OneMax problem, the number of ones in the solution is the objective to be maximized. Droste [9] interpreted this problem as maximizing the number of bits that match a given objective bit string. Based on this, he introduced the dynamic OneMax problem, in which dynamic changes happen on the objective bit-string over time. An extended version of this problem was defined by Kötzing et al. [21], where not only bit strings are allowed, but also each position can take on integer values in $\{0,\dots,r-1\}$ for $r\in\mathbb{N}_{\geq 2}$. The formal definition of the problem follows.

Let $[r]=\{0,\dots,r-1\}$ for $r\in\mathbb{N}_{\geq 2}$, and $x,y\in[r]$. Moreover, let the distance between x and y be

$$d(x,y)=\min\{(x-y) \mod r,(y-x) \mod r\}.$$

The extended OneMax problem, $\textsc{OneMax}_a : [r]^n \to \mathbb{R}$, where a is the objective string defining the optimum, is given as:

$$\textsc{OneMax}_a(x) = \sum_{i=1}^{n} d(a_i, x_i).$$

The goal is to find and maintain a solution with minimum value of $\textsc{OneMax}_a$.

Given a probability value p, the dynamism that is defined on this problem is that each component i, $1 \leq i \leq n$, of the optimal solution a changes independently as follows:

$$a_i = \begin{cases} (a_i + 1) \mod r & \text{with probability } p/2, \\ (a_i - 1) \mod r & \text{with probability } p/2, \\ a_i & \text{with probability 1-p.} \end{cases}$$

7.2.2 Linear Pseudo-Boolean Functions Under Dynamic Uniform Constraints

Linear pseudo-Boolean functions play a key role in the runtime analysis of evolutionary algorithms. Let $x = x_1 x_2 \ldots x_n$ be a search point in search space $\{0,1\}^n$, and w_i, $1 \leq i \leq n$ positive real weights. A linear pseudo-Boolean function $f(x)$ is defined as:

$$f(x) = w_0 + \sum_{i=1}^{n} w_i x_i.$$

For simplicity and as done in most studies, we assume $w_0 = 0$ in the following. The optimization of a linear objective function under a linear constraint is equivalent to the classical knapsack problem [20]. The optimization of a linear objective function together with a uniform constraint has recently been investigated in the static setting [15]. Given a bound B, $0 \leq B \leq n$, a solution x is feasible if the number of 1-bits of the search point x is at most B. The bound B is also known as the *cardinality bound*. We denote the number of 1-bits of x by $|x|_1 = \sum_{i=1}^{n} x_i$. The formal definition for maximizing a pseudo-Boolean linear function under a cardinality bound constraint is given by:

$$\max f(x)$$
$$\text{such that } |x|_1 \leq B.$$

The dynamic version of this problem, referred to as the problem with a *dynamic uniform constraint*, is defined in [33]. Here the cardinality bound changes from B to some new value B^*. Starting from a solution that is optimal for the bound B, the problem is then to find an optimal solution for B^*. The re-optimization time of an evolutionary algorithm is defined as the number of fitness evaluations that is required to find the new optimal solution.

7.2.3 Dynamic Vertex Cover Problem

The vertex cover problem is one of the best-known NP-hard combinatorial optimization problems. Given a graph $G = (V, E)$, where $V = \{v_1, \ldots, v_n\}$ is the set of vertices and $E = \{e_1, \ldots, e_m\}$ is the set of edges, the goal is to find a minimum subset of nodes $V_C \subseteq V$ that covers all edges in E, i.e. $\forall e \in E, e \cap V_C \neq \emptyset$. In the dynamic version of the problem, an edge can be added to or deleted from the graph.

As the vertex cover problem is NP-hard, it has been mainly studied in terms of approximations. The problem can be approximated within a worst case approximation ratio of 2 by various algorithms. One standard approach to obtain a 2-approximation is to compute a maximal matching and take all nodes adjacent to the chosen matching edges for the vertex cover. Starting from a solution that is a 2-approximation for the current instance of the problem, in the dynamic version of the problem the goal is to obtain a 2-approximate solution for that instance of the problem after one dynamic change. The re-optimization time for this problem refers to the time required for the investigated algorithm to find a 2-approximate solution for the new instance. This dynamic setting has been investigated in [30].

7.2.4 Dynamic Makespan Scheduling Problem

The makespan scheduling problem can be defined as follows. Given n jobs and their processing times $p_i > 0$, $1 \leq i \leq n$, the goal is to assign each job to one of two machines M_1 and M_2 such that the makespan is minimized. The makespan is the time that the busier machine takes to finish all assigned jobs. A solution is represented by a vector $x \in \{0,1\}^n$ which means that job i is assigned to machine M_1 if $x_i = 0$ and it is assigned to M_2 if $x_i = 1$, $1 \leq i \leq n$. With this representation, the makespan of a given solution x is given by

$$f(x) = \max\left\{\sum_{i=1}^{n} p_i(1-x_i), \sum_{i=1}^{n} p_i x_i\right\}$$

and the goal is to minimize f. In the dynamic version of this problem, the processing time of a job may change over time, but stays within a given interval. In [28], the setting $p_i \in [L, U]$, $1 \leq i \leq n$, where L and U are a lower and an upper bound on each processing time, was investigated. The analysis concentrates on the time evolutionary algorithms need to produce a solution where the two machines have a discrepancy of at most U. Dynamic changes to the processing times of the jobs were investigated in two different settings. In the first setting, an adversary is allowed to change the processing time of exactly one job. In the second setting, the job to be changed is picked by an adversary but the processing time of a job is altered randomly.

7.2.5 Stochastic Problems and Noise Models

We consider stochastic optimization problems where the fitness function is subject to some noise. There are two main noise models that have been studied in the area of the theoretical analysis of evolutionary computation. Noise that affects the solution before the evaluation is called *prior noise*. In this case, the fitness function returns the fitness value of a solution that may differ from the given solution because of the noise. Droste studied the effect of prior noise which flips one randomly chosen bit of the given solution with probability of p before each evaluation [10]. Note that the noise does not change the solution, but it causes the fitness function to evaluate a solution with a noisy bit flip. Other kinds of prior noise have also been considered. For example, prior noise which flips each bit with the probability of p or which sets each bit independently with a probability of p to 0 [17].

Another important type of noise is where the fitness of the solution is changed after evaluation. This type of noise is called *posterior noise* or *additive posterior noise*. The noise, which commonly comes from a defined distribution D, adds the value of a random variable sampled from D to the value coming from the original fitness function [12, 14, 17].

In a noisy environment, the problem of finding the optimal solutions is harder, as the noise misleads the search. The goal is to find an optimal solution for the original non-noisy fitness function by evaluating solutions on the fitness function affected by noise. However, it has been proven that simple evolutionary algorithms behave very well when facing this kind of problem. In addition to this, properties of stochastic settings that are hard for evolutionary algorithms to deal with have also been studied [12].

We concentrate on stochastic problems with a fixed known solution length that are subject to noise. We would like to mention, however, that there have also been studies investigating the performance of evolutionary algorithms with unknown solution length. This poses a different type of uncertainty which we will not capture in this chapter. We refer the interested reader to [2, 6].

Algorithm 7.1: RLS

```
1 The initial solution x is given;
2 while stopping criterion not met do
3     y ← flip one bit of x chosen uniformly at random;
4     if f(y) ≥ f(x) then
5         x ← y;
```

Algorithm 7.2: (1+1) EA

```
1 The initial solution x is given;
2 while stopping criterion not met do
3     y ← flip each bit of x independently with probability 1/n;
4     if f(y) ≥ f(x) then
5         x ← y;
```

7.2.6 Evolutionary Algorithms

Analyzing evolutionary algorithms often starts by investigating a standard randomized local search approach and a simple $(1+1)$ EA. Here we present these algorithms in addition to a population-based $(\mu+\lambda)$ EA for which results are summarized in Section 7.4.

A standard RLS (see Algorithm 7.1) starts with a bit string as the initial solution, makes a new solution by flipping one bit of the current solution uniformly at random at each iteration, and replaces the current solution with the new solution if the new one is better in terms of fitness. The algorithm repeats these steps, as long as the stopping criterion is not met.

The $(1+1)$ EA (see Algorithm 7.2) is a simple evolutionary algorithm in which the population consists of only one solution, and only one solution is generated at each time step. This algorithm is quite similar to RLS, except that multiple bit flips are allowed at each iteration. Instead of flipping one bit uniformly at random, in this algorithm all bits of the current solution are flipped with probability $1/n$, where n is the size of the solution.

A classical question in the area of evolutionary computation is whether populations help to achieve better results compared with algorithms working at each time step with a single solution. The $(\mu+\lambda)$ EA (Algorithm 7.3) is the population-based version of the $(1+1)$ EA. In this algorithm, μ denotes the size of the parent population. In each iteration, the algorithm creates λ offspring by mutating λ parents which have been chosen uniformly at random from the parent population. Finally, all the solutions from parents and offspring are evaluated and the μ best ones (in terms of fitness function) survive. They constitute the parent population of the next generation.

One of the questions raised by using a population is the effect of a crossover operator on the robustness. This has been investigated by Friedrich et

Algorithm 7.3: $(\mu+\lambda)$ EA

```
1  P is a set of μ uniformly chosen solutions;
2  while stopping criterion not met do
3      O ← ∅;
4      for i = 1 to λ do
5          pick x u.a.r. from P;
6          y ← flip each bit of x independently with probability 1/n;
7          O ← O ∪ y;
8      for x ∈ P ∪ O do
9          evaluate f(x);
10     P ← μ f-maximal elements from P ∪ O;
```

al. in [16]. To this end, they considered a framework consisting of one wide and many narrow parallel paths, with equal distances, for solutions to achieve the highest fitness value. The fitness grows more quickly along narrow paths and a solution which is not located in one of the paths does not survive. It was shown that algorithms with a higher recombination rate optimize through the wide path, while narrow paths are more favored by algorithms with zero recombination rate. A change that moves the framework along the x-axis will cause the extinction of solutions on the narrow paths, however, solutions on the wide path may survive. This shows the benefit of a crossover operation for the robustness of algorithms using a population.

When analyzing evolutionary algorithms with respect to their runtime behavior, one considers the number of solutions that are produced until a solution of the desired quality has been achieved. The expected time to reach this goal refers to the expected number of such solutions. The *expected optimization time* refers to the expected number of solutions that are produced until an optimal search point has been produced for the first time. When considering dynamic problems, we are often interested in the *expected re-optimization time* of an algorithm. Starting with a good (or even optimal) solution for the problem considered, the expected number of constructed solutions required to obtain a solution of the same quality after a dynamic change has occurred is analyzed.

7.3 Analysis of Evolutionary Algorithms on Dynamic Problems

In this section, we summarize recent theoretical analyses that have been performed on evolutionary algorithms dealing with dynamic optimization problems. In [33, 34], the efficiency of evolutionary algorithms for solving linear pseudo-Boolean functions with a dynamic linear constraint was investigated.

Particular attention was paid to the OneMax problem. OneMax has been the center of attention in some other related studies as well [9, 21]. We first present the investigations that have been performed on this problem, and then we give a summary of the results that have been obtained for linear pseudo-Boolean functions under dynamic uniform constraints. Furthermore, in this section we explain the analysis that has been carried out for the dynamic vertex cover problem and the makespan scheduling problem. Another problem which has been investigated in the context of dynamic optimization is the Maze problem for which evolutionary algorithms and ant colony optimization algorithms have been theoretically studied [22, 25, 26]. The results on evolutionary algorithms and ACO algorithms for this problem are presented in Section 7.3.5 and 7.5, respectively.

7.3.1 OneMax *Under Dynamic Uniform Constraints*

The first runtime analysis of evolutionary algorithms for a dynamic discrete problem was presented by Droste [9]. In that article, the OneMax problem was considered and the goal is to find a solution which has the minimum Hamming distance to an objective bit string. A dynamic change, in that publication, is changing one bit of the objective bit string, which happens at each time step with probability p' and results in the dynamic changes of the fitness function over time. Droste found the maximum rate of dynamic changes such that the expected optimization time of the $(1+1)$ EA remains polynomial for the problem studied. More precisely, he has proved that the $(1+1)$ EA has a polynomial expected runtime if $p' = O(\log(n)/n)$, while for any substantially larger probability the runtime becomes super polynomial. It is worth noting that the results of that article hold even if the expected re-optimization time of the problem is larger than the expected time until the next dynamic change happens.

Using drift analysis, Kötzing et al. [21] reproved some of the results in [9]. Furthermore, they carried out theoretical investigations for the extended dynamic OneMax problem (see Section 7.2.1), in which each variable can take on more than two values. They also carried out an *anytime analysis* (introduced in [19]) and showed how closely the algorithm that they investigated could track the dynamically moving target over time.

The optimization time of evolutionary algorithms for OneMax and the general class of linear pseudo-Boolean function, under the dynamic uniform constraint given in Section 7.2.2 was analyzed in [33, 34]. For now, we concentrate on OneMax with with a dynamic uniform constraint. The authors of [33, 34] analyzed a standard $(1+1)$ EA (Algorithm 7.2) and three other evolutionary algorithms, which are presented in Algorithms 7.4, 7.5, and 7.6. The results of their investigations are summarized in Table 7.1. The $(1+1)$ EA analyzed, uses the fitness function

Table 7.1 Upper bounds on the expected re-optimization times of evolutionary algorithms on the OneMax problem with a dynamic uniform constraint.

(1+1) **EA**	**MOEA**	**MOEA-S**	**MOGA**	
$O\left(n\log\left(\frac{n-B}{n-B^*}\right)\right)$	$O\left(nD\log\left(\frac{n-B}{n-B^*}\right)\right)$	$O\left(n\log\left(\frac{n-B}{n-B^*}\right)\right)$	$O\left(\min\{\sqrt{n}D^{\frac{3}{2}}, D^2\sqrt{\frac{n}{n-B^*}}\}\right)$	if $B < B^*$
$O\left(n\log\left(\frac{B}{B^*}\right)\right)$	$O\left(nD\log\left(\frac{B}{B^*}\right)\right)$	$O\left(n\log\left(\frac{B}{B^*}\right)\right)$	$O\left(\min\{\sqrt{n}D^{\frac{3}{2}}, D^2\sqrt{\frac{n}{B^*}}\}\right)$	if $B > B^*$

$$f_{(1+1)}(x) = f(x) - (n+1)\cdot\max\{0, |x|_1 - B^*\}$$

introduced earlier in [15]. This gives a large penalty to infeasible solutions by subtracting a term of $(n+1)$ for each unit of constraint violation. This implies that every infeasible solution is worse than any feasible one. The penalty of this fitness function guides the search towards the feasible region and does not allow the (1+1) EA to accept an infeasible solution after a feasible solution has been found for the first time.

Shi et al. [33, 34] used multiplicative drift analysis [8] to investigate the behavior of the algorithms that they studied. The potential function that they used for analyzing the (1+1) EA on OneMax with a dynamic uniform constraint is $|x|_0$, when $B \leq B^*$. Here, the initial solution, denoted by x_{org}, is feasible, and the algorithm needs to increase the number of ones of the solution until the cardinality bound B^* is reached. In this situation, the drift on $|x|_0$ is $\Omega(|x|_0/n)$ for the (1+1) EA. Using multiplicative drift analysis, the expected number of generations to reach a solution x^* with $|x^*|_0 = n - B^*$ is

$$O\left(n\log\left(\frac{|x_{\text{org}}|_0}{|x^*|_0}\right)\right) = O\left(n\log\left(\frac{n-B}{n-B^*}\right)\right).$$

For the situation where $B \geq B^*$, the initial solution is infeasible and the number of ones of the solution needs to decrease (and possibly increase again, if the last move to the feasible region has decreased $|x|_1$ to less than B^*). The potential function considered in this situation is $|x|_1$ and the drift on that is $\Omega(|x|_1/n)$, giving an expected re-optimization time of $O\left(n\log\left(\frac{B}{B^*}\right)\right)$.

The second algorithm that the authors investigated is the Multi-Objective Evolutionary Algorithm (MOEA) (see Algorithm 7.4). Here, dominance of solutions is defined with respect to the vector-valued fitness function

$$f_{\text{MOEA}}(x) = (|x|_1, f(x)).$$

A solution y dominates a solution z with respect to f_{MOEA} $(y \succeq z)$ iff $|y|_1 = |z|_1$ and $f(y) \geq f(z)$. Furthermore, y strictly dominates z $(y \succ z)$ iff $y \succeq z$ and $f(y) > f(z)$. The algorithm keeps at most one individual for each Hamming

Algorithm 7.4: MOEA; Assuming $B \leq B^*$. [33]

```
1 P ← an initial solution;
2 while stopping criterion not met do
3     Choose x ∈ P uniformly at random;
4     Obtain y from x by flipping each bit of x with probability 1/n;
5     if (B* ≥ |y|_1 ≥ B) ∧ (∄w ∈ P: w ≽_MOEA y) then
6         P ← (P ∪ {y}) \ {z ∈ P | y ≻_MOEA z};
```

Algorithm 7.5: MOEA-S; Assuming $B \leq B^*$. [33]

```
1 P ← an initial solution;
2 while stopping criterion not met do
3     Choose x ∈ P uniformly at random;
4     Obtain y from x by flipping bit one bit x_i, i ∈ {1,...,n} chosen u.a.r.;
5     if ∀z ∈ P: y ||_MOEA-S z then
6         P ← P ∪ {y}
7     if (B* ≥ |y|_1 ≥ B) ∧ (∃z ∈ P: y ≽_MOEA-S z) then
8         z ← y;
```

weight between B and B^*. Let $D = |B^* - B|$, then the size of the population P is at most $D+1$. The analysis shows that this population size slows down the re-optimization process for the OneMax problem. For the case where $B < B^*$ and $B > B^*$, the potential functions that Shi et al. [33] used for analyzing this algorithm were $M = \min_{x \in P} |x|_0$ and $M = \max_{x \in P} |x|_1$, respectively. The analysis is similar to their analysis of the $(1+1)$ EA, except that the drift on M is $\Omega(\frac{M}{n \cdot D})$. The D in the denominator comes from the fact that selecting the individual x with minimum $|x|_0$ for $B < B^*$ (or minimum $|x|_1$ for $B > B^*$) from the population happens at each iteration with probability at least $\frac{1}{D+1}$. Using multiplicative drift analysis, they obtained an upper bound of $O\left(nD \log\left(\frac{n-B}{n-B^*}\right)\right)$ for $B < B^*$ and an upper bound of $O(nD \log(\frac{B}{B^*}))$ for $B > B^*$.

The third algorithm investigated is a variant of MOEA named MOEA-S shown in Algorithm 7.5. In this algorithm only single-bit flips are allowed and a different definition for dominance is used. The new notion of dominance does not let the population size grow to a size larger than 2. If $B \leq B^*$, for two bit strings $y, z \in \{0,1\}^n$ we have:

- y dominates z, denoted by $y \succcurlyeq_{\text{MOEA-S}} z$ if *at most one* value among $|y|_1$ and $|z|_1$ equals B^* or $B^* - 1$, and $(|y|_1 > |z|_1) \vee (|y|_1 = |z|_1 \wedge f(y) \geq f(z))$;
- y dominates z, denoted by $y \succcurlyeq_{\text{MOEA-S}} z$ if both $|y|_1, |z|_1 \in \{B^*, B^* - 1\}$, and $|y|_1 = |z|_1 \wedge f(y) \geq f(z)$.

This implies that y and z are incomparable, denoted by $y \parallel_{\text{MOEA-S}} z$, iff $|y|_1 = B^*$ and $|z|_1 = B^* - 1$ or vice versa.

Algorithm 7.6: MOGA; Assuming $B \leq B^*$ [33], Concept from [5].

```
1  P ← {x}, x an initial solution;
2  while stopping criterion not met do
       /* Mutation phase.                                          */
3      Choose x ∈ P uniformly at random;
4      Choose ℓ according to Bin(n,p);
5      for i = 1 to λ do
6          x^(i) ← mutate_ℓ(x);
7      V = {x^(i) | x^(i) is valid};
8      if V ≠ ∅ then
9          Choose x' ∈ V uniformly at random;
10     else x' ← x;
       /* Crossover phase.                                         */
11     for i = 1 to λ do
12         y^(i) ← cross_c(x,x');
13     M = {y^(i) | y^(i) is ≽_MOEA-maximal ∧ |y^(i)|_1 = |x|_1 + 1};
14     if M = {y} then
15         y' ← y;
16     else y' ← x;
       /* Selection phase.                                         */
17     if (B* ≥ |y'|_1 ≥ B) ∧ (∄w ∈ P: w ≽_MOEA y') then
18         P ← (S ∪ {y'}) \ {z ∈ S | y' ≻_MOEA z};
```

For $B > B^*$, a similar definition of dominance is given by switching the dependency of $|y|_1 \geq |z|_1$ on the number of 1-bits to $|y|_1 \leq |z|_1$. The results of MOEA-S are obtained by observing that this algorithm behaves like RLS on ONEMAX. It was shown that the expected re-optimization time for ONEMAX with a dynamic uniform constraint is $O\left(n \log\left(\frac{n-B}{n-B^*}\right)\right)$ if $B < B^*$ and $O\left(n \log\left(\frac{B}{B^*}\right)\right)$ if $B > B^*$.

Shi et al. [33] have also introduced a multi-objective variant of the $(1+(\lambda+\lambda))$ GA [5] (MOGA, Algorithm 7.6), which is the fourth algorithm that they analyzed for ONEMAX with a dynamically changing uniform constraint. In this algorithm, the same notion of dominance as in MOEA is used, and the population size can grow to $D+1$. Having a solution x, at each iteration λ offspring are generated by the mutation operator, which flips $l = \mathrm{Bin}(n,p)$ random bits of x, where p is the mutation probability. The offspring that have a 0 flipped to 1 (or a 1 flipped to 0) are considered to be valid for $B^* > B$ (or for $B^* < B$, respectively). One of the valid offspring (if it exists), x', is then used in the crossover phase, in which it is recombined with the parent solution λ times. For a crossover probability c, the crossover operator creates a bit string $y = y_1 y_2 \cdots y_n$, where each bit $y_i, 1 \leq i \leq n$, is chosen to be x_i with probability c, and x'_i otherwise. The algorithm selects the best solution y with Hamming weight one larger than the Hamming weight of x. The solution y is added to the population if it meets the cardinality constraint and is not dominated by any other solution in the population.

It was proved that this algorithm solves the ONEMAX problem with a dynamically changing uniform constraint in expected time

$$O\left(\min\left\{\sqrt{n}D^{\frac{3}{2}}, D^2\sqrt{\frac{n}{n-B^*}}\right\}\right)$$

if $p = \frac{\lambda}{n}$, $c = \frac{1}{\lambda}$, $\lambda = \sqrt{n/(n-|x|_1)}$ for $B^* > B$, and in expected time

$$O\left(\min\left\{\sqrt{n}D^{\frac{3}{2}}, D^2\sqrt{\frac{n}{B^*}}\right\}\right)$$

if $\lambda = \sqrt{n/|x|_1}$ for $B^* < B$ [34]. The key argument behind these results is to show that there is a constant probability of producing a valid offspring in the mutation phase, and then show that there is a constant probability of generating a solution y in the crossover phase that is the same as x except for one bit, which is flipped from 0 to 1 for $B^* > B$ and from 1 to 0 for $B^* < B$.

7.3.2 Linear Pseudo-Boolean Functions Under Dynamic Uniform Constraints

The classical (1+1) EA and three multi-objective evolutionary algorithms were investigated in [33] for re-optimizing linear functions under dynamic uniform constraints. The general class of linear constraints on linear problems leads to exponential optimization times for many evolutionary algorithms [15, 37]. Shi et al. [33] considered the dynamic setting given in Section 7.2.2 and analyzed the expected re-optimization time of the evolutionary algorithms investigated. This section includes the results that they obtained, in addition to the proof ideas of their work.

The algorithms that were investigated in their work, have already been presented in Section 7.3.1 and the results are summarized in Table 7.2. The (1+1) EA (Algorithm 7.2) uses the following fitness function which was introduced by Friedrich et al. [15] (and is similar to the fitness function for ONEMAX in Section 7.3.1):

$$f_{(1+1)}(x) = f(x) - (nw_{\max} + 1) \cdot \max\{0, |x|_1 - B^*\}$$

Here, $w_{\max} = \max_{i=1}^{n} w_i$ denotes the maximum weight, and the large penalty for constraint violations guides the search towards the feasible region.

Table 7.2 Upper bounds on the expected re-optimization time of evolutionary algorithms on linear functions with a dynamic uniform constraint.

(1+1) **EA**	**MOEA**	**MOEA-S**	**MOGA**
$O(n^2 \log(B^* w_{\max}))$	$O(nD^2)$	$O(n \log D)$	$O(nD^2)$

Shi et al. [33] investigated this setting similarly to the analysis of OneMax under dynamic uniform constraints (Section 7.3.1). The main difference is that for a non-optimal solution with B^* 1-bits, an improvement is not possible by flipping a single bit. A 2-bit flip that flips a 1 and a 0 may be required, resulting in an expected re-optimization time of $O\big(n^2 \log(B^* w_{\max})\big)$.

The second investigated algorithm, MOEA, uses the fitness function f_{MOEA} and the notion of dominance defined in Section 7.3.1. Unlike the re-optimization time of this algorithm for the OneMax problem, whose upper bound is worse than the upper bound for the (1+1) EA; for general linear functions the upper bounds obtained for MOEA are smaller than the ones obtained for the (1+1) EA. The reason is that the algorithm is allowed to keep one individual for each Hamming weight between the two bounds in the population. This avoids the necessity for a 2-bit flip. To reach a solution that is optimal for cardinality $A+1$, the algorithm can use the individual that is optimal for cardinality A and flip the 0-bit whose weight is maximal. This happens in an expected number of at most $en(D+1)$ iterations, where $D = |B^* - B|$. As there are $D+1$ different cardinality values between the two bounds, the expected time to reach the optimal solution with cardinality B^* is $O(nD^2)$.

MOEA-S (Algorithm 7.5) was also analyzed for linear functions with a dynamically changing uniform constraint. It uses single bit flips and the population includes at most two solutions: one with Hamming weight at most $B^* - 1$ and one with Hamming weight B^*. With this setting, long waiting times for selecting a certain individual of the population are avoided. The algorithm starts with one solution in the population. It was shown that in time $O(n \log D)$ the population consists of one solution with Hamming weight $B^* - 1$ and one with Hamming weight B^*. Then the authors of [33] used a potential function to measure the difference of the current population from an optimal solution with Hamming weight B^*. The potential is given by the number of 0-bits in the two solutions that need to be set to 1 in order to obtain an optimal solution. Using multiplicative drift analysis with respect to the potential function, they proved that the expected re-optimization time of the algorithm is $O(n \log D)$.

The fourth algorithm that was analyzed in [33] was MOGA (Algorithm 7.6). The authors showed that, when an optimal solution of Hamming weight $A < B^*$ was chosen for reproduction, an optimal solution for Hamming weight $A+1$ is produced with probability $\Omega(n^{-1/2})$ in the next generation, if $p = \frac{\lambda}{n}$, $c = \frac{1}{\lambda}$ and $\lambda = \sqrt{n}$. Since there are $D+1$ different Hamming weights to consider, and each iteration of the algorithm constructs $O(\lambda) = O(\sqrt{n})$ solutions, the expected re-optimization time is upper bounded by $O(nD^2)$.

7.3.3 The Vertex Cover Problem

The common representation for solving the vertex cover problem by means of evolutionary algorithms is the node-based representation [11, 23, 29, 31]. A different representation, the edge-based representation, was suggested and analyzed in [18] for the static vertex cover problem. In this representation a search point is a bit string $x \in \{0,1\}^m$, where m denotes the number of edges in the given graph $G = (V, E)$. For a given search point x, $E(x) = \{e_i \in E \mid x_i = 1\}$ is the set of edges chosen. The cover set induced by x, denoted by $V_C(x)$, is the set of all nodes that are adjacent to at least one edge in $E(x)$.

Three variants of RLS and the (1+1) EA were investigated. These included one node-based approach and two edge-based approaches. The node-based approach and one of the edge-based approaches use a standard fitness function,

$$f(s) = |V_C(s)| + (|V|+1) \cdot |\{e \in E | e \cap V_C(s) = \emptyset\}|,$$

in which each uncovered edge obtains a large penalty of $|V|+1$. In [18], an exponential lower bound for finding a 2-approximate solution to the static vertex cover problem with these two approaches using the fitness function f was shown. Furthermore, considering the dynamic vertex cover problem, Pourhassan et al. [30] proved that there exist classes of instances of bipartite graphs where dynamic changes on the graph lead to a bad approximation behavior.

The third variant of an evolutionary algorithm that Jansen et al. [18] investigated, was an edge-based approach with a specific fitness function. The fitness function f_e has a very large penalty for common nodes among selected edges. It is defined as

$$\begin{aligned} f_e(s) = \quad & |V_C(s)| + (|V|+1) \cdot |\{e \in E \mid e \cap V_C(s) = \emptyset\}| \\ & + (|V|+1) \cdot (m+1) \cdot |\{(e,e') \in E(s) \times E(s) \mid e \neq e', e \cap e' \neq \emptyset\}|. \end{aligned}$$

This fitness function guides the search towards a matching, and afterwards to a maximal matching. In other words, whenever the algorithms find a matching, then they do not accept a solution that is not a matching, and whenever it finds a matching that induces a node set with k uncovered edges, then it does not accept a solution with $k' > k$ uncovered edges. It is well known that taking all the nodes belonging to the edges of a maximal matching for a given graph results in a 2-approximate solution to the vertex cover problem.

The variant of RLS and the (1+1) EA work with the edge-based representation and the fitness function f_e. Note that search points are bit strings of size m, and the probability of flipping each bit in the (1+1) EA is $1/m$. Jansen et al. [18] have proved that RLS and (1+1) EA with the edge-based approach find a maximal matching which induces a 2-approximate solution to the vertex cover problem in expected time $O(m \log m)$, where m is the number of edges.

The behavior of RLS and the (1+1) EA with this edge-based approach was investigated for the dynamic vertex cover problem (see Section 7.2.3) in [30]. It was proved in [30] that, starting from a 2-approximate solution for a current instance of the problem, in expected time $O(m)$ RLS finds a 2-approximate solution after a dynamic change of adding or deleting an edge. The authors of that paper investigated the situations for adding an edge and removing an edge separately. For adding an edge, they showed that the new edge either is already covered and the maximal matching stays a maximal matching, or is not covered by the current edge set and the current edge set is a matching that induces a solution with one (the new edge) uncovered edge. Since the number of uncovered edges does not grow in this approach and the algorithm selects the only uncovered edge with probability $1/m$, a maximal matching is found in expected m steps. This argument also holds for the (1+1) EA, but the probability of selecting the uncovered edge and having no other mutations with this algorithm is at least $1/(em)$. Therefore, the expected re-optimization time for the (1+1) EA after a dynamic addition is also $O(m)$.

When an edge is deleted from the graph, if it was selected in the solution, a number of edges can be uncovered in the new situation. All these uncovered edges were covered by the two nodes of the removed edge, and can be partitioned into two sets, U_1 and U_2, such that all edges in each set share a node. Therefore, if the algorithm selects one edge from each set (if any exist), the induced node set becomes a vertex cover again. It will again be a maximal matching and therefore a 2-approximate solution. On the other hand, no other one-bit flips in this situation can be accepted, because they either increase the number of uncovered edges, or make the solution become a non-matching. With RLS, in which only one-bit flips are possible, the probabilities of selecting one edge from U_1 and U_2 at each step are $\frac{|U_1|}{m}$ and $\frac{|U_2|}{m}$, respectively. Therefore, in expected time $O(m)$ one edge in each set is selected by the algorithm.

The analysis for the (1+1) EA dealing with a dynamic deletion is more complicated, because multiple-bit flips can happen. In other words, it is possible to deselect an edge and uncover some edges in the same step as when an edge in U_1 or U_2 is being selected to cover some other edges. An upper bound of $O(m \log m)$ was shown in [30] for the expected re-optimization time for the (1+1) EA after a dynamic deletion, which is the same as the expected time to find a 2-approximate solution with that algorithm, starting from an arbitrary solution.

7.3.4 Makespan Scheduling

Makespan scheduling is another problem which has been considered in a dynamic setting [28]. It is assumed that the processing time of job i, for $1 \le i \le n$, is $p_i \in [L, U]$, where L and U are lower and upper bounds, respectively, on the processing time of a job. The ratio between the upper bound and the lower bound is denoted by $R = U/L$. The runtime performance of the (1+1) EA and RLS was studied in terms of finding a solution with a good discrepancy and it was assumed that there is no stopping criterion for the algorithms except achieving such a solution. The discrepancy $d(x)$ of a solution x is defined as

$$d(x) = \left| \left(\sum_{i=1}^{n} p_i(1 - x_i) \right) - \left(\sum_{i=1}^{n} p_i x_i \right) \right|.$$

Note that a solution that has a smaller discrepancy also has a smaller makespan. Moreover, the proofs benefit from an important observation about the fuller machine (the machine which is loaded more heavily and determines the makespan). The observation is about the minimum number of jobs of the fuller machine in terms of U and L:

- every solution has at least $\lceil (P/2)/U \rceil \ge \lceil (nL/2)/U \rceil = \lceil (n/2)(L/U) \rceil = \lceil (n/2)R^{-1}) \rceil$ jobs on the fuller machine, where $P = \sum_{i=1}^{n} p_i$.

Two dynamic settings were studied for this problem. The first one is called the *adversary model.* In this model, a strong adversary is allowed to choose at most one arbitrary job i in each iteration and change its processing time to a new $p_i \in [L, U]$. It was proven that, independently of initial solution and the number of changes made by the adversary, RLS obtains a solution with discrepancy at most U in an expected time of $O(n \min\{\log n, \log R\})$. In the case of RLS, the number of jobs on the fuller machine increases only when the fuller machine is switched. Otherwise, the solution increases the makespan and will not be accepted by the algorithm. This fact is the basis of the proof. It was proved that if the fuller machine switches (either because of an RLS step, which moves a single job between machines, or because of a change that the adversary makes), then a solution with discrepancy at most U is found in steps before and after the switch.

The proof for the (1+1) EA is not as straightforward as for RLS, since the (1+1) EA may switch multiple jobs between the machines in one mutation step. However, it was shown that the number of incorrect jobs on the fuller machine, which should be placed on the other machine to decrease the makespan, has a drift towards zero. Using this argument, it was shown that the (1+1) EA will find a solution with discrepancy at most U in expected time $O(n^{3/2})$. Whether better upper bounds such as $O(n \log n)$ are possible is still an open question.

In the same dynamic setting, recovering a discrepancy of at most U was also studied for both RLS and the (1+1) EA. It was assumed that the algorithm had already achieved or had been initialized with a solution with a discrepancy of at most U and the processing time of a job changed afterwards. By applying the multiplicative drift theorem to the changes in the discrepancy and using the fact that the discrepancy will change by at most $U - L$, it was proven that the (1+1) EA and RLS recover a solution with a discrepancy of at most U in an expected time of $O(\min\{R, n\})$.

The makespan scheduling problem was also studied in another dynamic setting. In this model, which was called the *random model*, it is assumed that all job sizes are in $\{1, \ldots, n\}$. In each dynamic change, the adversary chooses one job i and its value changes from p_i to $p_i - 1$ or $p_i + 1$, each with a probability of 1/2. The only exceptions are $p_i = n$ and $p_i = 1$, for which the value changes to $p_i = n - 1$ and $p_i = 2$, respectively. Overall, this setting has less adversarial power than the *adversary model* owing to the randomness and the changes by only 1 involved.

Let the random variable X_i denote the random processing time of job i at any point in time. The following lemma proves that no large gap exists in the values of processing times which are randomly chosen for jobs.

Lemma 7.3.1 (Lemma 4 in [28]). *Let* $\phi(i) := |\{X_j \mid X_j = i \wedge j \in \{1, \ldots, n\}\}|$ *where* $i \in \{1, \ldots, n\}$, *be the frequency of jobs of size* i. *Let*

$$G := \max\{l \mid \exists i : \phi(i) = \phi(i+1) = \cdots = \phi(i+l) = 0\}$$

be the maximum gap size, i.e. the maximum number of intervals with zero frequency everywhere. Then, for some constant $c > 0$,

$$\Pr(G \geq l) \leq n 2^{-cl}.$$

This lemma states that, for any constant $c > 0$ and gap size $G \geq c' \log n$ with a sufficiently large c', there is no gap of size G with probability at least $1 - n \cdot n^{-c-1} = 1 - n^{-c}$. This probability was counted as a high probability in this study.

When the discrepancy is larger than G, it was proven that it decreases by at least one if two jobs swap between the fuller and the emptier machine. Furthermore, the maximum possible discrepancy for an initial solution is n^2 when all jobs have a processing time of n and are placed on one machine. Finally, it was proven that regardless of the initial solution, the (1+1) EA obtains with high probability a discrepancy of at most $O(\log n)$ after a one-time change in time $O(n^4 \log n)$.

The previous result considered the worst-case initial solution. However, it was proven that if the initial solution is generated randomly, then its expected discrepancy is $\Theta(n\sqrt{n})$ and it is $O(n\sqrt{n}\log n)$ with high probability. Thus, with a random initial solution, the (1+1) EA obtains a discrepancy of $O(\log n)$ after a one-time change in time $O(n^{3.5}\log^2 n)$ with high probability.

The two results on the (1+1) EA and in the random model are for a one-time change. In the extreme case, however, the processing time of a job may increase or decrease by one in each step, which makes it hard to obtain a discrepancy of $O(\log n)$, unlike the other results in this setting. Although, by using the results for the adversary model and assuming that $R = U = n$, it is possible to find a solution with a discrepancy of at most n. In the final theorem of this study, it was proven that independently of the initial solution and the number of changes, the (1+1) EA and RLS obtain a solution with discrepancy of at most n in expected time $O(n^{3/2})$ and $O(n\log n)$, respectively. In addition, it was shown that the expected ratio between the discrepancy and the makespan is $6/n$. This was done by considering that a solution of discrepancy at most n is obtained together with a lower bound on the makespan. The expected sum of all processing times is $n(n+1)/2$ and it is at least $n^2/3+n$ with a probability of $1-2^{-\Omega(n)}$. Hence, the expected makespan is at least $n^2/6+n/2$. Furthermore, if the sum of processing times is less than $n^2/3+n$, then the ratio is at least n/n since the processing times are at least one. Hence, if n is not too small, the ratio is bounded from above by

$$\frac{6}{n} - \frac{3}{n} + 2^{-\Omega(n)} \leq \frac{6}{n}.$$

7.3.5 The MAZE Problem

The dynamic pseudo-Boolean function MAZE, proposed in [22], consists of $n+1$ phases of $t_0 = kn^3\log n$ iterations. During the first phase, the function is equivalent to ONEMAX. In the next n phases, all bit strings except two, still have a value equivalent to ONEMAX. The two different bit strings, for each phase p, are 0^p1^{n-p} and $0^{p-1}1^{n-p+1}$, which have fitness values with an oscillating pattern: for two iterations out of three, these two bit strings are assigned values $n+2$ and $n+1$, respectively, and at the third iteration, this assignment is reversed. Note that during the last phase, MAZE behaves similarly to TRAP. The formal definition of MAZE follows:

$$\textsc{Maze}(x,t) = \begin{cases} n+2 & \text{if } t > (n+1)\cdot t_0 \ \wedge x = 0^n, \\ n+2 & \text{if } t > t_0 \ \wedge x = \mathrm{OPT}(t), \\ n+1 & \text{if } t > t_0 \ \wedge x = \mathrm{ALT}(t), \\ \textsc{OneMax}(x) & \text{otherwise,} \end{cases}$$

where

$$\mathrm{OPT}(t) = \begin{cases} \mathrm{OPT}_{\lfloor t/t_0 \rfloor} & \text{if } t \neq 0 \mod 3, \\ \mathrm{ALT}_{\lfloor t/t_0 \rfloor} & \text{otherwise,} \end{cases}$$

$$\mathrm{ALT}(t) = \begin{cases} \mathrm{ALT}_{\lfloor t/t_0 \rfloor} & \text{if } t \neq 0 \mod 3, \\ \mathrm{OPT}_{\lfloor t/t_0 \rfloor} & \text{otherwise,} \end{cases}$$

$$\mathrm{OPT}_p = 0^p 1^{n-p} \text{ for } p \leq n,$$

$$\mathrm{ALT}_p = 0^{p-1} 1^{n-p+1} \text{ for } p \leq n.$$

While it was shown in [22] that a $(1+1)$ EA loses track of the optimum for this problem and requires with high probability an exponential amount of time to find the optimum, Lissovoi and Witt [25] proved that the optimum of the MAZE function extended to finite alphabets, can be tracked by a $(\mu+1)$ EA when the parent population size μ is chosen appropriately and a genotype diversity mechanism is used.

In another publication [26], the behavior of parallel evolutionary algorithms was studied on the MAZE problem. In this analysis, it was proved that both the number of independent sub-populations (or islands), λ, and the length of the migration intervals, τ, influence the results. When τ is small, particularly for $\tau = 1$, migration occurs too often, and the algorithm behaves similarly to the $(1+\lambda)$ EA and fails to track MAZE efficiently for $\lambda = O(n^{1-\varepsilon})$, where ε is an arbitrary small positive constant. But, with a proper choice of τ, more precisely $\tau = t_0$, where t_0 is the number of iterations in each phase in the definition of the MAZE problem, and a choice of $\lambda = \Theta(\log n)$, the algorithm is able to track the optimum of MAZE efficiently.

The analysis of the $(\mu+1)$ EA and parallel evolutionary algorithms on the MAZE problem shows that both these algorithms have limitations on tracking the optimum. The $(\mu+1)$ EA not only exploits the small number of individuals among which the optimum oscillates, but also requires genotype diversity and a proper choice of μ. On the other hand, the positive results obtained for parallel evolutionary algorithms on the MAZE problem depend on a careful choice of migration frequency. But, on the plus side, with parallel evolutionary algorithms, the problem can be extended to a finite-alphabet version.

7.4 Analysis of Evolutionary Algorithms on Stochastic Problems

The performance of the $(1+1)$ EA in noisy environment was considered by Droste for the first time [10]. He proved that for prior noise which flips a randomly chosen bit with probability p, the (1+1) EA is able to deal with the noisy ONEMAX problem in polynomial time if and only if $p=O(\log(n)/n)$. Otherwise, the optimization time for $p=\omega(\log(n)/n)$ is super polynomial.

Recently, Gießen and Kötzing [17] considered the (1+1) EA together with population-based evolutionary algorithms in different noisy environments. They also reproved the results of Droste with new basic theorems and studied other types of prior and posterior noise against the (1+1) EA on the ONEMAX problem.

The new prior noise models that have been analyzed recently are

(a) noise which flips each bit independently with a probability of p;
(b) noise which assigns 0 to each bit independently with a probability of p.

The (1+1) EA is able to find the optimal solution of ONEMAX for both noise models in polynomial time only if $p=O(\log(n)/n^2)$, and the optimization time grows super polynomially if $p=\omega(\log(n)/n^2)$.

The study also covered the impact of two types of posterior noise on the performance of the (1+1) EA. It was stated that the ONEMAX problem with additive posterior noise from a random variable D with a variance of σ^2 is tractable in polynomial time with the (1+1) EA if $\sigma^2=O(\log(n)/n)$. In another case, if D is exponentially distributed with parameter 1, the (1+1) EA is able to find the optimum only in super polynomial time. Furthermore, an analysis of the behavior of the (1+1) EA on ONEMAX with posterior noise coming from a random variable with a Gaussian distribution $D\sim\mathcal{N}(0,\sigma^2)$ showed that it is able to deal with this noise in polynomial time if $\sigma^2\leq 1/(4\log n)$. But if $\sigma^2\geq c/(4\log n)$ for any $c>1$, then the (1+1) EA finds the optimal solution of the noisy ONEMAX problem in super polynomial time.

Following the studies in [10, 17], Dang-Nhu et al. recently proposed a more general approach to analyzing the behavior on noisy problems [4]. The generality of their approach comes from the elements that they considered to analyze the behavior of the algorithm. An apparent argument is that as the process gets closer to the optimal point, the effect of the noise will become more distracting in comparison to the contribution of the algorithm. The first part of their analysis was to calculate the time to achieve the equilibrium point where the contribution of the algorithm and the noise damage counterbalance in expectation. Afterwards, they considered the progress of the undisturbed algorithm, from the equilibrium point to the optimum, in the short period in which the noise does not impact on the results of comparisons. Using this new approach, they found more precise results for the performance of the (1+1) EA and $(1+\lambda)$ EA on noisy ONEMAX and LEADINGONES.

In addition to runtime analysis, there are other measurements for ranking the behavior of algorithms on noisy problems. The concept of regret, for example, considers the progress of algorithms in approximating noisy optimal solutions. Different definitions of regret and how they describe the performance of algorithms have been discussed in [27]. To measure the approximate solution achieved by an algorithm, Simple Regret (SR_n) uses the solutions of the nth iteration while Approximate Simple Regret (ASR_n) considers the closest solution to the optimum which has been produced up to the nth iteration. Hence, algorithms that do not use elitism may have better performance in terms of ASR in comparison to the SR measurement.

This section continues by considering studies of the influence of using a population in evolutionary algorithms for problems with noisy fitness functions. After this, some results on the performance of a population-based evolutionary algorithm against different types of noises are presented. Finally, we introduce an approach that modifies the algorithms by increasing the number of fitness evaluations to deal with noise.

7.4.1 Influence of the Population Size

In this section, we consider the impact of using populations in evolutionary algorithms for noisy problems. Gießen and Kötzing [17] studied this matter by considering the $(\mu+1)$ EA and $(1+\lambda)$ EA on the noisy OneMax problem. We assume a noisy function f, and let $(X_k)_{k\leq n}$ be a random variable taking on the value of the noisy function f for a solution with exactly k ones. It is also assumed that, $\forall j: 0<j<k<n$, we have $\Pr(X_k < X_{k+1}) \leq \Pr(X_j < X_{k+1})$. This means that when solutions are close, we are more likely to observe confusion caused by noise. The analysis of the performance of the $(\mu+1)$ EA on OneMax with prior noise is based on the following theorem.

Theorem 7.4.1 (Theorem 12 in [17]). *Let μ be given, and suppose that for each $k \leq n$, $X_k \in [k-1, k+1]$. For each $k<n$, let A_k be the event that when μ independent copies of X_k and one copy of X_{k+1} are drawn and then sorted, with ties being broken uniformly, the value of X_{k+1} does not come out least. If there is a positive constant $c \leq 1/15$ such that*

$$\forall k, n/4 < k < n: \Pr(A_k) \geq 1 - c\frac{n-k}{n\mu},$$

then the $(\mu+1)$ EA optimizes f in an expected number of $O(\mu n \log n)$ iterations.

The proof of this theorem is based on the definitions of two events to show that there is a positive drift on the number of ones in the best solution which has k ones in the current step of the algorithm. The first event, E_0, is the event that the new solution has at least one bit with a value one more than that in the current solution and it is not dominated by any other solutions, even when the noise is considered. The other event, E_1, is the situation where the new solution has fewer ones than the current best solution, the current best solution is unique, and it is ignored because of the noisy function. For this case to happen, the best solution with k ones must be evaluated to have $k-1$ ones, and all other solutions must have at least $k-2$ ones and be evaluated to have at least $k-1$ ones because of the noisy function. After this event, the number of ones in the best solution decreases at most by 2. Considering the probability of each event, the drift on the number of ones in the best solution is at least

$$\frac{n-k}{e\mu n} - 3\frac{c(n-k)}{n\mu} - \frac{2c}{n\mu}.$$

Finally, since $c \leq 1/15$, and using multiplicative drift analysis, the theorem is proven.

The above theorem was used to prove a corollary about the performance of the $(\mu+1)$ EA on noisy OneMax the prior noise, i.e. noise which flips a bit uniformly at random with probability p. It was proven that if $\mu \geq 12\log(15n)/p$, then the $(\mu+1)$ EA finds the optimum of OneMax in an expected number of $O(\mu n \log n)$ iterations. To be more specific, $\mu = 24\log(15n)$ is adequate to achieve such an expected time for $p = 1/2$.

Gießen and Kötzing also considered the performance of the $(1+\lambda)$ EA as another population-based evolutionary algorithm on the noisy OneMax problem. In the $(1+\lambda)$ EA, there exists an offspring population. The algorithm produces an offspring with size λ by mutating the current best solution λ times. Then, it chooses the best solution among the offspring and the parent as the next best solution. Gießen and Kötzing proved an important theorem to achieve their results on this topic. The theorem is as follows.

Theorem 7.4.2 (Theorem 14 in [17]). *Let $\lambda \geq 24\log n$ and, for each $k < n$, let Y_k denote the maximum over λ observed values of X_k (belonging to inferior individuals) and let Z_k denote the maximum over at least $\lambda/6$ observed values of X_k (belonging to better individuals). Suppose there is a $q < 1$ such that*

$$\forall k < n : \Pr(Y_k < X_{k+1}) \geq q \tag{7.4.1}$$

and

$$\forall k < n : \Pr(Y_{k-1} < Z_k) \geq 1 - \frac{q}{5}\frac{l\lambda}{en+l\lambda}. \tag{7.4.2}$$

Then the $(1+\lambda)$ EA optimizes f in $O((\frac{n\log n}{\lambda} + n)/q)$ iterations and needs $O((n\log n + n\lambda)/q)$ fitness evaluations.

In this theorem, l is the number of zeros in the current best solution. To prove the theorem, similar to Theorem 7.4.1, it was shown that the drift on the number of ones is positive and equal to $(q - \frac{4q}{5})\frac{l\lambda}{en+l\lambda}$. Furthermore, this theorem gives the sufficient conditions 7.4.1 and 7.4.2) on the noise to demonstrate that it is tractable with the $(1+\lambda)$ EA in a guaranteed expected number of iterations.

As a corollary, for prior noise which flips a bit uniformly at random with probability p, it was proven that the $(1+\lambda)$ EA with $\lambda \geq \max\{12/p, 24\} n \log n$ optimizes OneMax in expected time $O((\frac{n^2 \log n}{\lambda} + n^2)/p)$. Let $q = p/n$. To show that Equation 7.4.1 holds, it is enough to consider the event that the solution with k ones (the current best solution) is evaluated to have more ones. The complement of this event is a set of events where either there is no noisy bit flip or the noise flips a zero bit, which leads to

$$\Pr(Y_k < X_{k-1}) \geq 1 - \left(1 - p + \frac{pk}{n}\right) \geq \frac{p}{n} = q.$$

With a similar consideration about the probability of improving at least one of $\lambda/6$ of the solutions, it is observed that Equation 7.4.2 also holds and the corollary is correct.

The other corollary of Theorem 7.4.2 is about non-positive additive posterior noise D that is evaluated to greater than -1 with a non-zero probability p. It was proven that with this noise, for $\lambda \geq \max\{10e, \frac{-6\log(n/p)}{\log(1-p)}\}$, the $(1+\lambda)$ EA optimizes OneMax in time $O((n \log n + n\lambda)/p)$. The proof of this corollary is a bit more tricky. Since $D \leq 0$, we have $Y_k \leq k$ and

$$\Pr(Y_k \geq X_{k+1}) \leq \Pr(k \geq X_{k+1}) = \Pr(D \leq -1) = 1 - p.$$

This means that $\Pr(Y_k < X_{k+1}) \geq p$ which fulfills the first condition of Theorem 7.4.2. To consider the second condition, a similar complement technique is used, i.e., calculating the probability of $Y_{k-1} \geq Z_k$ which is the complement of $Y_{k-1} < Z_k$. To satisfy $Y_{k-1} \geq Z_k$, all of the $\lambda/6$ solutions which have k ones must be affected by noise with a value less than -1; thus $\lambda \geq \frac{-6\log(n/p)}{\log(1-p)}$ leads to the conclusion that $\Pr(Y_{k-1} \geq Z_k) \leq (1-p)^{\lambda/6} \leq p/n$. Finally, since $\lambda \geq 10e$, the second condition of $\Pr(Y_{k-1} < Z_k)$ has been proven to be satisfied; therefore, Theorem 7.4.2 holds.

7.4.2 Influence of Different Noise Distributions on the Performance of Population-Based Evolutionary Algorithms

Friedrich et al. [12, 14] considered the $(\mu+1)$ EA and additive posterior noise with different distributions. They introduced the concept of "graceful scaling" to determine the performance of an algorithm against noise. An algorithm scales gracefully with noise if there exists a parameter setting for the algorithm such that it finds the optimum of the real fitness function when evaluating the noisy one, in polynomial time.

They also introduced a sufficient condition for the $(\mu+1)$ EA not to be able to deal with noise and to be unable to find the real optimum through the noisy function. The condition is: if there are $\mu+1$ different values $\{d_1,\dots,d_{\mu+1}\}$ from the random variable D, Y is the minimum of $\{d_1,\dots,d_\mu\}$ and

$$\Pr(Y > d_{\mu+1}+n) \geq \frac{1}{2(\mu+1)},$$

then the $(\mu+1)$ EA, with a polynomially bounded μ, will not evaluate the optimum with high probability.

This theorem was used to analyze the performance of the $(\mu+1)$ EA with $\mu = \omega(1)$ and bounded from above by a polynomial, on noisy OneMax problems with different distributions. In this study it was proven that if the noise comes from a Gaussian distribution with $\sigma^2 \geq (na)^2$, for some $a = \omega(1)$, the $(\mu+1)$ EA will not find the optimum in polynomial time with high probability.

Furthermore, other noise distributions have were studied in [12]. The authors analyzed a random variable D with a distribution that decays exponentially. Here, the probability density function of D is as follows:

$$\begin{aligned} F(t) &:= \Pr(D < t) = \frac{1}{2}e^{ct} && \text{if } t \leq 0 \text{ and} \\ F(t) &:= 1 - \frac{1}{2}e^{-ct} && \text{if } t \geq 0, \end{aligned}$$

for some constant c and variable t. The probability mass function p of D is obtained by taking the derivative of F:

$$\begin{aligned} p(t) &= F'(t) = \frac{c}{2}e^{ct} && \text{if } t \leq 0 \text{ and} \\ p(t) &= \frac{c}{2}e^{-ct} && \text{if } t \geq 0. \end{aligned}$$

Note that this is a symmetric variant of an exponential distribution. It is observed that p is symmetric around 0 which implies that the distribution of D has mean 0.

The variance of D is calculated as follows:

$$\begin{aligned} \mathrm{Var}(D) &= \int_{-\infty}^{+\infty} t^2 p(t)\mathrm{d}t = \frac{c}{2}\left(\int_{-\infty}^{0} t^2 e^{ct}\mathrm{d}t + \int_{0}^{\infty} t^2 e^{-ct}\mathrm{d}t\right) \\ &= c\int_{-\infty}^{0} t^2 e^{ct}\mathrm{d}t = \left[\frac{(2-2ct+t^2c^2)e^{ct}}{c^3}\right]_{-\infty}^{0} \\ &= \frac{2}{c^2} =: \sigma^2. \end{aligned}$$

Now $F(t)$ can be rewritten in terms of σ:

$$F(t) := \frac{1}{2}e^{\sqrt{2}\frac{t}{\sigma}} \qquad \text{if } t \leq 0 \text{ and}$$
$$F(t) := 1 - \frac{1}{2}e^{-\sqrt{2}\frac{t}{\sigma}} \qquad \text{if } t \geq 0.$$

In this setting, it was proven that if the variance is large ($\sigma^2 = \omega(n^2)$), then the $(\mu+1)$ EA will not find the optimum of the noisy ONEMAX problem. The proof applies the condition mentioned above, and bounds $\Pr(Y > d_{\mu+1} + n)$. The idea is to consider a subset of events in which $Y > d_{\mu+1} + n$ holds. To this end, points t_0 and t_1 are defined such that $t_0 < t_1$ and we have $\Pr(D < t_0)$ and $\Pr(D < t_1)$ dependent on μ in such a way that $\Pr(D < t_0) < \Pr(D < t_1) < 1/2$. This definition leads to the following events:

- A: The event that $D < t_0 - n$ and $t_0 < Y$.
- B: The event that $t_0 - n < D < t_1 - n$ and $t_1 < Y$.

The fact that $\sigma^2 \geq (na)^2$ for $a = \omega(1)$ helps to find the lower bounds for the probabilities and results in $\Pr(Y > d_{\mu+1}) \geq \frac{1}{2(\mu+1)}$. Hence, the $(\mu+1)$ EA will not find the optimum of noisy ONEMAX if the noise comes from an exponential distribution as defined.

The other noise distributions which were studied by Friedrich et al. [12] were *Truncated Distributions*. It was proven that the $(\mu+1)$ EA scales gracefully with this kind of noise, which are generalizations of the uniform distribution.

Definition 7.4.3 (Definition 7 in [12]). Let D be a random variable. If there are $k, q \in \mathbb{R}$ such that $\Pr(D > k) = 0 \wedge \Pr(D \in (k-1, k]) \geq q$, then D is called upper q-truncated. Analogously, D is called lower q-truncated if there is are $k, q \in \mathbb{R}$ with $\Pr(D < k) = 0 \wedge \Pr(D \in [k, k+1)) \geq q$.

Using this definition, it was proven that the $(\mu+1)$ EA obtains the optimum of noisy ONEMAX with a lower $2\log(n\mu)/\mu$-truncated noise distribution in expected $O(\mu n \log n)$ iterations. The proof uses multiplicative drift and benefits from the fact that the best solution is never removed in the first $O(\mu n \log n)$ iterations if any other point is evaluated in the minimal bracket $[k, k+1)$. The first corollary of this result is that for an arbitrary lower q-truncated noise, the $(\mu+1)$ EA with $\mu \geq 3^{-1}\log(nq^{-1})$ evaluates the optimum of noisy ONEMAX after expected $O(\mu n \log n)$ iterations.

Finally the last corollary in this study considered a uniform distribution on $[-r, r]$, which is lower $1/2r$-truncated, as the noise function. In this manner, by using the previous results, it was proven that the $(\mu+1)$ EA scales gracefully on ONEMAX with additive posterior noise from a uniform distribution on $[-r, r]$.

7.4.3 Resampling Approach for Noisy Discrete Problems

In the previous sections, we have considered the behavior of evolutionary algorithms for noisy problems. This section presents another approach to dealing with such problems. Here, a modified version of an algorithm, which has a known performance on the noise-free case, is investigated. Akimoto et al. studied resampling methods to modify iterative algorithms and found upper bounds on its performance according to the proven performance of the known algorithm [1]. The framework that they presented suits EAs perfectly. In a resampling method, the evaluation of the noisy fitness function for each solution is repeated k times. The algorithm then takes the average of the k noisy values as the fitness value of the solution. Let *Opt* and *k-Opt* denote the original algorithm and the resampling modified version, respectively. The parameter k can be fixed, or be adapted during the optimization process. In [1], discrete optimization problems were classified into two different categories: Either there is a known algorithm available that finds the optimal solution in expected $r(\delta)$ fitness evaluations with probability $1-\delta$, or no such algorithm is known. In the first case, k is chosen according to $r(\delta)$, and in the second case, its value is set adaptively in each iteration.

We assume that additive Gaussian noise with variance σ^2 is applied to the fitness function. In the pre-known runtime case, it was proven that if we fix

$$k_g = \max\left(1, \left\lceil 32\sigma^2 \left[\ln(2) - \ln\left(1-(1-\delta)^{1/r(\delta)}\right)\right]\right\rceil\right),$$

k_g-*Opt* finds the optimum of the noisy function with probability at least$(1-\delta)^2$ and the expected running time is

$$O\left(r(\delta)\max\left(1,\sigma^2\ln\left(\frac{r(\delta)}{\delta}\right)\right)\right).$$

Let p denote the probability of the ratio of the noisy and real fitness values of point x being at least $1/4$. The proof first determines $p \leq 2\exp\left(-\frac{k/4^2}{2\sigma^2}\right)$. This leads to the probability of situations in which the noisy fitness value is sufficiently close to the real one.

On the other hand, suppose there is no algorithm to solve the noise-free problem. However, suppose that there is a known algorithm *Opt′* which satisfies the criterion *Opt* with probability at least $1-\delta$ after n total fitness evaluations in iteration n.

Let $\beta > 1$, and let

$$k_m = \left\lceil 32\sigma^2\left(\frac{2(n+1)\ln(n+1)^\beta}{\delta}\left(\sum_{i=2}^{\infty}\frac{1}{i\ln(i)^\beta}\right)\right)\right\rceil$$

be the number of resamplings of k-*Opt'* in iteration m. It was proven that k-*Opt'* satisfies Q for the noisy problem with probability at least $(1-\delta)^2$ after n iterations and the total number of fitness evaluations is

$$\sum_{m=1}^{n} k_m = O(n \ln n).$$

A more general scenario (called the heavy tail scenario) was also considered. Here, there is no assumption about the noise distribution, except that the variance σ^2 is finite and the distribution has mean zero. It was proven that if we fix

$$k_h = \max(1, \lceil 16\sigma^2/(1-(1-\delta)^{r(\delta)}) \rceil),$$

k_h-*Opt* solves the noisy problem with probability at least $(1-\delta)^2$ and the expected runtime is

$$O\left(r(\delta)\max\left(1, \sigma^2 \frac{r(\delta)}{\delta}\right)\right).$$

When no algorithm is known to find the optimum, then by using the definition of Q, k-*Opt'* with

$$k_m = \left\lceil \frac{16\sigma^2(n+1)\ln(n+1)^\beta}{\delta} \sum_{i=2}^{\infty} \frac{1}{i \ln(i)^\beta} \right\rceil,$$

where $\beta > 1$, satisfies Q for any heavy tail noisy function with probability at least $(1-\delta)^2$. The total number of fitness evaluation up to iteration n is

$$\sum_{m=1}^{n} k_m = O(n^2 \ln(n)^\beta).$$

Other studies have also considered the size of the resampling. Qian et al. [32] showed that increasing the resampling size does not always decrease the runtime of the algorithm. These authors proved that it takes expected exponential time for the (1+1) EA to solve the noisy LEADINGONES problem with $p = 1$, while using the resampling method with $k = 4n^4 \log n/15$ reduces it to an expected polynomial time. Furthermore, they proved that if the resampling method is applied with $k = n^5$, then the expected running time grows exponentially again.

Furthermore, the authors also investigated the ONEMAX problem with segmented noise, which applies different types of noise according to the size of the solution. They proved that the expected running time to solve the noisy ONEMAX problem for the (1+1) EA with a fixed size resampling and the $(\mu+1)$ EA with $\mu \in poly(n)$ is exponential. However, using the (1+1) EA with an adaptive resampling method solves the ONEMAX problem with segmented noise in expected polynomial running time. For two individuals, this adaptive

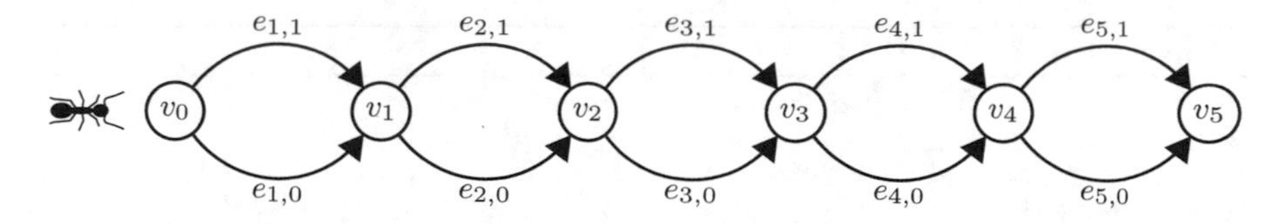

Fig. 7.1 Construction graph for pseudo-Boolean optimization with $n = 5$ bits.

resampling method uses the values of the noisy fitness functions to do the comparison only if the difference between the fitness values is within $[3n, n^4]$. Otherwise, it uses the average of n^5 fitness evaluations for each individual to compare them.

7.5 Ant Colony Optimization

After having investigated evolutionary algorithms for dynamic and stochastic problems, we now give a summary of the results obtained in the context of ant colony optimization. ACO algorithms construct solutions for a given optimization problem by performing random walks on a so-called construction graph. The construction graph frequently used in pseudo-Boolean optimization is shown in Figure 7.1. This random walk is influenced by so-called pheromone values on the edges of the graph. At each time step, the pheromone values induce a probability distribution on the underlying search space which is used to sample new solutions for the given problem. Pheromone values are adapted over time such that good components of solutions obtained during the optimization process are reinforced. The idea is that this reinforcement then leads to better solutions during the optimization process. An algorithm which is frequently studied in theoretical investigations of pseudo-Boolean maximization is MMAS (see Algorithm 7.7). This is a simplification of the Max-Min Ant System introduced in [35]. The algorithm, which is given in Algorithm 7.7, only uses one ant in each iteration. However, variants of MMAS called λ-MMAS, where λ ants are used in each iteration, have also been studied in the literature. Pheromone values are chosen within the interval $[\tau_{\min}, \tau_{\max}]$ where $\tau_{\min}$ and $\tau_{\max}$ are lower and upper bounds used in MMAS. Furthermore, the update strength ρ plays an important role in the runtime analysis of ACO algorithms. For MMAS, a large update strength such as $\rho = 1$ often makes the MMAS algorithms considered similar to simple evolutionary algorithms such as $(1+1)$ EA. The considered algorithms are usually analyzed with respect to the number of solutions until a given goal has been achieved. As in the case of runtime analysis of evolutionary algorithms, one is often interested in the expected number of solutions to reach the desired goal.

Algorithm 7.7: MMAS

```
1 Set τ_(u,v) = 1/2 for all (u,v) ∈ E.;
2 Construct a solution x*.;
3 Update pheromones w.r.t. x*.;
4 for forever do
5     Construct a solution x;
6     if f(x) ≥ f(x*) then
7         x* := x;
8         Update pheromones w.r.t. x*
```

7.5.1 Dynamic Problems

Kötzing and Molter [22] compared the behavior of the $(1+1)$ EA and MMAS on the MAZE problem. The MAZE problem has an oscillating behavior of different parts of the function and these authors have shown that MMAS is able to track this oscillating behavior if ρ is chosen appropriately, i.e. $\rho = \theta(1/n)$, whereas $(1+1)$ EA loses track of the optimum with probability close to 1.

In the case of dynamic combinatorial optimization problems, dynamic single-source shortest path problems have been investigated in [24]. Given a destination node $t \in V$, the goal is to compute for any node $v \in V \setminus t$ a shortest path from v to t. The set of these single-source shortest paths can be represented as a tree with root t, and the path from v to t in that tree gives a shortest path from v to t. The authors investigated different types of dynamic change for variants of MMAS. They first investigated MMAS and showed that this algorithm can effectively deal with one time changes, and built on investigations in [36] for the static case. They showed that the algorithm is able to recompute single-source shortest paths in an expected number of $O(\ell^*/\tau_{\min} + \ell \ln(\tau_{\max}/\tau_{\min})/\rho)$ iterations after a one change has happened. The parameter ℓ denotes the maximum number of arcs in any shortest path to node t in the new graph and $\ell^* = \min\{\ell, \log n\}$. The result shows that MMAS is able to track dynamic changes if they are not too frequent. Furthermore, these authors presented a lower bound of $\Omega(\ell/\tau_{\min})$ in the case where $\rho = 1$ holds. Afterwards, periodic local and global changes were investigated. In the case of the local changes investigated, λ-MMAS with a small λ is able to track periodic local changes for a specific setting. For global changes, a setting with oscillation between two simple weight functions was introduced, where an exponential number of ants would be required to make sure that an optimal solution is sampled with constant probability in each iteration.

7.5.2 Stochastic Problems

In stochastic environments, ACO algorithms have been analyzed for example benchmark functions and the stochastic shortest paths problem.

Thyssen and Sudholt [3] started the runtime analysis of ACO algorithms in stochastic environments. They investigated the single-destination shortest path (SDSP) problem where edge weights are subject to noise. For each edge e, the noise model returns a weight of $(1+\eta(e,p,t))\cdot w(e)$ instead of the exact real weight $w(e)$. This implies that the weight $w(e)$ of each edge e is increased according to the noise model. They considered a variant of MMAS for the shortest path problem introduced in [36]. They started by characterizing a noise model for which the algorithm can discover shortest paths efficiently. In the general setting, they examined algorithms in terms of approximations. The results depend on how much non-optimal paths differ at least from optimal ones. More precisely, they showed that if for each vertex $v \in V$ and some $\alpha > 1$ it holds that every non-optimal path has length at least $(1+\alpha \cdot E(\eta(opt_v)))\cdot opt_v$, then the algorithm finds an α-approximation in time proportional to $\alpha/(\alpha-1)$ and other standard ACO parameters such as pheromone bounds and pheromone update strengths. Here, opt_v denotes the value of a shortest path from v to the destination t and $E(\eta(opt_v))$ denotes the expected random noise on all edges of this path. Furthermore, for independent gamma distributions along the edges, they have showed that the algorithm may need exponential time to find a good approximation. Doerr et al. [7] extended these investigations of the stochastic SDSP problem. They considered a slight variation of MMAS for the stochastic SDSP which always reevaluates the best so-far solution when a new solution for comparison is obtained. This allows the MMAS version to easily obtain shortest paths in the stochastic setting.

Friedrich et al. [13] considered MMAS with a fitness-proportional pheromone update rule on linear pseudo-Boolean functions. They showed that the algorithm scales gracefully with noise, i.e., the runtime depends only linearly on the variance of the Gaussian noise. In contrast to this many of the noise settings considered are not solvable by simple evolutionary algorithms such as the $(1+1)$ EA [10]. This points out a clear benefit of using ant colony optimization for stochastic problems.

7.6 Conclusions

Evolutionary algorithms have been extensively used to deal with dynamic and stochastic problems. We have given an overview of recent results regarding the theoretical foundations of evolutionary algorithms for dynamic and stochastic problems in the context of rigorous runtime analysis. Various results for dynamic problems in the context of combinatorial optimization for problems such as makespan scheduling and minimum vertex cover have been summarized, and the benefits of different approaches to dealing with such dynamic problems have been pointed out. In the case of stochastic problems, the impact of the amount of noise and how population-based approaches are beneficial for coping with noisy environments have been summarized.

While all these studies have greatly contributed to the understanding of the basic working principles of evolutionary algorithms and ant colony optimization in dynamic and stochastic environments, analyzing the behavior of these algorithms on complex problems remains highly open. Furthermore, uncertainties often change over time and are therefore dynamic. Therefore, it would be very interesting to analyze the behavior of evolutionary algorithms for problems where uncertainties change over time. For future research, it would also be interesting to examine environments that are both dynamic and stochastic, as many real-world problems have both properties at the same time.

References

[1] Akimoto, Y., Morales, S.A., Teytaud, O.: Analysis of runtime of optimization algorithms for noisy functions over discrete codomains. Theor. Comput. Sci. **605**, 42–50 (2015)

[2] Cathabard, S., Lehre, P.K., Yao, X.: Non-uniform mutation rates for problems with unknown solution lengths. In: H. Beyer, W.B. Langdon (eds.) Foundations of Genetic Algorithms, 11th International Workshop, FOGA 2011, Schwarzenberg, Austria, January 5-8, 2011, Proceedings, pp. 173–180. ACM (2011)

[3] Coello, C.A.C., Cutello, V., Deb, K., Forrest, S., Nicosia, G., Pavone, M. (eds.): Parallel Problem Solving from Nature - PPSN XII - 12th International Conference, Taormina, Italy, September 1-5, 2012, Proceedings, Part I, *Lecture Notes in Computer Science*, vol. 7491. Springer (2012)

[4] Dang-Nhu, R., Dardinier, T., Doerr, B., Izacard, G., Nogneng, D.: A new analysis method for evolutionary optimization of dynamic and noisy objective functions. In: H.E. Aguirre, K. Takadama (eds.) Proceedings of the Genetic and Evolutionary Computation Conference, GECCO 2018, Kyoto, Japan, July 15-19, 2018, pp. 1467–1474. ACM (2018)

[5] Doerr, B., Doerr, C., Ebel, F.: From black-box complexity to designing new genetic algorithms. Theor. Comput. Sci. **567**, 87–104 (2015)

[6] Doerr, B., Doerr, C., Kötzing, T.: Solving problems with unknown solution length at (almost) no extra cost. In: S. Silva, A.I. Esparcia-Alcázar (eds.) Proceedings of the Genetic and Evolutionary Computation Conference, GECCO 2015, Madrid, Spain, July 11-15, 2015, pp. 831–838. ACM (2015)

[7] Doerr, B., Hota, A., Kötzing, T.: Ants easily solve stochastic shortest path problems. In: T. Soule, J.H. Moore (eds.) Genetic and Evolutionary Computation Conference, GECCO '12, Philadelphia, PA, USA, July 7-11, 2012, pp. 17–24. ACM (2012)
[8] Doerr, B., Johannsen, D., Winzen, C.: Multiplicative drift analysis. Algorithmica **64**(4), 673–697 (2012)
[9] Droste, S.: Analysis of the (1+1) EA for a dynamically changing OneMax-variant. In: Evolutionary Computation, 2002. CEC '02. Proceedings of the 2002 Congress on, vol. 1, pp. 55–60 (2002)
[10] Droste, S.: Analysis of the (1+1) EA for a noisy onemax. In: K. Deb, R. Poli, W. Banzhaf, H. Beyer, E.K. Burke, P.J. Darwen, D. Dasgupta, D. Floreano, J.A. Foster, M. Harman, O. Holland, P.L. Lanzi, L. Spector, A. Tettamanzi, D. Thierens, A.M. Tyrrell (eds.) Genetic and Evolutionary Computation - GECCO 2004, Genetic and Evolutionary Computation Conference, Seattle, WA, USA, June 26-30, 2004, Proceedings, Part I, *Lecture Notes in Computer Science*, vol. 3102, pp. 1088–1099. Springer (2004)
[11] Friedrich, T., He, J., Hebbinghaus, N., Neumann, F., Witt, C.: Approximating covering problems by randomized search heuristics using multi-objective models. Evolutionary Computation **18**(4), 617–633 (2010)
[12] Friedrich, T., Kötzing, T., Krejca, M.S., Sutton, A.M.: Graceful scaling on uniform versus steep-tailed noise. In: J. Handl, E. Hart, P.R. Lewis, M. López-Ibáñez, G. Ochoa, B. Paechter (eds.) Parallel Problem Solving from Nature - PPSN XIV - 14th International Conference, Edinburgh, UK, September 17-21, 2016, Proceedings, *Lecture Notes in Computer Science*, vol. 9921, pp. 761–770. Springer (2016)
[13] Friedrich, T., Kötzing, T., Krejca, M.S., Sutton, A.M.: Robustness of ant colony optimization to noise. Evolutionary Computation **24**(2), 237–254 (2016)
[14] Friedrich, T., Kötzing, T., Krejca, M.S., Sutton, A.M.: The compact genetic algorithm is efficient under extreme gaussian noise. IEEE Trans. Evolutionary Computation **21**(3), 477–490 (2017)
[15] Friedrich, T., Kötzing, T., Lagodzinski, G., Neumann, F., Schirneck, M.: Analysis of the (1+1) EA on subclasses of linear functions under uniform and linear constraints. In: C. Igel, D. Sudholt, C. Witt (eds.) Proceedings of the 14th ACM/SIGEVO Conference on Foundations of Genetic Algorithms, FOGA 2017, Copenhagen, Denmark, January 12-15, 2017, pp. 45–54. ACM (2017)
[16] Friedrich, T., Kötzing, T., Sutton, A.M.: On the robustness of evolving populations. In: J. Handl, E. Hart, P.R. Lewis, M. López-Ibáñez, G. Ochoa, B. Paechter (eds.) Parallel Problem Solving from Nature - PPSN XIV - 14th International Conference, Edinburgh, UK, September 17-21, 2016, Proceedings, *Lecture Notes in Computer Science*, vol. 9921, pp. 771–781. Springer (2016)

[17] Gießen, C., Kötzing, T.: Robustness of populations in stochastic environments. Algorithmica **75**(3), 462–489 (2016)
[18] Jansen, T., Oliveto, P.S., Zarges, C.: Approximating vertex cover using edge-based representations. In: F. Neumann, K.A. De Jong (eds.) Foundations of Genetic Algorithms XII, FOGA '13, Adelaide, SA, Australia, January 16-20, 2013, pp. 87–96. ACM (2013)
[19] Jansen, T., Zarges, C.: Evolutionary algorithms and artificial immune systems on a bi-stable dynamic optimisation problem. In: D.V. Arnold (ed.) Genetic and Evolutionary Computation Conference, GECCO '14, Vancouver, BC, Canada, July 12-16, 2014, pp. 975–982. ACM (2014)
[20] Kellerer, H., Pferschy, U., Pisinger, D.: Knapsack problems. Springer (2004)
[21] Kötzing, T., Lissovoi, A., Witt, C.: (1+1) EA on generalized dynamic OneMax. In: J. He, T. Jansen, G. Ochoa, C. Zarges (eds.) Proceedings of the 2015 ACM Conference on Foundations of Genetic Algorithms XIII, Aberystwyth, United Kingdom, January 17 - 20, 2015, pp. 40–51. ACM (2015)
[22] Kötzing, T., Molter, H.: ACO beats EA on a dynamic pseudo-boolean function. In: C.A.C. Coello, V. Cutello, K. Deb, S. Forrest, G. Nicosia, M. Pavone (eds.) Parallel Problem Solving from Nature - PPSN XII - 12th International Conference, Taormina, Italy, September 1-5, 2012, Proceedings, Part I, *Lecture Notes in Computer Science*, vol. 7491, pp. 113–122. Springer (2012)
[23] Kratsch, S., Neumann, F.: Fixed-parameter evolutionary algorithms and the vertex cover problem. Algorithmica **65**(4), 754–771 (2013)
[24] Lissovoi, A., Witt, C.: Runtime analysis of ant colony optimization on dynamic shortest path problems. Theor. Comput. Sci. **561**, 73–85 (2015)
[25] Lissovoi, A., Witt, C.: MMAS versus population-based EA on a family of dynamic fitness functions. Algorithmica **75**(3), 554–576 (2016)
[26] Lissovoi, A., Witt, C.: A runtime analysis of parallel evolutionary algorithms in dynamic optimization. Algorithmica **78**(2), 641–659 (2017)
[27] Morales, S.A., Cauwet, M., Teytaud, O.: Analysis of different types of regret in continuous noisy optimization. In: T. Friedrich, F. Neumann, A.M. Sutton (eds.) Proceedings of the 2016 on Genetic and Evolutionary Computation Conference, Denver, CO, USA, July 20 - 24, 2016, pp. 205–212. ACM (2016)
[28] Neumann, F., Witt, C.: On the runtime of randomized local search and simple evolutionary algorithms for dynamic makespan scheduling. In: Q. Yang, M. Wooldridge (eds.) Proceedings of the Twenty-Fourth International Joint Conference on Artificial Intelligence, IJCAI 2015, Buenos Aires, Argentina, July 25-31, 2015, pp. 3742–3748. AAAI Press (2015)
[29] Oliveto, P.S., He, J., Yao, X.: Analysis of the (1+1)-EA for finding approximate solutions to vertex cover problems. IEEE Trans. Evolutionary Computation **13**(5), 1006–1029 (2009)

[30] Pourhassan, M., Gao, W., Neumann, F.: Maintaining 2-approximations for the dynamic vertex cover problem using evolutionary algorithms. In: S. Silva, A.I. Esparcia-Alcázar (eds.) Proceedings of the Genetic and Evolutionary Computation Conference, GECCO 2015, Madrid, Spain, July 11-15, 2015, pp. 903–910. ACM (2015)

[31] Pourhassan, M., Shi, F., Neumann, F.: Parameterized analysis of multi-objective evolutionary algorithms and the weighted vertex cover problem. In: J. Handl, E. Hart, P.R. Lewis, M. López-Ibáñez, G. Ochoa, B. Paechter (eds.) Parallel Problem Solving from Nature - PPSN XIV - 14th International Conference, Edinburgh, UK, September 17-21, 2016, Proceedings, *Lecture Notes in Computer Science*, vol. 9921, pp. 729–739. Springer (2016)

[32] Qian, C., Bian, C., Yu, Y., Tang, K., Yao, X.: Analysis of noisy evolutionary optimization when sampling fails. In: H.E. Aguirre, K. Takadama (eds.) Proceedings of the Genetic and Evolutionary Computation Conference, GECCO 2018, Kyoto, Japan, July 15-19, 2018, pp. 1507–1514. ACM (2018)

[33] Shi, F., Schirneck, M., Friedrich, T., Kötzing, T., Neumann, F.: Reoptimization times of evolutionary algorithms on linear functions under dynamic uniform constraints. In: P.A.N. Bosman (ed.) Proceedings of the Genetic and Evolutionary Computation Conference, GECCO 2017, Berlin, Germany, July 15-19, 2017, pp. 1407–1414. ACM (2017)

[34] Shi, F., Schirneck, M., Friedrich, T., Kötzing, T., Neumann, F.: Reoptimization time analysis of evolutionary algorithms on linear functions under dynamic uniform constraints. Algorithmica (2018)

[35] Stützle, T., Hoos, H.H.: MAX-MIN ant system. Future Generation Comp. Syst. **16**(8), 889–914 (2000)

[36] Sudholt, D., Thyssen, C.: Running time analysis of ant colony optimization for shortest path problems. J. Discrete Algorithms **10**, 165–180 (2012)

[37] Zhou, Y., He, J.: A runtime analysis of evolutionary algorithms for constrained optimization problems. IEEE Trans. Evolutionary Computation **11**(5), 608–619 (2007)

Chapter 8
The Benefits of Population Diversity in Evolutionary Algorithms: A Survey of Rigorous Runtime Analyses

Dirk Sudholt

Abstract Population diversity is crucial in evolutionary algorithms to enable global exploration and to avoid poor performance due to premature convergence. This chapter reviews runtime analyses that have shown benefits of population diversity, either through explicit diversity mechanisms or through naturally emerging diversity. These analyses show that the benefits of diversity are manifold: diversity is important for global exploration and the ability to find several global optima. Diversity enhances crossover and enables crossover to be more effective than mutation. Diversity can be crucial in dynamic optimization, when the problem landscape changes over time. And, finally, it facilitates the search for the whole Pareto front in evolutionary multiobjective optimization.

The analyses presented rigorously quantify the performance of evolutionary algorithms in the light of population diversity, laying the foundation for a rigorous understanding of how search dynamics are affected by the presence or absence of population diversity and the use of diversity mechanisms.

8.1 Introduction

Evolutionary algorithms (EAs) are popular general-purpose metaheuristics inspired by the natural evolution of species. By using operators such as mutation, recombination, and selection, a multiset of solutions — the *population* — is evolved over time. The hope is that this artificial evolution will explore vast regions of the search space and yet use the principle of "survival of the fittest" to generate good solutions for the problem at hand. Countless applications as well as theoretical results have demonstrated that these algorithms are effective on many hard optimization problems.

Dirk Sudholt
Department of Computer Science, University of Sheffield, Sheffield, United Kingdom

B. Doerr, F. Neumann (eds.), *Theory of Evolutionary Computation*,
Natural Computing Series, https://doi.org/10.1007/978-3-030-29414-4_8

A key distinguishing feature from other approaches such as local search or simulated annealing is the use of a *population* of candidate solutions. The use of a population allows evolutionary algorithms to explore different areas of the search space, facilitating global exploration. It also enables the use of recombination, where the hope is to combine good features of two solutions.

A common problem in evolutionary algorithms is *premature convergence*: the population collapses to copies of the same genotype, or more generally, a set of very similar genotypes, before the search space has been explored properly. In this case there is no benefit from having a population; in the worst case, the evolutionary algorithm may behave like a local search algorithm, but with an additional overhead from maintaining many similar solutions.

What we want instead is a *diverse* population that contains dissimilar individuals to promote exploration. The benefits of diversity are manifold:

Global exploration. A diverse population is generally well suited for global exploration, as it can explore different regions of the search space, reducing the risk of the whole population converging to local optima of low fitness.

Facilitating crossover. Often, a diverse population is required for crossover to work effectively. Crossing over two very similar solutions will result in an offspring that is similar to both parents, and this effect can also be achieved by mutation. Many problems where crossover is essential do in fact require a diverse population.

Decision making. A diverse population provides a diverse set of solutions for a decision maker to choose from. This is particularly important in multi-objective optimization as there are often trade-offs between different objectives, and the goal is to provide a varied set of solutions for a decision maker.

Robustness. A diverse population reduces the risk of getting stuck in a local optimum of bad quality. It is also robust with regard to uncertainty, such as noisy fitness evaluations or changes to the fitness function in cases where the problem changes dynamically. A diverse population may be able to track moving optima efficiently or to maintain individuals on different peaks, such that when the global optimum changes from one peak to another, it is easy to rediscover the global optimum.

In the long history of evolutionary computation, many solutions have been proposed to maintain or promote diversity. These range from controlling diversity through balancing exploration and exploitation via careful parameter tuning and designing selection mechanisms carefully, to explicit diversity-preserving mechanisms that can be embedded in an evolutionary algorithm. The latter include techniques such as eliminating duplicates, using subpopulations with migration as in island models, and niching techniques that try to establish niches of similar search points and prevent niches from going extinct. Niching techniques include fitness sharing (where similar individuals are forced to "share" their fitness, i.e., their real fitness is reduced during selection), clearing (where similar individuals can be "cleared" away by setting their fitness to a minimum value), and deterministic crowding (where offspring compete directly against their parents), to name just a few.

For more extensive surveys of diversity-preserving mechanisms we refer the reader to recent surveys by Shir [53], Črepinšek, Liu, and Mernik [10], and Squillero and Tonda [54]. Details of the diversity mechanisms surveyed here will be presented in the respective sections.

Many of these techniques work either on the genotypic level, i.e., trying to create a diverse set of bit strings, or on a phenotypic level, trying to obtain different phenotypes, taking into consideration some mapping from genotypes to phenotypes. For instance, for functions of unitation (functions that depend only on the number of 1-bits in the bit string), the genotype is a bit string, but the phenotype is given by the number of 1-bits. Diversity mechanisms can focus on genotypic or phenotypic diversity.

Given the plethora of mechanisms to be applied, it is often not clear what the best strategy is. Which diversity mechanisms work well for a given problem, which do not, and, most importantly, *why*? In particular, the effects such mechanisms have on search dynamics and performance are often not well understood.

This chapter reviews rigorous theoretical runtime analyses of evolutionary algorithms where diversity plays a key role, in order to address these questions and to develop a better understanding of the search dynamics in the presence or absence of diversity.

The goal of runtime analysis is to estimate the random or expected time until an evolutionary algorithm has met a particular goal, by rigorous mathematical study. Goals can include finding a global optimum, finding a diverse set of optima, or, specifically in the context of multi-objective optimization, finding the Pareto front of Pareto-optimal solutions. The results help us to get insight into the search behavior of evolutionary algorithms in the presence or absence of diversity, and how parameters and explicit diversity mechanisms affect performance. They highlight in particular which diversity mechanisms are effective for particular problems, and which are ineffective. More importantly, they explain *why* diversity mechanisms are effective or ineffective, and how to design the most effective evolutionary algorithms for the problems considered.

The presentation of these results is intended to combine formal theorems with informal explanations in order to make it accessible to a broad audience, while maintaining mathematical rigor. Instead of presenting formal proofs, we focus on key ideas and insights that can be drawn from these analyses. The reader is referred to the original papers for rigorous proofs and further details. In many cases we present only selected results from the papers surveyed, or results that are simplified towards special cases for reasons of simplicity, and the original publications contain further and/or more general results.

The outline of this chapter is as follows. After some preliminaries in Section 8.2, we first review the use of diversity-preserving mechanisms for enhancing global exploration in mutation-based evolutionary algorithms in Section 8.3. Section 8.4 then reviews the benefits of diversity for the use of crossover in genetic algorithms. Section 8.5 reviews the benefits of diversity mechanisms in dynamic optimization, where only a few results are available to date. Section 8.6 presents a recent, novel approach, using diversity metrics to design parent selection mechanisms that speed up evolutionary multiobjective optimization by picking parents that are most effective for spreading the population on the Pareto front. We finish with conclusions in Section 8.7.

This chapter is not meant to be comprehensive; in fact, there are several further runtime analyses not surveyed in this chapter. These include island models with crossover [47, 61], achieving diversity through heterogeneous island models and how this helps for SetCover [45], or achieving diversity through a tailored population model as in the case of all-pairs shortest paths [16]. Another popular diversity mechanism is ageing: restricting the lifespan of individuals to promote diversity. There are several runtime analyses of ageing mechanisms [26, 31, 32, 49]; however, these are being reviewed in Zarges' chapter [65], in Section 10.3. This chapter focuses on single-objective optimization, though diversity is a very important topic in multi-objective optimization (see, e.g., the work of Horoba and Neumann [27]). We present only one recent study of multiobjective optimization in Section 8.6, and refer to Brockhoff's survey [3] for a review of further theoretical results.

There are also a few very recent papers that appeared after this chapter was written. These include runtime analyses of probabilistic crowding and restricted tournament selection on TwoMax [9] and an empirical comparison of diversity mechanisms on multimodal example problems [8]. The interested reader is referred to Covantes Osuna's PhD thesis [5].

8.2 Preliminaries

The rigorous runtime analysis of randomized search heuristics is a challenging task, as these heuristics have not been designed to support an analysis. For that reason, we mostly consider bare-bones algorithms to facilitate a theoretical analysis. Furthermore, enhancing a bare-bones evolutionary algorithm with a diversity mechanism allows us to compare different diversity mechanisms in a clear-cut way, keeping the baseline algorithm as simple as possible.

The algorithms presented here do not use a specific stopping criterion, as we are interested in the random time for achieving a set goal. This time is generally referred to as the runtime, or running time. The most common goal is finding a global optimum, and the first hitting time of a global optimum is called the optimization time. In other cases we aim to find all global optima, or the whole Pareto front in a multi-objective setting.

Gao and Neumann [23] provided an alternative approach using rigorous runtime analysis: they considered the task of maximizing diversity amongst all search points with a given minimum quality. Although this is a very interesting study that has inspired subsequent runtime analyses [12], it is considered out of the scope of this survey.

Unless mentioned otherwise, we assume a binary search space where n denotes the problem size, that is, the length of the bit string. We use 0^n and 1^n to indicate the all-zeros and the all-ones strings, respectively, and, more generally, blocks of bits of some value. Further, $|x|_1$ denotes the number of ones in a bit string x, and $|x|_0$ denotes the number of zeros.

In the following, we say that an event occurs *with high probability* if its probability is at least $1-n^{-\Omega(1)}$, and an event occurs *with overwhelming probability* if its probability is at least $1-2^{-\Omega(n^{\varepsilon})}$ for some $\varepsilon > 0$.

8.3 How Diversity Benefits Global Exploration

We first review runtime analyses where explicit diversity mechanisms were used to improve the ability of evolutionary algorithms to explore the search space and to find global optima. The first such study was presented by Friedrich, Hebbinghaus, and Neumann [21], who compared a genotypic and a phenotypic diversity mechanism on an artificially constructed problem. Here, we review results from subsequent work [7, 22, 50] in Section 8.3.1 that focuses on the bimodal problem TwoMax instead. TwoMax has a simple structure, but it is very challenging to evolve a population that contains both optima. The diversity mechanisms considered for TwoMax include avoiding genotype duplicates, avoiding fitness duplicates, deterministic crowding, fitness sharing (in two variants), and clearing.

In Section 8.3.2, we review theoretical analyses of island models [39, 40]: these use subpopulations that communicate via migration, and can be run effectively on parallel hardware.

8.3.1 Diversity Mechanisms on TwoMax*: A Simple Bimodal Function*

The function TwoMax (see Fig. 8.1) is a function of unitation, that is, the fitness value depends only on the number of 1-bits:

$$\text{TwoMax}(x) := \max\left\{\sum_{i=1}^{n} x_i, n - \sum_{i=1}^{n} x_i\right\}.$$

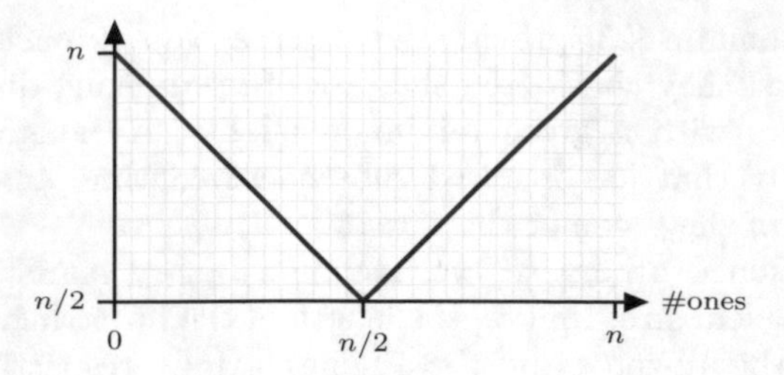

Fig. 8.1 Sketch of the function TwoMax with $n = 30$.

The fitness landscape consists of two hills with symmetric slopes, one for maximizing the number of ones and the other for maximizing the number of zeros. These sets are also referred to as branches. In [22], an additional fitness value for 1^n was added to distinguish between a local optimum 0^n and a unique global optimum. Here we use the original function with two global optima as also used in [7, 50], and measure the time needed in order to find both optima. The presentation of the results in [22] has been adapted to reflect this change.

TwoMax is an ideal benchmark function for studying diversity mechanisms as it is simply structured, hence facilitating a theoretical analysis, and it is hard for EAs to find both optima, as they have the maximum possible Hamming distance. The TwoMax function appears in well-known combinatorial optimization problems. For example, the VertexCover bipartite graph analyzed by Oliveto, He, and Yao [48] consists of two branches, one leading to a local optimum and the other to the minimum cover. Another function with a similar structure is the MinCut instance analyzed by Sudholt [58].

8.3.1.1 No Diversity Mechanism

In order to obtain a fair comparison of different diversity-preserving mechanisms, we keep one algorithm fixed as much as possible. The basic algorithm, the $(\mu+1)$ EA shown in Algorithm 8.1, has already been investigated by Witt [64].

The $(\mu+1)$ EA uses uniform random parent selection and elitist selection for survival. As parents are chosen uniformly at random, the selection pressure is quite low. Nevertheless, the $(\mu+1)$ EA is not able to maintain individuals on both branches for a long time. We now show that if μ is not too large, the individuals on one branch typically become extinct before the top of the branch is reached. Thus, the $(\mu+1)$ EA is unlikely to find both optima and the expected time for finding both optima is very large.

Theorem 8.3.1 (adapted from Theorem 1 in [22]). *The probability that the $(\mu+1)$ EA with no diversity-preserving mechanism and $\mu = o(n/\log n)$*

Algorithm 8.1: (μ+1) EA

```
1  t ← 0
2  Initialize P_0 with μ individuals chosen uniformly at random.
3  while termination criterion not met do
4      Choose x ∈ P_t uniformly at random.
5      Create y by flipping each bit in x independently with probability 1/n.
6      Choose z ∈ P_t with worst fitness uniformly at random.
7      if f(y) ≥ f(z) then
8          P_{t+1} = P_t \ {z} ∪ {y}
9      else
10         P_{t+1} = P_t
11     t ← t+1
```

finds both optima of TwoMax *in time* n^{n-1} *is* $o(1)$. *The expected time for finding both optima is* $\Omega(n^n)$.

The idea of the proof is to consider the first point in time where one optimum is found. Without loss of generality, we assume that this is 0^n. From there, one of two possible events may happen: another individual with genotype 0^n can enter the population or the other optimum, 1^n, can be found. The former event can occur if an individual with genotype 0^n is selected as a parent and no bit is flipped during mutation. The more copies of the 0^n optimum are contained in the population, the larger the probability of this event becomes. On the other hand, in order to create 1^n, a mutation has to flip all 0-bits in the parent (and no 1-bits). If the population size is small, copies of 0^n tend to take over the whole population before a mutation can create 1^n. If this happens, the (μ+1) EA has to flip n bits to create 1^n from 0^n, which has probability n^{-n}. Even considering n^{n-1} generations, the probability of this enormous jump happening is still $o(1)$, that is, it converges to 0.

8.3.1.2 Avoiding Genotype Duplicates

A simple way to enforce diversity within the population is not to allow genotype duplicates. We study a mechanism used by Storch and Wegener [55], where, in the "if" statement of the (μ+1) EA, the condition "and $y \notin P_t$" is added; see Algorithm 8.2. Note that here and in the following we only show the main loop, as the initialization steps are the same for all (μ+1) EA variants.

This mechanism ensures that the population always contains μ *different* genotypes (modulo possible duplicates occurring during initialization). However, this mechanism is not powerful enough to explore both branches of TwoMax.

Algorithm 8.2: (μ+1) EA avoiding genotype duplicates

```
1 while termination criterion not met do
2     Choose x ∈ P_t uniformly at random.
3     Create y by flipping each bit in x independently with probability 1/n.
4     Choose z ∈ P_t with worst fitness uniformly at random.
5     if f(y) ≥ f(z) and y ∉ P_t then
6         P_{t+1} = P_t \ {z} ∪ {y}
7     else
8         P_{t+1} = P_t
9     t ← t+1
```

Theorem 8.3.2 (adapted from Theorem 2 in [22]). *The probability that the (μ+1) EA with genotype diversity and $\mu = o(n^{1/2})$ finds both optima of* TwoMax *in time n^{n-2} is at most $o(1)$. The expected time for finding both optima is $\Omega(n^{n-1})$.*

The idea of the proof is similar to that of Theorem 8.3.2; however, we cannot rely on copies of one optimum, 0^n, taking over the population, as duplicates of 0^n are prevented from entering the population. The algorithm can still generate individuals similar to 0^n, for example by choosing 0^n as a parent and flipping a single 0-bit to 1. As mutations can flip any of n bits, the algorithm can easily create a population containing 0^n and many search points with only a single 1-bit that are at least as fit as the current best search points on the other branch. If the population size is $\mu = o(n^{1/2})$, the population is likely to be taken over by such search points before the other optimum, 1^n, is found. Note that our arguments rely on 0^n being selected as a parent, and there is only one individual with genotype 0^n. This leads to a more restrictive condition on μ ($\mu = o(n^{1/2})$) compared with the setting of no diversity mechanism ($\mu = o(n/\log n)$).

We conclude that avoiding genotype duplicates does create diversity in the population in the sense of different genotypes, but this kind of diversity is too weak for finding both optima of TwoMax, as there we need to evolve individuals on both branches.

8.3.1.3 Fitness Diversity

Avoiding genotype duplicates does not help much to optimize TwoMax as individuals from one branch are still allowed to spread on a certain fitness level and take over the population. A more restrictive mechanism is to avoid fitness duplicates, i.e., multiple individuals with the same fitness. Such a mechanism has been defined and analyzed by Friedrich, Hebbinghaus, and Neumann [21] for plateaus of constant fitness. In addition, this resembles the idea of fitness diversity proposed by Hutter and Legg [28].

Algorithm 8.3: (μ+1) EA with fitness diversity

```
1  while termination criterion not met do
2      Choose x ∈ P_t uniformly at random.
3      Create y by flipping each bit in x independently with probability 1/n.
4      if there exists z ∈ P_t such that f(y) = f(z) then
5          P_{t+1} = P_t \ {z} ∪ {y}
6      else
7          Choose z ∈ P_t with worst fitness uniformly at random.
8          if f(y) ≥ f(z) then
9              P_{t+1} = P_t \ {z} ∪ {y}
10         else
11             P_{t+1} = P_t
12     t ← t+1
```

The (μ+1) EA with fitness diversity avoids the situation where multiple individuals with the same fitness are stored in the population. If at some time t a new individual x is created with the same fitness value as a preexisting one $y \in P_t$, then x replaces y (see Algorithm 8.3).

The following theorem proves that if the population is not too large, then with high probability the individuals climbing one of the two branches will be extinguished before any of them reaches the top.

Theorem 8.3.3 (adapted from Theorem 3 in [22]). *The probability that the (μ+1) EA with fitness diversity and $\mu \le n^{O(1)}$ finds both optima*[1] *of* TwoMax *in time 2^{cn}, $c > 0$ being an appropriate constant, is at most $o(1)$. The expected time for finding both optima is $2^{\Omega(n)}$.*

The intuitive reason why this mechanism fails is that, once the population has reached one optimum, without loss of generality 0^n, the population has a tendency to spread on the branch leading to that optimum, until the whole population is contained on that branch. During this time there may be competition between the two branches: whenever one branch creates an offspring on the same branch, it may remove an individual of the same fitness on the opposite branch. This competition is biased towards the branch leading to 0^n, though, as that branch can use "downhill" mutations, that is, mutations flipping only one of the many 0-bits, whereas the opposite branch may have to rely on much rarer "uphill" mutations, that is, flipping only one of the rare 0-bits. The proof defines a potential function that captures the progress in this competition and shows that the branch that reaches its optimum first is likely to make individuals on the other branch go extinct.

[1] Owing to the fitness diversity mechanism, and since, in contrast to [22], we consider here a TwoMax variant with two global optima, the population can never actually contain both optima. To set a meaningful target, here we also consider cases where the union of the current population and a new offspring contains both optima.

Algorithm 8.4: (μ+1) EA with deterministic crowding

```
1 while termination criterion not met do
2     Choose x ∈ P_t uniformly at random.
3     Create y by flipping each bit in x independently with probability 1/n.
4     if f(y) ≥ f(x) then
5         P_{t+1} = P_t \ {x} ∪ {y}
6     else
7         P_{t+1} = P_t
8     t ← t+1
```

8.3.1.4 Deterministic Crowding

The main idea behind deterministic crowding is that offspring directly compete with their parents. In genetic algorithms with crossover, pairs of parents are formed, recombined, and mutated, and then the resulting offspring competes with one of its parents, replacing it if it is no worse.

We consider this mechanism in the absence of crossover, where offspring compete with their only parent. Then the population contains μ lineages that evolve independently (see Algorithm 8.4).

For sufficiently large populations, the algorithm can easily reach both global optima.

Theorem 8.3.4 (adapted from Theorem 4 in [22]). *The (μ+1) EA with deterministic crowding and $\mu \le n^{O(1)}$ reaches on* TwoMax *a population consisting of only global optima in expected time $O(\mu n \log n)$. In that case the population contains both global optima with probability at least $1 - 2^{-\mu+1}$.*

The probability of $1 - 2^{-\mu-1}$ follows from the fact that all μ lineages evolve independently, and that for each, once a global optimum is found, 0^n and 1^n are each found with probability 1/2. So, when one lineage that reaches a global optimum is fixed, the probability that the other $\mu - 1$ lineages all reach the same optimum is $2^{-\mu+1}$. The time bound is not immediate, as the (μ+1) EA picks a lineage to evolve further uniformly at random, so different lineages may receive different numbers of mutation steps. However, it is not difficult to show that the mutation steps are fairly concentrated around their expectation, leading to an upper time bound of $O(\mu n \log n)$.

8.3.1.5 Fitness Sharing

Fitness sharing [44] derates the real fitness of an individual x by an amount that represents the similarity of x to other individuals in the population. The similarity between x and y is measured by a so-called *sharing function* $\mathrm{sh}(x,y) \in [0,1]$, where a large value corresponds to large similarities and a value of 0 implies no similarity. The idea is that if there are several copies of

Algorithm 8.5: (μ+1) EA with fitness sharing

```
1  while termination criterion not met do
2      Choose x ∈ P_t uniformly at random.
3      Create y by flipping each bit in x independently with probability 1/n.
4      Choose z ∈ P_t with worst fitness uniformly at random.
5      Let P'_t := P_t ∪ {y}.
6      if f(y, P'_t) ≥ f(z, P'_t) then
7          P_{t+1} = P_t \ {z} ∪ {y}
8      else
9          P_{t+1} = P_t
10     t ← t+1
```

the same individual in the population, these individuals have to share their fitness. As a consequence, selection is likely to remove such clusters and to keep the individuals apart. We define the *shared fitness* of x in the population P and the fitness $f(P)$ of the population as

$$f(x,P) = \frac{f(x)}{\sum_{y \in P} \mathrm{sh}(x,y)} \qquad \text{and} \qquad f(P) = \sum_{x \in P} f(x,P),$$

respectively. It is common practice to use a so-called *sharing distance* σ such that individuals only share fitness if they have a distance less than σ. Given some distance function d, a common formulation for the sharing function is

$$\mathrm{sh}(x,y) = \max\{0, 1-(d(x,y)/\sigma)^\alpha\},$$

where α is a positive constant that regulates the shape of the sharing function. We use the standard setting $\alpha = 1$ and, following Mahfoud [44], we set the sharing distance to $\sigma = n/2$ as this is the smallest value allowing discrimination between the two branches. As TwoMax is a function of unitation, we allow the distance function d to depend on the number of ones: $d(x,y) := \big||x|_1 - |y|_1\big|$. Such a strategy is known as *phenotypic sharing* [44]. Our precise sharing function is then

$$\mathrm{sh}(x,y) = \max\left\{0,\ 1-2\frac{\big||x|_1 - |y|_1\big|}{n}\right\}.$$

There are different ways of performing selection according to the shared fitness, differing in the way the reference population P in the shared fitness $f(x,P)$ is chosen. In the following, we will review runtime analyses for two different variants of fitness sharing. The most common usage of fitness sharing is to consider the shared fitness according to the union of parents and offspring; see Algorithm 8.5.

Oliveto, Sudholt, and Zarges [50] showed that a population size of $\mu = 2$ is not sufficient to find both optima, and that the performance is even worse than for deterministic crowding with $\mu = 2$. The following theorem states that, with a probability greater than 1/2, the (2+1) EA will end up with both individuals in the same optimum, leading to an exponential running time from there. This performance is worse than for deterministic crowding, for which the probability of finding both optima is exactly 1/2 (see Theorem 8.3.4).

Theorem 8.3.5 (Theorems 1 and 2 in [50]). *The (2+1) EA with fitness sharing will, with probability* $1/2+\Omega(1)$*, reach a population with both members in the same optimum, and then the expected time for finding both optima from there is* $\Omega(n^{n/2})$*.*

However, with probability $\Omega(1)$ *the algorithm will find both optima in time* $O(n \log n)$*.*

The reason for the failure probability of $1/2 + \Omega(1)$ is that the algorithm typically gets stuck on one branch if both initial search points are on the same branch (which happens with probability around 1/2) or if the search points are initialized on different branches but one search point has a much higher fitness than the other. In that case the effect of fitness sharing is not strong enough, and the less fit individual will be replaced if the fitter one creates an offspring similar to itself.

If the population is initialized with two search points on different branches and with similar fitness, fitness sharing ensures that, with high probability, individuals on both branches survive. The reason is that whenever one parent creates an offspring on its branch, fitness sharing derates the fitness of both parent and offspring in such a way that the less fit one will be removed and the individual on the opposite branch survives.

For population sizes $\mu \geq 3$, fitness sharing becomes much more effective.

Theorem 8.3.6 (Theorem 3 in [50]). *For any population size* $\mu \geq 3$*, the* $(\mu+1)$ *EA with fitness sharing will find both optima of* TwoMax *in expected time* $O(\mu n \log n)$*.*

The analysis reveals a very interesting behavior. If all search points are initialized on one branch, the population starts to climb up that branch. But once a sufficiently large overall fitness value has been obtained (at the latest when two individuals have found an optimum), then these high-fitness individuals develop a sufficiently large critical mass that the effect of fitness sharing starts to become evident, and the population shows a very different behavior. From this point in time on, the population starts expanding towards lower fitness values and the individuals with the smallest and the largest numbers of 1-bits always survive. While the whole population may start to climb up one branch, at some point in time the individual with the lowest fitness starts to be repelled and makes its way back down, eventually reaching the other branch and climbing up to find the other optimum.

We can conclude that fitness sharing works for the (μ+1) EA with population sizes $\mu \geq 3$, but when larger offspring populations[2] are considered, it can have undesirable effects: if a cluster of individuals creates too many offspring, sharing decreases the shared fitness of all individuals in the cluster, and the cluster may go extinct.

In a similar vein, the population can even lose all global optima. In a (2+λ) EA with $\lambda \geq 6$, if the population contains two copies of the same global optimum, and then a generation creates $\lambda - 2$ clones and two individuals with Hamming distance 1 to the optimum, the latter two individuals will have a higher shared fitness and form the new population.

The following result shows that even with a small offspring population size of $\lambda = 2$, the (μ+λ) EA can fail.

Theorem 8.3.7 (Theorem 4 in [50]). *With probability* $1 - o(1)$ *the (2+2) EA with fitness sharing will, at some point in time, reach a population with both members in the same optimum. The expected time for finding both optima from there is* $\Omega(n^{n/2})$.

In order to avoid these problems, early runtime analyses of fitness sharing [20, 22, 57] used fitness sharing in a different sense. They set up a competition between *populations* instead of individuals: the (μ+1) EA variant considers the union of the parent population and the offspring population, P'_t and then selects the subset $P^* \subset P'_t$ of size μ that maximizes $f(P^*)$. This makes sense as the goal is to evolve a population of high population fitness.

Friedrich et al. [22] showed that the (μ+1) EA with a population-level implementation of fitness sharing can efficiently find both optima on TwoMax.

Theorem 8.3.8 (adapted from Theorem 5 in [22]). *The (μ+1) EA with fitness sharing and* $\mu \geq 2$ *finds both optima on* TwoMax *in expected time* $O(\mu n \log n)$.

The reason for this efficiency is as follows. If we imagine all parents and the new offspring on a scale of the number of 1-bits, the individuals with the smallest and the largest number of ones have the largest distance to all individuals in the population. Therefore, fitness sharing makes these outer individuals very attractive in terms of shared fitness, and hence these individuals are taken over to the next generation. This holds even if an outer individual has the worst fitness in the population; the best possible population that can be formed from parents and offspring will create individuals with a minimum and a maximum number of ones.

Hence the minimum number of ones in the population can never increase, and the maximum number of ones can never decrease. Both quantities can be improved whenever the outer individuals perform a hill-climbing step towards

[2] The (μ+λ) EA is a variant of the (μ+1) EA that creates λ offspring in parallel and then selects the μ best according to $f(\cdot, P'_t)$, where P'_t is the union of all μ parents and λ offspring, breaking ties towards preferring offspring.

Algorithm 8.6: Clearing procedure

input : A population P
output : Fitness values after clearing f' for all $x \in P$.

1. Let $f' = f$.
2. Sort P according to decreasing f' values.
3. **for** $i := 1$ **to** $|P_t|$ **do**
4.   **if** $f'(P[i]) > 0$ **then**
5.   winners $\leftarrow 1$.
6.   **for** $j := i+1$ **to** $|P|$ **do**
7.   **if** $f'(P[j]) > 0$ **and** $d(P[i], P[j]) < \sigma$ **then**
8.   **if** winners $< \kappa$ **then**
9.   winners $\leftarrow$ winners $+1$
10.   **else**
11. $f'(P[j]) \leftarrow 0$

their respective optima. Performing a hill-climbing task towards both 0^n and 1^n yields the expected time of $O(\mu n \log n)$.

A drawback of this design is that to find a population P^* that maximizes the population fitness, one needs to consider up to $\binom{\mu+\lambda}{\mu}$ different candidate populations of size μ. In the case of $\lambda = 1$ this is $\mu + 1$ combinations, but for large μ and λ this strategy is prohibitive.

8.3.1.6 Clearing

Clearing is a niching method that uses a similar principle to fitness sharing. While fitness sharing can be regarded as sharing resources evenly between similar individuals, clearing assigns these resources only to the best individual in each niche. Such an individual is referred to as a *winner*. All other individuals have their fitness set to 0 (or, more generally, to a value lower than the lowest fitness value in the search space).

Niches are established as in fitness sharing by using a clearing radius σ that determines up to which distance individuals will be considered to belong to the same niche. Each niche supports up to κ winners, where κ is a parameter called the *niche capacity*. The decision about which individuals are winners and which of them have their fitness cleared is made in a greedy procedure, shown in Algorithm 8.6. The individuals are first sorted in order of decreasing fitness. Then the clearing procedure processes individuals in this order. For each individual, if it has not been cleared, it is declared a winner. The procedure iterates through all remaining individuals (i.e., those with lower or equal fitness) that have not been cleared yet and that are within a clearing distance of σ, adding them to its niche until κ winners have been found, and clearing all remaining such individuals.

Algorithm 8.7: (μ+1) EA with clearing

```
1  while termination criterion not met do
2      Choose x ∈ P_t uniformly at random.
3      Create y by flipping each bit in x independently with probability 1/n.
4      Compute the fitness f' after clearing of all individuals in P_t ∪ {y} according
         to the clearing procedure.
5      Choose z ∈ P_t with worst fitness after clearing uniformly at random.
6      if f'(y) ≥ f'(z) then
7          P_{t+1} = P_t \ {z} ∪ {y}
8      else
9          P_{t+1} = P_t
10     t ← t+1
```

Clearing is a powerful mechanism, as it allows both exploitation and exploration: it allows winners to find fitness improvements, while at the same time enabling cleared individuals to tunnel through fitness valleys. In fact, cleared individuals are agnostic to the fitness landscape as they always have the worst possible fitness. Hence cleared individuals can explore the landscape by performing random walks. As we will show, this allows the algorithm to escape from local optima with even very large basins of attraction. The (μ+1) EA with clearing is shown in Algorithm 8.7.

Covantes Osuna and Sudholt [7] considered the performance of the (μ+1) EA with clearing for two choices of the dissimilarity measure d. When d is chosen to be the Hamming distance, we refer to this as *genotypic clearing*. When we choose the phenotypic distance as the difference in the number of ones, $d(x,y) := ||x|_1 - |y|_1|$, this strategy is referred to as *phenotypic clearing*.

With phenotypic clearing and a clearing radius $\sigma = 1$, every number of ones represents its own niche. If the population size is large enough to contain all niches, the population can easily spread throughout all niches. In the case of TwoMax, this means that both optima will have been found. In fact, this argument even extends to finding an optimum for *all* functions of unitation, as one of the niches will contain all global optima. The expected time for the population to spread across all niches is $O(\mu n \log n)$, which is the same time bound as for fitness sharing and deterministic crowding.

Theorem 8.3.9. *Let f be a function of unitation and let $\sigma = 1$, $\mu \geq (n+1) \cdot \kappa$. Then the expected optimization time of the (μ+1) EA with phenotypic clearing on f is $O(\mu n \log n)$.*

For genotypic clearing we have to consider larger niches, as otherwise each niche just consists of a single search point, and genotypic clearing essentially amounts to avoiding duplicates in the population (see Theorem 8.3.2). The most natural choice is $\sigma = n/2$ as for fitness sharing, as this is allows us to distinguish the two branches. For this setting, we have the following performance guarantee.

Theorem 8.3.10. *The expected time for the (μ+1) EA with genotypic or phenotypic clearing, with $\mu \geq \kappa n^2/4$, $\mu \leq n^{O(1)}$, and $\sigma = n/2$ to find both optima on* TwoMax *is $O(\mu n \log n)$.*

The idea behind the proof is to consider the situation after one of the optima has been reached, and once the population contains κ copies of that optimum. It is easy to show that the expected time until this happens, or both optima are found, is bounded by $O(\mu n \log n)$.

We then consider a potential function that describes the state of the current population: the sum of the Hamming distances of all individuals to the optimum. Note that the phenotypic and genotypic distances to an optimum (0^n or 1^n) are the same, and hence the analysis holds for both phenotypic and genotypic clearing. Imagine the situation when all individuals are close to the optimum. Then any mutation is likely to create an offspring that is further away from the optimum. Thus, mutation has a tendency to increase the potential.

Selection will then remove one of the non-winner individuals uniformly at random. There is a small bias, introduced by selection, towards remaining close to the winner. This is down to the fact that losers in the population do not evolve in complete isolation. The population always contains κ copies of the winner that may create offspring and may prevent the population from venturing far away from it. In other words, there is a constant influx of search points descended from winners.

All in all, mutation and selection yield opposite biases. The bias induced by selection decreases as the fraction of winners κ/μ decreases. If the population size μ is large enough with respect to κ and n, i.e., $\mu \geq \kappa n^2/4$, the potential shows a positive expected change until it reaches a value from which, by the pigeon-hole principle, we can conclude that at least one individual must have reached a distance at least $n/2$ from the winners.

From there, a new niche is created, and the other optimum can easily be found by hill climbing. The overall time is bounded by $O(\mu n \log n)$.

Note that the condition $\mu \geq \kappa n^2/4$ is a sufficient condition, and a quite steep requirement compared with that for fitness sharing, which works with constant population sizes μ. On the other hand, clearing works with genotypic distances, whereas fitness sharing has only been proved to work with phenotypic distances. A further advantage of clearing is that it also works on variants of TwoMax with different slopes, whereas the analysis of fitness sharing is sensitive to the absolute fitness values.

8.3.2 Diversity in Island Models

The presentation in this subsection is partly taken from this author's theory-flavored survey of parallel evolutionary algorithms [60].

Algorithm 8.8: Scheme of an island model with migration interval τ

1 Initialize a population made up of subpopulations or islands, $P^{(0)} = \{P_1^{(0)}, \dots, P_m^{(0)}\}$.
2 $t \leftarrow 1$.
3 **while** *termination criterion not met* **do**
4   **for** *each island* i **do in parallel**
5   **if** $t \bmod \tau = 0$ **then**
6   Send selected individuals from island $P_i^{(t)}$ to selected neighboring islands.
7   Receive immigrants $I_i^{(t)}$ from islands for which island $P_i^{(t)}$ is a neighbor.
8   Replace $P_i^{(t)}$ by a subpopulation resulting from a selection among $P_i^{(t)}$ and $I_i^{(t)}$.
9   Produce $P_i^{(t+1)}$ by applying reproduction operators and selection to $P_i^{(t)}$.
10   $t \leftarrow t+1$

Island models are popular ways of parallelizing evolutionary algorithms: they consist of subpopulations that may be run on different cores, and that coordinate their searches by using migration, communicating selected search points, or copies thereof, to other islands. These solutions are then considered for inclusion on the target island in a further selection process. Island models communicate on a communication topology, a directed graph that connects the islands, and migration involves sending solutions to all neighboring islands. Often periodic migration is used: migration happens every τ iterations, where τ is a parameter called the *migration interval.*

This way, islands can communicate and compete with one another. Islands that have got stuck in low-fitness regions of the search space can be taken over by individuals from more successful islands. This helps to coordinate the search, focus on the most promising regions of the search space, and use the available resources effectively. The islands also act as an implicit diversity mechanism: between migrations, islands evolve independently, and the flow of genetic information in the whole system is slowed down, compared with having one large population. This can help to increase diversity and to prevent or at least delay premature convergence. Note that the flow of information can be tuned by tuning the migration interval τ, the migration topology, and other parameters such as the number of individuals to be migrated or the policies (selection schemes) for emigration and immigration. Algorithm 8.8 shows the general scheme of a basic island model.

Common topologies include unidirectional rings (rings with directed edges in only one direction), bidirectional rings, torus or grid graphs, hypercubes, scale-free graphs [14], random graphs [24], and complete graphs. Fig. 8.2 sketches some of these topologies. An important characteristic of a topology $T = (V, E)$ is its *diameter*: the maximum number of edges on any shortest path between two vertices. Formally, $\mathrm{diam}(T) = \max_{u,v \in V} \mathrm{dist}(u, v)$ where $\mathrm{dist}(u, v)$ is the graph distance, the number of edges on a shortest path from u to v. The diameter gives a good indication of the time needed to propagate information throughout the topology. Rings and torus graphs have large diameters, while hypercubes, complete graphs, and many scale-free graphs have small diameters.

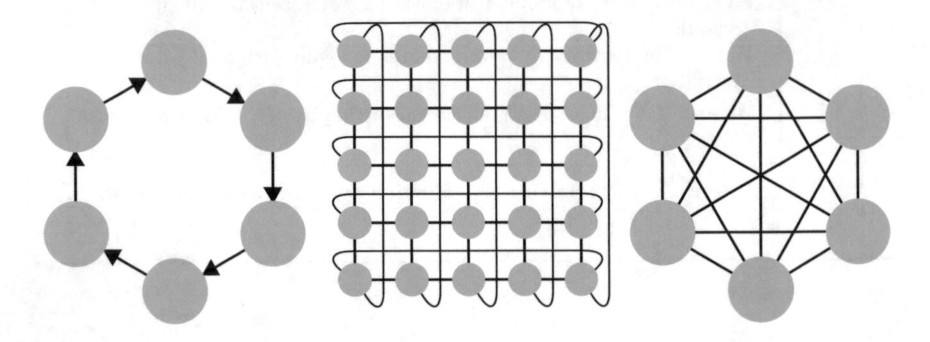

Fig. 8.2 Sketches of common topologies: a unidirectional ring, a torus, and a complete graph. Other common topologies include bidirectional rings, where all edges are undirected, and grid graphs, where the edges wrapping around the torus are removed.

8.3.2.1 A Royal Road for Island Models

Lässig and Sudholt [38, 39] presented a first example where communication makes the difference between exponential and polynomial running times, in a typical run. They constructed a family of problems called $\mathrm{LOLZ}_{n,z,b,\ell}$, where a simple island model, with all islands running (1+1) EAs, finds the optimum in polynomial time with high probability. This holds for a proper choice of the migration interval and any migration topology that is not too sparse. In contrast, both a single, large population as in the (μ+1) EA and independent islands (each running a (1+1) EA, or even when they also run a (μ+1) EA) need exponential time with high probability.

The basic idea of this construction is as follows. First, imagine a bit string where the fitness describes the length of the longest prefix of bits with the same value. Generally, a prefix of i leading ones yields the same fitness as a prefix of i leading zeros; for example, 111010 and 000110 both have fitness 3. However, the maximum possible fitness that can be attained by leading zeros is capped at some threshold value z. This means that, in the long run, gathering leading ones is better than gathering leading zeros. The former leads to an optimal value, while the latter leads to a local optimum that is hard to escape from.

The effect on an EA is as follows. In the beginning, the EA typically has to make a decision whether to collect leading ones (LO) or leading zeros (LZ). This holds not only for the (1+1) EA but also for a (not too large) panmictic population, as genetic drift will lead the whole population to either leading ones or leading zeros. After a significant prefix has been gathered, this decision gradually becomes irreversible, as many bits in the prefix need to be flipped at the same time to switch from leading ones to leading zeros or vice versa. So, with probability close to 1/2, the EA will end up finding an optimum by gathering leading ones, and again with probability close to 1/2 its population will get stuck in a hard local optimum.

To further increase the difficulty for EAs, this construction is repeated on several blocks of the bit string that need to be optimized one-by-one. Each block has length ℓ. Only if the right decision towards leading ones is made for the first block can the block be filled with further leading ones. Once the first block contains only leading ones, the fitness depends on the prefix in the second block, and a further decision between leading ones and leading zeros needs to be made. Only if the EA makes all decisions correctly can it find a global optimum. Table 8.1 illustrates the definition of the problem.

Table 8.1 Examples of solutions for the function LOLZ with four blocks and $z = 3$, along with their fitness values. All blocks have to be optimized from left to right. The sketch shows in red all bits that are counted in the fitness evaluation. Note how, in x_3, in the third block only the first $z = 3$ zeros are counted. Further 0-bits are ignored. The only way to escape from this local optimum is to flip all z 0-bits in this block simultaneously

x_1	11110011 11010100 11010110 01011110	$\mathrm{LOLZ}(x_1) = 4$
x_2	11111111 11010100 11010110 01011110	$\mathrm{LOLZ}(x_2) = 10$
x_3	11111111 11111111 00000110 01011110	$\mathrm{LOLZ}(x_3) = 19$

So, the problem requires an EA to make several decisions in succession. The number of blocks, b, is another parameter that determines how many decisions need to be made. Panmictic populations will sooner or later make a wrong decision and get stuck in some local optimum. If b is not too small, the same holds for independent runs. The results presented in [39] are summarized as follows; the second statement follows from the first one and the union bound.

Theorem 8.3.11. *Consider the* (μ+1) *EA with* $\mu \leq cn/(\log n)$ *for an arbitrary constant* $c > 0$ *on* $\mathrm{LOLZ}_{n,z,b,\ell}$ *with* $z = \omega(\log b)$, $b\ell \leq n$, *and* $z < \ell$. *With probability at least* $1 - e^{-\Omega(z)} - 2^{-b}$ *the* (μ+1) *EA does not find a global optimum within* $n^{z/3}$ *generations.*

The same holds when s *independent subpopulations, each running a* (1+1) *EA or* (μ+1) *EA as specified above, are considered; then the probability bound becomes* $1 - se^{-\Omega(z)} - s2^{-b}$.

However, an island model can effectively communicate the right decisions about blocks to other islands. Islands that have got stuck in a local optimum can be taken over by other islands that have made the correct decision. These dynamics make up the success of the island model, as it can be shown to find global optima with high probability. A requirement is, though, that the migration interval is carefully tuned so that migration transmits only the right information. If migration happens before the symmetry between leading ones and leading zeros is broken, it might be that islands with leading zeros take over islands with leading ones. We also need the topology to be able to spread the right information quickly enough. A topology G is called *well-expanding* if there is a constant $\varepsilon > 0$ such that the following holds: for every subset $V' \subseteq V$ with $|V'| \leq |V|^{\varepsilon}$, we have $|N(V')| \geq (2+\varepsilon)|V'|$. Lässig and Sudholt [39] gave the following result.

Theorem 8.3.12. *Consider an island model where each island runs a (1+1) EA with migration on a well-expanding migration topology with* $\tau = n^{5/3}$ *and* $\mu = n^{\Theta(1)}$ *subpopulations, accepting a best search point from all immigrants and the resident individual. Let the function* $\mathrm{LOLZ}_{n,z,b,\ell}$ *be parameterized according to* $\ell = 2\tau/n = 2n^{2/3}$, $z = \ell/4 = n^{2/3}/2$, *and* $b \leq n^{1/6}/16$. *If the migration counter* t *starts at* $\tau/2 = n^{5/3}/2$, *then with overwhelming probability the algorithm finds a global optimum within* $O(b\ell n) = O(n^2)$ *generations.*

The analysis is quite technical, but the main ideas can be summarized as follows. All islands optimize LOLZ by fixing bits from left to right, and at approximately the same pace. The migration interval is tuned such that between two migrations all islands will be starting to optimize the same new block, excluding islands that have got stuck on previous blocks. Assume for the moment that islands make decisions independently about the new block (we will discuss this assumption below). Then, in expectation, half the islands will get stuck in local optima, reducing the number of "good" islands still on track towards finding the global optimum. Once this number has dropped below $|V|^{\varepsilon}$, the properties of the topology ensure that these "good" islands propagate their information to sufficiently many other islands (many of which will be stuck in local optima) to ensure that a critical mass of "good" islands always survives, until a global optimum is found eventually.

An interesting finding is how islands do in fact make independent or nearly independent decisions about new blocks. After all, during migration, genetic information about all future blocks is transmitted. Hence, after migration, many islands share the same genotype on all future blocks. This is a real threat, as this dependence might imply that all islands make the same decision after moving on to the next block, compromising diversity.

However, under the conditions for the migration interval, there is a period of independent evolution following migration, before any island moves on to a new block. During this period of independence, the genotypes of future blocks are subjected to random mutations, independently for each island.

After some time, the distribution of bits on these future blocks will resemble a uniform distribution. This shows that independence can be gained by the use of periods of independent evolution. One could say that the island model combines the advantages of two worlds: independent evolution and selection pressure through migration. The island model is only successful because it can use both migration and periods of independent evolution.

8.3.2.2 Island Models for Eulerian Cycles

We now also give a simple and illustrative example from combinatorial optimization to show how island models can be beneficial through providing diversity. Lässig and Sudholt [40] considered island models for the Eulerian cycle problem. Given an undirected Eulerian graph, the task is to find an Eulerian cycle, i.e., a traversal of the graph on which each edge is traversed exactly once. This problem can be solved efficiently by tailored algorithms, but it has served as an excellent test bed for studying the performance of evolutionary algorithms [17–19, 46].

Instead of bit strings, the representation of the problem used by Neumann [46] is based on permutations of the edges of the graph. Each such permutation gives rise to a *walk*: starting with the first edge, a walk is the longest sequence of edges such that two subsequent edges in the permutation share a common vertex. The walk encoded by the permutation ends when the next edge does not share a vertex with the current one. A walk that contains all edges represents an Eulerian cycle. The length of the walk gives the fitness of the current solution.

Neumann [46] considered a simple instance that consists of two cycles of equal size, connected by one common vertex v^* (see Fig. 8.3). This instance is interesting, as it represents a worst case for the time until an improvement is found. This is with respect to randomized local search (RLS) working on this representation. RLS works like the (1+1) EA, but it uses only local mutations. As the mutation operator, it uses jumps: an edge is selected uniformly at random and then it is moved to a (different) target position chosen uniformly at random. All edges in between the two positions are shifted accordingly.

On the instance considered, RLS typically starts constructing a walk within one of these cycles, either by appending edges to the end of the walk or by prepending edges to the start of the walk. When the walk extends to v^* for the first time, a decision needs to be made. RLS can extend the walk to the opposite cycle (see Fig. 8.3). In this case RLS can simply extend both ends of the walk until an Eulerian cycle is formed. The expected time until this happens is $\Theta(m^3)$, where m denotes the number of edges.

But, if another edge in the same cycle is added at v^*, the walk will evolve into one of the two cycles that make up the instance. It is not possible to add further edges to the current walk unless the current walk starts and ends in v^*. However, the walk can be rotated so that the start and end vertex of

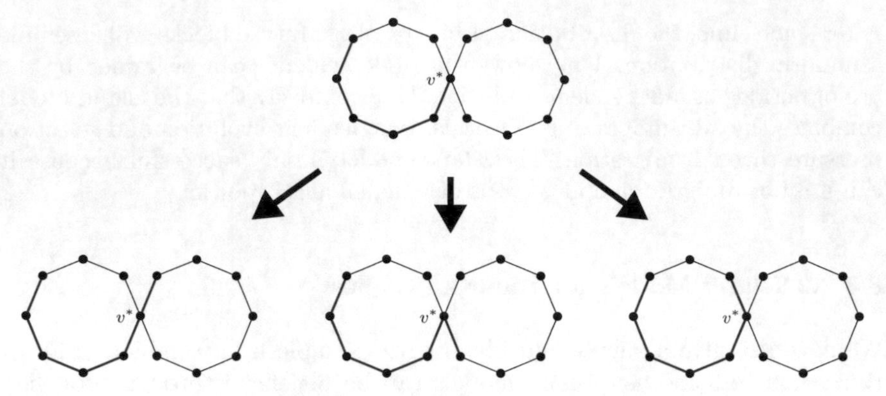

Fig. 8.3 Sketch of the graph G'. The top shows a configuration where a decision at v^* has to be made. The three configurations below show the possible outcomes. All these transitions happen with equal probability, but only the one on the bottom right leads to a solution where rotations are necessary.

the walk is moved to a neighboring vertex. Such an operation takes expected time $\Theta(m^2)$. Note that the fitness after a rotation is the same as before. Rotations that take the start and end closer to v^* are as likely as rotations that move it away from v^*. The start and end of the walk hence performs a fair random walk, and $\Theta(m^2)$ rotations are needed on average in order to reach v^*. The total expected time for rotating the cycle is hence $\Theta(m^4)$.

Summarizing, if RLS makes the right decision, then an expected time $\Theta(m^3)$ suffices in total. But, if rotations become necessary, the expected time increases to $\Theta(m^4)$. Now consider an island model with λ islands running RLS. If islands evolve independently for at least $\tau \geq m^3$ generations, all the decisions above are made independently, with high probability. The probability of making a wrong decision is $1/3$; hence, with λ islands, the probability that all islands make the wrong decision is $3^{-\lambda}$, leading to the following result.

Theorem 8.3.13. *The island model running RLS on $\lambda \leq m^{O(1)}$ islands, with $\tau \geq m^3$ and an arbitrary topology, optimizes G' in expected $O(m^3 + 3^{-\lambda} \cdot m^4)$ generations.*

The choice $\lambda := \log_3 m$ yields an expectation of $\Theta(m^3)$, and every value up to $\log_3 m$ leads to a superlinear and even exponential speedup, compared with the time $\Theta(m^4)$ for a single island running RLS.

Interestingly, this good performance only holds if migration is used rarely, or if independent runs are used. If migration is used too frequently, the island model rapidly loses diversity. If T is any strongly connected topology and $\text{diam}(T)$ is its diameter, we have the following.

Theorem 8.3.14. *Consider the island model with an arbitrary strongly connected topology T running RLS with jumps on each island. If $\tau \cdot \text{diam}(T) \cdot$*

$\lambda = O(m^2)$, then the expected number of generations on G' is at least $\Omega(m^4/(\log \lambda))$.

If $\tau \cdot \mathrm{diam}(T) \cdot \lambda = O(m^2)$, then there is a constant probability that the island that first arrives at a decision at v^* propagates this solution throughout the whole island model, before any other island can make an improvement. This results in an expected running time of $\Omega(m^4/\log(\lambda))$. This is almost $\Theta(m^4)$, even for very large numbers of islands. The speedup is therefore logarithmic in the number of islands at best, or even worse.

This natural example shows that the choice of the migration interval can make a difference between exponential and logarithmic speedups.

8.4 How Diversity Benefits Crossover

Now we look at examples where diversity enhances the use of crossover in evolutionary algorithms. In a population where all individuals are very similar, crossover is unlikely to be effective, as it will create an offspring that is similar to both of its parents. This effect can also be achieved by mutation. Therefore, many examples where crossover is essential require some form of diversity mechanism for crossover to work effectively.

In the following, we review several of these examples, from the very first constructed examples to problems from combinatorial optimization, and even simple hill-climbing problems, where crossover provides a noticeable speedup. In some cases the diversity is due to explicit diversity mechanisms; in others, diversity can emerge naturally from independent variations.

Most of the algorithms discussed in this section fit into the scheme described in Algorithm 8.9. Unless stated otherwise, parent selection is performed uniformly at random. With a crossover probability p_c, crossover is performed; this can be uniform crossover or k-point crossover. In any case, mutation is performed with a mutation rate of p, which is assumed to be the default value $p = 1/n$ unless stated otherwise. In the replacement selection, μ individuals with the best fitness are selected for survival. If there are ties, a specific tie-breaking rule can be used; the default is to break ties uniformly at random.

8.4.1 Real Royal Road Functions for Crossover

Jansen and Wegener [30] were the first to provide an example function for which it could be rigorously proved that a simple genetic algorithm (GA) with crossover takes expected polynomial time, whereas all $(\mu+\lambda)$ evolutionary algorithms using only standard bit mutation need exponential time with overwhelming probability.

Algorithm 8.9: Scheme of a $(\mu+\lambda)$ GA with mutation rate p and crossover with crossover probability p_c for maximizing $f\colon \{0,1\}^n \to \mathbb{R}$

```
1  Initialize population P of size μ ∈ N uniformly at random.
2  while optimum not found do
3      Let P' = ∅.
4      for i = 1,...,λ do
5          Choose p ∈ [0,1] uniformly at random.
6          if p ≤ p_c then
7              Select two parents x_1, x_2.
8              Let y := crossover(x_1, x_2).
9          else
10             Select a parent y.
11         Flip each bit in y independently with probability p.
12         Add y to P'.
13     Let P contain the μ best individuals from P ∪ P'; break ties according to a
         specified tie-breaking rule.
```

Their steady-state GA with population size μ can be regarded as a special case of Algorithm 8.9 with $\lambda = 1$, referred to as the $(\mu+1)$ GA in the following, using a tie-breaking rule that eliminates an individual with the largest number of duplicates in $P \cup P'$.[3] The latter is equivalent to breaking ties towards including individuals with the fewest duplicates in $P \cup P'$.

Jansen and Wegener [30] defined two classes of functions that they called *real royal road functions*: one for one-point crossover and one for uniform crossover. We focus on the one for one-point crossover as it is conceptually simpler. Denoting by $b(x)$ the length of the longest block consisting of ones only (e.g., $b(100111011) = 3$), the function class R_n is defined as (assuming $n/3 \in \mathbb{N}$)

$$R_n(x) = \begin{cases} 2n^2 & \text{if } x = 1^n, \\ n|x|_1 + b(x) & \text{if } |x|_1 \le 2n/3, \\ 0 & \text{otherwise.} \end{cases}$$

The function contains a strong gradient in the region of search points with at most $2n/3$ ones. The function also contains a fitness valley of fitness 0 that needs to be crossed to reach the optimum, 1^n. Moreover, the function encourages an evolutionary algorithm to evolve search points with $2n/3$ ones and a maximum block length of $2n/3$. This is to allow crossover to combine two such blocks to create the optimum 1^n, for instance by crossing over two parents $1^{2n/3}0^{n/3}$ and $0^{n/3}1^{2n/3}$ that have large blocks in different positions.

We give a simplified version of Jansen and Wegener's result, as their work includes an additional parameter that specifies the length m of the fitness

[3] In [30], the replacement selection stops without altering P if the fitness of the offspring is smaller than the fitness of the worst individual in P. Our algorithm is equivalent as in this case the offspring will be added to and immediately removed from the population.

valley, which is fixed at $m = n/3$ here. Note that the population size μ and the crossover probability p_c can be functions of the problem size n.

Theorem 8.4.1 (simplified from Theorem 3 in [30]). *Let $p_c \le 1-\varepsilon$ for some $0<\varepsilon<1$, and $\mu \ge 2n/3+1$. Then the expected optimization time of the (μ+1) GA breaking ties towards including individuals with the fewest duplicates in $P \cup P'$ on R_n is $O(\mu n^3 + (\mu \log \mu)n + \mu^2/p_c)$. For the typical case where p_c is a positive constant and $\mu = O(n)$, the bound is $O(n^4)$.*

The proof is a beautiful application of the so-called method of typical runs [62, Section 11], where a run is divided into phases that reflect the typical behavior of the algorithm. Then the expected time for completing each phase is estimated separately, using arguments most appropriate to that phase. Jansen and Wegener [30] showed that in expected time $O(\mu n)$ the population reaches a state where all search points have $2n/3$ ones (or the optimum is found beforehand). From there, the algorithm can focus on maximizing the maximum block length $b(x)$. In the next expected $O(n^2 \log(n) + (\mu \log \mu)n)$ generations the algorithm evolves a population where all search points x have the maximum block length $b(x) = 2n/3$ (or the optimum has been found).

Once such a population has been reached, we can rely on the diversity mechanism taking effect: since then all search points have the same fitness, selection for replacement is based solely on the number of duplicates in the population. There are only $2n/3+1$ different genotypes with $2n/3$ ones and a block length of $2n/3$:

$$1^{2n/3}0^{n/3},\ 01^{2n/3}0^{n/3-1},\ 0^2 1^{2n/3}0^{n/3-2},\ \dots\ ,0^{n/3}1^{2n/3}.$$

The population size μ is large enough to be able to store all these search points. Hence, once a particular genotype is created, the population will always retain such a genotype until the optimum is found. Using appropriate 2-bit flips, it is possible to create novel genotypes. The expected time until the population contains all the above genotypes is $O(\mu n^3)$.

Once the population contains the genotypes $1^{2n/3}0^{n/3}$ and $0^{n/3}1^{2n/3}$, if these are selected as parents, one-point crossover can easily create 1^n by choosing a cutting point in the middle third of the bit string. The expected time for this event is $O(\mu^2/p_c)$, and summing all expected times yields the claimed bound.

8.4.2 Coloring Problems

Fischer and Wegener [20] presented another example where a diversity mechanism enhances crossover for a combinatorial problem. They considered a simple variant of the Ising model, a well-known model of ferromagnetism that is NP-hard to solve in its general case. Here we consider an easy special case where vertices of an undirected graph can have one of two states,

0 and 1 (also referred to as "colors"), and it is beneficial to color two neighboring vertices with the same color. Then the fitness function corresponds to the number of monochromatic edges, and all colorings where all connected components have the same color are global optima.

The problem is an interesting test bed for evolutionary algorithms because subgraphs of the same color can be regarded as "building blocks" of optimal solutions. The inherent symmetry in the problem implies that competing building blocks may emerge, and evolutionary algorithms can get stuck in difficult local optima, depending on the graph.

For bipartite graphs, the problem is equivalent to the well-known GRAPH COLORING problem or, more specifically, to the 2-coloring problem where the goal is to color the graph with 2 colors such that no two adjacent vertices have the same color, and the fitness function is the number of correctly colored vertices. The reason is that there is a simple bijection between the Ising model and GRAPH COLORING: flipping all colors of one set of the bipartition turns all monochromatic edges into bichromatic edges, hence turning a solution for the Ising model into a GRAPH COLORING solution of the same fitness, and vice versa. All results derived for the Ising model variant described above also hold for the 2-coloring problem, if the underlying graph is bipartite.

Fischer and Wegener [20] studied ring graphs (or cycle graphs) where the i-th vertex is a neighbor of vertices $i-1$ and $i+1$ (identifying vertex 0 with vertex n), and observed that the fitness landscape contains a large number of plateaus. A search point such as 0001111000 contains blocks of bits with the same color, for example a block of four 1-bits. A mutation flipping only the first or the last bit of such a block can shorten the block; a mutation flipping only the last bit before the block or the first bit following the block can enlarge it. Those mutations are fitness-neutral (i.e., do not change the fitness) unless some block disappears, which leads to an increase in fitness. The main results from [20] are as follows. Note that rings with an even number of vertices are bipartite, allowing us to add statements about 2-coloring.

Theorem 8.4.2 (adapted from Theorems 4 and 5 in [20]). *The expected optimization time for the (1+1) EA on the Ising model and the 2-coloring problem on rings with even n is $O(n^3)$. This bound is asymptotically tight when one starts with two blocks of length εn and $(1-\varepsilon)n$, $0 < \varepsilon < 1/2$ a constant.*

The main observation is that the length of any block follows a fair random walk (apart from boundary states), with a large self-loop probability as the probability of changing the length of the block considered is $\Theta(1/n)$. From the setting described above with two blocks, it takes $\Theta(n^2)$ of these changes for some block to disappear, which results in a global optimum.

The use of a simple GA with fitness sharing, however, is able to find a global optimum in expected time $O(n^2)$.

Theorem 8.4.3. *Consider a (2+2) GA as a variant of Algorithm 8.9 that with probability* $p_c = 1/2$ *applies two-point crossover once to create two offspring and uses fitness sharing with sharing radius* $\sigma = n$ *to select two search points from amongst parents and offspring that maximizes the shared fitness of the population. The expected number of fitness evaluations until this GA finds an optimum for the Ising model and the 2-coloring problem on rings with even* n *is bounded by* $O(n^2)$.

The main observation is that fitness sharing turns plateaus into gradients, as it rewards the creation of dissimilar individuals. The GA then efficiently creates two complementary individuals (e.g., through 1-bit flips), and then two-point crossover is able to invert whole blocks, provided that the cutting points are chosen between two blocks, for example turning 00**110**010 into 00001100 by replacing the bits in bold with the values from the complementary parent.

A similar but more drastic effect was also shown for coloring complete binary trees [57]. Here, subtrees of the same color represent building blocks of good solutions. The problem is much harder than coloring rings, as it contains difficult local optima, for example when the two subtrees of the root are colored with different colors.

The present author showed that all algorithms in a large class of $(\mu+\lambda)$ EAs, with arbitrary mutation rates, need at least expected time $2^{\Omega(n)}$ to find a global optimum. In contrast, the (2+2) GA with fitness sharing finds an optimum in expected polynomial time.

Theorem 8.4.4. *Consider the (2+2) GA with fitness sharing described in Theorem 8.4.3. The expected optimization time for the Ising model and the 2-coloring problem on a complete binary tree with* n *vertices is bounded by* $O(n^3)$.

The analysis shows that fitness sharing again encourages an increase in the Hamming distance between the two current search points, x and y. In the case where x and y are complementary, two-point crossover can effectively substitute subtrees to increase the fitness. The challenge lies in showing that complementary search points evolve, and how they do so. In contrast to rings, binary trees do not contain any plateaus, and hence it is not always possible to increase the Hamming distance without compromising on fitness. Interestingly, a case distinction according to the function $H(x,y) + f(x) + f(y)$ shows that if this function is small then there are accepted 1-bit flips that increase the real fitness, possibly at the cost of decreasing the Hamming distance $H(x,y)$. But, if the function is large, there are accepted 1-bit flips that increase the Hamming distance at the expense of the real fitness. There is a "gray area" in between, where more complex operations (mutation and/or crossover) are required; however, these steps have probability $\Omega(1/n^2)$, leading to the overall time bound of $O(n^3)$.

In this scenario, even though fitness sharing can maximize diversity at the expense of the real fitness, it turns out to be an effective strategy, as the diversity of complementary search points can be exploited efficiently by crossover.

8.4.3 Diversity and Crossover Speed Up Hill Climbing

Diversity and crossover also prove useful in a very natural and well-known setting, albeit with smaller speedups compared with the examples seen so far. The simple problem OneMax is the most studied problem in the theory of randomized search heuristics. It can be regarded as a simple hill-climbing task, as a mutation flipping a single 0-bit to 1 increases the fitness. It can also be seen as a problem where ones are "building blocks" of the global optimum, and the algorithm has to assemble all building blocks to find the optimum. This perspective is related to the so-called "building block hypothesis," an attempt to explain the advantage of crossover, as GAs with crossover can combine building blocks of good solutions. Yet it has been surprisingly hard to come up with natural examples and rigorous proofs to cement or refute this hypothesis.

This author [56, 59] showed that the (μ+λ) GA with the duplicate-based tie-breaking rule is twice as fast as the fastest evolutionary algorithm using only standard bit mutation (modulo small-order terms).

Theorem 8.4.5 (simplified from Theorems 1 and 4 in [56][4]). *Let $n \geq 2$ and let $c > 0$ be a constant. Every evolutionary algorithm that uses only standard bit mutation with mutation rate $p = c/n$ to create new solutions has an expected optimization time of at least $\frac{e^c}{c} \cdot n \ln n \cdot (1 - o(1))$ on* OneMax *and every other function with a unique optimum.*

The expected optimization time of the (μ+λ) GA breaking ties towards including individuals with the fewest duplicates in $P \cup P'$, with $0 < p_c < 1$ constant, mutation probability $p = c/n$, and $\mu, \lambda = o((\log n)/(\log \log n))$, on OneMax *is at most $\frac{e^c}{c \cdot (1+c)} \cdot n \ln n \cdot (1 + o(1))$.*

Modulo small-order terms, this is a speedup of $1 + c$, which is 2 for $c = 1$, reflecting the default mutation rate $p = 1/n$.

The idea behind the proof is to make a case distinction for all possible populations, according to the current best-so-far fitness and the diversity in the population, and then to upper bound the expected time spent in all these cases.

If a population contains individuals of different fitness values, the individuals of current best-so-far fitness i quickly take over the population (or an improvement of the best-so-far fitness is found). Owing to our restrictions on

[4] We remark that the results in [56] hold for much larger ranges of the mutation rate p and arbitrary parent selection mechanisms that do not disadvantage individuals with higher fitness.

the population sizes μ, λ, the total time across all best-so-far fitness values is $o(n \log n)$. If the population consists of μ identical genotypes of fitness i, this state will be left for good if either a fitness improvement is found, or a different individual with the same fitness is created. In the latter case, the diversity mechanism in the tie-breaking rule ensures that this diversity will never get lost (unless an improvement is found).

This diversity can be created by a fitness-neutral mutation that flips the same number of 0-bits to 1 as it flips 1-bits to 0. Such a multi-bit flip would be irrelevant for mutation-only evolutionary algorithms. But when crossover is used, it can exploit the diversity created in this way by choosing two parents with equal fitness but different genotypes, and creating a surplus of ones on the bit positions where the two parents differ. Creating such a surplus is very likely; the probability of such an event is at least 1/4, irrespective of the Hamming distance between the two parents. The time the algorithm spends evolving a diverse population is negligible compared with the time spent in a state where all individuals are identical.

The expected time is thus dominated by the time spent trying to leave states where all genotypes are identical. Compared with mutation-based evolutionary algorithms, the creation of diversity offers another route towards fitness improvements as crossover rapidly exploits this diversity, creating improvements almost instantly.

Corus and Oliveto [4] recently showed that the choice of tie-breaking rule is important for getting the above-mentioned speedup: when it is replaced with a uniform tie-breaking rule, we still get a constant-factor speedup, but the constant is worse.

Theorem 8.4.6 (simplified from Theorem 9 in [4]). *The expected optimization time of the (μ+1) GA with uniform tie-breaking, $p_c = 1$, mutation probability $p = c/n$, and $3 \leq \mu = o((\log n)/(\log\log n))$, on* OneMax *is at most* $\frac{e^c}{c \cdot (1+c/3)} \cdot n \ln n \cdot (1+o(1))$.

For the standard mutation rate of $1/n$, the previous speedup of 2 in Theorem 8.4.5 now becomes a factor of 4/3. This constant is the best possible under mild assumptions [4, Theorem 11].

8.4.4 Overcoming Fitness Valleys with Naturally Emerging Diversity and Crossover

In Section 8.4.3 we have seen that diversity and crossover can speed up hill climbing on OneMax by a constant factor. Now we consider the task of overcoming fitness valleys in order to solve multimodal problems. We specifically focus on the problem class Jump_k, the first example function where crossover was proven to be beneficial [29]:

$$\text{Jump}_k(x) = \begin{cases} k+|x|_1 & \text{if } |x|_1 = n \text{ or } |x|_1 \leq n-k, \\ n-|x|_1 & \text{otherwise.} \end{cases}$$

In this problem, GAs have to overcome a fitness valley such that all local optima have $n-k$ ones and thus a Hamming distance k to the global optimum, 1^n. Jansen and Wegener [29] showed that while mutation-only algorithms such as the (1+1) EA require expected time $\Theta(n^k)$, a simple (μ+1) GA with crossover needs only time $O(\mu n^2 k^3 + 4^k/p_c)$. This time is $O(4^k/p_c)$ for large k, and hence significantly faster than for mutation-only GAs.

The factor $4^k/p_c$ results from the fact that, if the population contains pairs of parents that do not share a common 0-bit, then uniform crossover can set all the $k+k=2k$ bits where exactly one parent has a 1 to 1 in the offspring, with probability $2^{-2k} = 4^{-k}$. Hence the expected time for a successful crossover that creates the optimum is bounded by $4^k/p_c$. Note that two such parents have the largest possible Hamming distance, $2k$, between local optima, and hence populations typically achieve the maximum possible diversity between many pairs of parents. A drawback of the analysis in [29] is that it requires an unrealistically small crossover probability $p_c \leq 1/(ckn)$, for a large constant $c > 0$.

Kötzing, Sudholt, and Theile [37] later refined these results towards a crossover probability $p_c \leq k/n$, which is still unrealistically small. Both approaches focus on creating a maximum Hamming distance between local optima through a sequence of lucky mutations, relying on crossover to create the optimum once sufficient diversity has been created. The arguments break down if crossover is applied frequently. Hence, these analyses do not reflect the typical behavior in GA populations with constant crossover probabilities $p_c = \Theta(1)$ as used in practice.

We now review recent results from Dang et al. [11] where realistic crossover probabilities were considered, at the cost of a smaller (but still significant) speedup. Previous work [29, 37] relied on independent mutations providing diversity, and regarded crossover as potentially harmful, as the effect of crossover on diversity was not well understood. This led to a worst-case perspective on crossover: previous proofs considered mutation to build up diversity over time, like a house of cards, with the worst-case assumption being that one unexpected application of crossover would destroy the buildup of diversity, collapsing the house of cards, and the buildup of diversity had to restart from scratch. This view is backed up by a negative result [37, Theorem 8], showing that if only crossover with $p_c = \Omega(1)$ is used, but no mutation following crossover, diversity reduces quickly, leading to inefficient running times for small population sizes ($\mu = O(\log n)$).

In [11] a different perspective was offered, an approach loosely inspired by population genetics: the paper showed that crossover, when followed by mutation, can actually be very beneficial in creating diversity. Note that the perspective of crossover creating diversity is common in population genetics [34, 63]. A frequent assumption is that crossover mixes all alleles in a population, leading to a situation called *linkage equilibrium*, where the state of a population is described by the frequency of alleles [1].

The main result can be stated as follows.

Theorem 8.4.7 (Theorem 6 in [11], simplified for $k \geq 3$). *The expected optimization time of the (μ+1) GA with $p_c = 1$ and $\mu \leq \kappa n$, for some constant $\kappa > 0$, on* Jump_k, *$3 \leq k = o(n)$, is $O(n^k/\mu + n^{k-1}\log(\mu))$.*

For $\mu = \kappa n$, the bound simplifies to $O(n^{k-1}\log n)$, a speedup of order $\Omega(n/\log n)$ compared with the expected time of $\Theta(n^k)$ for the (1+1) EA [29].

The analysis shows that, on Jump_k, diversity emerges naturally in a population: the interplay of crossover, followed by mutation, can serve as a catalyst for creating a diverse range of search points out of few different individuals. Consider the situation where all individuals in the population are local optima with $n-k$ ones, and assume pessimistically that there is no diversity: all individuals are identical. In the following, we refer to a collection of identical individuals with $n-k$ ones as a *species*. Mutation is able to create a new species, for instance by flipping a single 0-bit and a 1-bit. This new species can grow in size, or become extinct over time.

Crossing over two individuals from different species can easily create a surplus of ones, where the offspring has $n-k+1$ ones. The following mutation now creates a local optimum if it flips a 1-bit back to 0. Note that here there are $n-k+1$ ones to choose from, each leading to a different species. This means that, once mutation has created a small amount of diversity, crossover and mutation can work together in this way to create a burst of diversity that has a good chance of prevailing for a long time, before the population loses all diversity or the global optimum is found.

In the proof of Theorem 8.4.7, the size of the largest species is taken as a potential function: if the size of the largest species is μ, there is no diversity, but if it is bounded away from μ, it is easy to select two parents from different species with uniform parent selection. The size of the largest species behaves like an almost fair random walk, and the population has a good chance of spending long periods of time in states where the size of the largest species is small. In these situations, when two parents from different species are selected, crossover has a chance to create a surplus of 1-bits, and then the global optimum can be found by flipping the at most $k-1$ remaining 0-bits to 1.

This argument also shows that speedups can be achieved from small amounts of diversity; in contrast to previous work [29, 37] it is not necessary to rely on a maximum Hamming distance between parents emerging.

A further finding in [11] is that increasing the mutation rate to $p = (1+\delta)/n$ for an arbitrarily small constant $\delta > 0$ turns the almost fair random walk describing the size of the largest species into an unfair random walk that is biased towards increased diversity. In other words, larger mutation rates facilitate the emergence and maintenance of diversity in this setting. This

Table 8.2 Overview of the main results of [11, 12] (Theorems 8.4.7 and 8.4.8), restricted to $3 \leq k = o(n)$. Most bounds come with mild restrictions on k, μ, or p_c; see [11, 12] for details. The second column shows simplified bounds, assuming a choice of μ and p_c that yields the best possible upper bound

Mechanism	General μ, p_c	Best μ, p_c
None, $p = 1/n$	$O(n^k/\mu + n^{k-1}\log\mu)$	$O(n^{k-1}\log n)$
None, $p = (1+\delta)/n$	$O(\mu^2 + n^{k-1})$	$O(n^{k-1})$
Duplicate elimination	$O(\mu^2 n + n^{k-1})$	$O(n^{k-1})$
Duplicate minimization	$O(\mu n + n^{k-1})$	$O(n^{k-1})$
Deterministic crowding	$O(\mu n + n\log n + ne^{5k}\mu^{k+2})$	$O(n\log n + ne^{5k}2^k)$
Convex hull max.	$O(\mu n^2 \log n + 4^k/p_c)$	$O(n^2\log n + 4^k)$
Hamming distance max.	$O(n\log n + \mu^2 kn\log(\mu k) + 4^k/p_c)$	$O(n\log n + nk\log k + 4^k)$
Fitness sharing	$O(n\log n + \mu^2 kn\log(\mu k))$	$O(n\log n + nk\log k + 4^k)$
Island model	$O(n\log n + \mu^2 kn + \mu^2 4^k)$	$O(n\log n + kn + 4^k)$

leads to the following improved upper bound, which for reasonably small μ gives a speedup of order n over the expected time of the (1+1) EA.

Theorem 8.4.8 (Theorem 10 in [11], simplified for $k \geq 3$). *The (μ+1) GA with mutation rate $(1+\delta)/n$, for a constant $\delta > 0$, and a population size $\mu \geq ck\ln(n)$ for a sufficiently large constant $c > 0$, has for $3 \leq k = o(n)$ an expected optimization time $O(\mu^2 + n^{k-1})$ on Jump_k.*

8.4.5 Speeding Up Fitness Valley Crossing with Explicit Diversity Mechanisms

The performance of the (μ+1) GA on Jump_k can be further improved by using explicit diversity mechanisms in the tie-breaking rule of the (μ+1) GA. This was studied by Dang et al. [12], and the main results are summarized in Table 8.2. For comparison, the table also contains results reviewed in Section 8.4.4 for uniform tie-breaking, where no diversity mechanism is used.

The different mechanisms (except for the island model) appear only in the tie-breaking rule; they are described as follows, along with the main ideas behind their analysis.

Duplicate elimination always chooses an individual for removal that has duplicates in the population, if duplicates exist. Otherwise, it removes an individual uniformly at random. The analysis shows that after $O(\mu^2 n)$ generations in expectation, there will only be $(1 - \Omega(1))\mu$ duplicates in the population, and this property will be maintained forever. Then the probability of picking nonidentical parents is $\Omega(1)$. Then, as argued in Section 8.4.4, crossover followed by mutation can find the optimum with probability $\Omega(n^{k-1})$, as crossover creates a surplus of 1-bits with probability $\Omega(1)$ and then mutation has to flip at most $k-1$ bits to reach the optimum.

Duplicate minimization is the familiar rule that breaks ties towards including individuals with the fewest duplicates in $P \cup P'$. Here it is easy to show that the size of the largest species decreases to $(1-\Omega(1))\mu$ in expected time $O(\mu n)$. Then we apply the trail of thought for duplicate elimination.

Deterministic crowding, in the case of fitness ties, always removes the parent if the offspring was created by mutation only, or one of the two parents chosen uniformly at random if the offspring was created by crossover and mutation. The analysis follows the approach of [37], relying on a sequence of events that evolves in a pair of search points that have a maximum Hamming distance of $2k$. Then there is a reasonable chance that uniform crossover will create the optimum by crossing over these parents and setting all differing bits to 1.

Convex hull maximization breaks ties towards maximizing the convex hull of the population, which is the set of search points that can be produced from uniform crossover of any two parents. More precisely, we maximize the convex hull by maximizing the number of bit positions where the population contains both a 0 and a 1 in some individual. The analysis, similar to that in [23], shows that in expected time $O(\mu n^2 \log n)$ a maximum amount of diversity is created, where all of the $\mu k \leq n$ zeros in the population occupy different bit positions. Then any two (different) parents have maximum Hamming distance $2k$, and the optimum can be constructed with probability $p_c \cdot 4^{-k}$ (as argued earlier).

Hamming distance maximization breaks ties towards maximizing the total Hamming distance between all pairs of search points. Similarly to convex hull maximization, we reach a population of maximum diversity in expected time $O(\mu^2 kn \log(\mu k))$. Repeating the arguments from there yields the claimed bound.

Fitness sharing, with a sharing radius of $\delta \geq 2k$ in the setting of populations with equal fitness, turns out to be equivalent to maximizing the total Hamming distance between pairs of search points, and hence the previous analysis carries over.

The island model uses a particular topology called the *single-receiver model* [61], where μ islands run a (1+1) EA independently, and there is a single receiver island that in every generation chooses two islands uniformly at random, copies their current search points, and performs a uniform crossover on these. The analysis shows that, when we fix any two islands, these islands will either have zeros in different positions or will, in expectation, reduce the number of bit positions where they have a zero in common. Once the islands have no zero in common, the receiver island has a good chance to create the optimum when crossing over individuals from these two islands.

The island model with the single-receiver topology was introduced in [61], and the paper relied on this diversity mechanism to prove that a constructed royal road function with a building-block structure could be solved efficiently by crossover. It was also used in [47], where it was shown that crossover during migration can be effective for constructed functions as well as for instances of the Vertex Cover problem. We refer the reader to [47, 61] and the survey [60, Section 46.5.4] for details.

8.5 How Diversity Benefits Dynamic Optimization

Another very important use of population diversity is to ensure good performance in dynamic optimization, where the problem can change over time. Diversity can ensure that the population is able to keep track of global optima, or to rediscover global optima when different local optima change their fitness and another local optimum becomes the new global optimum.

The runtime analysis of dynamic evolutionary optimization is still in its very infancy, with only a few results available (e.g., [13, 33, 35, 36, 42, 51, 52]).

8.5.1 Diversity Mechanisms for BALANCE

Oliveto and Zarges [51] considered diversity mechanisms for the dynamic function BALANCE [52].

Definition 8.5.1 (BALANCE [52]). Let $a, b \in \{0,1\}^{n/2}$ and $x = ab \in \{0,1\}^n$. Then

$$\text{BALANCE}(x) = \begin{cases} n^3 & \text{if } \text{LO}(a) = n/2, \text{else} \\ |b|_1 + n \cdot \text{LO}(a) & \text{if } n/16 < |b|_1 < 7n/16, \text{else} \\ n^2 \cdot \text{LO}(a) & \text{if } |a|_0 > \sqrt{n}, \text{else} \\ 0 & \text{otherwise,} \end{cases}$$

where $|x|_1 = \sum_{i=1}^{n/2} x_i$, $|x|_0 = n/2 - |x|_1$ is the number of zeros, and $\text{LO}(x) := \sum_{i=1}^{n/2} \prod_{j=1}^{i} x_j$ counts the number of leading ones.

For the majority of search points, the function gives hints to maximize the number of ones in the suffix (also referred to as the ONEMAX part), and even stronger hints to maximize the number of leading ones in the prefix (the leading-ones part). All search points with a maximum of $n/2$ leading ones are global optima; however, the function also contains two traps and a fitness valley of fitness 0 that separates the traps from the region of global optima.

The upper trap contains all search points with more than $7n/16$ ones, and the lower trap contains all search points with fewer than $n/16$ ones.

The function was used in a dynamic framework where, every τ generations, for a change frequency parameter τ, the roles of zeros and ones in the suffix are reversed, so that the fitness gradient switches between maximizing and minimizing the number of ones in the ONEMAX part. Unless stated otherwise, the arguments given below assume that the number of ones is maximized.

Oliveto and Zarges [51] showed that a (μ+1) EA with no diversity mechanism tends to fail on BALANCE, as the whole population is likely to run into one of the traps.

Theorem 8.5.2. *If $\tau > 20\mu n$ and $\mu \leq n^{1/2-\varepsilon}$, then the expected time for the (μ+1) EA to optimize* BALANCE *is at least $n^{\Omega(\sqrt{n})}$. If $\tau > 38\mu n^{3/2}$ and $\mu \leq n^{1/2-\varepsilon}$, then the ($\mu$+1) EA requires at least $n^{\Omega(\sqrt{n})}$ steps with overwhelming probability.*

The intuitive reason for this poor performance is that it is easier to optimize ONEMAX than it is to maximize the number of leading ones, and the algorithm only needs to come moderately close to the ONEMAX (or ZEROMAX) optimum to fall into a trap. With low frequencies of change, this is very likely to happen.

The authors of [51] investigated how far this poor performance can be mitigated by using diversity-preserving mechanisms such as the ones studied for TWOMAX in [22]. The main results are explained in the following.

Genotype diversity, that is, preventing genotype duplicates from being accepted, is too weak to affect the main search behavior; the (μ+1) EA still tends to run into traps.

Theorem 8.5.3. *For the (μ+1) EA with genotype diversity (Algorithm 8.2), the results in Theorem 8.5.2 apply.*

Deterministic crowding does not help: recall that deterministic crowding is based on offspring competing against their direct parents, and hence (since no crossover is used) the (μ+1) EA evolves μ independent lineages. Each lineage still has a high probability of running into a trap; hence, for polynomial population sizes and low frequencies of change, there is a high probability that the whole population will be led into a trap.

Theorem 8.5.4. *With overwhelming probability, the (μ+1) EA using deterministic crowding and $\mu \leq n^{O(1)}$ requires exponential time to optimize* BALANCE *if $\tau > 8e\mu n$.*

Fitness diversity as in Algorithm 8.3 turns out to perform a lot better: it can find the optimum efficiently for all values of τ. This is surprising as this mechanism showed the worst performance for TWOMAX [22].

Theorem 8.5.5. *Let $\mu > n-2(\sqrt{n}-1)$. Then, with overwhelming probability, the (μ+1) EA with fitness diversity optimizes* BALANCE *in time $O(\mu n^3)$ for arbitrary $\tau \geq 0$.*

The proof shows that, as the population size is quite large, the (μ+1) EA is able to "fill up" both traps in the sense that the algorithm will eventually contain individuals representing all fitness values inside a trap, and then no other point in the trap will be accepted. This then allows the algorithm to evolve a lineage leading to the global optimum, avoiding the trap.

Finally, Oliveto and Zarges [51] considered a variant of the (μ+1) EA with population size $\mu = 2$ combining fitness sharing (with $\alpha = 1$ and sharing radius $\sigma = n$) and deterministic crowding: in the selection step, the shared fitness of the current population is compared against the shared fitness of the population where the offspring replaces its parent, and the latter population is selected if its shared fitness is no smaller. Instead of standard bit mutation, local mutations are used that flip exactly one bit chosen uniformly at random, as done in RLS. The resulting algorithm is referred to as the (2+1) RLS.

The (2+1) RLS is efficient with probability close to 1/2.

Theorem 8.5.6. *With probability at least* $1/2 - e^{-\Omega(n)}$, *the (2+1) RLS with fitness sharing and crowding finds the optimum of* BALANCE *in time* $O(n^2)$ *for arbitrary* $\tau \geq 0$.

The analysis observes, similarly to [57] reviewed earlier in Section 8.4.2, that the function $H(x,y) + f(x) + f(y)$ for current search points x, y decides whether $H(x,y)$ can be increased at the expense of fitness, or whether the fitness can be increased at the expense of the Hamming distance. Bit flips in the ONEMAX part of BALANCE only change the fitness by 1. If $H(x,y) + f(x) + f(y) > 2n$, such bit flips are accepted if and only if they increase the Hamming distance. If $H(x,y) + f(x) + f(y) < 2n$, such bit flips are accepted if and only if they increase the fitness.

Now, a fitness larger than $2n$ is easily achieved if, at initialization, x and y have a total of at least two leading ones. This happens with probability at least 1/2, and then the (2+1) RLS will always have a fitness larger than $2n$. Then any bit flips in the ONEMAX part will be accepted only if they increase the Hamming distance $H(x,y)$. With high probability there will be many bit positions i where $x_i = 1, y_i = 0$ and many bit positions j where $x_j = 0, y_j = 1$. These values will never change, and hence the ONEMAX part of any search point will never meet the extreme values corresponding to a trap. The leading-ones part will be optimized, as it has a much larger impact on the fitness, leading to a global optimum in the claimed time.

However, the algorithm can also fail badly with constant probability, getting stuck in a local optimum from which there is no escape.

Theorem 8.5.7. *Let* $\tau > 12n + 1$. *With a probability bounded from below by a constant, the (2+1) RLS with fitness sharing and crowding requires infinite time to optimize* BALANCE.

This statement can be shown by observing that with constant probability, the fitness will remain below $2n$, hence maximizing the number of ones in the ONEMAX part, while one of the search points reaches the upper trap.

Then the fitness will always be larger than $2n$, which makes the algorithm maximize the Hamming distance and hence drives the other search point into the lower trap. Here the algorithm gets stuck, as the traps cannot be left and local mutations cannot create the global optimum from a trap.

An interesting conclusion when contrasting the performance of diversity mechanisms on TwoMax [22] and Balance [51] is that mechanisms that perform well on one function may not perform well on the other. Fitness diversity shows the worst performance guarantees for TwoMax, but it performs the best on Balance. Deterministic crowding performs well on TwoMax, but performs poorly on Balance. Fitness sharing performs the best on TwoMax, but is only effective on Balance with constant probability, and otherwise fails badly.

8.5.2 Island Models for the Maze Function

Lissovoi and Witt [43] presented another example where diversity mechanisms prove useful in dynamic optimization. They showed that island models can help to optimize the dynamic function Maze, introduced earlier by Kötzing and Molter [36]. The function Maze changes in phases of t_0 steps. In the first phase, the function is equivalent to OneMax. In the next n phases, higher fitness values are assigned to two search points on a shortest Hamming path from 1^n to 0^n, in an oscillating pattern. Every two iterations out of three, 0^i1^{n-i} receives fitness $n+2$, while the previous point on the path, $0^{i-1}1^{n-i+1}$, receives fitness $n+1$. Every three iterations, the fitness values of these two points are reversed. In every phase, the index i increases by 1. All other search points always retain their OneMax value; hence, whenever an algorithm loses track of the path, it is likely to be led back into 1^n. The optimum can only be reached if an algorithm tracks the moving optimum on the whole Hamming path, eventually reaching 0^n after n phases. We refer to [36, 43] for formal definitions of Maze.

The (1+1) EA fails badly on Maze [36] and the same holds for a (1+λ) EA with a moderate offspring population size, as shown in the following theorem. The reason is that in every phase there is a constant probability that the algorithm will maintain the previous point on the path, $0^{i-1}1^{n-i+1}$, and will fall off the path once the next phase starts.

Theorem 8.5.8. *The (1+λ) EA with $\lambda = O(n^{1-\varepsilon})$, for any constant $\varepsilon > 0$, will with high probability lose track of the optimum of* Maze, *i.e., with high probability it will require an exponential number of iterations to construct the final optimum.*

In sharp contrast, a simple island model running λ (1+1) EAs is effective on Maze, even with a much smaller number λ of offspring created in each generation.

Theorem 8.5.9. *An island model with $\lambda = c \log n$ islands, where c is a sufficiently large constant, with each island running a (1+1) EA, and migration on a complete topology occurring during the first iteration of every phase (i.e., with migration interval $\tau = t_0$), is able to find the optimum of* MAZE *with phase length $t_0 = kn^3 \log n$ in polynomial time with high probability.*

The intuitive reason is that each island on the path has a constant probability of ending the phase in $0^{i-1}1^{n-i+1}$ and a constant probability of ending it in 0^i1^{n-i}. In the latter case, these islands will still be on the path once the index i has increased at the start of the next phase. There is a high probability that at least one island will still be on the path, and its fitness will be no less than that of all the other islands. Hence migration will ensure that all islands that may have fallen off will be put back onto the path.

Note that the choice of the migration interval aligns with the time interval t_0 for dynamic changes, such that at the time of migration, the search points further up on the path have a higher fitness. If migration occurs at other points in time, island models with $O(\log n)$ islands may still fail on MAZE [43, Theorem 14].

8.6 Diversity-Based Parent Selection

All the results surveyed so far use diversity mechanisms in the environmental selection, i.e., to decide which search points are allowed to survive to the next generation. Here we present recent work by Covantes Osuna et al. [6], who suggested using diversity mechanisms in the parent selection in the context of evolutionary multiobjective optimization.

Well-established multiobjective evolutionary algorithms (MOEAs) such as NSGA-II [15], SPEA2 [2], and IBEA [66] have two basic principles driven by selection. First of all, the goal is to push the current population close to the "true" Pareto front. The second goal is to "spread" the population along the front such that it is well covered. The first goal is usually achieved by dominance mechanisms between the search points or by indicator functions that prefer nondominated points. The second goal involves the use of diversity mechanisms. Alternatively, indicators such as the hypervolume indicator play a crucial role in obtaining a good spread of the different solutions of the population along the Pareto front.

In the context of evolutionary multi-objective optimization (EMO), parent selection is usually uniform, whereas offspring selection is based on dominance and the contribution of an individual to the diversity of the population. The paper [6] shows that diversity mechanisms can also be highly beneficial when embedded into the parent selection mechanisms in EMO. The goal is to speed up the optimization process of an EMO algorithm by selecting individuals that have a high chance of producing beneficial offspring. The idea is to use a diversity metric, such as the hypervolume contribution or the crowding distance contribution, and to preferably select parents with a higher diversity score. The hypervolume describes the area that is dominated by points in the population; the hypervolume contribution describes the contribution a search point x makes to the hypervolume, i.e., the difference between the hypervolume of the whole population P and that of the population $P \setminus \{x\}$ without x. The crowding distance is a well-known measure from NSGA-II; it is based on the distances in objective space to the search points with the closest objective values, considering each objective separately.

The main assumption is that individuals with a high diversity score are located in poorly explored or less dense areas of the search space, so that the chances of creating new nondominated individuals are better than in areas where there are several individuals. In this sense the new parent selection schemes focus on individuals where the neighborhood is not fully covered and, in consequence, force reproduction in those areas and lead to the spread of the population over the search space.

We consider two well-known pseudo-Boolean functions $\{0,1\}^n \to \mathbb{N}^2$ with two objectives. For

$$\text{ONEMINMAX}(x_1,\ldots,x_n) := \left(n - \sum_{i=1}^{n} x_i, \sum_{i=1}^{n} x_i\right),$$

the aim is to maximize the numbers of zeros and ones at the same time. For

$$\text{LOTZ}(x_1,\ldots,x_n) := \left(\sum_{i=1}^{n}\prod_{j=1}^{i} x_j, \sum_{i=1}^{n}\prod_{j=i}^{n}(1-x_j)\right),$$

the goal is to simultaneously maximize the numbers of leading ones and trailing zeros.

OneMinMax has the property that every single solution represents a point in the Pareto front and that no search point is strictly dominated by another one. The goal is to cover the whole Pareto front, i.e., to compute a set of individuals that contains for each i, $0 \le i \le n$, an individual with exactly i ones. In the case of LOTZ, all non-Pareto-optimal decision vectors only have Hamming neighbors that are better or worse, but never incomparable to them. This fact facilitates the analysis of population-based algorithms, which certainly cannot be expected from other multiobjective optimization problems. Note that the Pareto front for LOTZ is given by the set of $n+1$ search points $\{1^i 0^{n-i} \mid 0 \le i \le n\}$.

We consider the Simple Evolutionary Multiobjective Optimizer (SEMO), shown in Algorithm 8.10, which is popular for theoretical analyses owing to its simplicity. The paper [6] also contains results for a variant, GSEMO, which uses standard bit mutations instead of local mutations. For simplicity, we present only results for SEMO in this survey.

Algorithm 8.10: SEMO

1 Choose an initial solution $s \in \{0,1\}^n$ uniformly at random.
2 Determine $f(s)$ and initialize $P := \{s\}$. **while** *not stopping* **do**
3 Choose s uniformly at random from P.
4 Choose $i \in \{1,\dots,n\}$ uniformly at random.
5 Define s' by flipping the i-th bit of s.
6 **if** *s' is not dominated by any individual in P* **then**
7 Add s' to P, and remove all individuals weakly dominated by s' from P.

The following theorem summarizes results from [6, 25, 41] on the performance of SEMO.

Theorem 8.6.1. *The expected time for SEMO to cover the whole Pareto front on* OneMinMax *and* LOTZ *is $\Theta(n^2 \log n)$ and $\Theta(n^3)$, respectively.*

The expected time is larger by a factor of $\Theta(n)$ than the expected time of the (1+1) EA for optimizing any single objective. The reason for SEMO being slower is that, once the Pareto front has been reached, only the search points with a maximum objective value, when chosen as parents, can expand the Pareto front further. All other choices of parents lead to the creation of an offspring whose objective values are already represented in the population. Once the population has grown to a linear size $\mu = \Theta(n)$, the probability of selecting a parent that allows SEMO to progress is only $\Theta(1/n)$, i.e., most steps are wasted. This leads to the additional factor of order n compared with the (1+1) EA.

Diversity-based parent selection using either the hypervolume contribution or the crowding distance contribution can improve these running times. The parent selection mechanisms considered use one of these diversity metrics to select parents according to processes that favor higher diversity: sorting the population according to the ranks of the diversity metric and picking the i-th ranked individual with probability proportional to 2^{-i} (exponential scheme) or $1/i^2$ (power-law scheme), or using tournament selection based on the diversity score with tournament size μ (i.e., the current size of the population). Here, the tournament is picked with replacement; hence search points can be picked multiple times or be excluded from the tournament.

Theorem 8.6.2. *Consider diversity-based parent selection using either the hypervolume contribution or the crowding distance contribution and selecting parents according to the exponential or power-law scheme, or according to a tournament with tournament size μ.*

Then the expected time for SEMO with diversity-based parent selection to cover the whole Pareto front on OneMinMax *and* LOTZ *is $O(n \log n)$ and $O(n^2)$, respectively.*

The proofs show that the expected time for SEMO is bounded from above by $O((n \log n)/p_{\text{good}})$ and $O(n^2/p_{\text{good}})$, respectively, where p_{good} is (a lower bound on) the probability of selecting a parent that has a Hamming neighbor whose objective vector is on the Pareto front, but not yet represented in the population. The diversity score assigns the highest values to the search points with maximum objective values. However, the extreme points 0^n and 1^n may themselves not be "good" search points; if the Pareto front has reached one "end" of the search space, SEMO still may need to expand in the other direction. All of the parent selection mechanisms mentioned have a probability of $\Omega(1)$ of selecting the individual with the highest diversity rank, but they also have a probability of $\Omega(1)$ of selecting the second best (and third best) individual. Hence, even if the population does contain 0^n or 1^n, the parent selection is still able to find a "good" parent to expand the Pareto front efficiently. Hence $p_{\text{good}} = \Omega(1)$ and the claimed bounds follow.

8.7 Conclusions

Maintaining and promoting diversity in evolutionary algorithms is a very important task. Surveys of diversity mechanisms [10, 53, 54] reveal a multitude of approaches to enhancing and promoting diversity, yet it is often unclear which of these mechanisms perform well, and why.

We have surveyed rigorous runtime analyses of evolutionary algorithms with explicit diversity mechanisms, ranging from avoiding genotype or fitness duplicates, deterministic crowding, fitness sharing, and clearing to island models. Other studies have shown that diversity can also emerge naturally, without any explicit mechanisms, through independent mutations, phases of independent evolution in the context of island models, or, in the case of Jump_k, through the interplay of different operators such as crossover followed by mutation and selection.

We have seen that diversity can be highly beneficial for enhancing the global exploration capabilities of evolutionary algorithms. It can enable crossover to work effectively, it can improve performance and robustness in dynamic optimization, and it is vital for evolutionary multiobjective optimization. In many cases diversity mechanisms can be highly effective for the problems considered, speeding up the expected or typical optimization time by constant factors, polynomial factors, or even exponential factors.

Comparing results for TwoMax, Jump_k, and Balance, we found that diversity mechanisms that are effective for one problem may be ineffective for other problems, and vice versa. The analyses have rigorously quantified performance to demonstrate these effects. More importantly, they have laid the foundation for a rigorous understanding of how search dynamics are affected by the presence or absence of population diversity and the introduction of diversity mechanisms.

Acknowledgements The author would like to thank Edgar Covantes Osuna and an anonymous reviewer for helpful comments. This work originated from the Dagstuhl seminar 17191, "Theory of Randomized Optimization Heuristics"; the author would like to thank the organizers and participants for inspiring discussions.

References

[1] N. Barton and T. Paixão. Can quantitative and population genetics help us understand evolutionary computation? In *Proceedings of the Genetic and Evolutionary Computation Conference (GECCO '13)*, pages 1573–1580, 2013.

[2] S. Bleuler, M. Brack, L. Thiele, and E. Zitzler. Multiobjective genetic programming: reducing bloat using SPEA2. In *Proceedings of the 2001 Congress on Evolutionary Computation (CEC 2001)*, volume 1, pages 536–543, 2001.

[3] D. Brockhoff. Theoretical aspects of evolutionary multiobjective optimization. In *Theory of Randomized Search Heuristics–Foundations and Recent Developments*. World Scientific Publishing, 2011.

[4] D. Corus and P. S. Oliveto. Standard steady state genetic algorithms can hillclimb faster than mutation-only evolutionary algorithms. *IEEE Transactions on Evolutionary Computation*, 22(5):720–732, 2017.

[5] E. Covantes Osuna. *Theoretical and Empirical Evaluation of Diversity-preserving Mechanisms in Evolutionary Algorithms: On the Rigorous Runtime Analysis of Diversity-preserving Mechanisms in Evolutionary Algorithms*. PhD thesis, University of Sheffield, 2018.

[6] E. Covantes Osuna, W. Gao, F. Neumann, and D. Sudholt. Speeding up evolutionary multi-objective optimisation through diversity-based parent selection. In *Proceedings of the Genetic and Evolutionary Computation Conference (GECCO '17)*, pages 553–560. ACM, 2017.

[7] E. Covantes Osuna and D. Sudholt. Analysis of the clearing diversity-preserving mechanism. In *Proceedings of Foundations of Genetic Algorithms (FOGA 2017)*, pages 55–63. ACM Press, 2017.

[8] E. Covantes Osuna and D. Sudholt. Empirical analysis of diversity-preserving mechanisms on example landscapes for multimodal optimisation. In *Parallel Problem Solving from Nature (PPSN '18)*, pages 207–219. Springer, 2018.

[9] E. Covantes Osuna and D. Sudholt. Runtime analysis of probabilistic crowding and restricted tournament selection for bimodal optimisation. In *Proceedings of the Genetic and Evolutionary Computation Conference (GECCO 2018)*, pages 929–936. ACM, 2018.

[10] M. Črepinšek, S.-H. Liu, and M. Mernik. Exploration and exploitation in evolutionary algorithms: A survey. *ACM Computing Surveys*, 45(3):35:1–35:33, 2013.

[11] D.-C. Dang, T. Friedrich, T. Kötzing, M. S. Krejca, P. K. Lehre, P. S. Oliveto, D. Sudholt, and A. M. Sutton. Escaping local optima using crossover with emergent diversity. *IEEE Transactions on Evolutionary Computation*, 22(3):484–497.

[12] D.-C. Dang, T. Friedrich, M. S. Krejca, T. Kötzing, P. K. Lehre, P. S. Oliveto, D. Sudholt, and A. M. Sutton. Escaping Local Optima with Diversity-Mechanisms and Crossover. In *Proceedings of the Genetic and Evolutionary Computation Conference (GECCO 2016)*, pages 645–652. ACM Press.

[13] D.-C. Dang, T. Jansen, and P. K. Lehre. Populations can be essential in tracking dynamic optima. *Algorithmica*, 78(2):660–680, 2017.

[14] M. De Felice, S. Meloni, and S. Panzieri. Effect of topology on diversity of spatially-structured evolutionary algorithms. In *Proceedings of the Genetic and Evolutionary Computation Conference (GECCO '11)*, pages 1579–1586. ACM, 2011.

[15] K. Deb, A. Pratap, S. Agarwal, and T. Meyarivan. A fast and elitist multiobjective genetic algorithm: NSGA-II. *IEEE Transactions on Evolutionary Computation*, 6(2):182–197, 2002.

[16] B. Doerr, E. Happ, and C. Klein. Crossover can provably be useful in evolutionary computation. *Theoretical Computer Science*, 425(0):17–33, 2012.

[17] B. Doerr, N. Hebbinghaus, and F. Neumann. Speeding up evolutionary algorithms through asymmetric mutation operators. *Evolutionary Computation*, 15:401–410, 2007.

[18] B. Doerr and D. Johannsen. Adjacency list matchings—an ideal genotype for cycle covers. In *Proceedings of the Genetic and Evolutionary Computation Conference (GECCO '07)*, pages 1203–1210. ACM Press, 2007.

[19] B. Doerr, C. Klein, and T. Storch. Faster evolutionary algorithms by superior graph representation. In *First IEEE Symposium on Foundations of Computational Intelligence (FOCI '07)*, pages 245–250. IEEE, 2007.

[20] S. Fischer and I. Wegener. The one-dimensional Ising model: Mutation versus recombination. *Theoretical Computer Science*, 344(2–3):208–225, 2005.

[21] T. Friedrich, N. Hebbinghaus, and F. Neumann. Rigorous analyses of simple diversity mechanisms. In *Proceedings of the Genetic and Evolutionary Computation Conference (GECCO '07)*, pages 1219–1225. ACM Press, 2007.

[22] T. Friedrich, P. S. Oliveto, D. Sudholt, and C. Witt. Analysis of diversity-preserving mechanisms for global exploration. *Evolutionary Computation*, 17(4):455–476, 2009.

[23] W. Gao and F. Neumann. Runtime analysis for maximizing population diversity in single-objective optimization. In *Proc. of GECCO '14*, pages 777–784, 2014.

[24] M. Giacobini, M. Tomassini, and A. Tettamanzi. Takeover time curves in random and small-world structured populations. In *Proceedings of the Genetic and Evolutionary Computation Conference (GECCO '05)*, pages 1333–1340. ACM Press, 2005.

[25] O. Giel and P. K. Lehre. On the effect of populations in evolutionary multi-objective optimisation. *Evolutionary Computation*, 18(3):335–356, 2010.

[26] C. Horoba, T. Jansen, and C. Zarges. Maximal age in randomized search heuristics with aging. In *Proceedings of the Genetic and Evolutionary Computation Conference (GECCO '09)*, pages 803–810, 2009.

[27] C. Horoba and F. Neumann. *Approximating Pareto-Optimal Sets Using Diversity Strategies in Evolutionary Multi-Objective Optimization*, pages 23–44. Springer Berlin Heidelberg, 2010.

[28] M. Hutter and S. Legg. Fitness uniform optimization. *IEEE Transactions on Evolutionary Computation*, 10:568–589, 2006.

[29] T. Jansen and I. Wegener. On the analysis of evolutionary algorithms—a proof that crossover really can help. *Algorithmica*, 34(1):47–66, 2002.

[30] T. Jansen and I. Wegener. Real royal road functions—where crossover provably is essential. *Discrete Applied Mathematics*, 149:111–125, 2005.

[31] T. Jansen and C. Zarges. Analyzing different variants of immune inspired somatic contiguous hypermutations. *Theoretical Computer Science*, 412(6):517–533, 2011.

[32] T. Jansen and C. Zarges. On the role of age diversity for effective aging operators. *Evolutionary Intelligence*, 4(2):99–125, 2011.

[33] T. Jansen and C. Zarges. Evolutionary algorithms and artificial immune systems on a bi-stable dynamic optimisation problem. In *Proceedings of the 2014 Annual Conference on Genetic and Evolutionary Computation (GECCO '14)*, pages 975–982. ACM, 2014.

[34] N. L. Komarova, E. Urwin, and D. Wodarz. Accelerated crossing of fitness valleys through division of labor and cheating in asexual populations. *Scientific Reports*, 2012.

[35] T. Kötzing, A. Lissovoi, and C. Witt. (1+1) EA on generalized dynamic onemax. In *Proceedings of the 2015 ACM Conference on Foundations of Genetic Algorithms (FOGA '15)*, pages 40–51. ACM, 2015.

[36] T. Kötzing and H. Molter. ACO beats EA on a dynamic pseudo-boolean function. In *Parallel Problem Solving from Nature (PPSN XII)*, pages 113–122. Springer Berlin Heidelberg, 2012.

[37] T. Kötzing, D. Sudholt, and M. Theile. How crossover helps in pseudo-Boolean optimization. In *Proceedings of the 13th Annual Genetic and Evolutionary Computation Conference (GECCO 2011)*, pages 989–996. ACM Press, 2011.

[38] J. Lässig and D. Sudholt. The benefit of migration in parallel evolutionary algorithms. In *Proceedings of the Genetic and Evolutionary Computation Conference (GECCO 2010)*, pages 1105–1112. ACM Press, 2010.

[39] J. Lässig and D. Sudholt. Design and analysis of migration in parallel evolutionary algorithms. *Soft Computing*, 17(7):1121–1144, 2013.

[40] J. Lässig and D. Sudholt. Analysis of speedups in parallel evolutionary algorithms and (1+λ) EAs for combinatorial optimization. *Theoretical Computer Science*, 551:66–83, 2014.

[41] M. Laumanns, L. Thiele, and E. Zitzler. Running time analysis of multiobjective evolutionary algorithms on pseudo-boolean functions. *IEEE Transactions on Evolutionary Computation*, 8(2):170–182, 2004.

[42] A. Lissovoi and C. Witt. On the utility of island models in dynamic optimization. In *Proceedings of the 2015 Annual Conference on Genetic and Evolutionary Computation*, GECCO '15, pages 1447–1454, New York, NY, USA, 2015. ACM.

[43] A. Lissovoi and C. Witt. A runtime analysis of parallel evolutionary algorithms in dynamic optimization. *Algorithmica*, 78(2):641–659, 2017.

[44] S. W. Mahfoud. Niching methods. In T. Bäck, D. B. Fogel, and Z. Michalewicz, editors, *Handbook of Evolutionary Computation*, pages C6.1:1–4. Institute of Physics Publishing and Oxford University Press, Bristol, New York, 1997.

[45] A. Mambrini, D. Sudholt, and X. Yao. Homogeneous and heterogeneous island models for the set cover problem. In *Parallel Problem Solving from Nature (PPSN 2012)*, volume 7491 of *LNCS*, pages 11–20. Springer, 2012.

[46] F. Neumann. Expected runtimes of evolutionary algorithms for the Eulerian cycle problem. *Computers & Operations Research*, 35(9):2750–2759, 2008.

[47] F. Neumann, P. S. Oliveto, G. Rudolph, and D. Sudholt. On the effectiveness of crossover for migration in parallel evolutionary algorithms. In *Proceedings of the Genetic and Evolutionary Computation Conference (GECCO 2011)*, pages 1587–1594. ACM Press, 2011.

[48] P. S. Oliveto, J. He, and X. Yao. Population-based evolutionary algorithms for the vertex cover problem. In *Proceedings of the IEEE Congress on Evolutionary Computation (CEC '08)*, pages 1563–1570, 2008.

[49] P. S. Oliveto and D. Sudholt. On the runtime analysis of stochastic ageing mechanisms. In *Proceedings of the Genetic and Evolutionary Computation Conference (GECCO 2014)*, pages 113–120. ACM Press, 2014.

[50] P. S. Oliveto, D. Sudholt, and C. Zarges. On the runtime analysis of fitness sharing mechanisms. In *13th International Conference on Parallel Problem Solving from Nature (PPSN 2014)*, volume 8672 of *LNCS*, pages 932–941. Springer, 2014.

[51] P. S. Oliveto and C. Zarges. Analysis of diversity mechanisms for optimisation in dynamic environments with low frequencies of change. *Theoretical Computer Science*, 561:37–56, 2015.

[52] P. Rohlfshagen, P. K. Lehre, and X. Yao. Dynamic evolutionary optimisation: an analysis of frequency and magnitude of change. In *Proceedings*

of the 2009 Genetic and Evolutionary Computation Conference (GECCO '09), pages 1713–1720. ACM Press, 2009.

[53] O. M. Shir. Niching in evolutionary algorithms. In G. Rozenberg, T. Bäck, and J. N. Kok, editors, *Handbook of Natural Computing*, pages 1035–1070. Springer, 2012.

[54] G. Squillero and A. Tonda. Divergence of character and premature convergence: A survey of methodologies for promoting diversity in evolutionary optimization. *Information Sciences*, 329:782–799, 2016. Special issue on Discovery Science.

[55] T. Storch and I. Wegener. Real royal road functions for constant population size. *Theoretical Computer Science*, 320:123–134, 2004.

[56] D. Sudholt. How crossover speeds up building-block assembly in genetic algorithms. *Evolutionary Computation*, 25(2):237–274.

[57] D. Sudholt. Crossover is provably essential for the Ising model on trees. In *Proceedings of the Genetic and Evolutionary Computation Conference (GECCO '05)*, pages 1161–1167. ACM Press, 2005.

[58] D. Sudholt. Hybridizing evolutionary algorithms with variable-depth search to overcome local optima. *Algorithmica*, 59(3):343–368, 2011.

[59] D. Sudholt. Crossover speeds up building-block assembly. In *Proceedings of the Genetic and Evolutionary Computation Conference (GECCO 2012)*, pages 689–696. ACM Press, 2012.

[60] D. Sudholt. Parallel evolutionary algorithms. In J. Kacprzyk and W. Pedrycz, editors, *Handbook of Computational Intelligence*, pages 929–959. Springer, 2015.

[61] R. A. Watson and T. Jansen. A building-block royal road where crossover is provably essential. In *Proceedings of the Genetic and Evolutionary Computation Conference (GECCO '07)*, pages 1452–1459. ACM, 2007.

[62] I. Wegener. Methods for the analysis of evolutionary algorithms on pseudo-Boolean functions. In R. Sarker, X. Yao, and M. Mohammadian, editors, *Evolutionary Optimization*, pages 349–369. Kluwer, 2002.

[63] D. B. Weissman, M. W. Feldman, and D. S. Fisher. The rate of fitness-valley crossing in sexual populations. *Genetics*, 186:1389–1410, 2010.

[64] C. Witt. Runtime analysis of the $(\mu+1)$ EA on simple pseudo-Boolean functions. *Evolutionary Computation*, 14(1):65–86, 2006.

[65] C. Zarges. Theoretical foundations of immune-inspired randomized search heuristics for optimization. In B. Doerr and F. Neumann, editors, *Theory of Evolutionary Computation – Recent Developments in Discrete Optimization*. Springer, 2019.

[66] E. Zitzler and S. Künzli. Indicator-based selection in multiobjective search. In *Proceedings of the Parallel Problem Solving from Nature - PPSN VIII*, pages 832–842. Springer Berlin Heidelberg, 2004.

Chapter 9
Theory of Estimation-of-Distribution Algorithms

Martin S. Krejca and Carsten Witt

Abstract Estimation-of-distribution algorithms (EDAs) are general metaheuristics used in optimization that represent a more recent alternative to classical approaches such as evolutionary algorithms. In a nutshell, EDAs typically do not directly evolve populations of search points but build probabilistic models of promising solutions by repeatedly sampling and selecting points from the underlying search space. Recently, significant progress has been made in the theoretical understanding of EDAs. This chapter provides an up-to-date overview of the most commonly analyzed EDAs and the most recent theoretical results in this area. In particular, emphasis is put on the runtime analysis of simple univariate EDAs, including a description of typical benchmark functions and tools for the analysis. Along the way, open problems and directions for future research are described.

9.1 Introduction

Optimization is one of the most important fields in computer science, with many problems being NP-hard and thus not necessarily easy to solve. Hence, *heuristics* play a major role, i.e., optimization algorithms that try to yield solutions of good quality in a reasonable amount of time. Research over the past decades has resulted in many good heuristics being developed for classical NP-hard problems. Unfortunately, these heuristics are tailored with specific problems in mind and exploit certain problem-specific properties in order to

Martin S. Krejca
Hasso Plattner Institute, University of Potsdam, Potsdam, Germany
e-mail: martin.krejca@hpi.de

Carsten Witt
DTU Compute, Technical University of Denmark, Kgs. Lyngby, Denmark
e-mail: cawi@dtu.dk

B. Doerr, F. Neumann (eds.), *Theory of Evolutionary Computation*,
Natural Computing Series, https://doi.org/10.1007/978-3-030-29414-4_9

save computation time. Thus, they cannot be used for problems that do not feature these specific properties.

One alternative to problem-specific heuristics is *general-purpose* heuristics. The information about the problem to be optimized that these algorithms have access to is fairly limited, up to the point that they are only able to compare the quality of different solutions relatively. This has the advantage that the problem itself does not have to be formalized but only the quality of a solution, as the problem formalization is communicated implicitly via the quality measure to the algorithm. In turn, this results in great reusability of these algorithms for different problems.

One such class of general-purpose heuristics is *evolutionary algorithms* (EAs) [34]. EAs are characterized by creating new solutions from already generated solutions. Oftentimes, many solutions are stored and only changed (*evolved*) locally, preferably discarding bad solutions and saving good ones. Such algorithms are EAs in the classical sense [60].

The concept of EAs can be broadened if we are less restrictive about what is being evolved. A similar approach to changing solutions directly is to instead change the procedure that generates the solutions in the first place. Thus, a solution-generating mechanism is evolved. Algorithms following this approach are called *estimation-of-distribution algorithms* (EDAs) [29, 37, 54, 55]. They are not EAs in the classical sense but can be considered EAs in the broad sense, as just described.

EDAs have been used very successfully in real-world applications [29, 37, 54, 55] and have recently gathered momentum in the theory community analyzing EAs [10, 20, 22, 36, 38, 63, 66]. The aim of theoretically analyzing EAs is to provide guarantees for the algorithms and to gain insights into their behavior in order to optimize the algorithms themselves. Common guarantees include the expected time until an algorithm finds a solution of sufficient quality, the probability of doing so after a certain time, and the fact that the algorithm is even able to find desired solutions.

In this chapter, we provide a state-of-the-art overview of the theoretical results on EDAs for *discrete* domains, as that is their main field of application. To the best of our knowledge, while continuous EDAs exist, no detailed theoretical analyses have been conducted so far. We present the most commonly investigated EDAs and give an outline of the history of their analyses, providing deep insights into some of the latest results. After reading this, the reader should be familiar with EDAs in general, the current state of theoretical research, and common tools used for the analyses.

In Section 9.2, we go more into detail about how EDAs work, we introduce the scenario used in most theoretical papers, and we provide different ways of classifying EDAs, stating the most commonly analyzed algorithms. Further, we mention some tools that are often used when deriving results for EDAs. Then, in Section 9.3, we give a short overview of the most commonly considered objective functions. In Section 9.4, we discuss the historically older results of convergence analyses on EDAs. After that, in Section 9.5, we present more recent results on EDAs, which consider the actual runtime of an algorithm. We end this article in Section 9.6 with some conclusions and open problems.

9.2 Estimation-of-Distribution Algorithms

In general, EDAs are problem-agnostic optimization algorithms that store a probabilistic model over the solution space. This model is the core part of these algorithms. It implies a probability distribution over the solution space and is iteratively refined, using samples. Ideally, the model converges to a state that produces only optimal solutions.

Since EDAs make use of sample sets – called *populations* – they are quite similar in this respect to EAs. However, the main difference is that EAs exclusively store a population and progress using solely this information, by varying samples – called *individuals* – from the population. Thus, they have quite a local view of the solution space and advance locally. In contrast, the probabilistic model of an EDA models most of the time the entire solution space. Updates to the model are done using the old model as well as a population. Hence, EDAs employ a more general view of the solution space than do classical EAs.

The probabilistic model of an EDA is used as an *implicit* probability distribution over the solution space, instead of an explicit distribution. This is usually done by constraining the distributions that can be modeled and by factorizing them, i.e., by writing the distribution as a product of marginal probabilities. Hauschild and Pelikan [29] distinguish between many different classes of EDAs with respect to how strongly constrained the models are. An advantage of factorizing a distribution is that it saves a lot of memory, since an explicit distribution would make it necessary to store a probability for each solution, which is not feasible. With a factorization, only the factors have to be stored in memory. However, even then it is possible for the model to grow to sizes exponential in the input [25].

As mentioned above, EDAs also use populations, like EAs, sampled from their probabilistic model, in order to update that model. It is up to the EDA to decide what to do with its population. However, all EDAs theoretically analyzed so far have in common the fact that they always discard their population after every iteration, valuing the model higher than the population.

In the following, we first state the optimization domain for the EDAs that we consider in this chapter. Then we discuss different classifications of EDAs and name various algorithms that fall into the various classes. Last, we mention the tools that are commonly used in the current theoretical research on EDAs.

9.2.1 Scenario

As in the theory of EAs, theoretical analyses of EDAs consider mainly pseudo-Boolean optimization, i.e., optimization of a function $f\colon \{0,1\}^n \to \mathbb{R}$, often referred to as the *fitness function*. Conventionally, the function value of a bit string $\boldsymbol{x}$ is called the *fitness of* $\boldsymbol{x}$.

The aspect of an EDA being a general-purpose solver is modeled as a classical black-box setting, where the algorithm gains problem-specific information only from querying the fitness function by inputting bit strings and receiving their respective fitness. In this setting, mostly two different scenarios have been of major interest.

Convergence analyses. In this historically older topic, EDAs have been analyzed with respect to the convergence of their probabilistic model, i.e., if they succeed at all in optimizing certain fitness functions. We discuss this scenario in more depth in Section 9.4.

Runtime analyses. A more recent trend is the analysis of an EDA's runtime on certain functions. In this scenario, the focus is on the number of queries needed until an optimum or a solution of sufficient quality is sampled, i.e., the first hitting time of an algorithm sampling such a solution. Although sampling a desired solution can happen by chance, the analyses usually entail that the probabilistic model of an EDA makes it very likely for such a solution to be sampled again. Section 9.5 goes into detail about this topic.

9.2.2 Classifications of EDAs

Arguably, the most straightforward way of classifying EDAs is with respect to the power of their underlying probabilistic model. *Univariate* algorithms use only a single variable in their model per problem variable.[1] In contrast, *multivariate* algorithms use more than a single variable to model a problem variable. Thus, univariate EDAs are not able to capture dependencies between problem variables, whereas multivariate EDAs are explicitly constructed to do so.

Pelikan et al [55] give a more fine-grained classification of EDAs, differentiating multivariate EDAs even further with respect to how many dependencies can be captured by the underlying probabilistic model.

Note that the classification into univariate and multivariate EDAs does not constrain the populations at all.

[1] In our setting of pseudo-Boolean optimization, a *problem variable* is a position in a bit string, i.e., one dimension of a hypercube.

9.2.2.1 Univariate Algorithms

When optimizing a pseudo-Boolean function, univariate EDAs assume independence of all of the n different bit positions to be optimized. Under this assumption, every probability distribution can be factorized into a product of n different probabilities $\boldsymbol{p}_i$, collected together in a vector $\boldsymbol{p}$ of length n. A bit string $\boldsymbol{x}$ is then sampled by choosing each bit $\boldsymbol{x}_i$ to be 1 with probability $\boldsymbol{p}_i$ and 0 otherwise. Since each $\boldsymbol{p}_i$ determines how frequently, in expectation, a 1 is sampled at position i, we call these probabilities *frequencies*, following the common naming convention [22]. The vector $\boldsymbol{p}$ is then consequently called the *frequency vector*.

n-Bernoulli-λ-EDA

Although the class of univariate EDAs does not limit the populations of the algorithms in any way, the most commonly considered univariate EDAs discard their entire population after every iteration. Thus, from a theoretical point of view, a run of a univariate EDA can be modeled as a series $(\boldsymbol{p}^{(t)})_{t\in\mathbb{N}_0}$ of frequency vectors over the number of iterations t. Usually, $\boldsymbol{p}^{(0)}$ models the uniform distribution by satisfying the condition that $\boldsymbol{p}_i^{(0)} = 1/2$ for each i. Friedrich et al [22] capture this class of univariate EDAs in a framework called the *n-Bernoulli-λ-EDA* (Algorithm 9.1).

The n-Bernoulli-λ-EDA samples λ individuals in each iteration and performs an update to its frequency vector, using the current frequency vector as well as all of the just-sampled individuals and their respective fitnesses. The function performing this update is called the *update scheme* and fully characterizes the algorithm.

Note that we do not specify a termination criterion. In fact, determining what a good criterion is may vary between different use cases of the algorithm. When considering the expected runtime of these algorithms (Section 9.5), we are interested in the number of fitness function evaluations until an optimal solution is sampled for the first time.

In many EDAs, if a frequency is either 0 or 1, all bits sampled at the respective position will be 0 or 1, respectively, and the update scheme will not change the frequency anymore. To prevent this, the algorithm is usually modified such that each frequency is only allowed to take values in an interval $[m, 1-m] \subset [0,1]$, where $m \in (0, 1/2]$ is called a *margin*; the values m and $1-m$ are called *borders*. Usually, a margin of $1/n$ is chosen [7, 10, 50]. In a scenario with a margin, line 8 of Algorithm 9.1 can be modified as follows:

foreach $i \in \{1, \ldots, n\}$ **do**

$$\boldsymbol{p}_i^{(t+1)} \leftarrow \max\Big\{m,\ \min\{1-m,\ \varphi\big(\boldsymbol{p}^{(t)}, (\boldsymbol{x}, f(\boldsymbol{x}))_{\boldsymbol{x}\in D}\big)_i\}\Big\};$$

[2] Note that D is a multiset, that is, we allow duplicates.

Algorithm 9.1: n-Bernoulli-λ-EDA with a given update scheme φ, optimizing f

1 $t \leftarrow 0$;
2 **foreach** $i \in \{1,\dots,n\}$ **do** $\boldsymbol{p}_i^{(t)} \leftarrow \frac{1}{2}$;
3 **repeat**
4 $\quad D \leftarrow \emptyset$;
5 $\quad$ **foreach** $j \in \{1,\dots,\lambda\}$ **do**
6 $\quad\quad \boldsymbol{x} \leftarrow$ offspring sampled with respect to $\boldsymbol{p}^{(t)}$;
7 $\quad\quad D \leftarrow D \cup \{\boldsymbol{x}\}$; [2]
8 $\quad \boldsymbol{p}^{(t+1)} \leftarrow \varphi\big(\boldsymbol{p}^{(t)}, \big(\boldsymbol{x}, f(\boldsymbol{x})\big)_{\boldsymbol{x} \in D}\big)$;
9 $\quad t \leftarrow t+1$;
10 **until** termination criterion met;

We will continue to give an overview of the most commonly theoretically analyzed univariate EDAs and show how they fit into the n-Bernoulli-λ-EDA framework. We present the algorithms without a margin although they are commonly analyzed with a margin of $1/n$.

Since many of the following examples do not make use of the entire population of size λ (the *population size*) but select a certain number μ (the *effective population size*) of individuals according to their fitness values, we denote the k-th-best individual as $\boldsymbol{x}^{(k)}$, where $1 \leq k \leq \mu$; ties are broken *uniformly at random*. Thus, $\boldsymbol{x}^{(1)}$ denotes an individual with the best fitness.

UMDA

The arguably easiest update scheme is given by the *univariate marginal distribution algorithm* (UMDA; Algorithm 9.2) [49]. It samples λ individuals in each iteration, of which μ of the best are chosen. Then, each frequency $\boldsymbol{p}_i$ is set to the relative frequency of 1s at position i in the set of the μ best individuals, regardless of the current frequency.

The update scheme of UMDA allows it to go from any valid frequency to any other in a single step if not stuck. Thus, the difference of two consecutive frequencies $\boldsymbol{p}_i^{(t)}$ and $\boldsymbol{p}_i^{(t+1)}$ can only be trivially bounded by roughly 1. We call such a difference the *step size* of the algorithm.

PBIL

A variant of UMDA that has an adjustable step size is the *population-based incremental learning* algorithm (PBIL; Algorithm 9.3) [4]. A frequency is updated in a way similar to UMDA, but the new frequency is a convex combination with parameter ρ of the current frequency and the relative frequencies

Algorithm 9.2: UMDA with population size λ, effective population size μ, optimizing f

```
1  t ← 0;
2  foreach i ∈ {1,...,n} do p_i^(t) ← 1/2 ;
3  repeat
4      D ← ∅;
5      foreach j ∈ {1,...,λ} do
6          x ← offspring sampled with respect to p^(t);
7          D ← D ∪ {x};
8      foreach i ∈ {1,...,n} do p_i^(t+1) ← (1/μ) Σ_{k=1}^{μ} x_i^(k) ;
9      t ← t+1;
10 until termination criterion met;
```

of 1s at that position. Thus, the step size is now bounded by ρ, and UMDA is a special case of PBIL with $\rho = 1$.

Algorithm 9.3: PBIL with population size λ, effective population size μ, and learning rate ρ, optimizing f

```
1  t ← 0;
2  foreach i ∈ {1,...,n} do p_i^(t) ← 1/2 ;
3  repeat
4      D ← ∅;
5      foreach j ∈ {1,...,λ} do
6          x ← offspring sampled with respect to p^(t);
7          D ← D ∪ {x};
8      foreach i ∈ {1,...,n} do p_i^(t+1) ← (1 − ρ)p_i^(t) + (ρ/μ) Σ_{k=1}^{μ} x_i^(k) ;
9      t ← t+1;
10 until termination criterion met;
```

$\mathrm{MMAS_{ib}}$

Another important univariate EDA is the *max-min ant system with iteration-best update* ($\mathrm{MMAS_{ib}}$; Algorithm 9.4) [50], which is a special case of PBIL where we set $\mu = 1$, i.e., where we consider only the best individual in each iteration. $\mathrm{MMAS_{ib}}$ also falls into the general class of *ant colony optimization* (ACO) algorithms [16]. Although ACO spans an entire research topic independent of EDAs and is typically not considered to be an EDA, the process of how it produces solutions iteratively can be viewed as refining a probabilistic model. Thus, we view ACO as an EDA here.

ACO considers graphs whose edges are weighted with probabilities, called *pheromones*. Additionally, the algorithm uses agents – called *ants* – that traverse the graph and thus construct paths. At each vertex v, if a path needs to be extended, an ant chooses an edge with a certain probability with respect to the pheromones on all of the outgoing edges of v. After the data of all ants has been collected, all pheromones decrease (they *evaporate*) and then some are increased afterward, usually the ones that are part of the best solutions constructed.

When pseudo-Boolean optimization is considered, a graph for ACO can be modeled as a multigraph with $n+1$ vertices from 0 to n, each vertex having exactly two outgoing edges to its direct successor (except for vertex n; see Fig. 9.1). One of these edges is interpreted as a 0, and the other one as a 1. Each solution is constructed by letting an ant traverse the graph starting at 0 and ending at n. The corresponding edges are then interpreted as a bit string of length n. Note how the probability of choosing an edge corresponding to a 1 is equal to an n-Bernoulli-λ-EDA's frequency for that respective position.

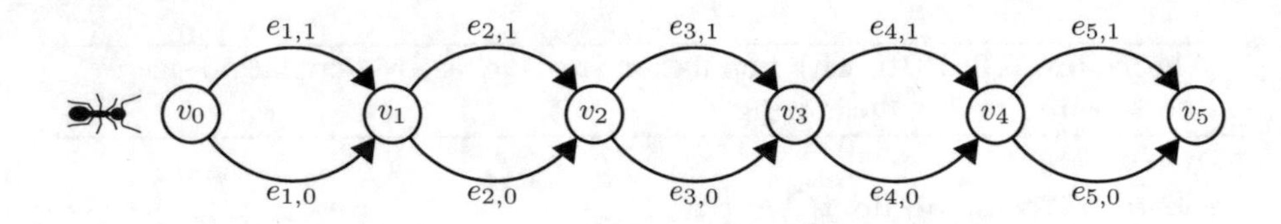

Fig. 9.1 The ACO graph for pseudo-Boolean optimization with $n = 5$ bits.

MMAS$_{\text{ib}}$ is a variant of the max-min ant system algorithm [61] that only makes an update with respect to the path of the best ant in each iteration, using a classical update rule in ACO.

Algorithm 9.4: MMAS$_{\text{ib}}$ with population size λ and evaporation factor ρ, optimizing f

```
1  t ← 0;
2  foreach i ∈ {1,...,n} do p_i^(t) ← 1/2 ;
3  repeat
4      D ← ∅;
5      foreach j ∈ {1,...,λ} do
6          x ← offspring sampled with respect to p^(t);
7          D ← D ∪ {x};
8      foreach i ∈ {1,...,n} do p_i^(t+1) ← (1 − ρ)p_i^(t) + ρx_i^(1) ;
9      t ← t + 1;
10 until termination criterion met;
```

cGA

An algorithm with a different approach is the *compact genetic algorithm* (cGA; Algorithm 9.5) [27]. It samples exactly two individuals in each iteration and compares their bit values componentwise. If the bits at position i are the same, the frequency $\boldsymbol{p}_i$ is left unchanged. Otherwise, the frequency is adjusted by $\pm 1/K$, where K is an algorithm-specific parameter, often referred to as the *population size*, such that the probability of sampling the bit value of the fitter individual is higher in the next iteration.

Algorithm 9.5: cGA with population size K, optimizing f

```
1  t ← 0;
2  foreach i ∈ {1,...,n} do p_i^(t) ← 1/2 ;
3  repeat
4      D ← ∅;
5      foreach j ∈ {1,2} do
6          x ← offspring sampled with respect to p^(t);
7          D ← D ∪ {x};
8      foreach i ∈ {1,...,n} do p_i^(t+1) ← max{0, min{1, p_i^(t) + (1/K)(x^(1) − x^(2))}};
9      t ← t+1;
10 until termination criterion met;
```

9.2.2.2 Multivariate Algorithms

The class of multivariate EDAs consists of all algorithms that can use multiple variables to model one problem variable and thus express dependencies. A compact representation of such dependencies can be modeled as a directed graph whose vertices are the variables and whose edges denote dependencies among the variables. For each vertex, the probability distribution conditional on all its adjacent vertices with an incoming edge (its *parents*) is stored. This results in a factorization of the problem space that respects the given dependencies. Multivariate EDAs can assume a certain dependency model and learn only the respective (conditional) probabilities of the factorization, or they can additionally try to learn a model that fits well to the samples.

The *factorized distribution algorithm* (FDA) [48] falls into the former category. It assumes a factorization according to a so called *additively decomposable function* (ADF), i.e., a function that is a sum of multivariate subfunctions. For each set of variables per subfunction, FDA creates a *metavariable*, and it expresses the objective function (the ADF) with respect to those metavariables. In each iteration, it samples solutions with respect to the factorization, selects a subset of them, and estimates the conditional probabilities based on these samples. Note that FDA is a generalization of the update of UMDA and coincides with it if no dependencies between the problem variables exist.

Another approach that also uses metavariables is the *extended compact genetic algorithm* (ECGA) [28]. Differently from FDA, a metavariable of ECGA represents multiple variables at once (i.e., it is assumed that such variables are strongly correlated). In each iteration, the algorithm starts by placing each problem variable into its own class. Then, it greedily merges two classes such that a certain metric (the so-called *Bayesian information criterion*) is maximized, using samples from the current model. If no further improvement can be made, the merging process stops and the algorithm uses the newly created model.

The easiest of the multivariate cases is the one where each variable can be at most dependent on one other variable, i.e., a bivariate setting, and the arguably easiest probabilistic model in such a setting is a path. This model is used in the *mutual-information-maximization input clustering* (MIMIC) algorithm introduced by De Bonet et al [11]. The idea of the underlying model is to construct a path that minimizes the Kullback–Leibler divergence with respect to the bivariate setting, i.e., to find a permutation that can explain the sample data best. However, since there are $n!$ possible permutations for n variables, the authors of [11] suggest a greedy approach that makes use of the empirical entropies, i.e., the entropies of the sample data. First, a variable with minimum entropy is chosen as the start vertex of the path. Then, the path is continued by choosing a node that has minimum conditional entropy with respect to the currently last vertex in the path.

The *bivariate marginal distribution algorithm* (BMDA) [52] uses a somewhat similar approach. However, it does not consider paths as its model for dependency graphs but rather a forest of rooted trees. In order to determine which variables are dependent on which other variables, the Pearson's chi-squared statistic is used as an indicator. If the indicator is too low, the corresponding variables are considered independent. The forest is then created greedily very similarly to regular algorithms for maximum spanning trees: iteratively, a vertex is added to one of the trees that has maximum Pearson chi-squared value.

The *Bayesian optimization algorithm* (BOA) [53] is a very general multivariate EDA and constructs an arbitrary dependency graph with respect to a metric of choice. If wanted, the degree of incoming edges, i.e., the number of dependent variables, can be limited. Pelikan et al [53] proposed the Bayesian Dirichlet metric as one possibility to determine the quality of a dependency graph, and they stated that the general problem of finding an optimal graph is NP-hard. Thus, they suggested greedy algorithms or heuristics for efficiently creating good graphs.

9.2.2.3 Other Classifications

Another approach to classifying EDAs is to differentiate them not by how many dependencies they can model but by certain invariances that their probabilistic models may have.

One such classification stems from the theory of EAs and was introduced by Lehre and Witt [39]. These authors considered a new black-box complexity known as *unbiased black-box complexity* in order to prove tighter lower bounds for commonly analyzed EAs. This definition is so general that it applies to any black-box algorithm optimizing pseudo-Boolean functions, thus including EDAs.

Unbiased black-box complexity considers black-box algorithms optimizing perturbations of the hypercube, where a perturbation is any isometric automorphism of the hypercube.[3] For example, cyclically shifting a bit string by one position to the right and changing the value of the first bit in the result is an isometric automorphism.

Given a fitness function and a perturbed variant of it, a black-box algorithm is said to be *unbiased* if the queries to the black box in the perturbed setting are the same as the queries in the unperturbed setting when inverted with respect to the perturbation. Thus, an unbiased algorithm does not favor certain positions over other positions or 1s over 0s, or vice versa, i.e., it has no bias in this respect.

When considering general-purpose algorithms, unbiasedness is a nice property to have, as it certifies that the algorithm has no bias with respect to the encoding of the search space. However, when considering certain problems, different values may have a strict, different meaning, such that unbiasedness with respect to those values does not make sense.

All of the EDAs presented in Section 9.2.2.1 are unbiased when uniform tie-breaking is used.

A seemingly similar but unrelated property that many EDAs feature is that their probabilistic model does not change, in expectation, if all samples have the same fitness, i.e., there is no signal from the fitness function. Friedrich et al [22] called this property *balanced*, with respect to the n-Bernoulli-λ-EDA. However, this property had already been considered before by Shapiro [58], albeit with different terminology.

Although balancedness seems beneficial at first glance, it actually leads to the probabilistic model converging to one of the corners of the hypercube [22, 58]. This is a general problem of martingales, i.e., random processes that do not change in expectation, with a bounded range, which will eventually end up at the bounds of their range. This means that balancedness implies a bias toward outer regions of the hypercube, also called *genetic drift* [3], as this is an inherent drift due to the genotypes of the sampled population. In order

[3] The isometric automorphisms of the hypercube are all isomorphisms that permute any positions and may change a value of x to $1 - x$ at any position.

to overcome this bias and optimize successfully, the drift due to selection introduced by the fitness function has to be larger than the genetic drift.

Different approaches have been suggested in order to prevent an EDA's probabilistic model from quickly converging to a corner of the hypercube. Shapiro [58] proposed to reject updates made to the probabilistic model with a probability equal to the ratio of going from one model to the other. This has the advantage that the resulting implicit distribution is the uniform distribution over the hypercube. However, the transition probabilities have to be known and computed in order to get the correct rejection probabilities. Another approach proposed by Shapiro [58] and also by Friedrich et al [22] is to introduce an artificial bias that counteracts the one introduced by the balancedness.

In the context of balancedness, Friedrich et al [22] introduced another concept, which they called *stable*. An n-Bernoulli-λ-EDA is stable if the limit distribution of each frequency, when no fitness signal is received, is unimodal with its maximum at $1/2$. This means that a stable n-Bernoulli-λ-EDA has a bias toward the center of the hypercube. These authors showed that this concept is mutually exclusive with an n-Bernoulli-λ-EDA being balanced, as such an EDA has a bias toward the corners of the hypercube. The *stable* property is similar to the concept of an EDA's limit distribution being the uniform distribution, as considered by Shapiro [58].

9.2.3 Tools for Analyzing EDAs Theoretically

Most of the theoretical results on EDAs consider univariate algorithms, as we explain in Section 9.5. Thus, tools that make use of independent events are commonly used. However, that does not limit the use of these tools to the univariate case. Especially, *drift analysis*, which we present later in this section, can be applied in any setting.

Many proofs make use of classical probabilistic concentration bounds, such as Markov's inequality, Chebyshev's inequality, or, most importantly, Chernoff bounds [44]. The latter are used very frequently, since the sampling process of a univariate EDA is usually done independently of the other samples. Thus, such a bound can be applied.

Since the theory of EDAs usually considers first hitting times, more specialized tools suited for that purpose are used as well. One such tool is the coupon collector problem [45], which gives highly concentrated first-hitting-time results if a certain number of events with low probability have to occur to reach the target. For EDAs, this can be thought of as a certain number of factors of the probabilistic model being at the wrong end of their spectrum, thus slowing down optimization, since they need to be changed for the optimization process to succeed.

Another tool for determining first hitting times, and the most prominent one when looking at the theory of EAs and EDAs in general, is *drift theory*. It is loosely akin to the potential method in complexity theory. To apply drift theory, one needs to define a potential that maps the stochastic process into the reals. Then, the expected difference between two consecutive steps of the process is considered: the *drift*. This can be thought of as the expected velocity of the process. If the drift can be bounded, the expected hitting time of the process reaching a target is easily deducible, i.e., if there is a known bias in the process toward a certain direction, the first hitting time can easily be bounded.

We now state the three most commonly used drift theorems. The most general theorem with respect to the prerequisites of the process – the additive drift theorem (Theorem 9.2.1) – was stated by He and Yao [30]. However, the ideas used date back to Wald's equation [65].

Theorem 9.2.1 (additive drift [30, 31]). *Let $(X_t)_{t \in \mathbb{N}_0}$ be random variables over a bounded space $S \subseteq \mathbb{R}_{\geq 0}$ containing 0, and let $T = \min\{t \mid X_t = 0\}$.*

If there is a constant $\delta > 0$ such that, for all $t < T$, $E[X_t - X_{t+1} \mid X_t] \geq \delta$, then

$$E[T \mid X_0] \leq \frac{X_0}{\delta}.$$

And if there is a $\delta > 0$ such that, for all $t < T$, $E[X_t - X_{t+1} \mid X_t] \leq \delta$, then

$$E[T \mid X_0] \geq \frac{X_0}{\delta}.$$

The additive drift theorem can be applied when the expected difference between two potentials is known. However, oftentimes it is easier to determine the expected difference conditional on the current potential, i.e., $E[X_t - X_{t+1} \mid X_t]$. Owing to the law of total expectation, a lower bound on the conditional expected value is also a lower bound on the unconditional one.

A theorem more suited to processes whose potential changes at least linearly with respect to the current potential is the following multiplicative drift theorem.

Theorem 9.2.2 (multiplicative drift [14]). *Let $(X_t)_{T \in \mathbb{N}_0}$ be nonnegative random variables over $\mathbb{R}$, each with finite expectation, and let $T = \min\{t \mid X_t < 1\}$.*

If there is a constant $\delta > 0$ such that, for all $t < T$, $E[X_t - X_{t+1} \mid X_t] \geq \delta X_t$, then

$$E[T \mid X_0] \leq \frac{1 + \ln X_0}{\delta}.$$

The multiplicative drift theorem is not well suited if the difference in potential is dependent on the current potential but not in a linear fashion. Such cases are covered by the following variable drift theorem. However, note that all these theorems assume that the difference in potential does not increase when one gets closer to the goal.

Theorem 9.2.3 (variable drift [35, 43]). *Let $(X_t)_{t \in \mathbb{N}_0}$ be nonnegative random variables over a bounded space $S \subseteq \mathbb{R}_{\geq 0}$ containing 1, each with finite expectation, and let $T = \min\{t \mid X_t < 1\}$.*

If there exists a monotonically increasing function $h\colon \mathbb{R}_{\geq 0} \to \mathbb{R}_{\geq 0}$ such that $1/h$ is integrable and, for all $t < T$, $E[X_t - X_{t+1} \mid X_t] \geq h(X_t)$, then

$$E[T \mid X_0] \leq \frac{1}{h(1)} + \int_1^{X_0} \frac{1}{h(x)}\,\mathrm{d}x.$$

The drift theorems above have been formulated in a simple, easy-to-read form that covers the most typical scenarios in which they are applied. However, more general drift theorems can be obtained [40, 41]; for example, to apply Theorem 9.2.1 in unbounded state spaces, to apply Theorems 9.2.2 and 9.2.3 with respect to arbitrary minimum states $s_{\min} > 0$ in the definition of T instead of state 1, and to allow processes adapted to arbitrary stochastic filtrations instead of the natural one implicit in the formulations above. These generalizations come partly at the cost of more complicated theorem statements, and sometimes require some additional technical assumptions about the underlying stochastic process.

9.3 Common Fitness Functions

The most commonly analyzed pseudo-Boolean functions for EDAs are OneMax [46] and LeadingOnes [56]. However, other functions have also been analyzed [6, 7], with BinVal being the most prominent one of them [17, 48].

OneMax counts the number of 1s in a bit string. Thus, the unique optimum is the all-1s bit string:

$$\textsc{OneMax}(\boldsymbol{x}) := \sum_{i=1}^{n} \boldsymbol{x}_i. \tag{9.1}$$

This function can be generalized to a class of functions, each having a target bit string $\boldsymbol{a}$ – which denotes the unique global optimum – and yielding the number of incorrectly set bits. Note that any unbiased algorithm, as introduced in Section 9.2.2.3, behaves on OneMax exactly as on the generalized version.

The OneMax function class is used to analyze how well an EDA performs as a hill climber. The usual expected runtime of an EDA on this function is $\Theta(n \log n)$ [36, 38, 63, 66].

Whereas OneMax is oftentimes considered to be the easiest pseudo-Boolean function, BinVal is said to be the hardest [17]. In contrast to OneMax, where all bits are equally weighted, BinVal uses exponentially scaled weights on its bit positions:

$$\textsc{BinVal}(\boldsymbol{x}) := \sum_{i=1}^{n} 2^{n-i}\boldsymbol{x}_i. \tag{9.2}$$

That means that BinVal represents value of a bit string interpreted as a binary unsigned integer.

Since the sum of all powers of 2 up to an exponent j is less than 2^j, BinVal can be interpreted as a lexicographic order on the hypercube, where lexicographically greater bit strings have a better fitness.

As with OneMax, in its general form, the global optimum of BinVal is any bit string $\boldsymbol{a}$, and the fitness of any bit string is the weight of the respective index if the bit value is the same as that of $\boldsymbol{a}$, and it is 0 otherwise.

LeadingOnes yields the number of consecutive 1s in a bit string, starting from the left:

$$\textsc{LeadingOnes}(\boldsymbol{x}) := \sum_{i=1}^{n}\prod_{j=1}^{i}\boldsymbol{x}_j. \tag{9.3}$$

As with OneMax, the unique global optimum is the all-1s bit string. In its general version, the function yields the number of consecutively correctly chosen bits with respect to a fixed permutation π and a target bit string $\boldsymbol{a}$.

LeadingOnes is used to analyze how an EDA copes with dependencies between the bits. The known expected runtime of certain EDAs on this function is $O(n^2)$ [10], which is compliant with the usual upper bound for EAs on this function [1].

9.4 Convergence Analyses

The earliest theoretical studies of EDAs focused mostly on their convergence, and were similar in style to the research that had been done for evolutionary algorithms in the 1990s [56, 64]. More precisely, it was studied how an algorithm behaves in the limit $t \to \infty$, i.e., if the algorithm is allowed to run for an arbitrary amount of time. If optimal solutions will be found in this limit, the algorithm is considered effective.

Almost all convergence analyses of EDAs consider univariate models. An early publication by Höhfeld and Rudolph [32] studied the vector of frequencies $\boldsymbol{p}^{(t)}$ in PBIL using a Markov chain model and rigorously proved that if $\mu = 1 < \lambda$ and $\rho > 0$, it will converge in expectation to some solution $\boldsymbol{x}^* = (\boldsymbol{x}_1^*, \dots, \boldsymbol{x}_n^*)$; more precisely, $E[\boldsymbol{p}_i^{(t)}] \to \boldsymbol{x}_i^*$ as $t \to \infty$. This solution need not be an optimal one but may correspond to a local optimum to which the search process is led in the very first steps. If the fitness function f is a linear pseudo-Boolean function, then in fact $E[\boldsymbol{p}_i^{(t)}] \to \boldsymbol{x}_i^*$ with respect to the optimal solution $\boldsymbol{x}^*$. This includes classical benchmark functions such as OneMax. However, as pointed out by Shapiro [58], convergence in expectation does not imply that PBIL eventually will sample the optimum of such functions. In fact, genetic drift may lock frequencies to values that make it impossible to sample the optimum.

PBIL was also theoretically analyzed by González et al [26] using a dynamical systems model. Convergence of the model to local optima of the fitness functions was proven for $\mu = 1$, and it was argued that the actual PBIL will resemble the model if ρ is chosen sufficiently close to 0. Hence, the approach does not make predictions for high learning rates ρ, in particular, it excludes the special case of $\rho = 1$ as used in UMDA.

Several subsequent publications have considered UMDA and its generalization FDA. Mühlenbein and Mahnig [47] also used an approach similar to dynamical systems theory to derive a quantitative statement about the behavior of the frequencies in FDA and UMDA over time. In fact, both fitness-proportionate and the usual truncation selection (take the best μ out of λ individuals) were considered. Specifically, for the classical UMDA on OneMax, they derived the result that, roughly,

$$\boldsymbol{p}_i^{(t+1)} \approx \boldsymbol{p}_i^{(t)} + \frac{I}{\sqrt{n}} \sqrt{\boldsymbol{p}_i^{(t)} \left(1 - \boldsymbol{p}_i^{(t)}\right)}\,, \tag{9.1}$$

where I is the so-called selection intensity, which is determined from the ratio μ/λ and can be thought of as being constant. By solving a differential equation, the formula can be turned into an approximation to the expected frequency at time t. Interestingly, (9.1) resembles a rigorous statement about the drift of the frequencies that was recently proven in [66] and is crucial for upper bounds on the runtime; see a more detailed discussion in Section 9.5.2.1.

A more comprehensive convergence study of FDA was done by Mühlenbein and Mahnig [48]. As a general assumption, the FDA is instantiated with the correct decomposition of an additively decomposable function $f(x) = \sum_{i=1}^{k} f(X_j)$, where $X_j \subset \{1,\ldots,n\}$, into its subfunctions. Then the algorithm will compute a probabilistic model, comprising unconditional and conditional frequencies from the sampled search points. Strong results are obtained if a fitness-proportionate selection scheme called Boltzmann selection is used. Under some assumptions about the initial population, the algorithm will converge to a distribution that is uniform on the set of optimal solutions. The drawback of this result is that Boltzmann selection is computationally very expensive. For the usual truncation selection, results building on simplifying assumptions were obtained. Moreover, using infinite-population models, the paper derived quantitative statements similar to (9.1) about the time for a frequency of UMDA to converge to its optimum value, regarding OneMax and BinVal.

In the 2000s, rigorous convergence proofs of FDA (including UMDA) with fitness-proportionate [72] and truncation selection [71] followed. To study the regions of convergence and their stability, this research was supplemented by a fixed-point analysis for UMDA and FDA with 2-tournament selection in [70]. It turns out that FDA, given an appropriate decomposition of a nonlinear function, converges under milder assumptions about the starting population than UMDA. Roughly, this indicates that a multivariate model, as used in FDA, can be superior to a univariate model, to which UMDA is restricted. However, the analyses in [70–72] also make the assumption of an infinite population size, which was very common in early convergence analyses of nature-inspired algorithms [64]. Infinite populations simplify the analysis, since certain stochastic effects leading to a deviation from the expected behavior, so-called fluctuations such as genetic drift, vanish under this assumption. Often this type of analysis has been accompanied by experiments, which support the validity of the statements for finite population sizes also. Theoretically motivated research often demands rigorous statements that also hold for finite populations, see the following sections on runtime analysis.

A more recent publication by Wu and Kolonko [68] presented a convergence analysis of a so-called generalized cross-entropy optimization algorithm. The algorithm generalizes PBIL by adding so-called feasibility information to elements of the search space. This information corresponds to the *heuristic information* used in ACO [15]. It was shown, for constant ρ and under different assumptions about the feasibility information, that the algorithm may stagnate in suboptimal points owing to genetic drift. However, for a time-dependent update scheme, almost sure convergence to a set of solutions that may include optimal points was proven. Finally, an initial runtime analysis on LeadingOnes was presented. However, this specific result has been superseded by more detailed analyses in a follow-up paper [69], discussed below.

To conclude this overview of convergence analyses, we mention a very recent publication by Ollivier et al [51]. They introduced the *information-geometric optimization* (IGO) algorithm, which is a very general EDA framework derived from three invariance properties: invariance under the parameterization of the search space, invariance under the parameterization of the probabilistic model, and invariance under monotone transformations of the fitness function. This means that IGO does not care about the encoding of the search space, the probabilistic model, or absolute fitness values. These authors showed that IGO results in a general EDA that encompasses PBIL and cGA when it is used on the discrete hypercube, considering Poisson binomial distributions. Further, they considered a time-continuous infinite-population version of IGO, which they called IGO flow, in the setting of linear pseudo-Boolean optimization and proved that it always converges to the optimum if the probabilistic model is not ill-initialized, i.e., none of the probabilities are initialized such that sampling the optimum is impossible.

9.5 Runtime Analyses

In contrast to convergence results as described in Section 9.4, the focus of runtime analyses is the number of iterations until an algorithm samples a solution of sufficient quality for the first time, usually an optimum. Normally, the analyses consider both the expected number of iterations and concentration results.

In this section, we first give an in-depth overview of the history of runtime analyses on EDAs, ending with a very detailed discussion of the most recent results. These results are summarized in Table 9.1. Then, we consider noisy scenarios, i.e., scenarios where the fitness function is perturbed by some kind of noise, usually as an additive term to the original fitness. In this setting, every time a solution is evaluated, the noise is drawn again and independently of any prior noise, and the goal is to optimize the underlying unperturbed function despite the noise.

9.5.1 Early Results

We start with a discussion of the first publications addressing runtime aspects of EDAs, which date back to the early 2000s. Although some of the runtime bounds proven in these publications can now be improved with state-of-the-art methods, the analyses already point out typical scenarios and challenges in the runtime behavior of EDAs, in particular regarding genetic drift. Also they give insights into fundamental properties of EDAs that distinguish them from other nature-inspired algorithms such as EAs.

9.5.1.1 First Steps Towards Runtime Analyses

As pointed out above, rigorous runtime analyes must avoid the infinite-population model and derive statements for populations of finite size. However, the finiteness comes at a cost: if very small populations are used, there is a high risk of genetic drift and premature convergence in suboptimal regions of the search space. In a series of publications, Shapiro [57–59] addressed sources of genetic drift in EDAs, quantified its impact, and proposed measures to avoid it. In [58], he pointed out that the probability distribution evolved by an EDA may converge to suboptimal points and, using a dynamical systems approach, determined $\sqrt{n}$ as the minimum population size for UMDA to avoid genetic drift on the OneMax problem, and even exponential sizes for Needle. Later, Sudholt and Witt [63] and Krejca and Witt [36] gave rigorous proofs of the fact that genetic drift can happen up to population sizes of $O(\sqrt{n}\log n)$ in cGA and UMDA. Alternatively, for PBIL, the learning rate ρ may be reduced to counteract genetic drift. Using a dynamical systems approach, Shapiro [57] derived the result that the learning rate should be $O(1/\sqrt{n})$ and $O(2^{-n})$ to avoid genetic drift on the OneMax and Needle functions, respectively.

In [59], Shapiro also gave a rigorous theorem on the speed at which genetic drift moves the probabilistic model belonging to a specific class of EDAs called SML-EDA (including UMDA) into suboptimal regions. Also, a rigorous bound $\Omega(2^{n/2}/\sqrt{n})$ was determined for the population size required to make genetic drift on Needle unlikely.

Finally, in [58, p. 115], early conjectures about the runtime of UMDA appeared. More precisely, the paper reported an experimental determination of a runtime of $\Theta(\lambda\sqrt{n})$ for UMDA on OneMax (given that λ is asymptotically larger than $\sqrt{n}$). This bound was rigorously proven in [66]. However, it should be noted that Shapiro's UMDA slightly differs from the standard.

9.5.1.2 First Runtime Analyses

The first rigorous runtime analysis of an EDA was given by Droste [17]. He considered cGA without borders and proved the general lower bound $\Omega(K\sqrt{n})$ for its expected runtime on all linear functions. Using classical drift analysis and Chernoff bounds, Droste also proved the bound $O(K\sqrt{n})$ for OneMax, using $K = \Omega(n^{1/2+\varepsilon})$, i.e., slightly above the threshold stated by Shapiro [58]. This bound becomes $O(n^{1+\varepsilon})$ for the smallest K covered by his analysis. Finally, Droste argued that BinVal is more difficult to optimize than OneMax and asymptotically most difficult within the class of linear functions by proving that cGA without borders takes time $O(Kn)$ with at least constant probability on this function if $K = \Omega(n^{1+\varepsilon})$, and expected time at least $\Omega(Kn)$. The upper bound is $O(n^{2+\varepsilon})$ for the smallest possible K allowed. However, the lower bound $\Omega(Kn)$ does not come with a minimum value for K.

The results for OneMax were recently refined by Sudholt and Witt [63], using more advanced tools. In particular, all of Droste's upper bounds apply Chernoff bounds to show that genetic drift is unlikely; more precisely, he showed that the probability of a frequency dropping below 1/3 during the optimization is superpolynomially small. Using a negative drift theorem, the upper bound was improved from $O(n^{1+\varepsilon})$ to $O(n \log n)$ in [63]. See Section 9.5.2.1 for more details. Regarding BinVal, a very recent analysis by Witt [67] proved Droste's conjecture that the function is harder to optimize than OneMax, since the expected optimization time of cGA on BinVal is $\Omega(n^2)$ no matter how K is chosen. The idea of the analysis is to show, for all $K = o(n)$, that genetic drift will lock many frequencies to 0 before the optimum can be found.

The results discussed in the previous two paragraphs are summarized in the following theorem.

Theorem 9.5.1 ([17, 67]). *Choosing $K = n^{1+\varepsilon}$ for some constant $\varepsilon > 0$, the runtime of cGA without borders on* BinVal *is bounded by $O(Kn)$ with probability $\Omega(1)$. Moreover, the expected runtime of cGA with and without borders on* BinVal *is bounded from below by $\Omega(\min\{n^2, Kn\})$.*

In the years following Droste's seminal work, runtime analysis focused more on UMDA and variants thereof. The first runtime analysis of a UMDA variant was given by Chen et al [5], who studied the LeadingOnes function and a modification called TrapLeadingOnes. An expected optimization time of $O(\lambda n)$ of UMDA on LeadingOnes was derived under the so-called *no-random-error* assumption, which is similar to an infinite-population model and basically eliminates genetic drift. These authors also showed that TrapLeadingOnes, which starts out in the same way as LeadingOnes but requires an almost complete change of the probabilistic model for the EDA to reach the global optimum, yields expected exponential optimization time for UMDA, using 2-tournament selection instead of the usual truncation selection. Moreover, a generalization of UMDA similar to PBIL was considered, but it turned out that the strongest bounds apply for UMDA.

Strictly speaking, Chen et al [5] only derived runtime bounds for a model of UMDA. In subsequent work [7], they therefore supplemented these with a rigorous proof of the fact that UMDA, when appropriate borders for the frequencies are used, with high probability requires superpolynomial time to optimize TrapLeadingOnes. Similarly to Droste's early work, Chernoff bounds were applied to show that the frequencies do not deviate much from their expected behavior, i.e., do not exhibit strong genetic drift. For the Chernoff bounds to be sufficiently strong, unusually large population sizes such as $\lambda = \Omega(n^{2+\varepsilon})$ are required.

This approach was successfully picked up and extended in a more comprehensive journal publication [8]. Using $\lambda = \Omega(n^{2+\varepsilon})$ again, the authors of [8] showed that UMDA without borders optimizes LeadingOnes in time $O(\lambda n)$ with overwhelming probability. Furthermore, the utility of appropriately set frequency borders was shown on a modification called BVLO, where the fitness landscape requires the frequency of the last bit to be changed from one extremal value to the other one. Here, UMDA with borders has expected polynomial runtime, whereas UMDA without borders will with overwhelming probability be stuck at nonoptimal solutions.

Finally, using similar proof techniques, in particular Chernoff bounds, Chen et al [6] presented a constructed example function called Substring, on which simple EDAs and simple evolutionary algorithms behave fundamentally differently. More precisely, it was proven that the $(1+1)$ EA with any mutation probability c/n, where $c > 0$ is constant, with overwhelming probability needs exponential time to find the optimum of the function, while UMDA using $\lambda = \Omega(n^{2+\varepsilon})$ and $\lambda/\mu = O(1)$ finds with very high probability the optimum in time $O(\lambda n)$. Specifically, it is beneficial for the optimization that UMDA can sample search points with high variances as long as all frequencies are close to $1/2$. The $(1+1)$ EA always samples with low variance in the vicinity of the best-so-far solution, which is detrimental with the specific example function.

9.5.2 Recent Advances

Only very few runtime analyses of EDAs were published in the years 2010–2014, most notably [50, 68]. Starting from 2015, this research area gained significant momentum again (see, e.g., [10, 20, 21]). We now discuss the latest results in runtime analysis of EDAs. They mostly consider the standard benchmark function for EAs: OneMax. Using and advancing the toolbox for the analysis, matching upper and lower bounds have been proven, giving a tight runtime result that allows a direct comparison of the performance of EDAs with other nature-inspired algorithms.

9.5.2.1 Upper Bounds for OneMax

Interestingly, early runtime analyses of EDAs focused more on variants of LeadingOnes instead of OneMax, which is the most commonly considered example function in evolutionary computation. In fact, the first runtime analysis of UMDA on OneMax was not published until 2015 [10]. A possible explanation is that the hierarchical structure of LeadingOnes makes it more accessible to a runtime analysis than OneMax: if the best-so-far LeadingOnes value is k and the frequencies of the first k bits all have attained their maximum value, it is likely to sample only 1s there, which is typically needed for an improvement of the best function value seen. In contrast, there is no direct relationship between the OneMax value and frequencies at specific bits. Also, modern runtime analyses of UMDA [10, 38] reveal that a proof of runtime bounds for LeadingOnes can be relatively short and simple once the case of OneMax has been understood.

Results for cGA and $\mathrm{MMAS_{ib}}$

Before we describe the advances made in the runtime analysis of UMDA in more detail, we discuss the state of the art for the simpler EDAs $\mathrm{MMAS_{ib}}$ and cGA. As mentioned above, Droste [17] showed that cGA typically optimizes OneMax in time $O(n^{1+\varepsilon})$, using $K = n^{1/2+\varepsilon}$. His variant of cGA does not use any borders on the frequencies, which is why he used a comparatively large K to make convergence of a frequency to 0 by genetic drift sufficiently unlikely. More recent analyses of cGA and also other EDAs such as UMDA mostly impose borders $\{1/n, 1 - 1/n\}$ on the frequencies, as mentioned in Section 9.2.2.1. Using a more careful analysis of the stochastic behavior of frequencies, the classical $O(n \log n)$ runtime can be obtained, as shown in the following summary of theorems.

Theorem 9.5.2 ([50, 63]). *If $\rho \le 1/(cn^{1/2}\log n)$ for a sufficiently large constant $c > 0$ and $\rho \ge 1/\text{poly}(n)$, then MMAS$_{ib}$ (with borders) optimizes* ONEMAX *in expected time $O(\sqrt{n}/\rho)$. For $\rho = 1/(cn^{1/2}\log n)$, the runtime bound is $O(n \log n)$.*

The expected optimization time of cGA (with borders) on ONEMAX *with $K \ge c\sqrt{n}\log n$ for a sufficiently large $c > 0$ and $K = \text{poly}(n)$ is $O(\sqrt{n}K)$. This is $O(n \log n)$ for $K = c\sqrt{n}\log n$.*

Theorem 9.5.2 makes statements for two slightly different EDAs but the proofs of these statements follow roughly the same structure. Crucially, the effect of genetic drift is bounded: in the given time bound, for example, $O(\sqrt{n}K)$ generations, the expected number of frequencies that drop below 1/3 is proven to be polynomially small, for example, $O(1/n^2)$. Such a statement is typically obtained from a negative drift theorem. Next, the drift of frequencies towards 1 induced by selection (the so-called bias) is analyzed. It turns out that this bias is at least proportional to the sampling variance of the EDA: roughly, each frequency $\boldsymbol{p}_i$ increases by an expected amount $O\big(\boldsymbol{p}_i(1-\boldsymbol{p}_i)/(K\sqrt{\sum_{j=1}^{n}\boldsymbol{p}_j})\big)$ in each generation. An analysis of this variable drift, using the variable drift theorem, then gives the desired runtime bound. (As variable drift analysis was not available to Neumann et al [50], a unified and simpler proof of the statement for MMAS$_{\text{ib}}$ was given in [63].) In the unlikely event that a frequency has reached the wrong border $1/n$ owing to genetic drift, an event of probability $\Omega(1/n)$ is sufficient to lift the frequency again, which is absorbed into the total runtime owing to the low expected number of such bad frequencies.

First Phase Transition Around $\sqrt{n}\log n$

Theorem 9.5.2 requires $K \ge c\sqrt{n}\log n$. Recent research reveals that cGA in fact exhibits a phase transition in the regime $\Theta(\sqrt{n}\log n)$, similarly to MMAS$_{\text{ib}}$. If $K \le c'\sqrt{n}\log n$ for a sufficiently small constant $c' > 0$, then genetic drift will outweigh the drift due to selection such that a significant number of frequencies will drop to the lower border. In this case, classical arguments about coupon collector processes show that the runtime must be at least $\Omega(n \log n)$; see more arguments below, in Section 9.5.2.2, on lower bounds. There are no upper bounds on the runtime of cGA and MMAS$_{\text{ib}}$ in the regime corresponding to $K \le c'\sqrt{n}\log n$, but it is conjectured that bounds resembling the existing ones for UMDA (see Theorems 9.5.3 and 9.5.4 below) can be obtained if $K \in [c_1 \log n, c_2\sqrt{n}\log n]$ for appropriate constants $c_1, c_2 > 0$.

Results for UMDA

We complete this discussion of upper bounds with a review of recent advancements for UMDA. As mentioned, Dang and Lehre [10] were the first to prove upper bounds for UMDA on OneMax. If $\lambda \ge c\log n$ for a sufficiently large constant $c > 0$ and $\lambda \ge 13e\mu/(1-c')$ for an arbitrarily small constant $c' > 0$, then the expected runtime of UMDA on OneMax is $O(n\lambda\log\lambda)$. Hence, plugging in the smallest value of λ allowed in the statement, the bound is $O(n\log n\log\log n)$, i.e., slightly above the $O(n\log n)$ bound discussed above with respect to cGA and MMAS_{ib}.

Dang and Lehre used a powerful proof technique to obtain their bound. Interestingly, the so-called level-based theorem [9], which was originally developed for the analysis of population-based evolutionary algorithms, can be applied in this context. It was shown how the truncation selection of the best μ out of λ individuals leads to a reasonable chance of improving the best-so-far OneMax value and allows one to satisfy the other conditions of the level-based theorem with certain parameter settings. As a side-result, using the same proof technique, the bound $O(n\lambda + n^2)$ with respect to the LeadingOnes function was also obtained. Somewhat unusually, these proofs mostly consider populations instead of analyzing the values of single frequencies. For this to work, it is necessary that a frequency vector can be translated more or less unambiguously back into the population from which it was computed. This is possible in UMDA but not even in the slight generalization PBIL, where a frequency vector depends on the history of previous populations.

Obviously, the proof of the above-mentioned $O(n\log n\log\log n)$ bound immediately raised the question of whether this was the best possible runtime of UMDA on OneMax. Recently, two independent improvements of the bound were presented. The first one, due to Lehre and Nguyen [38], builds on a refinement of the level-based analysis, carefully using properties of the Poisson–binomial distribution, and is summarized by the following theorem. We emphasize that UMDA always refers to the algorithm with borders $1/n$ and $1-1/n$ on the frequencies, i.e., Algorithm 9.2 extended by a step that narrows all frequencies down to the interval $[1/n, 1-1/n]$.

Theorem 9.5.3 ([38]). *For some constant $a > 0$ and any constant $c \in (0,1)$, UMDA (with borders) with a parent population size $a\ln n \le \mu \le \sqrt{n(1-c)}$ and an offspring population size $\lambda \ge (13e)\mu/(1-c)$ has expected optimization time $O(n\lambda)$ on OneMax.*

Hence, Theorem 9.5.3 proves that the runtime of UMDA is $O(n\log n)$ for an appropriate choice of the parameters. This is tight owing to the recent lower bound $\Omega(n\log n)$ discussed below in Section 9.5.2.2. Interestingly, the set of appropriate choices for the $O(n\log n)$ behavior is confined to $\lambda = \Theta(\log n)$, which corresponds to a parameter choice below the above-mentioned phase transition, i.e., a choice where the algorithm exhibits severe genetic drift. Also, the theorem includes a limit on μ, which is exactly in the regime of the phase transition. For greater values of μ and λ, Witt [66] independently derived runtime bounds (see the following theorem); this result also includes the regime covered by Lehre and Nguyen [38], albeit with an assumption about the ratio λ/μ.

Theorem 9.5.4 ([66]).

(a) Let $\lambda = (1+\beta)\mu$ for an arbitrary constant $\beta > 0$ and let $\mu \geq c\sqrt{n}\log n$ for some sufficiently large constant $c > 0$. Then the optimization time of UMDA, both with and without borders, on OneMax *is bounded from above by $O(\lambda\sqrt{n})$ with probability $\Omega(1)$. For UMDA with borders, the expected optimization time is also bounded in this way.*

(b) Let $\lambda = (1+\beta)\mu$ for an arbitrary constant $\beta > 0$ and let $\mu \geq c\log n$ for a sufficiently large constant $c > 0$ as well as $\mu = o(n)$. Then the expected optimization time of UMDA with borders on OneMax *is $O(\lambda n)$. For UMDA without borders, it is infinite with high probability if $\mu < c'\sqrt{n}\log n$ for a sufficiently small constant $c' > 0$.*

The two statements of Theorem 9.5.4 reflect the above-mentioned phase transition. For $\mu \geq c\sqrt{n}\log n$, as required in the first statement, the behavior is similar to that underlying Theorem 9.5.2 with respect to cGA and MMAS_{ib}. Frequencies move smoothly towards the upper border, and it is unlikely that frequencies will exhibit genetic drift towards smaller values than $1/3$. Hence, it is unlikely as well that UMDA without borders will get stuck with frequencies at 0. The runtime $O(n\log n)$ is obtained for $\lambda = c\sqrt{n}\log n$ for an appropriately large constant $c > 0$.

The second statement of Theorem 9.5.4 applies to a case where genetic drift is likely, but frequencies that have hit the lower border $1/n$ have a reasonable chance to recover in the given time span, which is $O(n\lambda)$ instead of only $O(\sqrt{n}\lambda)$ now. In fact, the analysis carefully considers the drift of frequencies from the lower towards the upper border and analyzes the probability that a frequency leaves its upper border again. To do so, a very careful analysis of the bias introduced by selecting the best μ individuals is required. Without such selection, a single frequency would correspond to a so-called martingale, but, owing to selection, there is a small drift upwards, similarly to what we described with respect to cGA above. Hence, the proof of Theorem 9.5.4 also gives insights into the stochastic process described by single frequencies. It is more involved than that for cGA since UMDA can change frequencies globally instead of only by $\pm 1/K$. The runtime $O(n\log n)$ can be obtained again, this time for $\lambda = c\sqrt{n}\log n$.

It is worth pointing out that Theorems 9.5.3 and 9.5.4 make nonoverlapping statements. Theorem 9.5.4 also applies to λ above the phase transition and describes a transition of $O(n\lambda)$ to $O(n\sqrt{\lambda})$ in the runtime. However, it crucially assumes $\lambda = (1+\Theta(1))\mu$ in both statements, an assumption that was also useful in earlier analyses of EDAs [59] but restricts the generality of the statements. In contrast, Theorem 9.5.3 applies to settings such as $\mu = 1$, $\lambda = c \log n$ and shows the $O(n \log n)$ bound also for this somewhat extreme choice of parameters.

We conclude this discussion of upper bounds by summarizing a recent study by Wu et al [69], who presented the first runtime analysis of PBIL (called the cross-entropy (CE) method in their paper). Using $\mu = n^{1+\varepsilon} \log n$ for some constant $\varepsilon > 0$ and $\lambda = \omega(\mu)$, they obtained the result that the runtime of PBIL on OneMax is $O(\lambda n^{1/2+\varepsilon/3}/\rho)$ with overwhelming probability. Hence, if $\rho = \Omega(1)$, including the special case $\rho = 1$, where PBIL collapses to UMDA, a runtime bound of $O(n^{3/2+(4/3)\varepsilon} \log n)$ holds, i.e., slightly above $n^{3/2}$. In light of the detailed analyses of UMDA presented above, one may conjecture that this bound is not tight even if $\rho < 1$ is used, i.e., PBIL actually uses its learning approach to include solutions from several previous generations in the probabilistic model. In addition to that, a bound of the type $O(n^{2+\varepsilon})$ on LeadingOnes is obtained if $\rho = \Omega(1)$, $\mu = n^{\varepsilon/2}$ and $\lambda = \Omega(n^{1+\varepsilon})$. Technically, Wu et al [69] used concentration bounds such as Chernoff bounds to bound the effect of genetic drift, as well as anti-concentration results, in particular for the Poisson–binomial distribution, to obtain their statements. All bounds hold with high probability only, since PBIL is formulated without borders. Probably, using a more detailed analysis of genetic drift and applying modern drift theorems, the bound for LeadingOnes can be improved to an expected $O(n^2)$ runtime for all $\rho = \Omega(1)$, provided that the classical borders $\{1/n, 1 - 1/n\}$ are used.

9.5.2.2 Lower Bounds for OneMax

Deriving lower bounds on the runtime of EDAs is often more challenging than deriving upper bounds. Roughly, most existing approaches show that the probabilistic model is not sufficiently adjusted towards the set of optimal solutions within a given time span. A relatively straightforward approach relates the runtime to the strength of updates in the algorithm. With respect to simple univariate algorithms such as cGA and UMDA, one can show that frequencies do not increase by more than $1/K$ (with probability 1) or $O(1/\mu)$ (in expectation, assuming $\lambda = (1 + \Theta(1))\mu$) in a step. This naturally leads to a lower bound of $\Omega(K)$ or $\Omega(\mu)$, respectively, on the runtime on OneMax. However, the bound is weak, as it pessimistically assumes that each generation changes frequencies in the right direction. More detailed analyses reveal that cGA, in the early phases of the optimization process, has only a probability of $O(1/\sqrt{n})$ of performing a step where the two offspring differ in fewer than two bits, i.e., the probability that the outcome of a certain bit is relevant for selection is then only $O(1/\sqrt{n})$ [63]. Similar results can be obtained for UMDA [36]. Thus, each bit only moves by up to an expected amount of $O(1/(K\sqrt{n}))$ or $O(1/(\mu\sqrt{n}))$, respectively, per generation. Then a drift analysis translates this into the lower bounds $\Omega(K\sqrt{n})$ and $\Omega(\mu\sqrt{n})$

that appear in the following theorems. The first bound was already known for cGA without borders from Droste's work [17].

Theorem 9.5.5 ([63]). *The optimization time of cGA (with borders) with $K \leq \text{poly}(n)$ on* OneMax *is $\Omega(K\sqrt{n}+n\log n)$ with high probability and in expectation.*

Theorem 9.5.6 ([36]). *Let $\lambda = (1+\beta)\mu$ for some constant $\beta > 0$ and $\lambda \leq \text{poly}(n)$. Then the expected optimization time of UMDA on* OneMax *is $\Omega(\mu\sqrt{n}+n\log n)$ (both with and without borders).*

Sudholt and Witt [63] also stated Theorem 9.5.5 in an analogous fashion for MMAS_{ib}, with the parameter K replaced by $1/\rho$. As its working principle is rather similar to that of cGA, we do not discuss MMAS_{ib} further in this section.

The lower bounds $\Omega(K\sqrt{n})$ and $\Omega(\mu\sqrt{n})$ we have illustrated so far are very weak if K and μ, respectively, are small. In fact, they can be even worse than the bounds $\Omega(n/\log n)$ that follow from black-box complexity [18]. Until 2016, it was not clear whether the runtime of these simple EDAs was also bounded by $\Omega(n\log n)$ or whether they could possibly optimize OneMax in $o(n\log n)$ time and hence be faster than simple evolutionary algorithms. A negative answer was given by the two above theorems, both of which also contain an $\Omega(n\log n)$ term.

The proof of the bound $\Omega(n\log n)$ is technically demanding. It relies on the following strategy:

(a) Show that with high probability several frequencies, for example $\sqrt{n}$ of them, reach the lower border before the optimum is sampled. This requires a detailed analysis of the stochastic behavior of several dependent, single frequencies instead of considering merely the sum $P_t := \sum_{i=1}^{n} \boldsymbol{p}_i^{(t)}$ of the frequencies, whose stochastic behavior is already quite well understood and can relatively easily be analyzed by drift analysis, as sketched in the paragraph following Theorem 9.5.2. In fact, in the detailed analysis of single frequencies, it is even required to show that some frequencies walk to the lower border while most other frequencies do not move up too far to the upper border; otherwise one cannot rule out with sufficiently high probability the possibility that the optimum is sampled in the meantime.
(b) Once polynomially many frequencies have reached the lower border $1/n$, a so-called coupon collector effect arises. A relatively straighforward generalization of the coupon collector theorem [44, 45] to the case where still polynomially many bits have to be corrected, where a correction is made with probability at most $1/n$, yields the following statement: *Assume cGA reaches a situation where at least $\Omega(n^{\varepsilon})$ frequencies attain the lower border $1/n$. Then, with high probability and in expectation, the remaining optimization time is $\Omega(n\log n)$.* The underlying modification of the coupon collector theorem may be called folklore in probability theory,

but is interesting for its own sake: collecting the last n^{ε} coupons takes asymptotically the same time as collecting them all.

A major effort is required to flesh out the behavior sketched in item (a) above. Roughly speaking, one exploits the fact that frequencies behave similarly to a martingale and can walk to the lower border owing to genetic drift. However, the effect of genetic drift is dependent on many factors. When all frequencies have reached a border, genetic drift is much less pronounced than in situations where many frequencies are close to the median value 1/2 (which is initially the case). To handle this dependency on time, it has to be shown that some frequencies move unusually fast, which means faster than the expected time, to the lower border while the majority of the frequencies is still at a medium value. More precisely, the proofs approximate the hitting time of the lower border by a normally distributed random variable, which is not sharply concentrated around the mean and exhibits exactly the desired reasonable probability of deviating from the mean. Additionally, the drift analysis features a novel use of potential functions that smooth out the variances of the movements of frequencies, which would be place-dependent and not applicable to the approximation by a normal distribution otherwise.

Second Phase Transition Around $\log n$

Not much research has been done on very small values of the population size λ and K in UMDA and cGA, corresponding to very large ρ in MMAS_{ib}. Neumann et al [50] gave an exponential bound on the runtime of MMAS_{ib} if $\rho \geq c/\log n$, indicating a second phase transition in behavior around $\log n$. Roughly speaking, if the set of possible values for a frequency becomes less than $\log n$, then the scale is too coarse for the probabilistic model to adjust slowly towards the set of optimal solutions. For example, even after a frequency has reached its maximum $1-1/n$ once, an unlucky step may lead to a drastic decline in frequency which, on average, cannot be recovered in polynomial time. It is conjectured that cGA and UMDA will not optimize OneMax in polynomial time either if $K \leq c\log n$ or if $\lambda \leq c\log n$, respectively, for a small constant $c > 0$.

Major Open Problems

Even if we ignore the values below $\log n$ corresponding to the second phase transition just mentioned, the lower bounds given in Theorems 9.5.5 and 9.5.6 still do not give a complete picture of the runtime of the algorithms on OneMax. For example, for μ in the medium regime between the phase transitions, i.e., when μ is both $\omega(\log n)$ and $o(\sqrt{n}\log n)$, it is not clear whether a lower bound of the kind $\Omega(\mu n)$ (which would match the upper bound given above in Theorem 9.5.4) or any other runtime $\omega(n\log n)$ holds. It is an open problem to prove tight bounds on the runtime of simple EDAs in this medium regime. As usual, we expect analyses to be harder for UMDA than for cGA, as the former algorithm can change frequencies in a global way, while the latter only changes them locally by $\pm 1/K$.

Some progress on the way to tight bounds has been made very recently by Lengler et al. [42], who proved a lower bound of $\Omega(K^{1/3}n + n\log n)$ on the expected optimization time of the cGA if $K = O(n^{1/2}/(\log n \log\log n))$. Hence, the expected optimization time will be $\Omega(n^{7/6}/(\log n \log\log n))$ for $K = O(n^{1/2}/(\log n \log\log n))$, while it is bounded from above by $O(n\log n)$ for $K = cn^{1/2}$ if c is chosen as a sufficiently large constant. Hence, the runtime seems to depend in a multimodal way on K. Nevertheless, this still remains a conjecture, since there are no upper bounds on the runtime of the cGA for $K = o(n^{1/2})$; there are only upper bounds for the UMDA if $\lambda = o(n^{1/2})$ that support this conjecture.

A summary of proven upper and conjectured bounds on the runtime of UMDA on OneMax is displayed in Fig. 9.2. We believe that similar results hold for cGA and MMAS_{ib}, with λ replaced by K and $1/\rho$, respectively.

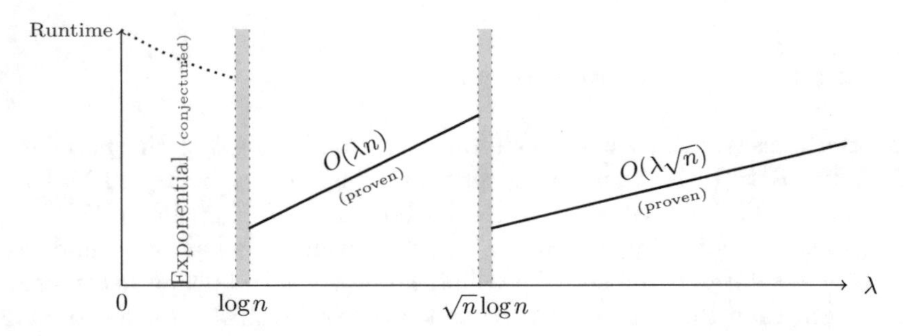

Fig. 9.2 Picture of runtime bounds with UMDA on OneMax, assuming $\lambda = (1+\Theta(1))\mu$.

We have carried out experiments for UMDA on OneMax to gain some empirical insights into the relationship between λ and the runtime. The algorithm was implemented in the C programming language using the WELL512a random number generator. The problem size was set to $n = 2000$, λ was increased from 14 to 350 in steps of size 2, μ was set to $\lambda/2$, and, owing to the high variance of the runs, especially for small λ, an average was taken over 3000 runs for every setting of λ. The left-hand side of Fig. 9.3 demonstrates that the runtime in fact shows a multimodal dependence on λ. Starting from very high values, it has a minimum at $\lambda \approx 20$ and then increases again up to $\lambda \approx 70$. Thereafter it falls again up to $\lambda \approx 280$, and finally increases rather steeply for the rest of the range. The right-hand side also illustrates that the number of times the lower border is hit seems to decrease exponentially with

λ. The phase transition where the behavior of frequencies turns from chaotic into stable is empirically located somewhere between 250 and 300.

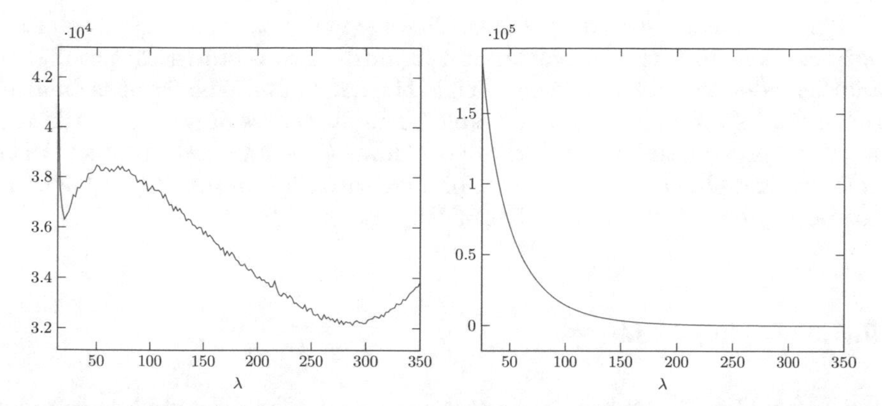

Fig. 9.3 Left-hand side: empirical runtime of UMDA on OneMax, right-hand side: number of hits of lower border; for $n = 2000$, $\lambda \in \{14, 16, \ldots, 350\}$, $\mu = \lambda/2$, and averaged over 3000 runs

9.5.2.3 New Advances in Tackling Genetic Drift

Genetic drift slows down optimization because it basically adds a random signal to the objective function. One reason why this impacts the algorithms is their myopic behavior: they have to perform an update to their frequencies based only on information from the current iteration. Especially if this sample size is small, such as for cGA or MMAS_{ib}, the amount of information gained during a single iteration may be too small to perform a sensible decision with respect to the update.

In order to counteract such ill-informed updates, Doerr and Krejca [12] proposed a new EDA that tries to reduce the number of incorrect frequency updates by relying not only on information from a single iteration but also information from multiple previous iterations. Their *significance-based cGA* (sig-cGA) stores a frequency vector, like an n-Bernoulli-λ-EDA, but additionally also stores a history H_i for each bit position i. In each iteration, only two offspring are sampled, and the bits of the better individual are saved in the respective histories. Then the algorithm checks, for each history, whether a *significance* occurs, that is, whether the number of 1s or 0s saved is drastically more than expected when assuming that each 1 occurs with probability $\boldsymbol{p}_i$. The level of confidence can be regulated by a parameter called ε. If a significance of 1s is detected at a position, the respective frequency is set to $1 - 1/n$; if a significance of 0s is detected, the frequency is set to $1/n$; otherwise, the frequency is left unchanged. Overall, the algorithm uses only three different frequency values: $1-1/n$, $1/2$, and $1/n$, where $1/2$ is used only as a starting value – if a frequency once takes a value different from $1/2$, it never returns.

This significance-based approach allows sig-cGA, at the beginning of an optimization, to keep frequencies at $1/2$ until there is statistical proof that another value would be more beneficial. Thus, it can be thought of as an algorithm that is both balanced and stable.[4] The usefulness of this approach was shown by proving that this algorithm optimizes OneMax and LeadingOnes both in time $O(n \log n)$ in expectation and with high probability, which has not been proven for any other EA or EDA before [12].

9.5.3 Noisy Settings

In real-world optimization, the evaluation of a solution often involves a degree of uncertainty due to inaccuracies in the evaluation process. We call this uncertainty in the fitness *noise*. Since EDAs, as general-purpose heuristics, build on this inaccurate information, it is interesting to analyze how they perform when faced with noise.

Most EDA scenarios with noise consider ACO variants on single-destination shortest-path problems, mostly not in the context of EDAs at all [13, 19, 33, 62]. However, some results have analyzed pseudo-Boolean optimization [23, 24].

9.5.3.1 Combinatorial Optimization

Horoba and Sudholt [33] considered an acyclic weighted graph and were interested in finding a shortest path from each vertex to a single given destination. The noise was modeled by drawing a random nonnegative value η per edge weight w, possibly dependent on the edge, and its new weight w' was determined by $w' = w(1+\eta)$. Thus, depending on the distribution of η, large weights increase more than small weights. The algorithm of interest was an ACO variant. This constructs paths from each node to the destination, using the perturbed weights and choosing an edge with a probability related to its pheromone value with respect to the pheromones of all competing edges. The algorithm compares each constructed path with the currently best-so-far solution per node without reevaluation. That means that the best-so-far solutions, as well as their possibly perturbed weights, are stored and used for lookup.

These authors provided instances in which the algorithm does not find a desired approximation within polynomial time with high probability. This is

[4] Since sig-cGA is not an n-Bernoulli-λ-EDA (owing to the histories that store data from multiple iterations), the actual definitions of *balanced* and *stable* do not apply.

Table 9.1 Expected runtimes (number of fitness evaluations) of various EDAs until they first find an optimum for the three functions OneMax (9.1), LeadingOnes (9.3), and BinVal (9.2)

Algorithm	OneMax	Constraints	LeadingOnes	Constraints	BinVal	Constraints
UMDA/PBIL[5]	$\Omega(\lambda\sqrt{n} + n\log n)$ (Thm. 9.5.6)	$\mu=\Theta(\lambda), \lambda=O\big(\text{poly}(n)\big)$	$O(n\lambda\log\lambda+n^2)$ [10]	$\lambda=\Omega(\log n)$, $\mu=\Theta(\lambda)$	Unknown	–
	$O(\lambda n)$ (Thm. 9.5.3 and 9.5.4)	$\mu=\Omega(\log n)\cap O(\sqrt{n}), \lambda=\Omega(\mu)$ or $\mu=\Omega(\log n)\cap o(n), \mu=\Theta(\lambda)$				
	$O(\lambda\sqrt{n})$ (Thm. 9.5.4)	$\mu=\Omega(\sqrt{n}\log n), \mu=\Theta(\lambda)$				
cGA/2-MMAS$_{\text{ib}}$	$\Omega\left(\frac{\sqrt{n}}{\rho} + n\log n\right)$ (Thm. 9.5.5)	$\frac{1}{\rho}=O\big(\text{poly}(n)\big)$	Unknown	–	$\Omega(\min\{n^2, Kn\})$ (Thm. 9.5.1)[6]	None
	$O\left(\frac{\sqrt{n}}{\rho}\right)$ (Thm. 9.5.2)	$\frac{1}{\rho}=\Omega(\sqrt{n}\log n)\cap O\big(\text{poly}(n)\big)$			$O(Kn)$ (Thm. 9.5.1)[6]	$K=n^{1+\varepsilon}$, $\varepsilon=\Theta(1)$
sig-cGA	$O(n\log n)$ [12]	$\varepsilon>12$	$O(n\log n)$ [12]	$\varepsilon>12$	Unknown	–

[5] The results shown for PBIL are the results for UMDA, since the latter is a special case of the former. Wu et al [69] also analyzed PBIL but with worse results.

[6] This result was only proven for cGA.

due to the best-so-far solution not being reevaluated. Thus, if a nonoptimal path is evaluated to be very good by chance, it will get reinforced many times, making it more unlikely that the algorithm will sample other paths that, additionally, have to be evaluated even better. However, these authors also proved that optimization will succeed if the noise follows the same distribution for every edge.

Sudholt and Thyssen [62] extended the results of Horoba and Sudholt [33] by considering a larger range of noise distributions, showing how long it takes to approximate optimal solutions or even when optimization succeeds.

Doerr et al [13] considered a similar scenario to the one analyzed by Horoba and Sudholt [33], the difference being that the weights of the graphs were purely random, i.e., there was no groundtruth to rely on. This setting makes it harder to define what an optimal solution actually is.

The authors of [13] first considered a multigraph consisting of two nodes with multiple edges between those nodes. They called an edge *preferred* if its probability of being shorter than any other edge from the same vertex was at least $1/2 + \delta$, where $\delta > 0$ is a constant, and they stated how this scenario relates to armed-bandit settings. Using the same ACO algorithm as Horoba and Sudholt [33] but reevaluating the best-so-far solution each iteration, they gave an upper bound on the expected time until the pheromone on the preferred edge was maximal. They then provided examples of weight distributions that result in an edge being preferred. The paper concludes with a more general graph setting that assumes that there exists an inductively defined set of edges S, starting at a given node, such that each edge that extends paths using edges from S is preferred. The authors of the paper gave an upper bound, in the case where S is a tree, on the expected time until the ACO variant considered maximizes the pheromones on all of the edges in S.

Feldmann and Kötzing [19] analyzed the same setting as Doerr et al [13] but investigated another ACO variant: MMAS-fp. This algorithm does not store best-so-far solutions but always makes an update with respect to the current samples; however, the update is done with respect to each sample's fitness. Thus, good solutions yield larger changes in the update than bad solutions. The authors of [19] explained the difference in this approach with respect to those of Horoba and Sudholt [33] and Doerr et al [13] by saying that MMAS-fp optimizes paths that are shortest in expectation. They proved this claim by providing upper bounds on the expected number of iterations until MMAS-fp finds expected shortest paths in graphs where, for each node, the difference between the expected lengths of different outgoing edges can be lower-bounded by a value $\delta > 0$, which influences the runtime.

9.5.3.2 Pseudo-Boolean Optimization

Friedrich et al [23] (conference version [21]) also considered MMAS-fp, just like Feldmann and Kötzing [19], but in the setting of optimizing linear pseudo-

Boolean functions. The noise was mostly modeled as Gaussian *additive posterior noise*, i.e., when evaluating the fitness of an individual, a normally distributed random variable is added to the fitness, every time anew and independently. Friedrich et al [23] showed that MMAS-fp *scales gracefully* in this scenario. That means that, for every polynomially bounded variance of the noise, there is a configuration of MMAS-fp such that the runtime is polynomially bounded as well. Since the runtime results hold with high probability, by performing an uninformed binary search using restarts, the correct variance of a problem with Gaussian noise can be guessed correctly within polynomial time. Thus, MMAS-fp can be modified such that the runtime is, with high probability, polynomial if the variance of the noise is.

Additionally, the authors of [23] extended their results to posterior noise other than Gaussian. Further, they considered a *prior noise* model where, before evaluating the fitness of an individual, a uniformly randomly chosen bit is flipped. In both of these settings, they proved that the algorithm scales gracefully.

Friedrich et al [24] (conference version [20]) also considered cGA under the Gaussian additive posterior noise model. As for MMAS-fp, they proved that the algorithm scales gracefully. Further, they showed that the $(\mu + 1)$ EA, a commonly analyzed EA, does not scale gracefully. Both of the results of Friedrich et al [23, 24] suggest that EDAs are inherently more tolerant to noise than standard EAs, as the EDAs did not need to be modified to cope with noise, except for choosing correct parameters. These authors also compared the restart version of cGA with an approach that uses resampling in order to basically remove the noise in the fitness, as described by Akimoto et al [2]. Since the number of resamples is closely tied to the noise's variance, the cGA variant using restarts instead of resampling emerges victorious.

9.6 Conclusions and Open Problems

We have given an overview of the state of the art in the theory of discrete EDAs, where the most recent research surpasses convergence analyses and instead deals with the runtime of especially simple univariate EDAs such as cGA, UMDA, and PBIL. In this domain, increasingly precise results have been obtained with respect to well-established benchmark problems such as OneMax, but, as we have emphasized in this chapter, there are several open problems even for this simple problem. In particular, a complete picture of the runtime of the simple EDAs depending on their parameters is still missing. We think that further results for benchmark functions will give insight into the right choice of specific EDAs, including the choice of parameters such as the population size and the borders on the frequencies depending on the problem characteristics. We also expect that this research will lead to runtime results and advice on the choice of algorithms and parameters with respect to more practically relevant combinatorial optimization problems. Here in particular, noisy settings or, more generally, optimization under uncertainty seem to represent scenarios where EDAs can outperform classical evolutionary algorithms. Also, the combinatorial structure may favor the application of multivariate EDAs, a type of EDA for which almost no theoretical results exist yet.

Acknowledgements Carsten Witt was supported by a grant from the Danish Council for Independent Research (DFF-FNU 4002-00542). Support by the COST Action 15140 "Improving Applicability of Nature-Inspired Optimisation by Joining Theory and Practice" (ImAppNIO) is also gratefully acknowledged.

References

[1] Afshani P, Agrawal M, Doerr B, Doerr C, Larsen KG, Mehlhorn K (2013) The query complexity of finding a hidden permutation. In: Space-Efficient Data Structures, Streams, and Algorithms – Papers in Honor of J. Ian Munro on the Occasion of His 66th Birthday, pp 1–11, DOI 10.1007/978-3-642-40273-9_1

[2] Akimoto Y, Astete-Morales S, Teytaud O (2015) Analysis of runtime of optimization algorithms for noisy functions over discrete codomains. Journal of Theoretical Computer Science 605:42:50, DOI 10.1016/j.tcs.2015.04.008

[3] Asoh H, Mühlenbein H (1994) On the mean convergence time of evolutionary algorithms without selection and mutation. In: Proc. of PPSN '94, pp 88–97

[4] Baluja S (1994) Population-based incremental learning: A method for integrating genetic search based function optimization and competitive learning. Tech. Rep. CMU-CS-94-163, Carnegie Mellon University, Pittsburgh, PA

[5] Chen T, Tang K, Chen G, Yao X (2007) On the analysis of average time complexity of estimation of distribution algorithms. In: Proc. of CEC '07, pp 453–460

[6] Chen T, Lehre PK, Tang K, Yao X (2009) When is an estimation of distribution algorithm better than an evolutionary algorithm? In: Proc. of CEC '09, pp 1470–1477

[7] Chen T, Tang K, Chen G, Yao X (2009) Rigorous time complexity analysis of univariate marginal distribution algorithm with margins. In: Proc. of CEC '09, pp 2157–2164

[8] Chen T, Tang K, Chen G, Yao X (2010) Analysis of computational time of simple estimation of distribution algorithms. IEEE Transactions on Evolutionary Computation 14(1):1–22

[9] Corus D, Dang DC, Eremeev AV, Lehre PK (2017) Level-based analysis of genetic algorithms and other search processes. IEEE Transactions on Evolutionary Computation 22(5):707–719
[10] Dang D, Lehre PK (2015) Simplified runtime analysis of estimation of distribution algorithms. In: Proc. of GECCO '15, pp 513–518
[11] De Bonet JS, Isbell CL Jr, Viola PA (1997) MIMIC: Finding optima by estimating probability densities. In: Proc. of NIPS '96, pp 424–430
[12] Doerr B, Krejca MS (2018) Significance-based estimation-of-distribution algorithms. In: Proc. of GECCO '18, pp 1483-1490
[13] Doerr B, Hota A, Kötzing T (2012) Ants easily solve stochastic shortest path problems. In: Proc. of GECCO '12, pp 17–24, DOI 10.1145/2330163.2330167
[14] Doerr B, Johannsen D, Winzen C (2012) Multiplicative drift analysis. Algorithmica 64(4):673–697, DOI 10.1007/s00453-012-9622-x
[15] Dorigo M, Stützle T (2004) Ant Colony Optimization. MIT Press
[16] Dorigo M, Maniezzo V, Colorni A (1991) Positive feedback as a search strategy. Tech. Rep. 91–016, Dipartimento di Elettronica, Politecnico di Milano, Milan, Italy
[17] Droste S (2006) A rigorous analysis of the compact genetic algorithm for linear functions. Natural Computing 5(3):257–283, preliminary version in GECCO '05
[18] Droste S, Jansen T, Wegener I (2006) Upper and lower bounds for randomized search heuristics in black-box optimization. Theory of Computing Systems 39:525–544
[19] Feldmann M, Kötzing T (2013) Optimizing expected path lengths with ant colony optimization using fitness proportional update. In: Proc. of FOGA '13, pp 65–74, DOI 10.1145/2460239.2460246
[20] Friedrich T, Kötzing T, Krejca MS, Sutton AM (2015) The benefit of recombination in noisy evolutionary search. In: Proc. of ISAAC '15, pp 140–150
[21] Friedrich T, Kötzing T, Krejca MS, Sutton AM (2015) Robustness of ant colony optimization to noise. In: Proc. of GECCO '15, pp 17–24
[22] Friedrich T, Kötzing T, Krejca MS (2016) EDAs cannot be balanced and stable. In: Proc. of GECCO '16, pp 1139–1146, DOI 10.1145/2908812.2908895
[23] Friedrich T, Kötzing T, Krejca MS, Sutton AM (2016) Robustness of ant colony optimization to noise. Evolutionary Computation 24(2):237–254
[24] Friedrich T, Kötzing T, Krejca MS, Sutton AM (2017) The compact genetic algorithm is efficient under extreme gaussian noise. IEEE Transactions on Evolutionary Computation 21(3):477–490
[25] Gao Y, Culberson J (2005) Space complexity of estimation of distribution algorithms. Evolutionary Computation 13(1):125–143, DOI 10.1162/1063656053583423

[26] González C, Lozano J, Larrañaga P (2000) Analyzing the PBIL algorithm by means of discrete dynamical systems. Complex Systems 12(4):465–479
[27] Harik G, Lobo FG, Goldberg DE (1998) The compact genetic algorithm. IEEE Transactions on Evolutionary Computation pp 523–528
[28] Harik GR, Lobo FG, Sastry K (2006) Linkage learning via probabilistic modeling in the extended compact genetic algorithm (ECGA). In: [54], pp 39–61
[29] Hauschild M, Pelikan M (2011) An introduction and survey of estimation of distribution algorithms. Swarm and Evolutionary Computation 1(3):111–128
[30] He J, Yao X (2001) Drift analysis and average time complexity of evolutionary algorithms. Artificial Intelligence 127(1):57–85, DOI 10.1016/S0004-3702(01)00058-3
[31] He J, Yao X (2004) A study of drift analysis for estimating computation time of evolutionary algorithms. Natural Computing 3(1):21–35, DOI 10.1023/B:NACO.0000023417.31393.c7
[32] Höhfeld M, Rudolph G (1997) Towards a theory of population-based incremental learning. In: Proc. of ICEC '97, pp 1–5
[33] Horoba C, Sudholt D (2010) Ant colony optimization for stochastic shortest path problems. In: Proc. of GECCO '10, pp 1465–1472, DOI 10.1145/1830483.1830750
[34] Janusz Kacprzyk WP (ed) (2015) Springer Handbook of Computational Intelligence. Springer, DOI 10.1007/978-3-662-43505-2
[35] Johannsen D (2010) Random combinatorial structures and randomized search heuristics. PhD thesis, Universität des Saarlandes, Saarbrücken, Germany and the Max-Planck-Institut für Informatik
[36] Krejca MS, Witt C (2017) Lower bounds on the run time of the univariate marginal distribution algorithm on OneMax. In: Proc. of FOGA '17, pp 65–79, DOI 10.1145/3040718.3040724
[37] Larrañaga P, Lozano JA (2002) Estimation of Distribution Algorithms: A New Tool for Evolutionary Computation, Genetic Algorithms and Evolutionary Computation, vol 2. Springer
[38] Lehre PK, Nguyen PTH (2017) Improved runtime bounds for the univariate marginal distribution algorithm via anti-concentration. In: Proc. of GECCO '17, pp 1383–1390, DOI 10.1145/3071178.3071317
[39] Lehre PK, Witt C (2010) Black-box search by unbiased variation. In: Proc. of GECCO '10, pp 1441–1448
[40] Lehre PK, Witt C (2014) Concentrated hitting times of randomized search heuristics with variable drift. In: Proc. of ISAAC '14, pp 686–697, DOI 10.1007/978-3-319-13075-0
[41] Lehre PK, Witt C (2017) General drift analysis with tail bounds, arXiv:1307.2559
[42] Lengler J, Sudholt D, Witt C (2018) Medium step sizes are harmful for the compact genetic algorithm. In: Proc. of GECCO '18, pp 1499–1506

[43] Mitavskiy B, Rowe JE, Cannings C (2009) Theoretical analysis of local search strategies to optimize network communication subject to preserving the total number of links. International Journal of Intelligent Computing and Cybernetics 2(2):243–284, DOI 10.1108/17563780910959893
[44] Mitzenmacher M, Upfal E (2005) Probability and Computing: Randomized Algorithms and Probabilistic Analysis. Cambridge University Press
[45] Motwani R, Raghavan P (1995) Randomized Algorithms. Cambridge University Press
[46] Mühlenbein H (1992) How genetic algorithms really work: Mutation and hillclimbing. In: Proc. of PPSN '92, pp 15–26
[47] Mühlenbein H, Mahnig T (1999) Convergence theory and applications of the factorized distribution algorithm. Journal of Computing and Information Technology 7:19–32
[48] Mühlenbein H, Mahnig T (1999) FDA – A scalable evolutionary algorithm for the optimization of additively decomposed functions. Evolutionary Computation 7(4):353–376, DOI 10.1162/evco.1999.7.4.353
[49] Mühlenbein H, Paass G (1996) From Recombination of Genes to the Estimation of Distributions I. Binary Parameters. In: Proc. of PPSN '96, pp 178–187
[50] Neumann F, Sudholt D, Witt C (2010) A few ants are enough: ACO with iteration-best update. In: Proc. of GECCO '10, pp 63–70
[51] Ollivier Y, Arnold L, Auger A, Hansen N (2017) Information-geometric optimization algorithms: A unifying picture via invariance principles. Journal of Machine Learning Research 18:1–65
[52] Pelikan M, Mühlenbein H (1999) The bivariate marginal distribution algorithm. In: Advances in Soft Computing, Springer, pp 521–535
[53] Pelikan M, Goldberg DE, Cantú-Paz E (1999) BOA: The bayesian optimization algorithm. In: Proc. of GECCO '99, pp 525–532
[54] Pelikan M, Sastry K, Cantú-Paz E (2006) Scalable Optimization via Probabilistic Modeling: From Algorithms to Applications, Studies in Computational Intelligence, vol 33. Springer
[55] Pelikan M, Hauschild M, Lobo FG (2015) Estimation of distribution algorithms. In: [34], pp 899–928, DOI 10.1007/978-3-662-43505-2_45
[56] Rudolph G (1997) Convergence properties of evolutionary algorithms. Verlag Dr. Kovač
[57] Shapiro JL (2003) The sensitivity of PBIL to its learning rate, and how detailed balance can remove it. In: Proc. of FOGA '02, pp 115–132
[58] Shapiro JL (2005) Drift and scaling in estimation of distribution algorithms. Evolutionary Computation 13(1):99–123, DOI 10.1162/1063656053583414
[59] Shapiro JL (2006) Diversity loss in general estimation of distribution algorithms. In: Proc. of PPSN '06, Springer, pp 92–101
[60] Simon D (2013) Evolutionary Optimization Algorithms. John Wiley & Sons

[61] Stützle T, Hoos HH (2000) MAX–MIN ant system. Future generation computer systems 16(8):889–914
[62] Sudholt D, Thyssen C (2012) A simple ant colony optimizer for stochastic shortest path problems. Algorithmica 64(4):643–672, DOI 10.1007/s00453-011-9606-2
[63] Sudholt D, Witt C (2016) Update strength in EDAs and ACO: How to avoid genetic drift. In: Proc. of GECCO '16, pp 61–68
[64] Vose MD (1999) The Simple Genetic Algorithm: Foundations and Theory. MIT Press
[65] Wald A (1944) On cumulative sums of random variables. The Annals of Mathematical Statistics 15(3):283–296, DOI 10.1214/aoms/1177731235
[66] Witt C (2017) Upper bounds on the runtime of the univariate marginal distribution algorithm on OneMax. In: Proc. of GECCO '17, pp 1415–1422, DOI 10.1145/3071178.3071216
[67] Witt C (2018) Domino convergence: why one should hill-climb on linear functions. In: Proc. of GECCO '18, ACM Press, to appear
[68] Wu Z, Kolonko M (2014) Asymptotic properties of a generalized cross-entropy optimization algorithm. IEEE Transactions on Evolutionary Computation 18(5):658–673, DOI 10.1109/TEVC.2014.2336882
[69] Wu Z, Kolonko M, Möhring RH (2017) Stochastic runtime analysis of the cross-entropy algorithm. IEEE Transactions on Evolutionary Computation 21(4):616–628, DOI 10.1109/TEVC.2017.2667713
[70] Zhang Q (2004) On stability of fixed points of limit models of univariate marginal distribution algorithm and factorized distribution algorithm. IEEE Transaction on Evolutionary Computation 8(1):80–93, DOI 10.1109/TEVC.2003.819431
[71] Zhang Q (2004) On the convergence of a factorized distribution algorithm with truncation selection. Complexity 9(4):17–23, DOI 10.1002/cplx.20013
[72] Zhang Q, Mühlenbein H (2004) On the convergence of a class of estimation of distribution algorithms. IEEE Transactions on Evolutionary Computation 8(2):127–136

Chapter 10
Theoretical Foundations of Immune-Inspired Randomized Search Heuristics for Optimization

Christine Zarges

Abstract Artificial immune systems are a class of nature-inspired algorithms based on the immune system of vertebrates. They have been used in a large number of different areas of application, most prominently learning, classification, pattern recognition, and (function) optimization. In the context of optimization, clonal selection algorithms are the most popular and constitute an interesting and promising alternative to evolutionary algorithms. While structurally similar, they offer very different features and capabilities. Over the last decade, significant progress has been made in the theoretical foundations of clonal selection algorithms. This chapter gives an overview of the state of the art in the theory of artificial immune systems with a focus on optimization. It provides pointers to corresponding articles where more details and proofs can be found.

10.1 Introduction

Artificial immune systems (AIS) are derived from various immunological theories, namely the clonal selection principle [4], immune network theory [46], and the danger theory [6]. Besides the natural tasks of anomaly detection and classification, they are often applied to function optimization. In the latter context, most immune-inspired randomized search heuristics are based on the clonal selection principle [4], a theory which describes the basic features of an adaptive immune response to invading pathogens (antigens). The most popular clonal selection algorithms to tackle optimization problems include CLONALG [7], Opt-IA [21], the B-cell algorithm [47], and MISA [10]. All

Christine Zarges
Department of Computer Science, Aberystwyth University, Aberystwyth, United Kingdom
e-mail: c.zarges@aber.ac.uk

B. Doerr, F. Neumann (eds.), *Theory of Evolutionary Computation*,
Natural Computing Series, https://doi.org/10.1007/978-3-030-29414-4_10

these algorithms are population-based. The input is usually represented by a population of antigens; a population of immune cells represents candidate solutions of the problem considered. Various aspects of clonal selection are used in these immune-inspired algorithms, for example mutations of different types of immune cells found in the immune system, resulting in a large number of very different approaches that share a common biological inspiration. Many of these algorithms resemble evolutionary algorithms from a structural point of view. However, their concrete implementations are usually very different.

This chapter provides an overview of the state of the art in the theory of immune-inspired randomized search heuristics in discrete search spaces. Most theoretical studies so far have concentrated on pseudo-Boolean optimization and classical example functions; however, some initial work on combinatorial optimization (vertex cover, longest common subsequence) exists. We discuss problem definitions, analytical frameworks, and common algorithms and operators in the corresponding sections. However, note that we consider bit strings of length n as the representation, $x \in \{0,1\}^n$, and that $x[i]$ denotes the i-th bit in x (with $i \in \{0,1,\ldots,n-1\}$, denoting the leftmost position in x by $x[0]$ and the rightmost position by $x[n-1]$).

The main part of this chapter will concentrate on performance analyses such as runtime analysis (where we are interested in the number of function evaluations required to locate an optimal solution, called the optimization time) and fixed-budget analysis (where we analyze the expected solution quality for a given budget of function evaluations). We therefore start with a brief overview of other related publications: early theoretical work was particularly concerned with Markov chain models of clonal selection algorithms and convergence analyses, i.e., the study of whether a given algorithm is guaranteed to converge to a global optimum for time $t \to \infty$. Based on Markov chain theory, Villalobos-Arias et al. [56, 57] proved convergence of the multi-objective clonal selection algorithm MISA [10] under the condition that the algorithm maintains an elitist memory throughout the search process. Later, using a similar approach, convergence results for the B-cell algorithm [47] based on a Markov model for contiguous hypermutations were presented [8, 9]. Cutello et al. [20] considered a more general framework of immune algorithms and examined conditions sufficient for their convergence. They provided some problem-independent upper bounds for their class of immune algorithms, but pointed out that such analyses should be related to some problem class and its characteristics in order to give useful insights. More recently, Hong and Kamruzzaman [29] used martingale theory to prove convergence for a class of elitist clonal selection algorithms. Timmis et al. [55] provided a survey of early theoretical advances and pointed out that runtime analysis would be much more useful than convergence analysis. The remainder of this chapter will therefore concentrate on these more advanced results.

The vast majority of work to date has concentrated on AIS for optimization. In this context, two defining aspects of AIS have been particularly considered: hypermutation operators (Section 10.2) and a diversity mechanism called aging (Section 10.3). However, more recently, insights into the interplay between different operators have allowed the first analyses of "complete" AIS as published in the literature (Section 10.4).

For the sake of completeness, we remark that theoretical studies in other subareas of artificial immune systems exist. In the context of classification, we refer the reader to the above-mentioned survey by Timmis et al. [55] and more recent work by Elberfeld and Textor [24, 54] and Gu et al. [27]. Moreover, a relatively large number of surveys provide a comprehensive view of the various application areas of artificial immune systems. They include general overviews and introductions to the field [6, 22, 23, 53], as well as more specialized surveys of, for example, optimization [1] and security [25].

10.2 Theoretical Analyses of Hypermutations

Mutation in AIS is very different from mutation in other randomized search heuristics, for example evolutionary algorithms. While in evolutionary algorithms generally moderate mutation probabilities are employed, AIS incorporates mutations at a high rate, so-called hypermutations. Different types of immune-inspired mutation operators with roots in different (processes of the) immune systems can be found in the literature. Among the most prominent classes are inversely fitness-proportional mutations (Section 10.2.1), contiguous hypermutations (Section 10.2.2), and hypermutations with mutation potential (Section 10.2.3). We will discuss different variants of these operators in the following.

All the results presented in this section consider immune-inspired mutation operators in minimalistic algorithmic frameworks to study them in as much isolation as possible. These frameworks include in particular the (1+1) framework shown in Algorithm 10.1 and a simple population-based $(\mu+1)$ framework as described in Algorithm 10.2. The performance of hypermutation operators has particularly been compared with local search (flipping exactly one random bit) and standard bit mutations (flipping each bit with probability $1/n$).

Algorithm 10.1: (1+1) framework

1 Choose $x \in \{0,1\}^n$ uniformly at random.
2 **repeat**
3   Create offspring $y := \text{mutate}(x)$.
4   **if** $f(y) \geq f(x)$ **then**
5   Set $x := y$.
6 **until** *some termination criterion is met*

Algorithm 10.2: (μ+1) framework

Parameters: Population size μ
1 Choose $x_1,\dots,x_\mu \in \{0,1\}^n$ independently, uniformly at random.
2 Let $P := \{x_1,\dots,x_\mu\}$.
3 **repeat**
4 Choose $x \in P$ uniformly at random.
5 Create offspring $y :=$ mutate(x).
6 Choose $z \in P$ with minimum fitness.
7 **if** $f(y) \geq f(z)$ **then**
8 Let $P = P \setminus \{z\} \cup \{y\}$.
9 **until** *some termination criterion is met*

10.2.1 Inversely Fitness-Proportional Mutations

The idea of inversely fitness-proportional mutations derives directly from the widely accepted clonal selection principle [4]. It aims to balance exploration and exploitation of the search space by using an individual mutation rate for each search point (immune cell) in the population, i.e., focusing on exploitation for good search points that are hopefully close to a local or global optimum and focusing on exploration otherwise. As a result, search points in "better" regions of the search space are only subject to small mutations, while for search points located far away from optimal regions larger mutation rates are used.

Inversely fitness-proportional mutation operators exist in both continuous and discrete versions, but to the best of our knowledge theoretical studies so far have concentrated on discrete settings or, more precisely, pseudo-Boolean optimization. Here, the mutation rate is a function that depends on the (normalized) fitness of a search point and determines the probability for each bit in a given bit string to be mutated (see Algorithm 10.3). The relationship between mutation rate and fitness is not required to be inversely proportional in a strict mathematical sense, but only needs to follow the rule that the higher the fitness the smaller the mutation rate, and vice versa.

Algorithm 10.3: Inversely fitness-proportional mutations

Input: Search point $x \in \{0,1\}^n$; fitness-dependent mutation rate $p(v)$
Output: Mutated search point $x \in \{0,1\}^n$
1 Let $v :=$ normalize$(f(x)) \in [0,1]$.
2 Let $y := x$.
3 **for** $i := 0$ **to** $n-1$ **do**
4 With probability $p(v)$ set $y[i] := 1 - y[i]$.

Inversely fitness-proportional mutations are a key ingredient of CLONALG [7]. For the case of maximization, the operator uses the inverse of an exponential function to establish a relationship between the mutation probability and the normalized fitness value v:

$$p_{\text{CLONALG}}(v) = \exp(-\rho \cdot v), \tag{10.1}$$

where ρ is a so-called decay parameter that controls the smoothness of the exponential function and needs to be set to a value appropriate for the problem considered. A similar operator with a slightly different parameterization is used in Opt-AiNet [5], an immune-inspired algorithm based on immune network theory [46]:

$$p_{\text{Opt-AiNet}}(v) = \exp(-v)/\rho. \tag{10.2}$$

Here, the parameter ρ is not incorporated into the exponent of the exponential function but rather used to scale its result, leading to a very different impact of ρ on the optimization process [61].

Zarges [59, 61] examined the role of the decay parameter ρ for these two operators on the OneMax problem in a simple (1+1) framework (see Algorithm 10.1). Constant decay values ρ as well as values logarithmic and linear in the length of the bit string were considered. Using, among others, drift arguments, it was shown that both operators are very sensitive to parameterization and, if parameterized inappropriately, very bad at hill-climbing.

For maximization problems, the standard normalization method [7] is to divide the fitness by the best fitness in the current population or by the best fitness seen so far. For a single search point, Zarges [59, 61] used the optimal value of the fitness function considered instead and argued that using an upper bound on the fitness would be appropriate if the optimal value was not known. However, it was noted that using an upper bound leads to generally larger mutation probabilities, while the use of the best current fitness reduces them.

For CLONALG mutations, it was proven that, with overwhelming probability, the algorithm using constant and linear settings of ρ is unable to locate the optimum of OneMax within a polynomial number of iterations. For a constant ρ, this is because the mutation probabilities grow much too large to be effective, while a linear ρ leads to exponentially small mutation rates, rendering the algorithm unable to perform any search at all. In addition to these two extreme cases, $\rho = \ln n$ was considered. It was noted that this parameterization yields reasonable values for the mutation rate between $1/2$ and $1/n$. While one can still observe a large negative drift for this setting (and consequently the algorithm is unable to optimize OneMax in polynomial time with high probability), the algorithm demonstrates much better performance in practice, as the probability of not finding the optimum within a polynomial number of iterations converges much more slowly to 1 than it does for constant ρ. In fact, in experiments the performance was comparable to that of standard bit mutations up to bit string lengths of about 10^5.

For Opt-AiNet mutations, a constant ρ results in roughly the same mutation rates as already seen for the case of CLONALG, and thus the algorithm is also not efficient on OneMax with this setting. However, using $\rho = \Theta(n)$ yields mutation rates of $\Theta(1/n)$ for all possible fitness values. Thus, we get an optimization time of $\Theta(n \log n)$ for this case, which can easily be shown by using fitness-level arguments and adapting previous analyses for standard bit mutations.

Later, the CLONALG hypermutation operator was analyzed in a $(\mu + 1)$ framework (see Algorithm 10.2), using the current best fitness value for normalization [60]. Setting $\rho = \ln n$ (as this leads to reasonable mutation rates for OneMax [59]) and using fitness-level arguments, it was shown that even a population size of 2 considerably improves the performance on the OneMax problem and, more generally, a class of smooth integer functions of unitation, i.e., a class of functions where the function value depends only on the number of 1-bits in the bit string and neighboring points in the search space have similar function values. A key insight for this result is that for this parameterization the behavior of the current best search point in the population mimics that of standard bit mutations.

A matching lower bound was proven by bounding the bandwidth of the fitness values in the current population using inductive arguments and by employing the technique of analyzing randomized family trees. Here, an important insight is that the rate of inversely fitness-proportional mutations depends directly on the structure of the population or, more precisely, the difference between the best and the worst fitness value. If this difference is small enough, the expected number of flipping bits during a single iteration can be bounded by a constant.

While fitness-dependent mutations are common in immune-inspired randomized search heuristics, they have only recently emerged in the context of evolutionary computation. For example, Oliveto et al. [51] analyzed a rank-based mutation operator, an alternative to using normalized function values to reduce the effect of large differences in absolute function values. Böttcher et al. [2] derived an optimal adaptive mutation rate for the LeadingOnes problem that has the form of an inversely fitness-proportional mutation rate, $p_{\text{simple}}(v) = 1/(v+1)$ with $v = f(x)$. The same mutation rate is efficient for the OneMax problem [61]. However, a Hamming-distance-based mutation rate maximizing the probability of finding the global optimum of OneMax in a single mutation step, i.e., $p_{\text{Hamming}}(v) = v/n$ with $v = n - f(x)$, yields exponential optimization time.

Jansen and Zarges [43] considered inversely fitness-proportional mutations in the context of fixed-budget analysis, where, instead of the expected time needed for optimization, the expected performance within a given time frame is analyzed. They showed that CLONALG mutations outperform local search at the beginning of the run, but are eventually overtaken later in the run. This insight was used to devise a hybrid algorithm that starts with CLONALG mutations and switches to local search when progress stagnates. These results are discussed in more detail in Section 5.5 of this book.

10.2.2 Contiguous Hypermutations

Contiguous hypermutations were introduced as part of the B-cell algorithm [47]. They were inspired by the observation that mutation of B-cell receptors (a type of immune cell) often focuses on specific regions of the receptor. To mimic this behavior, the mutation operator first selects a contiguous region of the search point's representation and restricts the mutation to this region. The use of contiguous hypermutations is limited to discrete search spaces and employs a bit string representation. Mutation flips all bits within the chosen region with a given probability $r \in [0,1]$ and does not change any bit that is outside of this region. It has been noted that, depending on its parameterization, this mutation operator can easily be trapped in local optima [9, 37]: if $r = 1$, there might not exist a mutation leading from a local to a global optimum. However, the following analyses consider this extreme case only.

Jansen and Zarges [34, 37] considered three different variants of this hypermutation operator which differ in the way the contiguous mutation region is determined. The original operator chooses a random starting position p and a random length l of an interval to be mutated (see Algorithm 10.4). It does not wrap around, and thus has a strong positional bias and strongly different mutation probabilities for mutations of single bits depending on their location: bits towards the end of the bit string have a higher probability of being mutated during a mutation. As such a bias is considered undesirable unless it suits known problem characteristics, this observation motivates the definition of a variant that wraps around and thus has no positional bias at all (see Algorithm 10.5). A third variant selects random start and end points for the mutation region (see Algorithm 10.6). Similarly to the original operator, this variant has a strong positional bias – here, bits towards the middle of the bit string have a higher probability of being mutated. All three variants have in common that, in expectation, they flip a linear number of bits. The probability of performing a single bit mutation is $\Theta(1/n^2)$ in all three cases (with the exception of the version that does not wrap around, where the probability of flipping only the last bit is $\Theta(1/n)$). This is considerably smaller than the corresponding probability for standard bit mutations and explains why contiguous hypermutations are also bad at simple hill-climbing.

However, they can have advantages when large mutations are needed: standard bit mutations perform specific b-bit mutations with probability $\Theta(1/n^b)$, while all three variants of contiguous hypermutation achieve this with probability $O(1/n^2)$. Jansen and Zarges [34, 37] investigated this in a rigorous way

Algorithm 10.4: Contiguous hypermutations, not wrapping around [8]

Input: Search point $x \in \{0,1\}^n$; mutation probability $r \in (0,1]$
Output: Mutated search point $x \in \{0,1\}^n$
1 Select $p \in \{0,1,\ldots,n-1\}$ uniformly at random.
2 Select $l \in \{0,1,\ldots,n\}$ uniformly at random.
3 **for** $k := 0$ *to* $\min\{l-1, n-1-p\}$ **do**
4 With probability r, invert the bit $x[p+k]$.

Algorithm 10.5: Contiguous hypermutations, wrapping around [34]

Input: Search point $x \in \{0,1\}^n$; mutation probability $r \in (0,1]$
Output: Mutated search point $x \in \{0,1\}^n$
1 Select $p \in \{0,1,\ldots,n-1\}$ uniformly at random.
2 Select $l \in \{0,1,\ldots,n\}$ uniformly at random.
3 **for** $i := 0$ **to** $l-1$ **do**
4 With probability r set $x[(p+i) \mod n] := 1 - x[(p+i) \mod n]$.

Algorithm 10.6: Contiguous hypermutations, two hotspots [34]

Input: Search point $x \in \{0,1\}^n$; mutation probability $r \in (0,1]$
Output: Mutated search point $x \in \{0,1\}^n$
1 Select $p_1 \in \{0,1,\ldots,n-1\}$ uniformly at random.
2 Select $p_2 \in \{0,1,\ldots,n-1\}$ uniformly at random.
3 **for** $k := \min\{p_1, p_2\}$ *to* $\max\{p_1, p_2\}$ **do**
4 With probability r, invert the bit $x[k]$.

by presenting different examples for functions where contiguous hypermutations are superior or inferior to the standard bit mutations typically used in evolutionary algorithms in a simple (1+1) framework (see Algorithm 10.1). Using fitness level and drift arguments, they showed that contiguous hypermutations can drastically outperform standard bit mutations for a previously know family of example functions that require mutations of many bits simultaneously,

$$\mathrm{CLOB}_{b,k}(x) = n \cdot \left(\sum_{h=1}^{k} \sum_{i=1}^{n/(bk)} \prod_{j=0}^{i \cdot b-1} x\left[(h-1) \cdot \frac{n}{k} + j\right] \right) - \textsc{OneMax}(x),$$

for $b,k,n \in \mathbb{N}$ with $n/k \in \mathbb{N}$, $n/(bk) \in \mathbb{N}$, and $x \in \{0,1\}^n$. Here, contiguous hypermutations yield an optimization time of $O(n^2 \log n)$, while standard bit mutations require time $\Theta(n^b(l/b + \ln k))$. In addition, it was shown that contiguous hypermutations do not necessarily lose a factor of $\Theta(n)$ on functions where mutations of single bits are responsible for optimization: while this is the case for OneMax, contiguous hypermutations lose at most a factor of $\log n$ on LeadingOnes.

Jansen and Zarges [34, 37] investigated the role of initialization for contiguous hypermutations and demonstrated that advantageous starting points with large blocks of contiguous 0s (e.g., the all-zero bit string 0^n) can speed up optimization for this operator while having nearly no impact on standard bit mutations. However, whether this advantage, which is big in the beginning where it is easy to make progress and decreases towards the end where making progress is much harder, is sufficient to yield an asymptotically smaller expected optimization time depends on how long this advantage can be preserved during the optimization process. It is important to note that all positive results rely on the extreme choice $r = 1$. Thus, it can be concluded that contiguous hypermutations can only play out their strength if r is set to some value at least close to 1.

Jansen and Zarges [43] later also considered contiguous hypermutations in the context of fixed-budget analysis (see Section 5.5 of this book). Revisiting negative results for immune-inspired hypermutations for the simple OneMax function and the observation that careful initialization can speed up the optimization process [37], they demonstrated that contiguous hypermutations can be much more efficient in the beginning of a run when progress is still easy to achieve and, thus, given a limited budget of function evaluations, such mutations can by far outperform a random local search operator. This is mainly due to the fact that hypermutations are able to accumulate many small steps into a single large one, while random local search needs to perform those steps one after each other and thus needs time to catch up. This insight helps to explain the success of seemingly inefficient mutation operators, as in practice the length of a run is usually limited.

The above theoretical results were put into practice by designing a more efficient hybrid search heuristic that applies contiguous hypermutations in the beginning when they can be expected to be more beneficial, and switches to local search when contiguous hypermutations start to become slow. Two strategies to switch the mutation operator were investigated, one directly based on the theoretical findings, the other using the expected progress of the two operators adaptively in each iteration. Experiments showed that both strategies yield noticeable improvements over simple local search if careful initialization is performed, but that the more sophisticated adaptive strategy does not yield any significant advantage. These results are discussed in more detail in Section 5.5 of this book.

Other work on the analysis of contiguous hypermutations includes a runtime and fixed-budget analysis for the highly multimodal HIFF (hierarchical if and only if) problem by Jansen and Zarges [44] and later by Xia and Zhou [58]. Jansen and Zarges [44] showed that under certain conditions contiguous hypermutations can be successful hill-climbers. Using fitness-level arguments, they showed that contiguous hypermutations in a simple (1+1) framework (see Algorithm 10.1) solve HIFF in time $O(n^3 \log n)$, while random local search does not find an optimum with overwhelming probability. Moreover, they demonstrated that contiguous hypermutations are not outperformed by random local search on HIFF for any given budget by performing a fixed-budget analysis (see Section 5.5 of this book): for small budgets both algorithms have roughly equal performance, but contiguous hypermutations outperform random local search for moderately large budgets. This result is somewhat counterintuitive as at the beginning, i.e., before reaching a local optimum, HIFF can be optimized by simple hill-climbing – something contiguous hypermutations are particularly bad at.

Xia and Zhou [58] additionally considered contiguous hypermutations on Trap functions and the max-cut and minimum s-t-cut problems, again using the simple (1+1) framework in Algorithm 10.1. They showed that Trap can be optimized in time $O(n^2 \log n)$ and considered a family of graphs for max-cut that can be efficiently optimized using contiguous hypermutations but not using standard bit mutations and a problem-specific local search operator. A similar result was shown for a family of graphs for the minimum s-t-cut problem.

Very recently, Corus et al. [11, 12] derived an easiest function for contiguous hypermutations. Again using the insight that contiguous hypermutations can have advantages on functions that require mutations of many bits simultaneously, they introduced the following fitness function.

Definition 10.2.1. Let $L_0 \dot{\cup} L_1 \dot{\cup} L_2 \dot{\cup} \cdots \dot{\cup} L_l = \{0,1\}^n$ be a partition and let

$$\textsc{MinBlocks}(x) = l - i \text{ for } x \in L_i,$$

with $l = \lfloor n/2 \rfloor + 1$, $L_0 = \{1^n\}$, $L_1 = \{0^n\}$, and $L_i = \{x \in \{0,1\}^n \mid x \text{ contains } i-1 \text{ different 1-blocks}\}$ for each $i \in \{2,3,\ldots,l\}$.

MinBlocks has a unique global optimum, 1^n, with fitness $\lfloor n/2 \rfloor + 1$. The second best bit string is 0^n, with fitness $l-1$. Corus et al. [11, 12] presented both runtime and fixed-budget analyses of contiguous hypermutations on this function and showed that MinBlocks is indeed an easiest function using a method introduced by He et al. [28]. The runtime of contiguous hypermutations embedded in a (1+1) framework (see Algorithm 10.1) on MinBlocks is $\Theta(n^2)$.

MinBlocks turns out to be an asymptotically hardest function for standard bit mutations. Owing to the symmetry of 0- and 1-bits, the (1+1) EA (Algorithm 10.1 using standard bit mutations) will reach 0^n (instead of 1^n) with probability 1/2. In this situation, it needs to flip all bits in a single mutation, resulting in an expected optimization time of at least $n^n/2$. This is only smaller by a factor of at most 2 than the expected optimization time of the (1+1) EA on its hardest function, Trap [28].

Finally, Corus et al. [11, 12] discussed a number of hybridizations of standard bit mutations and contiguous hypermutations. These allow one to combine the advantages of the two operators, and yield optimal asymptotic performance on both OneMax and MinBlocks.

Some analyses of contiguous hypermutations consider the mutation operator within the complete B-cell algorithm rather than a minimalistic framework. We discuss these results later in Section 10.4.1.

10.2.3 Hypermutations with Mutation Potential

Hypermutations with mutation potential were introduced as a mutation operator in Opt-IA [21]. The main idea behind this kind of mutation is to determine the number of local mutation steps by a given function, the so-called mutation potential. Mutation potentials exist in different flavors, for example static, fitness-proportional, and inversely fitness-proportional. Moreover, they can be restricted to certain regions of the bit string (hypermacromutation).

Algorithm 10.7 provides pseudocode of four different variants of this mutation operator, defined for minimization problems. For a number M of local mutation steps, the operator sequentially draws M not necessarily distinct positions in the bit string and flips them independently. We distinguish a TABU variant (where the operator is prevented from choosing a specific position two or more times) and a non-TABU variant (where bits can be flipped back in a later mutation step). Moreover, a mechanism often used in conjunction with mutation potentials is the so-called "stop at first constructive mutation" (FCM). Here, a fitness evaluation is performed after every single mutation step and the mutation stops if an improvement has been found (a so-called constructive mutation). It has been shown that the question of whether local mutation steps may undo each other is far less important than the use of the FCM mechanism [13, 18, 40]. We provide more detail in the following.

Algorithm 10.7: Mutation with mutation potential M (minimization)

```
   Input: Search point x ∈ {0,1}^n; flags TABU and FCM
   Output: Mutated search point x ∈ {0,1}^n
 1 Set y = x.
 2 repeat
 3 |   if TABU = 0 then
 4 |   |_ Select i ∈ {1,...,n} uniformly at random (u.a.r.).
 5 |   else
 6 |   |_ Select i ∈ {1,...,n} u.a.r., i not previously chosen.
 7 |   Invert the bit y[i].
 8 |   if (FCM = 1) AND (f(y) < f(x)) then
 9 |   |_ BREAK
10 until M times
```

10.2.3.1 Inversely Fitness-Proportional Mutation Potentials

The first analysis of hypermutations with mutation potential was presented by Jansen and Zarges [40]. They considered all four variants of this operator for an inversely fitness-proportional mutation potential $M_c(v)$, where $c \in]0,1[$ and f_{OPT} is the minimum function value for the fitness function considered, $f \colon S \to \mathbb{R}^+$:

$$M_c(v) = \lceil (1 - f_{\text{OPT}}/v) \cdot c \cdot n \rceil. \tag{10.3}$$

Using the simple (1+1) framework (see Algorithm 10.1), they showed that the FCM mechanism is crucial for the performance of mutation potentials even on very simple optimization problems such as ZEROMIN, the minimization variant of ONEMAX. The main reason for this is that hypermutations with mutation potential that do not make use of FCM basically perform a random walk of length equal to the mutation potential – such a random walk has hardly any chance of locating a specific search point (which can easily be proven by using results for the gambler's ruin problem). While adding FCM considerably improves the optimization time, it loses a factor of n in comparison with standard bit mutations: the expected optimization time of Algorithm 10.1 using hypermutations with mutation potential on ZEROMIN is $2^{\Omega(n)}$ without FCM and $\Theta(n^2 \log n)$ with FCM. The upper bounds for the algorithms with FCM were derived using fitness-layer arguments, while the lower bound was obtained by Chernoff bounds and arguments from the classical ballot theorem for the non-TABU variant, and a careful analysis of the underlying random walk for the TABU version.

Moreover, Jansen and Zarges [40] analyzed the ability of the original hypermutation operator with FCM to locate optima precisely and at a large distance from other promising regions of the search space by considering a previously introduced example function called Sp-Target (short path with target). The main idea of this function is that the vast majority of the search space guides the search heuristics towards a path with increasing function values starting from the all-zero bit string 0^n. The global optimum is a large area with a minimum Hamming distance to the path.

A typical run of a search heuristic finds and climbs the path before finding the optimal region. Thus, in order to be able to optimize this function, an algorithm requires the ability to "jump" from the path into the global optimum. It is known that standard bit mutations are unable to efficiently locate the global optimum if the distance is $\omega(\log n / \log\log n)$. Jansen and Zarges [40] proved that hypermutations with mutation potential yield an optimization time of $O(n^3)$ provided that c in $M_c(v)$ is chosen large enough.

More recently, Corus et al. [17] compared different variants of inversely fitness-proportional mutation potentials based on Hamming distance and fitness difference. They showed that a potential that increases exponentially with the Hamming distance to the optimum (called M_{expoHD}) is most promising and argued that using Hamming distance instead of fitness difference also comes with the advantage of robustness to scaling of the fitness function. In comparison with static mutation potentials, they showed a considerable speedup for all inversely fitness-proportional variants on standard unimodal example functions, for which the global optimum is known. In addition, M_{expoHD} was considered in situations, where the global optimum is unknown. Using the best found solution to estimate the mutation rate and combining M_{expoHD} with hybrid aging (see Section 10.3.2), it was demonstrated that the algorithm might not be able to identify new local optima on slopes that lead away from previous ones. As a consequence, a symmetric version of M_{expoHD} was introduced and shown to be effective on two well-known bimodal example functions, CLIFF_d and TWOMAX.

10.2.3.2 Static Mutation Potentials

Corus et al. [13, 18] presented a first detailed study of static mutation potentials. They proved for a static mutation potential (where the number of bits flipped is linear in the problem size, $c \cdot n$ for constant $c > 0$) and the TABU variant that unless the FCM mechanism is applied, hypermutations with mutation potential require exponential expected time to optimize any function with a polynomial number of optima. They argued that the search point created by such hypermutations is uniformly distributed over all search points with distance $c \cdot n$ to the search point it was derived from. Since there are exponentially many such search points, the probability of a specific outcome is exponentially small. Corus et al. [13, 18] pointed out that this result could easily be extended to other types of mutation potential such as inversely fitness-proportional [40] and fitness-proportional mutation potentials [19]. In [18], the authors additionally suggested that it may be beneficial to call a mutation "constructive" if the fitness is at least as good (instead of strictly better) to improve the algorithm's exploration capabilities.

Moreover, Corus et al. [13, 18] showed that the expected optimization time if FCM is used is at most larger by a linear factor than the upper bound obtained for random local search (with any neighborhood size) via fitness-level arguments. This demonstrates that it is sufficient to analyze random local search instead of hypermutations with such a mutation potential (which is often easier) to achieve a valid upper bound on the expected optimization time. Corus et al. [13, 18] showed that these bounds are tight for easy example functions such as ONEMAX and LEADINGONES.

Finally, Corus et al. [13, 18] compared hypermutations with mutation potential with standard mutation- and crossover-based evolutionary algorithms on the JUMP_k and CLIFF_d functions. They proved that their hypermutations operator could exponentially speed up the process of escaping from local optima, particularly in cases where the jump is hard to perform. However, the upper bound on the expected optimization time was still exponential in the distance between the local and the global optimum.

Very recently, Corus et al. [14, 16] presented an analysis for the NP-hard number partitioning problem. They showed that, due to its ability to escape from local optima, a simple artificial immune system using static hypermutations with mutation potential is able to efficiently solve a class of problem instances that are known to be hard for random local search and evolutionary algorithms using standard bit mutations. More importantly, they proved that such an artificial immune system is a randomised polynomial time approximation scheme (for $\varepsilon = \omega(n^{-1/2})$), i.e., it guarantees an approximation ratio of $(1+\varepsilon)$ for any problem instance in expected time polynomial in the problem size and exponential in $1/\varepsilon$. The authors pointed out that, to the best of their knowledge, this was the first time performance guarantees were proven for an artificial immune system on a classical combinatorial optimization problem.

10.2.3.3 Variants of Mutation Potentials

Based on some of the above findings, novel variants of hypermutations with mutation potential were introduced. Jansen and Zarges [40] proposed an improved version of the mutation potential based on ranks that allows one to parameterize the trade-off between efficiency in local search and the ability to perform huge changes in a single mutation:

$$\hat{M}_{c,\rho}(v_i) = \lceil (1 - n^{\rho}/(n^{\rho} + i - 1)) \cdot c \cdot n \rceil, \qquad (10.4)$$

where i is the rank of the fitness value considered among all fitness values in the search space, and ρ controls the degree of mutation aversion the hypermutation operator has. It was proven that for $\rho > 1$ the expected optimization time on ZEROMIN with FCM decreases to $\Theta(n \log n + n^{3-\rho})$.

More recently, Corus et al. [15] proposed a "fast" variant of hypermutations with mutation potential, where instead of deterministically performing a fitness evaluation after each bitflip, the fitness after the i-th bitflip is only evaluated with probability roughly $p_i \approx \gamma/i$. In doing so, fewer function evaluations are "wasted" during the hypermutation process, particularly for easy problems, for which local search strategies are efficient. The effectiveness of two variants of this operator coupled with and without FCM was demonstrated by analyzing problems that had been considered previously for hypermutations with mutation potential and Opt-IA (see Section 10.4.2) and recommendations for setting the parameter γ were provided. Moreover, it was demonstrated how upper bounds for "fast" hypermutations with mutation potential could be derived from upper bounds for random local search that were obtained via fitness-level arguments.

10.3 Theoretical Analyses of Aging Operators

Aging operators require that each search point in the population is equipped with an individual age that is increased by 1 in each iteration of the search heuristic. A maximum lifespan τ is introduced, and each search point with an age exceeding τ is removed from the current population, making room for new and perhaps more promising search points. The mechanism of aging is thought of as increasing the diversity of the population and it is hoped that it will be helpful for multimodal problems where simpler search heuristics may get stuck in local optima.

Different variants exist, and static pure aging and stochastic aging have both been used within Opt-IA [21]. Both strategies usually have in common that the initial age of a new search point is set to 0 only if its function value is strictly larger than the function value of the search point it was derived from (the parent); otherwise, it inherits the age of this search point. This scheme is intended to give an equal opportunity to each improving new search point to explore the landscape effectively. Alternatively, each new search point can be assigned age 0; however, this is more common in evolutionary computation [38].

In static pure aging, search points exceeding a predefined maximum lifespan (maximum number of iterations) τ are removed from the population. In stochastic aging, each search point x survives aging at the end of the iteration with a probability p_{die}. In order to keep the size of the population constant at a certain size μ, often new random search points with age 0 are introduced, if necessary.

The publications reviewed in the following subsections consider different kinds of aging in a minimal algorithmic framework by extending the $(\mu+1)$ framework introduced in Algorithm 10.2. The extended framework in Algorithm 10.8 again uses a population of size μ. It works in rounds, where in each round all search points grow older, one new search point is generated as a random variation of existing search points, its age is decided, search points that are too old are removed, and new randomly generated search points are introduced to keep the number of search points constant at μ.

10.3.1 Static Pure Aging

In static pure aging, offspring inherit by default the age of their parent and are only assigned age 0 if their function value is strictly larger than that of their parents (see Algorithm 10.9). At the end of an iteration, search points that exceed the maximum age τ are removed deterministically (see Algorithm 10.10).

Horoba et al. [30] were the first to present a rigorous runtime analysis of static pure aging in artificial immune systems and consider its most impor-

Algorithm 10.8: $(\mu+1)$ framework with aging

```
Parameters: Population size μ
1  Choose x_1,...,x_μ ∈ {0,1}^n independently, uniformly at random, and let
     P := {x_1,...,x_μ}.
2  for all x ∈ P do
3  |  Set x.age = 0.
4  repeat
5  |  for all x ∈ P do
6  |  |  set x.age = x.age + 1.                              /* Growing older */
7  |  Choose x ∈ P uniformly at random.
8  |  Create offspring y := mutate(x).                          /* Variation */
9  |  Decide about the age of y.                           /* (see details) */
10 |  Remove search points due to age.                     /* (see details) */
11 |  if |P| > μ then
12 |  |  Remove one z ∈ P with minimum fitness.                    /* Removal */
13 |  else
14 |  |  Keep all search points in P.
15 |  |  Fill up P with random points until |P| = μ.                 /* Birth */
16 until some termination criterion is met
```

Algorithm 10.9: Static pure aging: age of offspring

```
Input: Parent x; offspring y
1 if f(y) ≥ f(x) then
2 |  Set y.age := 0.
3 else
4 |  Set y.age := x.age.
```

Algorithm 10.10: Static pure aging: removal due to age

```
Input: Population P; maximum lifespan τ
1 for all x ∈ P do
2 |  if x.age > τ then
3 |  |  Set P := P \ {x}.
```

tant parameter: the maximum lifespan τ. They showed that the smaller the maximum age, the more the search process resembles pure random search and, thus, becomes ineffective. To be more precise, they demonstrated that τ needs to be large enough to allow the algorithm to create a better offspring. Considering a RIDGE-like function that includes a number of gaps of size k, they made their arguments more precise. They proved that, for such a function, τ needs to be sufficiently large in order for the algorithm to be successful, by presenting a proof that considers a typical run of a simple algorithm using static pure aging (see Algorithm 10.8) and standard bit mutations. Additionally, a common lower bound of $\tau = \omega(\mu n \log \mu)$ was derived – for smaller τ, the algorithm is unable to perform hill-climbing.

However, a maximum age that is too large severely limits the influence of the operator, as fewer search points are subject to removal by age. Moreover, there exist situations in which a small τ can prevent the algorithm under consideration from getting trapped in parts of the search space that keep it away from the global optimum. Again, this argument was made more precise by constructing an example function and proving that it can only be optimized efficiently using static pure aging if τ is sufficiently small.

Finally, it was shown that aging can be very sensitive to the maximum age of a search point and that it may be difficult to set this appropriately, i.e., the appropriate age can be within a very narrow range. Horoba et al. [30] demonstrated this by carefully devising a new example function by combining the two previous functions.

Building upon this work, Jansen and Zarges [33, 38] compared static pure aging with aging in evolutionary algorithms where new search points are always assigned age 0. It was shown that new random search points that are introduced during the birth phase typically have very low fitness and thus die out quickly. Using well-known example functions, they demonstrated that static pure aging is able to escape from local optima by recognizing stagnation and performing a kind of restart; however, when there are plateaus of constant function value, it mistakes the absence of progress in function values for stagnation and thus is not able to perform a random walk on the plateau. The performance of evolutionary aging is exactly opposite, i.e., it is not able to escape local optima but it can perform a random walk on a plateau. Based on these insights, a modified aging operator was introduced that provably shares the advantages of both aging mechanisms (see Algorithm 10.11): while the function values do not increase in both situations, being stuck in a local optimum additionally means that no new search points are created.

Algorithm 10.11: Genotypic aging: age of offspring

```
  Input: Parent x; offspring y
1 if f(y) ≥ f(x) and y ≠ x then
2 |  Set y.age := 0.
3 else
4 L  Set y.age := x.age.
```

Later, Jansen and Zarges [35, 36, 39] analyzed the interplay of static pure aging with the replacement strategy used. It was demonstrated that static pure aging can achieve performance improvements that go beyond what restarts can accomplish [35]. Since it is often stated in the literature that aging increases the diversity within the population of search points, this is an important step in understanding how and why aging can make an algorithm more efficient. In this context, crossover plays an important role, as the main effect shown is based on the recombination (k-point crossover) of a local optimum and a randomly generated search point [35]. However, given the original definition of static pure aging, it is unclear how the age of a new search point is set in the case of more than one parent. Different strategies, including setting the age to the age of the older parent and setting it to the age of the better/worse parent, were introduced and analyzed in [36], where it was pointed out that even subtle differences can have a huge impact on the performance of the algorithm.

In [39], Jansen and Zarges argued that static pure aging can be subdivided into an aging and a replacement strategy. While the aging strategy determines the age of a new search point, the replacement strategy decides how it is introduced into the population. Considering a number of different implementations for both strategies, their interplay was analyzed. It was shown that not only the maximum age but also diversity with respect to age plays a key role and can make a difference between efficient and inefficient optimization. Different strategies that use the age to remove one of the search points with the worst function value from the population were considered. To compare these aging and replacement strategies, an example function was constructed and the performance for all possible combinations of operators was analyzed. As age diversity is mainly determined by the interplay of the aging and the replacement strategies, a careful algorithm design and description was considered crucial to obtaining meaningful results.

Very recently, Corus et al. [14, 16] considered an artificial immune system using standard bit mutations and static pure aging in the context of the NP-hard number partitioning problem. Similarly to their analysis for hypermutations with mutation potential (see Section 10.2.3.2) they proved that their artificial immune system is also able to efficiently solve the same "hard" problem instances and constitutes a randomised polynomial time approximation scheme for the partitioning problem (for $\varepsilon \geq 4/n$).

10.3.2 Stochastic Aging

Stochastic aging usually uses the same mechanism as static pure aging to decide the age of an offspring (see Algorithm 10.9); however, search points are removed based on some probability p_{die} (see Algorithm 10.12).

Algorithm 10.12: Stochastic aging: removal due to age

Input: Population P; probability of dying p_{die}
1 **for** *all* $x \in P$ **do**
2 Set $P := P \setminus \{x\}$ with probability p_{die}.

Oliveto and Sudholt [52] presented the first theoretical analysis of stochastic aging. They showed that, just like static pure aging, stochastic aging can implicitly perform restarts but, more importantly, they also considered the question of what aging can achieve beyond performing standard restarts. They presented a framework for the analysis of stochastic aging using a given probability p_{die} and showed that stochastic aging can be effective in a natural setting (i.e., without crossover, which is not usually found in artificial immune systems) where restarts do not work.

Using the same classical example function that Jansen and Zarges [38] used for static pure aging, they provided guidance for parameterization and showed that stochastic aging as in Algorithm 10.12 is effective only for not too large population sizes, as the probability of performing a restart (i.e., the probability that all search points die roughly at the same time) is exponential in the population size μ. To tackle this problem, a hybrid aging operator (see Algorithm 10.13) combining ideas from static pure aging and stochastic aging was introduced. Like static pure aging, it protects a search point from dying for τ generations, but search points with an age larger than τ are removed from the population with probability p_{die}. The efficiency of this novel operator was demonstrated for example functions from the literature in both dynamic and static environments. These results hold for arbitrary population sizes. For the dynamic BALANCE function, it was shown that hybrid pure aging enables the algorithm to escape from a local optimum if all but one search point of the population die and, in the same iteration, the surviving search point moves out of the local optimum – something that an evolutionary algorithm using standard bit mutations is unable to achieve. Moreover, it was shown that static pure aging is inefficient in the dynamic setting considered. As a by-product of their analysis, Oliveto and Sudholt also remarked that the parameter setting for hybrid pure aging is at the opposite side of the spectrum from that for stochastic aging: while stochastic aging requires a high probability of surviving (close to 1), hybrid pure aging requires a low survival probability ($\Theta(1/\mu)$).

Algorithm 10.13: Hybrid pure aging: removal due to age

Input: Population P; lifespan τ; probability of dying p_{die}

1 **for** *all* $x \in P$ **do**
2   **if** $x.age > \tau$ **then**
3   Set $P := P \setminus \{x\}$ with probability p_{die}

More recently, Corus et al. [13, 18] extended the analysis of Oliveto and Sudholt [52] by considering the more general example function CLIFF_d. They demonstrated that hybrid aging can be very efficient when coupled with local as well as standard bit mutations. For local mutations, they proved an expected optimization time of $O(n \log n)$ if the gap has linear size, i.e., when the function is most difficult for evolutionary algorithms using standard bit mutations. It is noted, that this asymptotically matches the lower bound for all unbiased mutation-based randomised search heuristics to optimise any function with a unique optimum [48]. For standard bit mutations, Corus et al. [13, 18] proved an expected optimisation time of $O(n^{1+\varepsilon})$, $\varepsilon > 0$ constant, if the gap has linear size. The study was further expanded in [18] by adding a genotype diversity mechanism (as proposed in the original Opt-IA). Here, it was shown that the algorithm can still escape from the local optima provided that the population size is not too large.

10.4 Theoretical Analyses of Complete AIS

While most of the theoretical work so far has considered only specific ingredients of artificial immune systems from the literature, some work has analyzed complete AIS as used in applications. Early examples examined the B-cell algorithm [47]. More recently, a study of Opt-IA [21] was presented. These publications shed a more detailed light on the strengths and weaknesses of immune-inspired approaches compared with other randomized search heuristics such as evolutionary algorithms or random local search by examining the interplay between different mechanisms. The following sections present pseudocode for both algorithms considered in this way, and an overview of the results obtained.

10.4.1 The B-Cell Algorithm

The B-cell algorithm (BCA; see Algorithm 10.14) uses a population of size μ, generates λ clones for each member of the population, and applies standard bit mutations to one random clone of each member and somatic contiguous hypermutations to all clones. It applies plus-selection between each member of the population and its clones. Jansen et al. [31] proposed a variant of the BCA where contiguous hypermutations are only applied with some constant probability $0 < p < 1$ to the search point undergoing standard bit mutations. Thus, in this version, one of the offspring is subject to standard bit mutations with only probability $1 - p$. We call this variant BCA*.

10.4.1.1 Vertex Cover

The work of Jansen et al. [31] constitutes the first runtime analysis of a clonal selection algorithm from the literature without any simplifications. It

Algorithm 10.14: The B-cell algorithm (BCA)

```
Parameters: Population size μ; offspring population size λ;
            mutation probability r ∈ (0,1]
1  Choose x_1,...,x_μ ∈ {0,1}^n independently uniformly at random (u.a.r.)
2  repeat
3      for i ∈ {1,...,μ} do                          /* Clonal expansion */
4          for j ∈ {1,...,λ} do
5              Set y_{i,j} := x_i.
6      for i ∈ {1,...,μ} do                          /* Standard bit mutations */
7          Select j ∈ {1,...,λ} u.a.r.
8          Perform SBM(y_{i,j}).
9      for i ∈ {1,2,...,μ} do                        /* Contiguous hypermutations */
10         for j ∈ {1,2,...,λ} do
11             Perform CHM(y_{i,j}). // see Algorithm 10.5
12     for i ∈ {1,2,...,μ} do                        /* Selection */
13         if f(x_i) ≤ max{f(y_{i,1}),...,f(y_{i,λ})} then
14             Set x_i := y_{i,j}, where f(y_{i,j}) = max{f(y_{i,1}), ..., f(y_{i,λ})}, break ties
               u.a.r.
15 until some termination criterion is met
```

considers the performance of the BCA and BCA* on the vertex cover problem and compares it with known results for evolutionary algorithms.

In the vertex cover problem, we are given an undirected graph $G=(V,E)$ with a set V of $n=|V|$ vertices and a set E of $m=|E|$ edges. A cover is a subset of nodes, $V' \subseteq V$, such that each edge $e \in E$ is covered by at least one node in V', i.e., $e \cap V' \neq \emptyset$. One is interested in finding a small cover, i.e., a cover $V^* \subseteq V$ such that no smaller subset of V can be a cover, i.e., $\forall V' \subseteq V\colon (|V'|<|V^*|) \Rightarrow (\exists e \in E\colon e \cap V^* = \emptyset)$.

In their work, Jansen et al. [31] used the standard node-based representation for vertex cover that assumes that each $x \in \{0,1\}^n$ encodes the node selection $V(x)=\{v_i \in V \mid x[i]=1\}$. For a bit string x that encodes a cover $V(x)$, we use its size $|V(x)|$ as its fitness; otherwise, the number of edges that are not covered is used as a penalty term, yielding the following fitness function that is to be minimized:

$$f(x)=\begin{cases}|V(x)| & \text{if } \forall e \in E\colon V(x)\cap e \neq \emptyset,\\ (|V|+1)\cdot|\{e \in E \mid V(x)\cap e=\emptyset\}| & \text{otherwise.}\end{cases}$$

Since, for the BCA, it is easy to flip contiguous bits but difficult to simultaneously flip a few bits which are far apart, the mapping between a bit in x and a node in V can have a significant influence on the performance of the algorithm. Thus, Jansen et al. [31] suggested an ordering heuristic to determine a suitable mapping instead of making an arbitrary choice. The main idea behind this heuristic is that nodes that are close to each other and share many neighbors are likely to be ordered together (see [31] for a formal definition and illustrative examples).

Using this encoding, Jansen et al. [31] considered a sequence of increasingly complex instances of the vertex cover problem that have been studied as example instances for different kinds of randomized search heuristics to explore their limits. The simplest example (introduced by Friedrich et al. [26]) is a bipartite graph where a small set of nodes V_1 (with size $|V_1| = \varepsilon n$) is completely connected to a larger set of nodes V_2 (with size $|V_2| = (1-\varepsilon)n$). The number of nodes n and the parameter ε that defines the imbalance in the sizes of the two sets are parameters. Friedrich et al. [26] proved that the (1+1) EA is easily caught in the local optimum and therefore is very inefficient with respect to expected optimization time. The same holds for random local search. Both heuristics require the introduction of restarts when stuck in a local optimum to become efficient on this problem instance [49]. The BCA, with any polynomial population size μ and any not too large number of clones ($\lambda = O(1)$), has expected optimization time $O(\mu n^2 \log n)$ and does not require restarts.

To make the difference between the (1+1) EA and the BCA more pronounced, one can "amplify" the result by considering a number of copies of the bipartite graph and adding a small number of additional edges to make the graph connected. For this graph, the (1+1) EA has an exponential expected optimization time even if it is equipped with an optimal restart strategy [50]. For a number l of copies of the bipartite graph, where each copy has h nodes (so that the total graph has $n = h \cdot l$ nodes), the BCA has expected optimization time $O(\mu n^2 (l + \log(n)))$, again for any polynomial population size μ and any not too large number of clones ($\lambda = O(1)$).

The (1+1) EA is not a very typical evolutionary algorithm, since it employs neither a proper population of solutions nor crossover. For evolutionary algorithms making use of both, more complex vertex cover instances become solvable. Oliveto et al. [49] proved that an evolutionary algorithm with a population size of μ that applies crossover with a small probability p_c is able to find an optimal solution for a more complex vertex cover instance in polynomial time with very high probability, namely in time $O(\mu^2 n/p_c)$, where the population size is at least $\mu \geq n^{1+\varepsilon}$ and the probability of applying crossover is at most $p_c \leq 1/(\mu\sqrt{n}\log n)$. Note that the upper bound is $\omega(n^{4.5} \log n)$, which is far from being efficient from a practical point of view. The (1+1) EA is provably very inefficient on this problem instance. The BCA, on the other hand, finds an optimum for this instance in expected time $O(\mu n^3)$, which can be as small as $O(n^3)$ if the population size μ is small ($\mu = O(1)$).

As seen in Section 10.2.2, contiguous hypermutations are a rather inefficient hill-climber, and thus so is the "pure" BCA. Using BCA* instead of the BCA reduces the upper bound to $O(\mu n^2 \log n)$, which becomes $O(n^2 \log n)$ for small population sizes ($\mu = O(1)$). This modification improves the hill-climbing abilities of the BCA considerably without compromising its search capabilities in a significant way. We remark that for $\mu = O(1)$ and $\lambda = O(1)$, BCA* also improves the expected optimization times for OneMax and LeadingOnes to $\Theta(n \log n)$ and $\Theta(n^2)$, respectively.

10.4.1.2 Longest Common Subsequence

A comparison similar to the one presented by Jansen et al. [31] for the vertex cover problem has been performed for the longest common subsequence problem [41]. Here, Jansen and Zarges showed that the BCA outperforms a large class of evolutionary algorithms using mutation and crossover on previously introduced hard problem instances.

In the longest common subsequence problem, we are given a set of m sequences of potentially different lengths over a common finite alphabet Σ, i.e., $X_1, X_2, \ldots, X_m \subseteq \Sigma^*$. By $|Y|$ we denote the length of a sequence Y, i.e., $|Y| = l$ for $Y = y[1]y[2]\cdots y[l] \in \Sigma^l$. A sequence $Y = y[1]y[2]\ldots y[l] \in \Sigma^l$ is called a subsequence of a sequence $X = x[1]x[2]\cdots x[n] \in \Sigma^n$ if there are indices $0 < i_1 < i_2 < \cdots < i_l \leq n$ such that $y[j] = x[i_j]$ holds for all $j \in \{1,2,\ldots,l\}$. The sequence of indices proving that Y is a subsequence of X need not be unique. A sequence Y is a common subsequence of $X_1, X_2, \ldots, X_m$ if it is a subsequence of X_i for all $i \in \{1,2,\ldots,m\}$. It is a longest common subsequence if all common subsequences of $X_1, X_2, \ldots, X_m$ do not have greater length.

Jansen and Zarges [41] used $S = \{0,1\}^n$, where n is the length of a shortest sequence in the input, as the search space. Let $X_1 = x[1]x[2]\cdots x[n] \in \Sigma^n$ denote the letters in the sequence X_1. For a search point $s = s[1]s[2]\cdots s[n] \in \{0,1\}^n$, let $I_1 = \{i_1, i_2, \ldots, i_l\} \subseteq \{1,2,\ldots,n\}$ (with $i_1 < i_2 < \cdots < i_l$) denote the positions of 1-bits in s, i.e., $s[i] = 1$ for all $i \in I_1$ and $s[i] = 0$ for all $i \in \{1,2,\ldots,n\} \setminus I_1$. The search point s encodes the sequence $x[i_1]x[i_2]\cdots x[i_l]$, a subsequence of X_1. Let $c(s)$ denote the sequence encoded by s. If $c(s)$ is a subsequence of all $X_1, X_2, \ldots, X_m$, it encodes a feasible solution, otherwise $c(s)$ is infeasible. The all-zero bit string encodes a trivial empty solution.

We discuss only one of the three fitness functions considered in [41], as the other two are either very complicated or merely of theoretical interest. The function f_{MAX} determines the maximum length k of a prefix of $c(s)$ such that $c(s)_{(k)}$ is a common subsequence of $X_1, X_2, \ldots, X_m$. This length minus the length of the remaining suffix of $c(s)$ is the function value:

$$\begin{aligned}
&\text{MAX}(c(s), X_1, X_2, \ldots, X_m) \\
&\quad = \min\{\max\{k \mid c(s)_{(k)} \text{ is subsequence of } X_i\} \mid i \in \{1,\ldots,m\}\}, \\
&f_{\text{MAX}}(s) \\
&\quad = \text{MAX}(c(s), X_1, X_2, \ldots, X_m) - (|c(s)| - \text{MAX}(c(s), X_1, X_2, \ldots, X_m)).
\end{aligned}$$

Jansen and Zarges [41] considered four hard instances from the literature [32], two for the theoretically motivated fitness function omitted here and two for the other two (including the one defined above):

- E_{MAX}:

$$X_1 = 0^{(8/32)n}1^{(24/32)n} \text{ and } X_2 = 1^{(24/32)n}0^{(5/32)n}1^{(13/32)n},$$

  where n is a multiple of 32;
- A_{MAX}:

$$X_1 = 0^{(1/l)n}1^{((l-1)/l)n} \text{ and } X_2 = 1^{((l-1)/l)n}0^{(5/(8l))n}1^{((4l-3)/(8l))n},$$

  where $l := \lceil (3/\varepsilon) - (1/2) \rceil$ for some $\varepsilon > 0$ constant and n a multiple of $8l$.

It is known that a large class of evolutionary algorithms fails to locate an optimal solution of E_{MAX} efficiently; for A_{MAX}, this class even fails to approximate an optimal solution up to a factor of $2-\varepsilon$ for any constant $\varepsilon > 0$ [32].

Jansen and Zarges [41] proved that the BCA is not efficient if random initialization of the population is used; for A_{MAX}, it also fails to find a good approximation – just like evolutionary algorithms. However, they showed that the BCA is very efficient if started with trivial empty candidate solutions. For both E_{MAX} and A_{MAX}, the expected optimization time of the BCA is $O(\mu\lambda n^2 \log n)$ for all settings of $\mu = n^{O(1)}$, $\lambda = n^{O(1)}$ with $\mu\lambda = \omega(n \log n)$. The algorithm benefits from deterministic initialization because contiguous hypermutations are able to introduce a linear number of 1-bits into a region where they are needed in a single step. As a by-product of their analyses, Jansen and Zarges [41] noted that the concrete choices of μ and λ make no difference as long as $\mu \cdot \lambda$ remains unchanged – in evolutionary computation, these choices usually have a very different effect.

While empirical observations for the longest common subsequence problem indicate that evolutionary algorithms perform better if started with trivial empty candidate solutions, this is not the case for the instances considered, and deterministic initialization does not lead to an improved behavior.

10.4.1.3 Dynamic Optimization

The BCA and its variant BCA* have also been considered in the context of dynamic optimization [42, 45]. Here, Jansen and Zarges particularly discussed why fixed-budget analysis is more appropriate for dynamic environments, where the limited time budget refers to the generations directly after a change in the fitness landscape. They introduced a novel dynamic bistable example function that exhibits phases of stability and rapid change. Motivated by earlier results, they investigated whether artificial immune systems have an advantage in situations of rapid change. A large number of concrete theoretical results for different combinations of execution platforms and parameters of the fitness function were presented. The specific way the optimum moves in the nonstable phases tends to be helpful for contiguous hypermutations, but within the analytical framework no clear advantage could be observed. The concrete contributions of [42, 45] are discussed in more detail in Section 5.5 of this book.

10.4.2 Opt-IA

The name Opt-IA [21] encompasses several clonal selection algorithms following similar ideas and using roughly the same operators. Just like the B-cell algorithm, Opt-IA is mostly used in the context of optimization and uses a bit string representation. We give a description of the algorithm's bare bones in Algorithm 10.15. For each search point in the population, a large number of clones are created that are then subject to mutation. Usually a static cloning operator is used, i.e., the number of clones is independent of the fitness; however, some versions of Opt-IA employ some form of fitness-dependent cloning, where a search point is selected for cloning with a probability that is proportional to its fitness (see [3] for an overview). Depending on the specific variant used, Opt-IA uses two different types of mutation operator: hypermutations with mutation potential and a form of contiguous hypermutations (often called hypermacromutation). Usually, both mutation operators are applied independently and separately to the clones. Opt-IA additionally introduces the concept of aging to clonal selection algorithms, as discussed in Section 10.3. Aging operators aim at increasing the diversity within the population by removing "too old" search points. If aging results in too few search points in the population, the population is filled up with new random search points. Moreover, usually no duplicates are allowed in the population.

Corus et al. [13, 18] considered the first runtime analysis of Opt-IA using only hypermutations with a static mutation potential (where the number of bits flipped is linear in the problem size, $c \cdot n$ for constant $c > 0$), including the FCM mechanism and ensuring that only distinct bits are flipped (TABU variant). Moreover, they replaced the standard static pure aging operator by hybrid aging, as introduced by Oliveto and Sudholt [52] (see Algorithm 10.13).

After considering standard example functions such as ONEMAX, LEADINGONES, CLIFF_d, and JUMP_k for this variant of Opt-IA (with and without genotype diversity), the study highlights problems where the use of the complete Opt-IA variant is crucial. For a carefully constructed novel example function called HIDDENPATH, it was shown that Opt-IA with appropriate parameterization has an expected polynomial optimization time, while the algorithm missing either aging or hypermutations requires at least superpolynomial time. To give a complete picture, the extension in [18] introduced another class of functions (called HYPERTRAP_y), for which, with

Algorithm 10.15: Opt-IA

```
   Parameters: Population size μ; offspring population size λ;
               mutation flags H, M
 1 Choose P = {x_1,...,x_μ} independently, uniformly at random (u.a.r.).
 2 repeat
 3     for i ∈ {1,...,μ} do                    /* Clonal selection and expansion */
 4         Generate λ clones of x_i.
 5         Place the clones in a clonal pool C_i = {y_{i,1},...,y_{i,λ}}.
 6         C_i^H = ∅. C_i^M = ∅.
 7         for j ∈ {1,...,λ} do                                /* Affinity maturation */
 8             if H then
 9                 ŷ_{i,j} ← Apply hypermutations with mutation potential to y_{i,j}.
10                 Add ŷ_{i,j} to C_i^H.
11             if M then
12                 ỹ_{i,j} ← Apply contiguous hypermutations to y_{i,j}.
13                 Add ỹ_{i,j} to C_i^M.
14     Apply aging to P, C_i^H, and C_i^M.                          /* Metadynamics */
15     Set P = P ∪ C_1^H ∪ ... ∪ C_μ^H ∪ C_1^M ∪ ... ∪ C_μ^M.          /* Selection */
16     if |P| > μ then
17         Keep the μ best search points from P, breaking ties u.a.r. and removing
           duplicates.
18     else
19         Keep all search points in P.
20         Fill up P with random points until |P| = μ.
21 until some termination criterion is met
```

overwhelming probability, Opt-IA is inefficient, while the simple (1+1) EA using standard bit mutations is efficient.

Corus et al. [13, 18] also considered a simple trap function, as such a function was used when Opt-IA was originally introduced. They proved an expected optimization time of $O(\mu n^2 \log n)$ for $\tau = \Omega(n^2)$, $c = 1$, and $\lambda = 1$ and pointed out that this does not match the empirical results reported in [19], where Opt-IA was unable to optimize trap functions for $n > 50$. It was conjectured that this was due either to not using FCM or too small a time budget.

In [15], Corus et al. analyzed Opt-IA using "fast" hypermutations with mutation potential (see Section 10.2.3.3) on previously considered example functions such as HIDDENPATH and CLIFF_d. The authors particularly pointed out that, in order to effectively work with aging on CLIFF_d, it was crucial not to use hypermutations with FCM as the FCM mechanism does not allow worsening of the fitness value[1]. Thus, the operator performed all n mutation steps and returned the best sampled search point instead of the first improved one.

[1] This observation holds for both "classical" and "fast" hypermutations with mutation potential.

Later, Corus et al. [17] demonstrated that their inversely fitness-proportional mutation potential (see Section 10.2.3.1) together with aging was able to optimize CLIFF_d with $d = \Theta(n)$ in expected polynomial time even if FCM is used.

10.5 Summary

In this chapter, we have provided an overview of the state of the art in the theory of immune-inspired randomized search heuristics for optimization. In this context, most algorithms are inspired by the so-called clonal selection principle, which describes the basic features of the adaptive immune response. A large variety of different clonal selection algorithms have been introduced, and over the last decade some significant progress has been made on the theoretical foundations of such algorithms. Initially, most theoretical studies concentrated on two defining aspects of artificial immune systems: hypermutation operators (inversely fitness-proportional mutations, contiguous hypermutations, and hypermutations with mutation potential) and a diversity mechanism called aging (static pure aging and stochastic aging). More recently, insights into the interplay between different operators have allowed the first analyses of "complete" artificial immune systems as published in the literature – this particularly includes analyses of the B-cell algorithm and Opt-IA.

Theoretical analyses have contributed to significant insights into the working principles of immune-inspired operators and algorithms. For example, a common observation in the literature is that typical immune-inspired operators such as hypermutations and aging allow us to efficiently escape from local optima – particularly when compared to evolutionary algorithms – but may have difficulties during the exploitation phase. The introduction of fixed-budget analysis (discussed in more detail in Chapter 5 of this book) has particularly contributed to our understanding of their strengths and weaknesses. In many cases, these insights have contributed to the development of improved versions of the operators or hybrid variants that combine immune-inspired mechanisms with techniques used in evolutionary computation and other randomized search heuristics. However, more research into the strengths and weaknesses of immune-inspired algorithms is needed, particularly in the context of combinatorial optimization. It would be interesting to see on what kind of problems these algorithms excel over other nature-inspired randomized search heuristics such as evolutionary algorithms. It is also often argued that immune-inspired algorithms are especially suited for multimodal or dynamic optimization problems. Further investigations in these directions are promising directions for future research.

References

[1] Bernardino, H.S., Barbosa, H.J.C.: Artificial immune systems for optimization. In: R. Chiong (ed.) Nature-Inspired Algorithms for Optimisation, *Studies in Computational Intelligence*, vol. 193, pp. 389–411. Springer (2009)

[2] Böttcher, S., Doerr, B., Neumann, F.: Optimal fixed and adaptive mutation rates for the LeadingOnes problem. In: Proceedings of the 11th International Conference on Parallel Problem Solving from Nature (PPSN 2010), *Lecture Notes in Computer Science*, vol. 6238, pp. 1–10. Springer (2010)

[3] Brownlee, J.: Clonal selection algorithms. Tech. Rep. 070209A, Swinburne University of Technology, Victoria, Australia (2007)

[4] Burnet, F.M.: The Clonal Selection Theory of Acquired Immunity. Cambridge University Press (1959)

[5] de Castro, L.N., Timmis, J.: An artificial immune network for multimodal function optimization. In: Proceedings of the Congress on Evolutionary Computation (CEC 2002), pp. 699–704. IEEE Press (2002)

[6] de Castro, L.N., Timmis, J.: Artificial Immune Systems: A New Computational Intelligence Approach. Springer (2002)

[7] de Castro, L.N., Von Zuben, F.J.: Learning and optimization using the clonal selection principle. IEEE Transactions on Evolutionary Computation **6**(3), 239–251 (2002)

[8] Clark, E.B.: A framework for modelling stochastic optimisation algorithms with Markov chains. Ph.D. thesis, University of York (2008)

[9] Clark, E.B., Hone, A., Timmis, J.: A Markov chain model of the B-cell algorithm. In: Proceedings of the 4th International Conference on Artificial Immune Systems (ICARIS 2005), *Lecture Notes in Computer Science*, vol. 3627, pp. 318–330. Springer (2005)

[10] Coello, C.A.C., Cortés, N.C.: Solving multiobjective optimization problems using an artificial immune system. Genetic Programming and Evolvable Machines **6**(2), 163–190 (2005)

[11] Corus, D., He, J., Jansen, T., Oliveto, P.S., Sudholt, D., Zarges, C.: On easiest functions for somatic contiguous hypermutations and standard bit mutations. In: Proceedings of the Genetic and Evolutionary Computation Conference (GECCO 2015), pp. 1399–1406. ACM (2015)

[12] Corus, D., He, J., Jansen, T., Oliveto, P.S., Sudholt, D., Zarges, C.: On easiest functions for mutation operators in bio-inspired optimisation. Algorithmica **78**(2), 714–740 (2017)

[13] Corus, D., Oliveto, P.S., Yazdani, D.: On the runtime analysis of the opt-IA artificial immune system. In: Proceedings of the Genetic and Evolutionary Computation Conference (GECCO 2017), pp. 83–90. ACM (2017)

[14] Corus, D., Oliveto, P.S., Yazdani, D.: Artificial immune systems can find arbitrarily good approximations for the NP-hard partition problem. In:

Proceedings of the 15th International Conference on Parallel Problem Solving from Nature (PPSN 2018), Part II, *Lecture Notes in Computer Science*, vol. 11102, pp. 16–28. Springer (2018)
[15] Corus, D., Oliveto, P.S., Yazdani, D.: Fast artificial immune systems. In: Proceedings of the 15th International Conference on Parallel Problem Solving from Nature (PPSN 2018), Part II, *Lecture Notes in Computer Science*, vol. 11102, pp. 67–78. Springer (2018)
[16] Corus, D., Oliveto, P.S., Yazdani, D.: Artificial immune systems can find arbitrarily good approximations for the NP-hard number partitioning problem. Artificial Intelligence **274**, 180–196 (2019). In press. `https://doi.org/10.1016/j.artint.2019.03.001`
[17] Corus, D., Oliveto, P.S., Yazdani, D.: On inversely proportional hypermutations with mutation potential. In: Proceedings of the Genetic and Evolutionary Computation Conference (GECCO 2019), pp. 215–223. ACM (2019).
[18] Corus, D., Oliveto, P.S., Yazdani, D.: When hypermutations and ageing enable artificial immune systems to outperform evolutionary algorithms. Theoretical Computer Science (2019). In press. `https://doi.org/10.1016/j.tcs.2019.03.002`
[19] Cutello, V., Nicosia, G., Pavone, M.: Exploring the capability of immune algorithms: A characterization of hypermutation operators. In: Proceedings of the 3rd International Conference on Artificial Immune Systems (ICARIS 2004), *Lecture Notes in Computer Science*, vol. 3239, pp. 263–276. Springer (2004)
[20] Cutello, V., Nicosia, G., Romeo, M., Oliveto, P.S.: On the convergence of immune algorithms. In: Proceedings of the IEEE Symposium on Foundations of Computational Intelligence (FOCI 2007), pp. 409–415. IEEE (2007)
[21] Cutello, V., Pavone, M., Timmis, J.: An immune algorithm for protein structure prediction on lattice models. IEEE Transactions on Evolutionary Computation **11**(1), 101–117 (2007)
[22] Dasgupta, D. (ed.): Artificial Immune Systems and Their Applications. Springer (1998)
[23] Dasgupta, D., Niño, L.F.: Immunological Computation: Theory and Applications. Auerbach (2008)
[24] Elberfeld, M., Textor, J.: Negative selection algorithms on strings with efficient training and linear-time classification. Theoretical Computer Science **412**(6), 534–542 (2011)
[25] Fernandes, D.A.B., Freire, M.M., Fazendeiro, P.A.P., Inácio, P.R.M.: Applications of artificial immune systems to computer security: A survey. Journal of Information Security and Applications **35**, 138–159 (2017)
[26] Friedrich, T., He, J., Hebbinghaus, N., Neumann, F., Witt, C.: Approximating covering problems by randomized search heuristics using multi-objective models. Evolutionary Computation **18**(4), 617–633 (2010)

[27] Gu, F., Greensmith, J., Aickelin, U.: Theoretical formulation and analysis of the deterministic dendritic cell algorithm. Biosystems **111**(2), 127–135 (2013)

[28] He, J., Chen, T., Yao, X.: On the easiest and hardest fitness functions. IEEE Transactions on Evolutionary Computation **19**(2), 295–305 (2015)

[29] Hong, L., Kamruzzaman, J.: Convergence of elitist clonal selection algorithm based on martingale theory. Engineering Letters **21**(4), 181–184 (2013)

[30] Horoba, C., Jansen, T., Zarges, C.: Maximal age in randomized search heuristics with aging. In: Proceedings of the Genetic and Evolutionary Computation Conference (GECCO 2009), pp. 803–810. ACM (2009)

[31] Jansen, T., Oliveto, P.S., Zarges, C.: On the analysis of the immune-inspired B-cell algorithm for the vertex cover problem. In: Proceedings of the 10th International Conference on Artificial Immune Systems (ICARIS 2011), *Lecture Notes in Computer Science*, vol. 6825, pp. 117–131. Springer (2011)

[32] Jansen, T., Weyland, D.: Analysis of evolutionary algorithms for the longest common subsequence problem. Algorithmica **57**, 170–186 (2010)

[33] Jansen, T., Zarges, C.: Comparing different aging operators. In: Proceedings of the 8th International Conference on Artificial Immune Systems (ICARIS 2009), *Lecture Notes in Computer Science*, vol. 5666, pp. 95–108. Springer (2009)

[34] Jansen, T., Zarges, C.: A theoretical analysis of immune inspired somatic contiguous hypermutations for function optimization. In: Proceedings of the 8th International Conference on Artificial Immune Systems (ICARIS 2009), *Lecture Notes in Computer Science*, vol. 5666, pp. 80–94. Springer (2009)

[35] Jansen, T., Zarges, C.: Aging beyond restarts. In: Proceedings of the Genetic and Evolutionary Computation Conference (GECCO 2010), pp. 705–712. ACM (2010)

[36] Jansen, T., Zarges, C.: On the benefits of aging and the importance of details. In: Proceedings of the 9th International Conference on Artificial Immune Systems (ICARIS 2010), *Lecture Notes in Computer Science*, vol. 6209, pp. 61–74. Springer (2010)

[37] Jansen, T., Zarges, C.: Analyzing different variants of immune inspired somatic contiguous hypermutations. Theoretical Computer Science **412**(6), 517–533 (2011)

[38] Jansen, T., Zarges, C.: On benefits and drawbacks of aging strategies for randomized search heuristics. Theoretical Computer Science **412**(6), 543–559 (2011)

[39] Jansen, T., Zarges, C.: On the role of age diversity for effective aging operators. Evolutionary Intelligence **4**(2), 99–125 (2011)

[40] Jansen, T., Zarges, C.: Variation in artificial immune systems: Hypermutations with mutation potential. In: Proceedings of the 10th Interna-

tional Conference on Artificial Immune Systems (ICARIS 2011), *Lecture Notes in Computer Science*, vol. 6825, pp. 132–145. Springer (2011)

[41] Jansen, T., Zarges, C.: Computing longest common subsequences with the B-cell algorithm. In: Proceedings of the 11th International Conference on Artificial Immune Systems (ICARIS 2012), *Lecture Notes in Computer Science*, vol. 7597, pp. 111–124. Springer (2012)

[42] Jansen, T., Zarges, C.: Evolutionary algorithms and artificial immune systems on a bi-stable dynamic optimisation problem. In: Proceedings of the Genetic and Evolutionary Computation Conference (GECCO 2014), pp. 975–982. ACM (2014)

[43] Jansen, T., Zarges, C.: Reevaluating immune-inspired hypermutations using the fixed budget perspective. IEEE Transactions on Evolutionary Computation **18**(5), 674–688 (2014)

[44] Jansen, T., Zarges, C.: Understanding randomised search heuristics. lessons from the evolution of theory: A case study. In: Proceedings of the 20th International Conference on Soft Computing (MENDEL 2014), pp. 293–298 (2014)

[45] Jansen, T., Zarges, C.: Analysis of randomised search heuristics for dynamic optimisation. Evolutionary Computation **23**, 513–541 (2015)

[46] Jerne, N.: Towards a network theory of the immune system. Annals of Immunology **125C**(1–2), 373–389 (1974)

[47] Kelsey, J., Timmis, J.: Immune inspired somatic contiguous hypermutation for function optimisation. In: Proceedings of the Genetic and Evolutionary Computation Conference (GECCO 2003), *Lecture Notes in Computer Science*, vol. 2723, pp. 207–218. Springer (2003)

[48] Lehre, P.K., Witt, C.: Black-box search by unbiased variation. Algorithmica **64**(4), 623–642 (2012)

[49] Oliveto, P.S., He, J., Yao, X.: Analysis of population-based evolutionary algorithms for the vertex cover problem. In: Proceedings of the Congress on Evolutionary Computation (CEC 2008), pp. 1563–1570. IEEE Press (2008)

[50] Oliveto, P.S., He, J., Yao, X.: Analysis of the (1+1)-EA for finding approximate solutions to vertex cover problems. IEEE Transactions Evolutionary Computation **13**(5), 1006–1029 (2009)

[51] Oliveto, P.S., Lehre, P.K., Neumann, F.: Theoretical analysis of rank-based mutation – combining exploration and exploitation. In: Proceedings of the Congress on Evolutionary Computation (CEC 2009), pp. 1455–1462. IEEE (2009)

[52] Oliveto, P.S., Sudholt, D.: On the runtime analysis of stochastic ageing mechanisms. In: Proceedings of the Genetic and Evolutionary Computation Conference (GECCO 2014), pp. 113–120. ACM (2014)

[53] Silva, G.C., Dasgupta, D.: A survey of recent works in artificial immune systems. In: P.P. Angelov (ed.) Handbook on Computational Intelligence, chap. Chapter 15, pp. 547–586. World Scientific (2016)

[54] Textor, J.: Efficient negative selection algorithms by sampling and approximate counting. In: Proceedings of the 12th International Conference on Parallel Problem Solving from Nature (PPSN 2012), *Lecture Notes in Computer Science*, vol. 7491, pp. 32–41. Springer (2012)
[55] Timmis, J., Hone, A., Stibor, T., Clark, E.: Theoretical advances in artificial immune systems. Theoretical Computer Science **403**(1), 11–32 (2008)
[56] Villalobos-Arias, M., Coello, C.A.C., Hernández-Lerma, O.: Convergence analysis of a multiobjective artificial immune system algorithm. In: Proceedings of the 3rd International Conference on Artificial Immune Systems (ICARIS 2004), *Lecture Notes in Computer Science*, vol. 3239, pp. 226–235. Springer (2004)
[57] Villalobos-Arias, M., Coello, C.A.C., Hernández-Lerma, O.: Asymptotic convergence of some metaheuristics used for multiobjective optimization. In: Proceedings of the 8th International Workshop on Foundations of Genetic Algorithms (FOGA 2005), *Lecture Notes in Computer Science*, vol. 3469, pp. 95–111. Springer (2005)
[58] Xia, X., Zhou, Y.: On the effectiveness of immune inspired mutation operators in some discrete optimization problems. Information Sciences **426**, 87–100 (2018)
[59] Zarges, C.: Rigorous runtime analysis of inversely fitness proportional mutation rates. In: Proceedings of the 10th International Conference on Parallel Problem Solving from Nature (PPSN 2008), *Lecture Notes in Computer Science*, vol. 5199, pp. 112–122. Springer (2008)
[60] Zarges, C.: On the utility of the population size for inversely fitness proportional mutation rates. In: Proceedings of the 10th International Workshop on Foundations of Genetic Algorithms (FOGA 2009), pp. 39–46. ACM Press (2009)
[61] Zarges, C.: Theoretical foundations of artificial immune systems. Ph.D. thesis, TU Dortmund, Germany (2011)

Chapter 11
Computational Complexity Analysis of Genetic Programming

Andrei Lissovoi and Pietro S. Oliveto

Abstract Genetic programming (GP) is an evolutionary computation technique to solve problems in an automated, domain-independent way. Rather than identifying the optimum of a function as in more traditional evolutionary optimization, the aim of GP is to evolve computer programs with a given functionality. While many GP applications have produced human competitive results, the theoretical understanding of what problem characteristics and algorithm properties allow GP to be effective is comparatively limited. Compared with traditional evolutionary algorithms for function optimization, GP applications are further complicated by two additional factors: the variable-length representation of candidate programs, and the difficulty of evaluating their quality efficiently. Such difficulties considerably impact the runtime analysis of GP, where space complexity also comes into play. As a result, initial complexity analyses of GP have focused on restricted settings such as the evolution of trees with given structures or the estimation of solution quality using only a small polynomial number of input/output examples. However, the first computational complexity analyses of GP for evolving proper functions with defined input/output behavior have recently appeared. In this chapter, we present an overview of the state of the art.

Andrei Lissovoi
Rigorous Research, Department of Computer Science, University of Sheffield, Sheffield, United Kingdom.
e-mail: a.lissovoi@sheffield.ac.uk

Pietro S. Oliveto
Rigorous Research, Department of Computer Science, University of Sheffield, Sheffield, United Kingdom.
e-mail: p.oliveto@sheffield.ac.uk

B. Doerr, F. Neumann (eds.), *Theory of Evolutionary Computation*,
Natural Computing Series, https://doi.org/10.1007/978-3-030-29414-4_11

11.1 Introduction

Genetic programming (GP) is a class of evolutionary computation techniques to evolve computer programs popularized by Koza [20]. GP uses genetic algorithm mutation, crossover and selection operators adapted to work on populations of program structures. Program fitness is evaluated using a *training set* consisting of samples of program inputs and the corresponding correct outputs. The goal of a GP system is to construct a program which, as well as producing the correct outputs on the inputs included in the training set, generalizes well to other possible inputs.

In standard tree-based GP, as popularized by Koza, programs are expressed as syntax trees rather than lines of code, with variables and constants (collectively referred to as *terminals*) appearing as leaf nodes in the tree, and functions (such as +, *, and `cos`) appearing as internal nodes. New programs are produced by mutation (which makes some changes to a solution) or crossover (which creates new solutions by combining subtrees of two parent solutions). Several other variants of GP exist that use different representations than tree structures. Popular ones are Linear GP [1], Cartesian GP [31], and Geometric Semantic GP (GSGP) [33]. Since most of the available computational complexity analyses focus on tree-based GP, this is where we keep our focus in this chapter. Work on GSGP is an exception that we will also consider [35].

One of the main points regarding GP made by Koza is that a wide variety of different problems from many different fields can be recast as requiring the discovery of a computer program that produces some desired output when presented with particular inputs [20]. Ideally, this process of discovery could take place without requiring a human to explicitly make decisions about the size, shape, or structural complexity of the solutions in advance. Since GP systems provide a way to search the space of computer programs for one which solves (or approximates) the problem at hand, they are thus applicable to a wide variety of problems, including those in artificial intelligence, machine learning, adaptive systems, and automated learning. GP has produced human-competitive results and patentable solutions on a large number of diverse problems, including the design of quantum computing circuits [51], antennas [26], mechanical systems [24], and optical lens systems [22]. From these results, Koza observes that GP may be especially productive in areas where little information about the size or shape of the ultimate solution is known, while large amounts of data and good simulators are available to measure the performance of candidate solutions [21].

While there are many examples of successful applications of GP (see [21] for an overview), the understanding of how such systems work and on which problems they are successful is much more limited. Compared with traditional evolutionary algorithms for function optimization, GP applications are further complicated by two additional factors: the variable-length representation of candidate programs, and the difficulty of evaluating their quality efficiently, since it is prohibitive or even impossible to test programs on all possible inputs. Such difficulties, naturally, impact the runtime analysis of GP considerably, where space complexity also comes into play. As a result, while nowadays the analysis of standard elitist [3, 4] and nonelitist genetic algorithms [2, 39, 40] has finally become a reality, analyzing standard GP systems is far more prohibitive. Indeed, McDermott and O'Reilly [30] remarked that "due to stochasticity, it is arguably impossible in most cases to make formal guarantees about the number of fitness evaluations needed for a GP algorithm to find an optimal solution." Similarly to how the analysis of simplified evolutionary algorithms (EAs) has gradually led to the achievement of techniques that nowadays allow the analysis of standard EAs, Poli et al. suggested that "computational complexity techniques being used to model simpler GP systems, perhaps GP systems based on mutation and stochastic hill-climbing" [48].

Following this guideline the first runtime analyses laying the groundwork for better understanding of GP considered simplified algorithms primarily based on mutation and hill-climbing (i.e., the (1+1) GP algorithm introduced in [9]). However, further simplifications compared with applications of GP in practice were necessary to deal with the additional difficulties introduced by the variable length of GP solutions, the stochastic fitness function evaluations when dynamic training sets were used, and the neighborhood structure imposed by the GP mutation and crossover operations acting on syntax trees. Indeed, Goldberg and O'Reilly observed that "the methodology of using deliberately designed problems, isolating specific properties, and pursuing, in detail, their relationships in simple GP is more than sound; it is the only practical means of systematically extending GP understanding and design" [13]. To this end, the first runtime analyses of GP considered the time required to evolve particular tree structures rather than proper computer programs. In particular, solution fitness was evaluated based on the tree structure rather than by executing the evolved syntax tree. Problems belonging to this category are ORDER, MAJORITY [9] and SORTING [56]. Even in such simplified settings, the characteristic GP problem, bloat (i.e., the continuous growth of evolved solutions that is not accompanied by significant improvements in solution quality), may appear.

In GP applications generally, either the set of all possible inputs is too large to evaluate the exact solution quality efficiently, or not much of it is known (i.e., only a limited amount of information about the correct input/output behavior is available). As a result, the performance of the GP system is usually considered in the probably approximately correct (PAC) learning framework [54], to show that the solution produced by the GP system generalizes well to all inputs. Kötzing et al. isolated this issue when they presented the first runtime analysis of a GP system in this framework [18]. They considered the problem of learning the weights assigned to n bits of a pseudo-Boolean function (i.e., the IDENTIFICATION problem), and proved that a simple GP system can discover the weights efficiently even if a limited sample of the possible inputs is used to evaluate solution quality.

A more realistic problem where the program output, rather than structure, is used as the basis for determining solution quality is the MAX problem [19], originally introduced in [12]. The problem is to evolve a program which, given some mathematical operators and constants (the problem admits no variable inputs), outputs the maximum possible value subject to a constraint on program size.

Only recently, the time and space complexity of the (1+1) GP has been analyzed for evolving Boolean functions of arity n [25, 29]. Solution quality was evaluated by comparing the output of the evolved programs with the target function on all possible inputs, or on a polynomially sized training set. The analyses show that while conjunctions of n variables can be evolved efficiently (either exactly, using the complete truth table as the training set, or in the PAC learning framework when smaller training sets are used), parity functions of n variables cannot. These results represent the first rigorous complexity analysis of a tree-based GP system for evolving functions with actual input/output behavior.

We will also consider the theoretical work on GSGP, where the variation operators used by the GP system are designed to modify program semantics rather than program syntax.

This chapter presents an overview of the state of the art. It is structured as follows. In Section 11.2, we introduce the (1+1) GP, the GP system used for most of the available computational complexity analysis results. In Section 11.3, we present an overview of the analyses of GP systems for evolving tree structures with specific properties (the ORDER, MAJORITY, and SORTING problems). In Section 11.4, we present results where GP systems evolve programs with limited functionality: the MAX problem is considered in Subsection 11.4.1, and the IDENTIFICATION problem in Subsection 11.4.2. Section 11.5 presents results for GP evolving proper Boolean functions of arity n. Section 11.6 presents a brief overview of the computational complexity results available for GSGP algorithms. Finally, Section 11.7 presents a summary of the presented results and discusses the open directions for future work.

11.2 Preliminaries

In this chapter, we will primarily consider the behavior of the simple (1+1) GP algorithm (Algorithm 11.1), which represents programs using syntax trees and uses the HVL-Prime operator (Algorithm 11.2) to perform mutations. This algorithm maintains a population of one individual (initialized either with an empty tree, or with a randomly generated tree), and at each generation chooses between the parent and a single offspring generated

Algorithm 11.1: The (1+1) GP

```
1 Initialize a tree X
2 for t ← 1,2,... do
3     X' ← X
4     k ← 1 + Poisson(1)
5     for i ← 1,...,k do
6         X' ← HVL-Prime(X')
7     if f(X') ≤ f(X) then
8         X ← X'
```

Algorithm 11.2: The HVL-Prime mutation operator

```
   Data: A binary syntax tree X.
1  Choose op ∈ {INS, DEL, SUB} uniformly at random
2  if X is an empty tree then
3      Choose a literal l ∈ L uniformly at random
4      Set l to be the root of X
5  else if op = INS then
6      Choose a node x ∈ X uniformly at random
7      Choose f ∈ F, l ∈ L uniformly at random
8      Replace x in X with f
9      Set the children of f to be x and l, order chosen uniformly at random
10 else if op = DEL then
11     Choose a leaf node x ∈ X uniformly at random
12     Replace x's parent in X with x's sibling in X
13 else if op = SUB then
14     Choose a node x ∈ X uniformly at random
15     Choose a replacement l ∈ L, or f ∈ F uniformly at random
16     Replace x in X with l if x is a leaf node, or with f if x is an internal node
```

by HVL-Prime mutation. This simple algorithm had already been considered in early comparative work between standard tree-based GP and iterated hill-climbing versions of GP [42–44].

The HVL-Prime mutation operator, introduced in [9] and shown in Algorithm 11.2 here, is an updated version of the HVL (hierarchical variable length) mutation operator [42]. It is specialized to deal with binary trees and is designed to perform similarly to bitwise mutation in evolutionary algorithms. The original motivation for using the HVL-Prime operator was that of making the smallest alterations possible to GP trees while respecting the key properties of the GP tree search space: variable length and hierarchical structure.

A single application of HVL-Prime selects uniformly at random one of three suboperations – insertion, substitution, and deletion – to be applied at a location in the solution tree chosen uniformly at random, selecting additional functions or terminals from the sets F and L of all available functions

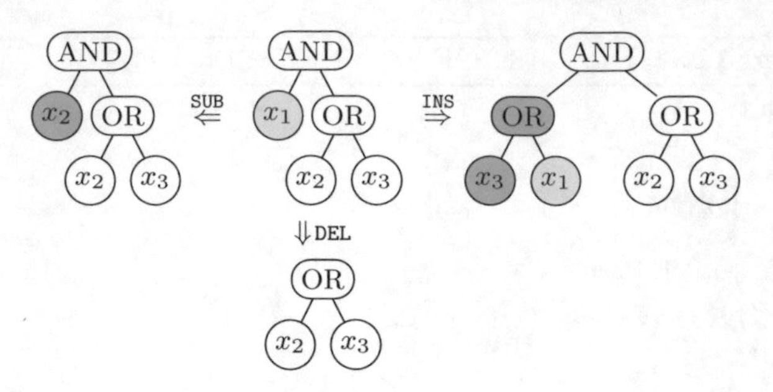

Fig. 11.1 HVL-Prime suboperations: substitution, insertion, and deletion.

and terminals as required. The suboperations are illustrated in Fig. 11.1: substitution can replace any node of the tree with another node chosen uniformly at random from the set of terminals or the set of functions (if the replaced node is a terminal or a function, respectively), insertion inserts a new leaf and function node at a random location in the tree, and deletion can remove a random leaf (replacing its parent with its sibling).

We note that for problems with trivial function or terminal sets (i.e., those that contain only one element), the substitution operator is typically restricted to select only from among those nodes which can be replaced with something other than their current content, avoiding the situation where the only option is to substitute a function or terminal node with a copy of itself. This restriction does not typically affect asymptotic complexity analysis results, as the only effect of allowing such substitutions is that approximately 1/6 of the HVL-Prime applications will not alter the current solution.

In this chapter, we refer to Algorithm 11.1, with $k = 1 + \text{Poisson}(1)$, as the $(1+1)$ GP, differentiating it from the simpler local search variant which always uses $k = 1$, which we call RLS-GP.[1]

$(1+1)$ GP algorithms do not use crossover or populations. Instead, larger changes to the current solution can be performed by multiple applications of the HVL-Prime operator without evaluating the fitness of the intermediate trees produced within an iteration. Since each application of HVL-Prime selects a location in the tree that it will modify independently, it is possible for this procedure to mutate the parent tree in several places, rather than only modifying a single subtree (which would be the case for the standard GP subtree mutation operator, which replaces a random subtree of the parent program with a randomly generated subtree [47]).

[1] In previous work, the name "$(1+1)$ GP" was used for both algorithms, relying either on explicitly specifying k or on using a suffix as in "$(1+1)$ GP-multi" and "$(1+1)$ GP-single" to distinguish between the two variants. Our notation matches the conventions for the runtime analysis of evolutionary algorithms [15, 41].

11.2.1 Bloat Control Mechanisms

Algorithm 11.1 depicts the nonstrictly elitist variant of the (1+1) GP, which accepts offspring as long as they do not decrease the fitness of the current solution. We use "(1+1) GP*" (and equivalently "RLS-GP*") to refer to the strictly elitist variant of the algorithm, which only accepts offspring which have strictly better fitness when compared with the current solution.

The difference between the elitist and nonelitist variants is often significant in how the algorithms cope with bloat problems. The (1+1) GP algorithm operates with a variable-length representation of its current solution: as mutations are applied, the number of nodes in the tree may increase or decrease. Poli et al. defined bloat as "program growth without (significant) return in terms of fitness" [47]. Bloat can reduce the effectiveness of GP, as larger programs are potentially more expensive to evaluate, can be hard to interpret, and may reduce the effectiveness of the GP operators in exploring the solution space. For example, if a large portion of the current solution is nonexecutable (perhaps inside an `if` statement with a trivially false condition), mutations applied inside that portion of the program would not alter its behavior, and hence are not helpful in attempting to improve the program.

Common techniques used to control the impact of bloat include modifying the genetic operators to produce smaller trees and considering additional nonfitness-related factors when determining whether an offspring should be accepted into the population. The latter can include imposing direct limits on the size of the accepted solutions (by imposing either a maximum tree depth or a maximum tree size limit), rejecting neutral solutions, or a parsimony pressure approach [47], which prefers smaller solutions when the fitness values of two solutions are equal.

Two bloat control approaches that frequently appear in theoretical analyses of GP algorithms are *lexicographic parsimony pressure* and *Pareto parsimony pressure* [27]. The former mechanism breaks ties between equal-fitness individuals (e.g., in line 7 of Algorithm 11.1) by preferring solutions of smaller size, whereas the latter treats fitness and solution size as equal objectives in a multiobjective approach to optimization, making the GP system maintain a population of individuals which do not Pareto-dominate each other.

11.2.2 Evaluating Solution Quality

In the GP problems analyzed in this chapter, the correct behavior of the target program is known for all possible inputs. Additionally, in most of the problems, the GP systems considered are able to evaluate program quality on all possible inputs efficiently. Both of these assumptions simplify the analysis, but may not be practical in real-world applications of GP: the correct output of the target function might only be known for a limited number of the possible inputs, and/or it might not be practical to evaluate the candidate solutions for all of the known inputs. Nevertheless, considering the performance of GP in this setting represents an important first step: systems which are unable to evolve a program with the desired behavior using a fitness function which considers all possible inputs are unlikely to fare better when using a limited approximation. Additionally, fully deterministic outcomes for solution fitness comparisons simplify the analysis of the GP systems, allowing their behavior to be described in greater detail.

When the exact fitness is not available, the performance of GP is analyzed in the PAC learning framework [54]. This considers the expected performance of the GP-evolved program on inputs it may not have encountered during the optimization process. In this framework, GP evaluates solution fitness by sampling input/output examples from a training set during the optimization process, and the goal is to produce a program with a low generalization error, i.e., with a good probability of producing correct output on any randomly sampled solution, including ones that have not been sampled during its construction. The number of samples used to compare the quality of solutions is an important parameter in this setting, potentially trading evaluation accuracy for time efficiency.

While a GP algorithm may evaluate solution fitness by relying on a static training set of polynomial size, for instance chosen at random from the set of all known inputs/outputs at the start of the optimization process, Poli et al. noted that in some circumstances doing so "may encourage the population to evolve into a cul-de-sac where it is dominated by offspring of a single initial program which did well on some fraction of the training cases, but was unable to fit the others" [47, Chapter 10]. To counteract this when the amount of training set data available is sufficient, GP systems can also opt to compare program quality on samples chosen from the available data for each comparison [11]. The complexity of these subset selection algorithms varies from simply selecting inputs/outputs at random (in the case of random subset selection), through attempting to identify useful inputs/outputs based on the current or previous GP runs (dynamic or historical subset selection), to hierarchical combinations of these approaches [5].

11.3 Evolving Tree Structures

In this section, we review the computational complexity results concerning the analysis of GP systems for the evolution of trees with specified properties, rather than the evolution of programs with inputs and outputs. The specific property that the evolved tree should satisfy depends on the problem class. The possibility of calculating the fitness of candidate solution trees without explicitly executing the program was regarded as a considerable advantage, since more realistic problems were deemed to be far too difficult for initial computational complexity analyses.

The earliest analysis of the evolution of tree structures considered two separable problems, called ORDER and MAJORITY. These problems, originally introduced by Goldberg and O'Reilly [13], were considered as "two much simplified, but still insightful, problems that exhibit a few simple aspects of program structure" [9]. Specifically, ORDER and MAJORITY were introduced as abstracted simplifications of the eliminative expression that takes place in conditional statements (where the presence or absence of some element may *eliminate* others from evaluation, e.g., by making it impossible for program execution to reach the body of an if statement with an always false condition), and of the accumulative expression present in many GP applications such as symbolic regression (where the GP system is able to *accumulate* information about the correct solution from the aggregate response of a large number of variables), respectively. In particular, the ORDER problem was meant to reflect conditional programs by making it impossible to express certain variables by inserting them at certain tree locations (representing portions of the program which might not ever be executed), while MAJORITY requires the identification of the correct set of solution components out of all possible sets. For both problems the fitness of a candidate solution is determined by an in-order traversal of its syntax tree.

Neumann additionally introduced weighted variants of the ORDER and MAJORITY problems. In WORDER and WMAJORITY, each pair of variables $x_i, \overline{x}_i$ has a corresponding weight w_i, which models the relative importance of the component to the correctness of the overall solution [37]. The idea behind these weighed variants to mimic the generalization of the complexity analysis of evolutionary algorithms from ONEMAX to the class of linear pseudo-Boolean functions [8, 41].

Another problem considered in the literature where the fitness of solutions depends on tree structure rather than program execution is SORTING. In the following three subsections, we review the state of the art concerning these problems.

The analyses of the toy problems considered in this section have two main aims. The first is to provide simplified settings that allow rigorous computational complexity analysis of GP systems by abstracting from the need of evaluating solution quality on a training set. The second is to evaluate to what extent bloat affects GP optimization on simplified problems with variable length representation. Since bloat seems to be a ubiquitous problem in GP, one expects it to appear also in the optimization process of the problems presented in this section.

11.3.1 The ORDER Problem

The ORDER problem, as originally introduced by Goldberg and O'Reilly [13], is defined as follows.

Problem 11.3.1 (ORDER) $F := \{J\}$, $L := \{x_1, \overline{x}_1, \ldots, x_n, \overline{x}_n\}$.
The fitness of a tree X is the number of literals x_i for which the positive literal x_i appears before the negative literal $\overline{x}_i$ in the in-order parse of X.

J (for "join") is the only available function in this problem, and the fitness of a tree is determined by an in-order parse of its leaf nodes; this reduces the importance of the tree structure in the analysis, making the representation somewhat similar to a variable-length list. For example, a tree X with in-order parse $(x_1, \overline{x_4}, x_2, \overline{x_1}, x_3, \overline{x_6})$ has fitness $f(X) = 3$ because x_1, x_2, and x_3 appear before their negations. Any tree that contains all the positive literals and in which each negative literal $\overline{x_i}$ that appears in the tree is preceded by the corresponding positive literal x_i has a fitness of n and is optimal.

ORDER was introduced as a simple problem that reflects the typical eliminative expressions that take place in conditional statements and other logical elements of computer programs, where the presence of an element determines the execution of one program branch rather than another. The overall idea is that the conditional execution path is determined by inspecting whether a literal or its complement appear first in the in-order leaf parse. The task of the GP algorithm is to identify and appropriately position the conditional functions to achieve the correct behavior.

Durrett et al. [9] proved that the (1+1) GP can optimize ORDER in expected time $O(nT_{\max})$, where $T_{\max}$ represents the maximum size the evolved tree reaches throughout the optimization process. The exact result is stated in the following theorem.

Theorem 11.3.2 ([9]). *The expected optimization time of the strictly and nonstrictly elitist cases of the RLS-GP and* $(1+1)$ *GP algorithms on* ORDER *is $O(nT_{\max})$ in the worst case, where n is the number of variables x_i and $T_{\max}$ denotes the maximum tree size at any stage during the execution of the algorithm.*

The proof idea uses standard fitness-based partition arguments. Given that at most k variables are expressed correctly (i.e., the positive literal appears before any instances of the corresponding negative literal in the in-order parse of the GP tree), a lower bound of $p_k = \Omega((n-k)^2/(n\max(T,n)))$ may be achieved on the probability of expressing an additional literal by an insertion operation given that the GP tree contains exactly T leaf nodes. Then, by standard waiting-time arguments, the expected number of iterations required to improve the solution is $1/p_k$, and the expected time until all literals are expressed is $\sum_{k=1}^{n} 1/p_k$.

The runtime bound stated in Theorem 11.3.2 depends on the tree size $T_{\max}$. If, as often happens in GP applications, a bound on the maximum size of the tree is imposed, then this bound is also a bound on $T_{\max}$. However, if no restriction on the maximum tree size is imposed, then bounding the maximum size of the tree is challenging. Nevertheless, if strict selection and local mutations are used, then it can be shown that the tree does not grow too much from its initialized size. The following corollary of Theorem 11.3.2, which states this result precisely, is slightly more general than the one presented in [9].

Corollary 11.3.3. *The expected optimization time of RLS-GP* on* ORDER *is $O(n^2 + nT_{\text{init}})$ if the tree is initialized with T_{init} terminals.*

Proof. RLS-GP* will accept only mutations which improve the fitness of the current solution, and, as there are only $n+1$ possible fitness values, at most n mutations can be accepted by the GP algorithm before the optimum is found.

A single application of HVL-Prime cannot increase the size of the tree by more than one leaf. Thus, $T_{\max} \leq T_{\text{init}} + n$, and applying Theorem 11.3.2 yields the desired runtime bound. □

It is still an open problem to bound $T_{\max}$ for the $(1+1)$ GP, or even for RLS-GP where nonstrict selection is used. It has been conjectured [9] that the same bound as in Corollary 11.3.3 should also hold for the $(1+1)$ GP*. In general, Durrett et al. noted that the acceptance of neutral moves on ORDER causes a "feedback loop that stimulates the growth of the tree" [9], as there is a slight bias towards accepting insertions rather than deletions in the problem, and larger trees create more opportunities for neutral insertions to take place.

A subsequent experimental analysis performed by Urli et al. led those authors to conjecture an $O(T_{\text{init}} + n \log n)$ upper bound on the runtime [53], which would imply, if correct, that the bound given in Corollary 11.3.3 is not tight.

As shown in the following subsection, by using bloat control mechanisms, more precise results have been achieved by exploiting more explicit control of the tree size.

11.3.1.1 Bloat Control

The performance of the $(1+1)$ GP with lexicographic parsimony pressure on ORDER has been considered by Nguyen et al. [38] and Doerr et al. [6]. This mechanism controls bloat by preferring trees of smaller size when ties amongst solutions of equal fitness are broken.

Nguyen et al. used a negative drift theorem to show that as long as the initial tree is not too large ($T_{\text{init}} < 19n$), it does not grow significantly in less than exponential time (i.e., $T_{\max} < 20n$ with high probability). With this bound on $T_{\max}$, it was then proven that the optimum is found in $O(n^2 \log n)$

iterations with high probability, showing that the solution can be improved up to n times via a cycle of shrinking it down to minimal size (containing no redundant copies of any variable) and then expressing a new variable (pessimistically assuming that this insertion also creates a large number of redundant terminals in the tree, requiring another round of shrinking to occur prior to the next insertion). Experimental results led to the conjecture of an $O(T_{\text{init}} + n \log n)$ bound [53].

A more precise analysis proves the bound and its tightness, as given in the following theorem [6].

Theorem 11.3.4 ([6]). *The* (1+1) *GP with lexicographic parsimony pressure on* ORDER *takes* $\Theta(T_{\text{init}} + n \log n)$ *iterations in expectation to construct the minimal optimal solution.*

The lower bound of the theorem is proven by using standard coupon collector and additive drift arguments. For the upper bound, the variable drift theorem [49] is applied using a potential function that takes into account both the number of expressed literals and the size of the tree.

Neumann considered the *Pareto parsimony pressure* approach to bloat control by introducing a multiobjective GP algorithm (SMO-GP), and using both the solution fitness and its size as objectives [37]. This approach was motivated by noting that GP practitioners can, when presented with a variety of solutions, gain insight into how solution complexity trades off against quality.

The SMO-GP algorithm maintains a population of solutions P, representing the current best approximation of the Pareto front. Similarly to the (1+1) GP, the algorithm produces a single offspring individual by applying the HVL-Prime operator k times to a parent individual chosen uniformly at random from P in each iteration. If the offspring is not strictly dominated by any solution already in P, it is added to the population, while any solutions in P that it weakly dominates are removed. Thus, the size of the population P can vary throughout the run. The theoretical analysis considers the number of iterations required to compute a population containing the entire Pareto front.

Theorem 11.3.5 ([37]). *The expected optimization time of SMO-GP, using either* $k = 1$ *or* $k = 1 + \text{Poisson}(1)$, *on* ORDER *is* $O(nT_{\text{init}} + n^2 \log n)$.

The result is proven by showing that it is possible for the GP algorithm to construct the empty tree in expected $O(nT_{\text{init}})$ iterations. Once a minimal solution with k expressed variables exists in the population, the minimal solution with $k+1$ expressed variables can be constructed from it with probability at least $\frac{1}{3e}\frac{1}{n+1}\frac{n-k}{2n}$ in each iteration, and hence an upper bound on the expected runtime may be achieved by using the fitness-based partition method.

Experiments have led to the unproven conjecture that the bound in Theorem 11.3.5 is tight [53].

11.3.2 The MAJORITY Problem

The MAJORITY problem, as originally introduced by Goldberg and O'Reilly [13], is defined as follows.

Problem 11.3.6 (MAJORITY) $F := \{J\}$, $L := \{x_1, \overline{x}_1, \ldots, x_n, \overline{x}_n\}$.
The fitness of a tree X is the number of literals x_i for which the positive literal x_i appears in X at least once, and at least as many times as the corresponding negative literal $\overline{x}_i$.

J (for "join") is the only available function in this problem, and the fitness of a tree is determined by an in-order parse of its leaf nodes; this reduces the importance of the tree structure in the analysis, making the representation somewhat similar to a variable-length list. For example, a tree with an in-order parse of $(\overline{x}_1, x_1, x_2, x_3, \overline{x}_3, \overline{x}_3)$ would have a fitness of 2, as only the literals x_1 and x_2 are expressed (while $\overline{x}_3$ outnumbers x_3 in the tree, and x_3 is therefore suppressed). Any optimal solution, expressing all n positive literals, has a fitness of n.

The fitness of solutions in MAJORITY is based on the number of literals x_i and $\overline{x}_i$ in the tree, with only the literal in greater quantity (the majority) being expressed and potentially contributing to the fitness value. This serves to model problems where solution fitness can be accumulated through additions of more nodes to the tree, regardless of their exact positions.

In contrast to ORDER, where there is always a position in the tree where an unexpressed literal x_i can be inserted to express x_i and improve the fitness of a solution, in MAJORITY there exist trees where no single insertion of an unexpressed x_i will lead to x_i being expressed and thus improving the fitness, even though all literals x_i can contribute to expressing x_i in aggregate regardless of their position. Thus, GP variants which do not accept neutral moves have been found to perform quite badly, with RLS-GP* shown to be capable of getting stuck in easily constructed local optima, and (1+1) GP* having an exponential expected optimization time to recover from a worst-case initialization [9]. On the other hand, GP variants using nonstrict selection may be efficient.

Theorem 11.3.7 ([9]). *Let $T_{\max}$ denote the maximum tree size at any stage during the execution of the algorithm. Then the expected optimization time of RLS-GP on* MAJORITY *is*

$$O(n \log n + D T_{\max} n \log \log n)$$

in the worst case, where $D := \max(0, \max_i(c(\overline{x}_i) - c(x_i)))$ and $c(x)$ is the number of times the literal x appears in the initial tree.

If the algorithm is initialized with a random tree containing $2n$ terminals selected uniformly at random from L, the expected optimization time of RLS-GP on MAJORITY *is $O(n^2 T_{\max} \log \log n)$.*

The bounds presented depend on D, the maximum deficit between the numbers of positive literals and negative literals of any variable in the tree (thus, a tree with a single copy of x_1 and two copies of $\overline{x}_1$ would have a deficit $D=1$). The worst-case result, assuming a deficit of D literals for all n variables, follows from a generalized variant of the coupon collector problem [36], requiring the collection of D copies of each coupon. For a uniform initialization with $T_{\text{init}}=2n$, a bound $D=O(\log n/\log\log n)$ was derived using the balls-into-bins model [32]. It was then proven that a variable which initially has a deficit of D becomes expressed after an expected $O(DT_{\max})$ mutations involving that variable (which occur with probability $\Theta(1/n)$) by showing that the GP system essentially performs a random walk that is at least fair with respect to decreasing the deficit.

For the $(1+1)$ GP, only a hypothetical worst-case analysis for the elitist variant was presented in [9], noting that if the last unexpressed variable has k more negative literals than positive literals in the tree, the final mutation will require at least $\Omega(n^{k/2})$ time, and thus, unless k can be shown to be constant, the expected runtime remains superpolynomial. However, no bounds on the probability that a superconstant k would actually occur were given.

The problem, including the dependence on $T_{\max}$ was recently solved, proving the following upper and lower bounds on the expected optimization time [6].

Theorem 11.3.8 ([6]). *When the algorithm is initialized with a tree containing T_{init} terminals, the expected optimization time of the RLS-GP and $(1+1)$ GP algorithms on* MAJORITY *is at least $\Omega(T_{\text{init}}+n\log n)$ and at most $O(T_{\text{init}}\log T_{\text{init}}+n\log^3 n)$.*

The lower bound is proven by an application of the multiplicative drift theorem with bounded step size, while the upper bound relies on showing that if $T_{\text{init}}\geq n\log^2 n$, the tree will grow by at most a constant factor in $O(T_{\text{init}}\log T_{\text{init}})$ generations before the optimal solution is constructed. As a result, bloat does not hinder the optimization process, i.e., the final tree may be at most larger by a multiplicative polylogarithmic factor than the optimal solution size.

From the analysis, an interesting alternative to bloat control emerges. If the HVL mutation probabilities were changed such that deletions were more likely than insertions, a drift towards smaller solutions would be observed, leading to smaller trees, and hence faster optimization. Such a suggestion was originally made by Durrett et al., albeit for the ORDER problem [9]. Concerning MAJORITY, theoretical evidence in support of this has emerged, though no formal proof is available [6].

11.3.2.1 Bloat Control

Applying lexicographic parsimony pressure mitigates the analysis problems that arise with GP systems for MAJORITY. With this bloat control mechanism, mutations which solely remove negated terminals are always accepted, as they reduce the size of the tree. Accepting such mutations eventually leads the GP system to a solution where fitness can be improved by inserting a positive literal, allowing the optimum to be reached efficiently.

Theorem 11.3.9 ([37]). *The expected optimization time of RLS-GP with lexicographic parsimony pressure on* MAJORITY, *when initialized with a tree containing T_{init} literals, is $O(T_{\text{init}} + n \log n)$.*

The result is proven by reasoning that it takes $O(T_{\text{init}})$ iterations to remove the T_{init} negated terminals provided by a worst-case initialization, and $O(n \log n)$ iterations to express all n variables by an application of the coupon collector argument.

A tight bound for the $(1+1)$ GP, showing that the larger Poisson mutations do not affect the asymptotic runtime, has recently been proven [6], confirming a previous conjecture [53].

Theorem 11.3.10 ([6]). *The expected optimization time of the* $(1+1)$ *GP with lexicographic parsimony pressure on* MAJORITY, *when initialized with a tree containing T_{init} literals, is $\Theta(T_{\text{init}} + n \log n)$.*

The lower bound of the theorem is proven by using standard coupon collector and additive drift arguments. For the upper bound, the variable drift theorem [49] is applied using a potential function that takes into account both the number of expressed literals and the size of the tree. Intuitively, the size of the tree is only allowed to increase if the MAJORITY fitness is also increased, which can only occur a limited number of times, and the magnitude of the increase is unlikely to be overly large owing to the Poisson distribution used to determine k.

It is still an open problem to prove that lexicographic parsimony pressure asymptotically improves the runtime of the $(1+1)$ GP or that the upper bound given in Theorem 11.3.8 is not tight (Urli et al. conjectured an upper bound of $O(T_{\text{init}} + n \log n)$ without bloat control, based on experimental data [53]).

Applying Pareto parsimony pressure and treating the size of the tree as an additional objective in the multiobjective SMO-GP algorithm allows the GP system to compute the Pareto front of solutions in terms of fitness/complexity.

Theorem 11.3.11 ([37]). *The expected optimization time of SMO-GP (with either $k = 1$ or $k = 1 + \text{Poisson}(1)$) on* MAJORITY, *initialized with a single tree containing T_{init} terminals, is $O(nT_{\text{init}} + n^2 \log n)$.*

The SMO-GP population will contain at most $n+1$ individuals, as there are only $n+1$ distinct fitness values for MAJORITY. Similarly to the situation for lexicographic parsimony pressure, SMO-GP is able to construct an initial solution on the Pareto front by repeatedly removing any duplicate or negated terminals from the initial solution. Once a solution on the Pareto front exists, the entire front can be constructed by repeatedly selecting a solution at the edge of the front and expressing an additional variable or deleting an expressed variable.

11.3.2.2 More Complex MAJORITY Variants

Given that the MAJORITY problem can be efficiently optimized by simple GP systems without bloat appearing as a problem, more sophisticated versions of the problem have been designed [17].

In the $+c$-MAJORITY problem, x_i is expressed if and only if the number of x_i literals in the tree exceeds the number of $\overline{x}_i$ literals by at least c. It has been proven that the RLS-GP is with high probability not able to find the optimal solution when $c > 1$ and lexicographic parsimony pressure is employed, but is able to do so in expected polynomial time when no bloat control mechanism is used. In this problem, the impact of bloat is limited, as the insertions of x_i and $\overline{x}_i$ are accepted with equal probability when x_i is not expressed, and the necessary margin to express x_i can be reached as a consequence of a fair random walk. On the other hand, lexicographic parsimony pressure prevents this random walk from taking place, as only mutations which increase the number of expressed variables or reduce the size of the tree would be accepted. Thus, RLS-GP with lexicographic parsimony pressure cannot express x_i unless at least $c-1$ copies of x_i are already present in the initial solution.

The opposite holds for the 2/3-SUPERMAJORITY problem, which provides a fitness reward of $2 - 2^{c(\overline{x}_i)-c(x_i)}$ for every variable x_i for which $c(x_i) > 2c(\overline{x_i})$, where $c(z)$ denotes the number of times the literal z appears in the tree. In particular, the RLS-GP without bloat control is with high probability not able to express all n variables, and thus cannot find solutions with fitness above a certain threshold.

Theorem 11.3.12 ([17]). *For any constant $\nu > 0$, consider the RLS-GP without bloat control on 2/3-SUPERMAJORITY on the initial tree with size $s_{\text{init}} = \nu n$. There is $\varepsilon = \varepsilon(\nu) > 0$ such that, with probability $1 - o(1)$, an ε-fraction of the variables will never be expressed. In particular, the algorithm will never reach a fitness larger than $(2 - 2\varepsilon)n$.*

The proof idea relies on showing that the size of the current solution increases over time (due to the fitness rewards for inserting additional copies of positive literals for expressed variables), which makes insertions of non-expressed variables more likely to occur than their deletions. This makes reaching the 2/3-majority threshold to express a variable difficult, requiring a significant deviation from the expected outcome of a fair random process. Lexicographic parsimony pressure, when employed, sidesteps this problem by gradually removing literals of non-expressed variables from the tree, and eventually allowing x_i to be expressed by a single insertion of its positive literal.

Kötzing et al. additionally proved that a memetic GP algorithm with a simple concatenation crossover mechanism and local search to remove redundant literals is able to efficiently solve both the $+c$-MAJORITY and 2/3-SUPERMAJORITY problems [17] if lexicographic parsimony pressure is employed. Hence they provide an example where incorporating a population and applying crossover allows a wider range of problems to be solved.

11.3.3 The SORTING Problem

The SORTING problem is the first classical combinatorial optimization problem for which computational complexity results have been obtained for discrete evolutionary algorithms. For the application of evolutionary algorithms Scharnow et al. defined SORTING as the problem of maximizing different measures of sortedness of a permutation of a totally ordered set of elements [50].

Wagner et al. analyzed the performance of GP for the problem, aiming to investigate the differences between different bloat control mechanisms for GP [56, 57]. For GP, the measures of sortedness were adapted to deal with incomplete permutations of the literal set.

Problem 11.3.13 (SORTING) $F := \{J\}, L := \{1,2,\dots,n\}$.

The fitness of a tree X is computed by deriving a sequence π of symbols based on their first appearance in the in-order parse of X, and considering one of the following five measures of sortedness of this sequence.

- INV(π) *Number of pairs of adjacent elements in the correct order (maximize to sort), with* $\mathrm{INV}(\pi) = 0.5$ *if* $|\pi| = 1$.
- HAM(π) *Number of elements in correct position (maximize to sort).*
- RUN(π) *Number of maximal sorted blocks (minimize to sort), plus the number of missing elements* $n - |\pi|$, *with* $\mathrm{RUN}(\pi) = n+1$ *if* $|\pi| = 0$.
- LAS(π) *Length of longest ascending sequence (maximize to sort).*
- EXC(π) *Smallest number of exchanges needed to sort the sequence (minimize to sort), plus* $1+n-|\pi|$ *if* $|\pi| < n$.

J (for "join") is the only available function in this problem, and the fitness of a tree is determined by an in-order parse of its leaf nodes drawn from a totally ordered set of terminals L. This reduces the importance of the tree structure in the analysis, making the representation somewhat similar to a variable-length list. Thus, for $n=5$, the fitness of a tree with an in-order parse of $(1,2,1,4,5,4,3)$, and hence $\pi=(1,2,4,5,3)$ is $\text{INV}(\pi)=3$, $\text{HAM}(\pi)=2$, $\text{RUN}(\pi)=2$, $\text{LAS}(\pi)=4$, and $\text{EXC}(\pi)=2$. The fitness value of optimal trees for the INV, HAM, and LAS measures is n, while for the RUN and EXC measures it is 0.

Unlike the ORDER and MAJORITY problems considered in the previous sections, the SORTING problem is not separable, meaning that it cannot be split into subproblems that could be solved independently. The dependencies between the subproblems can thus significantly impact the overall time needed to solve the optimization problem, and the variable-length representation of solutions can create local optima from which it is difficult for GP systems to escape. Wagner et al. additionally remarked that the task of evolving a solution is more difficult for the RLS-GP and (1+1) GP systems considered than for the permutation-based EA, which in expectation requires $O(n^2 \log n)$ iterations for the INV, HAM, LAS, or EXC sortedness measure, and exponential time when using the RUN sortedness measure [50].

Theorem 11.3.14 ([57]). *The expected optimization time for the RLS-GP* and (1+1) GP* algorithms on* SORTING *using* INV *as the sortedness measure is $O(n^3 T_{\max})$, where n is the number of elements to be sorted, and $T_{\max}$ is the maximum size of the tree during the run of the algorithm.*

For the HAM, RUN, LAS, *and* EXC *measures, there exist initial solutions with $O(n)$ terminals such that the expected optimization time of RLS-GP* is infinite, and the expected optimization time of (1+1) GP* is $e^{\Omega(n)}$.*

The positive statement is proven by applying the artificial fitness level method, observing that there are $n \cdot (n-1)/2+1$ possible fitness values, and with probability $\Omega(1/(nT_{\max}))$ a mutation inserts a literal which corrects at least one unsorted pair without introducing any additional unsorted pairs.

For the HAM, RUN, LAS, and EXC measures, trees which require large mutations to improve fitness exist, which causes the expected optimization time to be infinite for RLS-GP* and $e^{\Omega(n)}$ for the (1+1) GP*. In general, the problematic solutions contain a large number of copies of a single literal in an incorrect location and a large sorted sequence, requiring either all the incorrectly placed copies to be removed simultaneously or the sorted sequence to be moved in a single mutation.

11.3.3.1 Bloat Control

When bloat control mechanisms are applied, GP systems may reduce the size of the redundant components of the solution even if mutations which make progress in this direction do not alter the solution's sortedness measure.

The impact of applying lexicographic parsimony pressure for the (1+1) GP family of algorithms and of Pareto parsimony pressure for the

Table 11.1 Known expected runtimes for GP algorithms on SORTING using various sortedness measures and bloat control mechanisms.

$F(X)$	No bloat control		Parsimony pressure	
	RLS-GP*	(1+1) GP*	RLS-GP	SMO-GP
INV	$O(n^3 T_{\max})^a$	$O(n^3 T_{\max})^a$	$O(T_{\text{init}} + n^5)^a$	$O(n^2 T_{\text{init}} + n^5)^a$
LAS	∞^a	$\Omega\left(\left(\frac{n}{e}\right)^n\right)^a$	$O(T_{\text{init}} + n^2 \log n)^{a,b}$	$O(nT_{\text{init}} + n^3 \log n)^a$
HAM	∞^a	$\Omega\left(\left(\frac{n}{e}\right)^n\right)^a$	∞^c	$O(nT_{\text{init}} + n^4)^c$
EXC	∞^a	$\Omega\left(\left(\frac{n}{e}\right)^n\right)^a$	∞^c	$O(n^2 T_{\text{init}} + n^3 \log n)^c$
RUN	∞^a	$\Omega\left(\left(\frac{n}{e}\right)^n\right)^a$	∞^c	$O(n^2 T_{\text{init}} + n^3 \log n)^c$

[a] Shown in [57].
[b] Also holds with probability $1 - o(1)$ for the (1+1) GP.
[c] Shown in [56].

SMO-GP algorithms has been considered [56, 57]. We summarize the results in Table 11.1.

In general, the positive results are proven by showing that there exists a sequence of fitness-improving mutations leading the GP system to the global optimum (in the case of (1+1) GP algorithms), or, for SMO-GP, to a solution on the Pareto front from which other Pareto front solutions can be constructed efficiently.

The majority of the negative results rely on showing the existence of local optima for the sortedness measure, which limits the availability of results for nonstrictly elitist algorithms, and especially for the (1+1) GP, which is capable of performing larger mutations.

The results in Table 11.1 suggest that the variable-length representation can cause difficulties for RLS-GP even when parsimony pressure is applied, for some simple measures of sortedness, while even a simple multiobjective algorithm is able to find the entire Pareto front of the problem efficiently when using any of the five measures considered.

Experimental results have been presented that suggest that the (1+1) GP algorithm is efficient (i.e., able to find the optimum in polynomial time) using all of the sortedness measures considered except RUN, both with and without bloat control mechanisms: concerning the average-case complexity, an $O(n^2 \log n)$ bound has been conjectured for the INV and LAS measures, and an $O(n^4)$ bound for the EXC and HAM measures [57]. Providing a rigorous theoretical analysis of the behavior of these GP systems remains an open question.

11.3.4 Outlook

In this section, we have provided an overview of the computational complexity results for simple GP systems for toy problems where the evolved GP trees may grow to arbitrarily large sizes. The main aim behind the analyses is to shed light on how bloat affects the optimization process of GP. Surprisingly, bloat does not hinder the efficient optimization of the $(1+1)$ GP for any of the basic problems. Theorem 11.3.8 provides a rigorous proof of this for MAJORITY, while experimental work has lead to similar conjectures for ORDER and SORTING, although formal proofs are not yet available.

Recently, a toy problem has been designed where the RLS-GP provably requires exponential time with overwhelming probability due to bloat. To achieve this result, the design of 2/3-SUPERMAJORITY closely follows the definition of "bloat". Indeed, fitness increases slightly with the increase of the tree size, making it less and less likely that significantly beneficial mutations occur. Nevertheless, simple bloat control mechanisms, such as lexicographic parsimony pressure, effectively address the issue. Thus they allow the RLS-GP to efficiently optimize 2/3-SUPERMAJORITY. Overall, there is still a need to design benchmark functions that reflect the reported behavior of GP in practice, i.e., problems where bloat occurs and are difficult to solve with the use of bloat control techniques.

11.4 Evolving Programs of Fixed Size

In this section, we consider two more advanced applications compared with those in the previous section. For both problems, the fitness of an evolved program is computed by evaluating its output. While more realistic, these problems are still different from real-world GP applications. In the first problem, MAX, the program to be evolved has no input variables, and thus the GP system has to construct a program which always outputs the same constant value, subject to constraints on problem size and available operators. Concerning the second problem, IDENTIFICATION, the structure of the optimal solution is fixed (i.e., no tree structure has to be evolved), and the GP system is not allowed to deviate from it, but must instead learn the exact weights of a predefined linear function while evaluating program quality by comparing the program output with the target function on only a limited number of the possible function inputs.

The first toy problem, MAX, may reflect practical GP applications where bloat is avoided by setting a maximum limit on the size or height of the evolved trees. When such a limit is reached, large tree modifications may be required to make further progress. Such a problem occurs, for example, for GP evolving Boolean conjunctions with a function set comprising of AND and OR (see Theorem11.5.10 in Section 11.5.1.3). The second problem, IDENTIFICATION, models the issue that the true fitness of candidate solutions in GP is usually unknown, and their quality has to be estimated using a training set.

11.4.1 The MAX Problem

The MAX problem was originally introduced by Gathercole and Ross as a means of analyzing the limitations of crossover when applied to trees of fixed size [12]. The fitness of the program depends on the evaluation of the arithmetic expression represented by the tree. However, the problem contains no variable inputs, and thus the goal of the GP algorithm is simply to construct a tree that evaluates to the maximum possible value subject to restrictions on the size of the tree, and on the available functions and terminals.

Problem 11.4.1 (MAX) $F := \{+, \times\}$, $L := \{t\}$, $t > 0$ *a positive constant, and maximum tree depth* D.

The fitness of a tree X *is the value produced by evaluating the arithmetic expression represented by the tree if the tree is of depth at most* D*, and 0 if the tree is of larger depth.*

The optimal solution to MAX is a complete binary tree of depth D, with t at all the leaf nodes, and with the lowest $\lfloor 1/2 + 1/t \rfloor$ levels of internal (i.e., nonleaf) nodes containing $+$ and the remaining internal nodes containing $\times$. It has been noted that lower values of $t < 1$ make the problem more difficult for crossover-based GP systems [12].

The behavior of GP systems on the MAX problem was previously studied experimentally, with Langdon and Poli observing that MAX is hard for GP systems utilizing crossover owing to the interaction of deception with the depth bound on the tree making it difficult to evolve solutions. The GP system is essentially forced to perform randomized hill climbing in the later stages of the optimization process, and hence requires exponential time with respect to the maximum allowed depth of the tree [23].

A theoretical analysis of the (1+1) GP for the MAX problem was presented by Kötzing et al. [19], who proved that the runtime of the mutation-only algorithm is polynomial with respect to $n = 2^{D+1} - 1$, the maximum allowed number of nodes in the tree.

Theorem 11.4.2 ([19]). *The RLS-GP algorithm finds the optimal solution for the MAX problem for any choice of* $t > 0$*, in expected* $O(n \log n)$ *iterations, where* n *is the maximum allowed number of nodes in a tree subject to the depth limit* D*.*

The theorem is proven by showing that the GP algorithm can first construct a complete binary tree with depth D in a way that prevents any node from being deleted, and then use the substitution suboperation of HVL-Prime to correct internal nodes.

Concerning the (1+1) GP, a weaker bound on the expected runtime was proven.

Theorem 11.4.3 ([19]). *The expected time for the* (1+1) *GP to find the optimal solution for the MAX problem with* $t = 1$ *is* $O(n^2)$.

The theorem is proven using fitness-based partitions, exploiting the existence of at least one leaf in a tree of size n which could be selected by insertion to grow the tree. Experimental results suggesting that the true runtime of the (1+1) GP on MAX is also $O(n \log n)$ were also presented, and the authors of [19] noted that a more precise potential function based on the contents of the tree would be required to show this upper bound using drift analysis.

Additionally, a modification of the insertion operation in HVL-Prime to grow the tree in a more balanced fashion was considered: rather than selecting a location to insert a new leaf node uniformly at random from the entire tree, selection would pick a leaf at depth d with probability 2^{-d} to be replaced with a new function node, using the original leaf and an inserted terminal as its children. As well as balancing the growth of the tree between different branches, this reduces the probability that mutation attempts insertion operations which would be blocked by the tree depth limit. With this modified insertion operator, an $O(n \log n)$ bound on the expected runtime of the (1+1) GP on MAX with $F = \{+\}$ was proven [19].

Closing the gap between the $O(n^2)$ upper bound for the (1+1) GP on MAX with $F = \{+, \times\}$ and the $\Omega(n \log n)$ lower bound given by a coupon collector argument remains an open problem. Furthermore, theoretical time complexity analyses of the performance of crossover-based GP systems, for which the MAX problem was originally introduced, are still unavailable.

11.4.2 The Identification Problem and PAC Learning

It is generally not possible to evaluate the quality of the evolved programs on *all* possible inputs efficiently, as they usually are too numerous when the number or the domain of input variables is too large. The IDENTIFICATION problem was introduced by Kötzing et al. [18] to evaluate the learning capabilities of a simple evolutionary algorithm, an EA with a local mutation operator that evaluates program quality by considering only a polynomial number of inputs chosen uniformly at random in each iteration. This setting is the same as that of the PAC learning framework [54]. The idea is that while some problems cannot always be solved exactly (as there might be no known polynomial-time algorithm that produces an exact solution, as, e.g., for NP-hard problems), a good approximation, i.e., one that is correct on a random input with high probability, may be achieved. A large class of functions has been shown to be PAC learnable by designing appropriate evolutionary algorithms [10, 55]. Compared with those studies, Kötzing et al. considered a simplified setting [18]. Unlike the problems previously considered, the structure of the desired solution is known in advance by the algorithm, which has to identify the target function among a known class of linear functions. More precisely, the IDENTIFICATION problem is to learn the weights of a linear function f_{OPT} defined over bit strings $x \in \{0,1\}^n$,

$$f_{\mathrm{OPT}}(x) = \sum_{i=1}^{n} w_i x_i,$$

where $w_i \in \{-1, 1\}$.

The goal of the EA (called the Linear GP algorithm) is to identify whether each weight w_i is positive or negative. The algorithm changes a single weight w_i in each iteration, and determines whether the mutated offspring has better fitness than its parent using a multiset S constructed independently in each iteration by selecting the desired number of points uniformly at random (with replacement) from $\{0,1\}^n$. The error e_S of each solution f is computed as

$$e_S(f, f_{\mathrm{OPT}}) = \sum_{x \in S} |f(x) - f_{\mathrm{OPT}}(x)|,$$

and solutions with lower error are preferred.

Thus, the focus of the analysis is to measure the ability of the GP system to extract information from a limited view of the true fitness function: if S is too small, the sampled error function may be an unreliable indication of the true quality of the solution. On the other hand, if S is too large, more computational effort than necessary is expended for each fitness evaluation, which could result in worse performance with respect to the overall CPU time spent.

The following theorem shows that the Linear GP algorithm is able to learn f_{OPT} efficiently if the number of inputs sampled in each iteration is sufficiently large.

Theorem 11.4.4 ([18]). *If $|S| \geq c_0 n \log n$, c_0 a large enough constant, the expected number of generations until the best-so-far function found by Linear GP has an expected error $\leq \delta$ is $O(n \log n + n^2/\delta^2)$.*

If f_{OPT} also has a linear number of both 1 and -1 weights, the expected number of generations until such a solution is found is $O(n + n^2/\delta^2)$.

In this setting, $e_S \leq 1$ implies that an optimal solution has been found, and thus the theorem additionally provides an $O(n^2)$ bound on the expected number of generations required to learn f_{OPT} perfectly (by setting $\delta = 1$). The theorem is proven by showing that in $O(n \log n)$ generations, the numbers c_1 and c_{-1} of incorrect weights in f set to 1 and -1, respectively, become balanced (such that there is at most one more incorrect weight of one kind than the other) with high probability, and remain balanced throughout the rest of the process. When $c_1 = c_{-1}$, mutations that increase either value are rejected with high probability, while mutations that reduce either value are accepted with high probability (but can be undone by the GP system until a wrong weight of the opposite kind is corrected). Thus, c_1 and c_{-1} can be reduced permanently by performing the two reductions in sequence (which occurs with probability at least $(i/n)^2$ if, initially, $c_1 = c_{-1} = i$), and, by a coupon collector-like argument, the number of incorrect weights is reduced to an acceptable level in expectation after $O(n^2/\delta^2)$ generations.

Extending the analysis to broader function classes and algorithms, for example considering functions with more than two options for each coefficient, or a $(1+1)$ GP-like mutation operator capable of performing more than one change in each iteration, remains an open direction for further research. The PAC learning framework will also be used to analyze the performance of the $(1+1)$ GP family of algorithms on Boolean functions in the next section.

11.4.3 Outlook

The MAX problem is easy for mutation-based GP systems. Yet, the achievement of precise asymptotic bounds on their runtime is still prohibitive. On the other hand, the crossover-based GP algorithms used in practice do not achieve a significant benefit from crossover on MAX [12]. How this could be rigorously proven remains an open problem.

Small super-linear polynomial size training sets suffice to efficiently estimate the true fitness of candidate solutions for linear functions with {1,-1} weights. This allows the exact identification of the target function of the IDENTIFICATION problem. Generalization of this result to larger weight sets and function classes would support future analyses of realistic symbolic regression applications.

11.5 Evolving Proper Programs: Boolean Functions

In real-world applications of GP systems, the goal is to evolve a program with specific behavior. In most applications, the program accepts some inputs and produces one or more output values, and the quality of candidate programs is evaluated by executing them on a variety of possible inputs for which the correct output is known. The structure of the target program is typically not known in advance, and thus the GP systems may be given access to more components (both functions and terminals) than is strictly necessary to represent an optimal solution. Real-world applications of GP can exhibit all the challenges that the previously discussed problems modeled in isolation: the length and structure of the target program are not known to the GP system, there may be a variety of function and terminal nodes, and solution quality is evaluated by executing the program on some or all of the possible inputs.

Boolean functions, which take a number of binary inputs and produce a single binary output, have long been used as benchmarks in the field of GP [20, 23] and are a natural next step for the complexity analysis of GP systems, as they can combine all of these challenges. The problems of evolving some Boolean functions, such as conjunctions (AND) or parity (XOR), are also well understood in the PAC learning framework [55] – conjunctions are evolvable efficiently, while parity problems are not. Additionally, such problems form an interesting sanity check for the (1+1) GP algorithms: if the simple algorithms are not able to evolve relatively simple functions, it would be interesting to determine which components of the more complex GP algorithms enable these problems to be solved efficiently, i.e., to identify how much sophistication is required in the GP system for it to be efficient.

A complexity analysis of (1+1) GP algorithms for the AND and XOR problems, where the goal is to construct a conjunction or an even parity function, has recently been presented [29]. For these problems, the fitness of the evolved solutions was evaluated by comparing their output with that of the target function on either the entire truth table or a polynomial training subset.

Using the complete truth table (i.e., all possible inputs) as the training set is typically only feasible for Boolean functions if the size of the problem, in terms of the number of input variables, is relatively small (as there are 2^n possible inputs for n Boolean input variables, and evaluating each candidate program on an exponential number of inputs would require exponential time). However, benchmark problems with small n have been considered for GP systems, and may still occur in some settings. Additionally, a confirmation of whether a given GP system can evolve a given function given an exact fitness function (i.e., the complete truth table) is also useful for further analysis: if it cannot, it is likely that mechanisms more complex than random sampling of inputs would be required to evolve the function in polynomial time.

If an incomplete training set is used, the GP system may either choose it once at the beginning of the run (the static incomplete training set case, as considered in [29]), or choose a fresh subset dynamically in every iteration (as in [25]). Both approaches may be valid in different practical settings. If the complete truth table is known but is prohibitively large, it may be sampled to estimate the fitness of a solution, reducing the computational effort required to evaluate the quality of a program at the cost of introducing some uncertainty. On the other hand, if only a limited number of input/output examples are available, some may need to be reserved to validate the quality of the solution on inputs that it has not been trained on.

11.5.1 Evolving Conjunctions

For the AND problem, the target function that the GP system has to evolve is a conjunction of some number of variables. Conjunctions have an easy to understand input-to-output mapping simplifying the analysis, and are known to be efficiently evolvable [55]. However, unlike tailored learning algorithms, the GP systems do not necessarily know that the target function is a conjunction – and ideally, should be able to evolve conjunctions even with access to a variety of functions and terminals.

Problem 11.5.1 (AND) *Let $L \subseteq \{x_1,\ldots,x_n\}$ be the set of available terminals, and F be the set of available functions.*

The fitness of a tree X using a training set T selected from the rows of the complete truth table C is the number of training set rows on which the value produced by evaluating the Boolean expression represented by the tree differs from the output of the target function: the conjunction of all (or some) of the n inputs. This fitness value should be minimized; the optimal solution has a fitness of 0.

AND_n is used to refer to the variant of this problem where the target function is a conjunction of all n input variables, while the target of $AND_{n,m}$ is composed of an unknown subset of $m \leq n$ variables.

For example, when the complete truth table is used as the training set T, the fitness of a tree containing only a single leaf x_1 for the AND_n problem with $n=3$ is 3, while the fitness of the optimum is 0 (the fitness function represents the *error* of the solution on the training set). In general, a conjunction of a distinct variables has a fitness of $2^{n-a}-1$ on the complete truth table.

The initial complexity analysis results for this problem consider the minimal function set (i.e., $F = \{AND\}$) to simplify the analysis by forcing all solutions considered by the GP algorithms to be conjunctions. This simplification renders the fitness function unimodal, making the AND_n problem somewhat similar to the OneMax benchmark problem for evolutionary algorithms: the GP system simply has to collect all n distinct variables together in its solution, with the fitness of the current solution improving with each distinct variable that is added. In this minimal setting, initializing with larger trees makes the problem easier for the GP system, as fewer variables would need to be inserted into the tree to complete the conjunction. Thus, for complexity analysis results, the initial solution is typically an empty tree.

Building upon these results, the impact of using richer function (e.g., by introducing disjunctions [7] and negations) and terminal sets (via the $\text{AND}_{n,m}$ problem) has been also been analyzed.

11.5.1.1 Complete Truth Table, Minimal Terminal and Function Sets

Mambrini and Oliveto showed that the RLS-GP and RLS-GP* algorithms can efficiently construct the optimal solution for the AND_n problem when they use the complete truth table to evaluate solution fitness [29].

Theorem 11.5.2 ([29]). *The expected optimization time of RLS-GP and RLS-GP* with $F = \{AND\}$ and $L := \{x_1,\ldots,x_n\}$ on the AND_n problem* *using the complete truth table as the training set is $\Theta(n \log n)$. The solution produced by RLS-GP* contains exactly n terminals.*

The proof applies a coupon collector argument, showing that with probability $(n-i)/(3n)$ a new variable is added to the solution, and that no mutations decreasing the number of distinct variables are ever accepted. As all internal nodes are forced to be conjunctions, collecting all n variables in the tree produces an optimal solution.

The following theorem presents a fixed budget analysis of the RLS-GP and RLS-GP* algorithms, providing a relationship between the expected number of distinct variables in the solution and the time the algorithms are allowed to run.

Theorem 11.5.3 ([25]). *Let $v(x)$ denote the number of distinct variables in solution x, and let x_b^* or x_b be the solution produced by the RLS-GP* or RLS-GP algorithms, respectively, with $F = \{AND\}$ and $L := \{x_1, \ldots, x_n\}$, given a budget of b iterations on the AND_n problem using the complete truth table as the training set when initialized with an empty tree. Then,*

$$E(v(x_b^*)) = n - n(1 - 1/(3n))^b,$$

$$n - n(1 - 1/(3n))^b \leq E(v(x_b)) \leq n - n(1 - 2/(3n))^b.$$

The theorem is proven by following the techniques used to analyze Randomized Local Search (RLS) on the ONEMAX problem in [16]. The exact expectation is known for RLS-GP*, which never accepts solutions that do not improve fitness, and hence can never have a substitution suboperation increase the number of distinct variables in the solution. The upper and lower bounds on $E(v(x_b))$ for RLS-GP stem from trivial bounds on the probability of a substitution suboperation of HVL-Prime increasing the number of distinct variables in the solution. We note that although the relationship $f(x) = 2^{n-v(x)} - 1$ between the solution fitness ($f(x)$) and the number of distinct variables it contains ($v(x)$) is known, it is not possible to apply linearity of expectation to transform a bound on $E(v(x_b))$ into a bound on $E(f(x_b))$ (as could be done for ONEMAX).

The runtime analysis results have been extended to cover the $(1+1)$ GP algorithms, and show that the expected number of terminals in the constructed solution is $\Theta(n)$.

Theorem 11.5.4 ([25]). *The expected optimization time of the $(1+1)$ GP and the $(1+1)$ GP* with $F = \{AND\}$ and $L := \{x_1, \ldots, x_n\}$ on the AND_n problem using the complete truth table as the training set is $\Theta(n \log n)$. In expectation, the solution produced by these algorithms contains $\Theta(n)$ terminals.*

For the AND problem, there are many possible trees which encode the desired behavior (because repeating a variable multiple times in the conjunction does not negatively affect the behavior of the program) and it is therefore possible that a "correct" program could contain many redundant leaf nodes. The space complexity result in Theorem 11.5.4 shows that the considered GP systems construct a tree that in expectation contains just $O(n)$ leaf nodes. This is proven by showing that the number of leaf nodes that contain variables present in the solution multiple times does not grow fast enough to affect the asymptotic tree size bound in the $O(n \log n)$ iterations required to collect all n variables with high probability.

11.5.1.2 Incomplete Training Sets, Minimal Terminal and Function Sets

In practice, it may not be possible to evaluate the exact fitness of a candidate solution on all 2^n possible Boolean inputs when n is large. If this is the case, solution quality could instead be evaluated by executing the program on a sampled subset of possible inputs (the "training set"). Without assuming any specific knowledge of the target function class, the training set could be sampled uniformly at random.

When training sets of polynomial size sampled uniformly at random are used for the AND_n problem, a solution representing a conjunction of a logarithmic number of distinct variables will with high probability be correct on all of the inputs included in the training set. This causes the optimization process to end prior to finding a solution that is correct on all possible inputs [29]. The following result holds both when the training set is sampled once and for all at the beginning of the run (i.e., a static training set) and when at each generation a new training set is sampled (i.e., a dynamic training set).

Theorem 11.5.5 ([25, 29]). *Let $s = \text{poly}(n)$ be the size of a training set chosen from the truth table uniformly at random with replacement. With $F = \{AND\}$ and $L := \{x_1, \ldots, x_n\}$, both RLS-GP and RLS-GP* will fit the training set on the AND_n problem in expected time $O(\log s) = O(\log n)$, and the solution will contain at most $O(\log n)$ variables.*

This result is proven by observing that rows selected uniformly at random from the truth table are unlikely to assign more than $Y = n/2 + \varepsilon n$ input variables to true, and hence can be satisfied by inserting any one of a linear number of variables into the solution. After $\log_{n/Y}(2s)$ successful insertions, the probability that some row of the s-row training set is still not satisfied is at most $n/2$, and hence in expectation the process satisfies all rows after $2k = O(\log n)$ distinct variables have been successfully inserted into the tree.

Theorem 11.5.5 also yields a lower bound on the generalization error of the solution: if it contains at most $O(\log n)$ variables, the probability that its output is wrong on a truth table row sampled uniformly at random is $2^{-O(\log n)} = n^{-O(1)}$, i.e., it requires in expectation a polynomial number of samples taken uniformly at random from C before a divergence from the target function is discovered.

Theorem 11.5.5 has been extended to cover the (1+1) GP and (1+1) GP* algorithms, using a multiplicative drift theorem to provide a runtime bound on the expected time to fit a static polynomial-sized training set [25]. Additionally, a similar bound holds if, instead of a static training set, each iteration samples s independent rows of the complete truth table to compare the fitness of two solutions (using a dynamic training set).

Theorem 11.5.6 ([25]). *Let $s = n^{2c+\varepsilon}$ rows from the complete truth table of the AND_n problem be sampled with replacement and uniformly at random in each iteration (where $c > 0$ and $\varepsilon > 0$ are any constants). With $F = \{AND\}$ and $L := \{x_1, \ldots, x_n\}$, RLS-GP, RLS-GP*, (1+1) GP, and (1+1) GP* will construct a solution with a generalization error of at most n^{-c} in expected $O(\log n)$ iterations. In expected $O(\log^2 n)$ iterations, the nonstrictly elitist algorithms will construct a solution with a sampled error of 0.*

Here, the training set size s is chosen to be sufficiently large to ensure that solutions with a generalization error greater than n^{-c} are wrong on at least one training set row with high probability, preventing the GP system from terminating early with a bad solution, while the $O(\log^2 n)$ runtime bound stems from a random walk argument pessimistically considering the probabilities of accepting solutions that increase or decrease the number of distinct variables in the tree to be equal.

11.5.1.3 More Expressive Function and Terminal Sets

In practical applications of GP, it may not be known which functions or input variables are useful for evolving the target function, and thus a generic GP system is usually given access to a wide variety of functions and terminals. In the setting of evolving conjunctions, this may be modeled by introducing input variables not included in the target conjunction (the $\text{AND}_{n,m}$ problem), or giving the GP systems access to additional Boolean operators (such as negation or disjunction). The aim is to evaluate whether the systems are still able to evolve the target function efficiently.

The $\text{AND}_{n,m}$ problem is a variant of the AND problem in which the target function is a conjunction of $m \leq n$ distinct variables from the terminal set L. This is similar to the conjunction evolution problem considered by Valiant [55] and has been analyzed for RLS-GP algorithms in [25]. The RLS-GP and RLS-GP* algorithms (the latter only when disallowing the HVL-Prime substitution suboperation) are able to construct an optimal solution for the $\text{AND}_{n,m}$ problem using the complete truth table in an expected $O(n \log n)$ iterations, while the canonical RLS-GP* will with high probability fail to find the optimum.

Theorem 11.5.7 ([25]). *The RLS-GP algorithm and the RLS-GP* algorithm (without the HVL-Prime substitution suboperation) using $F = \{AND\}$*

and $L := \{x_1,\dots,x_n\}$ find the optimum for the $AND_{n,m}$ problem in expected $O(n\log n)$ iterations when using the complete truth table as the training set.

The RLS-GP algorithm (with the substitution suboperation) will with high probability fail to find the optimum for the $AND_{n,m}$ problem when $m = cn$ for any constant $0 < c < 1$ when using the complete truth table as the training set.*

The analysis relies on showing that, initially, inserting both variables that are present in the target function ("correct" variables) and those that are not ("incorrect" variables) is beneficial for the fitness value of the candidate solution, while removing incorrect variables only becomes beneficial after all correct variables are present in the current solution. With local search mutation and the substitution suboperation of HVL-Prime, it is possible for RLS-GP* to accept a solution which substitutes the last copy of some incorrect variable with another copy of a still-present incorrect variable in the solution. If this occurs, RLS-GP* will not be able to reach the global optimum, because a single application of HVL-Prime could only remove a single copy of an incorrect variable present multiple times in the current solution, which would not provide a fitness improvement.

It is conjectured that a similar bound also holds for the runtime of the (1+1) GP and (1+1) GP* algorithms, which are able to introduce and remove duplicate terminals in the solution using larger mutation operations.

A more realistic function set as used in practice should also include additional Boolean operators, such as OR or NOT, with the aim of giving the GP system the expressive power necessary to represent any Boolean function. Mambrini and Oliveto have shown that if the unary NOT operation is introduced (by extending the set of literals with negated versions of each variable, avoiding the need to modify the HVL-Prime mutation operator to deal with nonbinary functions), the RLS-GP algorithms are no longer able to efficiently construct the optimal solution of the AND problem using the complete truth table as the training set [29]. This result was extended by Lissovoi and Oliveto to cover the (1+1) GP algorithms [25].

Theorem 11.5.8 ([25, 29]). *The RLS-GP, RLS-GP*, (1+1) GP and (1+1) GP* algorithms on the AND_n problem with $L = \{x_1,\dots,x_n,\overline{x}_1,\dots,\overline{x}_n\}$ and $F = \{AND\}$ do not construct an optimal solution in polynomial time, with overwhelming probability, when using the complete truth table as the training set.*

The theorem follows from the observation that a conjunction that contains both a variable x_i and its negation $\overline{x}_i$ always evaluates to false, and hence has a nearly optimal fitness value of 1 (i.e., it is wrong on just one of 2^n possible inputs). Such a pair of literals was shown to be present in the current solution with overwhelming probability once it contains $n/2$ distinct literals. For the strictly elitist GP algorithms, reaching the global optimum would then require a large simultaneous mutation with an exponential waiting time, while the nonstrictly elitist GPs would essentially need to perform a random walk in $2n$ dimensions and reach a particular point while receiving little guidance from the fitness function.

Additionally, even if the GP systems could be prevented from accepting any solution containing a contradiction (for instance, by weighting the all-true variable assignment much higher than any other input), the RLS-GP and (1+1) GP algorithms would still require exponential time to find the global optimum, as nonoptimal solutions containing all n variables (in either the positive or the negated form) share the same fitness value ($2^n - 2$, i.e., they are wrong on the all-true input and the single assignment satisfying the solution but not the target function), and the closer the GP system is to having all n positive literals, the more likely it is to produce an offspring which replaces a positive literal with a negative one.

On the other hand, if a training set of polynomial size is used as in practical applications, the GP systems can still efficiently construct a solution which generalizes well (even if it is not optimal) on the AND_n problem, even in the presence of negations.

Corollary 11.5.9 ([25]). *The (1+1) GP using $F = \{AND\}$ and $L = \{x_1, \ldots, x_n, \overline{x_1}, \ldots, \overline{x_n}\}$, is able to find a solution on the AND_n problem with a generalization error of at most n^{-c} for any constant $c > 0$ in polynomial time, when comparing program quality using a sufficiently large training set of polynomial size chosen either uniformly at random from the complete truth table in each iteration or during the first iteration.*

Doerr et al. [7] have analyzed the behavior of the RLS-GP algorithm using $F = \{AND, OR\}$ for the AND_n problem. To allow the analysis in this setting, a limit on the maximum solution size was imposed; specifically, solutions containing more than $\ell \geq n$ leaf nodes were rejected regardless of their fitness. However, there exist solutions with ℓ leaf nodes which cannot be modified by HVL-Prime without detrimentally affecting fitness, and hence RLS-GP requires an expected infinite number of iterations to find an optimal solution. To address this issue, the HVL-Prime deletion sub-operation was modified to select a node uniformly at random and remove the subtree rooted at that node (replacing the node's parent with the node's sibling). Allowing subtree deletions brings the operator closer to the sort of large-scale modifications of candidate solutions that are produced by the mutation operators of practical GP systems [47]. With the two modifications, RLS-GP is able to find the global optimum in expected polynomial time with respect to the number of variables and the limit on the tree size imposed if the complete truth table is used.

Theorem 11.5.10 ([7]). *The RLS-GP algorithm with $F = \{AND, OR\}$ and $L := \{x_1, \ldots, x_n\}$, a tree size limit $\ell \geq (1+c)n$ leaf nodes for any $c \in \Theta(1)$, HVL-Prime with subtree deletion, finds the optimum for the AND_n problem in expected $O(\ell n \log^2 n)$ iterations when using the complete truth table as the training set.*

This result was proven by showing that within $\Omega(\ell n \log^2 n)$ iterations, the current solution of the RLS-GP contains fewer than ℓ leaf nodes, and thus progress can be made by inserting a conjunction with a useful variable at the root of the offspring solution. A super-multiplicative drift theorem was then applied to bound the expected runtime. Experimental results suggest that a tree size limit is not required in this setting, and that systems with larger tree size limits find the optimum in fewer iterations than those with tree size limits close to n [7].

When using incomplete training sets to evaluate solution quality, it was shown that with probability $1-O(\log^2(n)/n)$, RLS-GP avoids inserting any disjunctions before finding a solution which satisfies its termination condition and with high probability reaches the desired generalization ability.

Theorem 11.5.11 ([7]). *For any constant $c>0$, consider an instance of the RLS-GP algorithm with $F=\{AND,OR\}$, $L=\{x_1,\ldots,x_n\}$, a tree size limit $\ell \geq n$, using a training set of $s=n^c \lg^2 n$ rows sampled uniformly at random from the complete truth table in each iteration to evaluate solution quality, and terminating when the sampled error of the solution is at most $c' \lg n$, where c' is an appropriately large constant. On the AND_n problem, the algorithm will, with probability at least $1-O(\log^2(n)/n)$, terminate within $O(\log n)$ iterations, and return a solution with a generalization error of at most n^{-c}.*

Notably, the theorem does not require an *upper* limit on the size of the tree; $\ell \geq n$ simply ensures that the target function is representable within the tree size limit. The proof shows that a solution with the desired generalization error is found once $O(\log n)$ insertions occur, and thus the RLS-GP with high probability does not exceed any reasonable tree size limit in this setting. Experimental results additionally show that solutions with fewer undesired disjunctions could be constructed by terminating the GP system once it achieves a logarithmic error on the training set rather than waiting for an error of 0 to be observed [7].

11.5.1.4 Optimal Training Sets

While the target conjunctions are unlikely to be evolved exactly (with a generalization error of 0) when using a polynomial training set chosen uniformly at random, there do exist small training sets of $O(n)$ rows which allow the RLS-GP and (1+1) GP algorithms to find exact solutions efficiently. In general, identifying such training sets may be nontrivial.

Theorem 11.5.12 ([25]). *Let M be an n-row training set, where row i sets x_i to false and all x_j (where $j \neq i$) to true, and let M' be a $2n+1$-row training set containing all the rows of M and $n+1$ copies of the row setting all inputs to true. The RLS-GP and (1+1) GP algorithms with $F=\{AND\}$ using the training sets M and M', respectively are able to find the exact solution* *of* AND_n *and* $AND_{n,m}$ *with* $F = \{AND\}$, $L = \{x_1, \ldots, x_n\}$ *(or* AND_n *with* $F = \{AND\}$ *and* $L = \{x_1, \ldots, x_n, \overline{x}_1, \ldots, \overline{x}_n\}$*) in expected* $O(n \log n)$ *fitness evaluations (or* $O(n^2 \log n)$ *training set row evaluations).*

For $L = \{x_1, \ldots, x_n, \overline{x}_1, \ldots, \overline{x}_n\}$, a variant of the $(1+1)$ GP which maintains and randomly selects from a population of μ individuals subject to a diversity mechanism prohibiting multiple solutions with identical outputs on the training set was proven to find an optimal solution in $O(\mu n \log n)$ iterations on an $n+1$-row training set (consisting of all the inputs in M and an input where all the n variables are set to true) [25]. Effectively, this uses the explicit diversity mechanism to avoid including multiple copies of the all-true row in the training set as in Theorem 11.5.12.

11.5.2 Evolving Parity

The XOR problem asks the GP system to evolve an exclusive disjunction of all n input variables. Unlike conjunctions, exclusive disjunctions are known to not be evolvable in the PAC learning framework [55].

Problem 11.5.13 (XOR) *Let* $L \subseteq \{x_1, \ldots, x_n\}$ *be the set of available terminals, and* F *be the set of available functions.*

The fitness of a tree X *using a training set* T *selected from the rows of the complete truth table* C *is the number of training set rows on which the value produced by evaluating the Boolean expression represented by the tree differs from the output of the exclusive disjunction of all* n *inputs.*

When $F = XOR$ and the complete truth table is used as the training set, the fitness of any nonoptimal solution is 2^{n-1}, while the fitness of the optimal solution is 0. Thus, using the complete truth table as the training set on XOR is similar to the Needle benchmark problem; Langdon and Poli noted that "the fitness landscape is like a needle-in-a-haystack, so any adaptive search approach will have difficulties" [23].

Predictably, the RLS-GP and $(1+1)$ GP algorithms are not able to optimize XOR efficiently. Strictly elitist variants will only move from their initial solution if the optimum is constructed as a mutation of that solution, which occurs in expected infinite time for RLS-GP* (as the optimum is not reachable by a single HVL-Prime mutation from many possible points), and in expected exponential time for the $(1+1)$ GP* (which essentially needs to construct the complete function in one mutation; if initialized with an empty tree, this mutation needs to perform at least n HVL-Prime insertion suboperations). When the complete truth table set is used as the training set, the expected optimization time for RLS-GP is exponential in the problem size, because the algorithm accepts any and all mutations, while reaching the optimal solution requires all n variables to appear an odd number of times in the solution [29].

Theorem 11.5.14 (Theorem 4, [29]). *RLS-GP using* $F = \{XOR\}$, $L = \{x_1, \ldots, x_n\}$, *and using the complete truth table as the training set to evolve* XOR_n *requires more than* $2^{\Omega(n/\log n)}$ *iterations with probability* $p > 1 - 2^{-\Omega(n/\log n)}$ *to reach the optimum.*

The theorem is proven by an application of the simplified negative drift theorem, showing that when the number of variables that appear in the current solution an odd number of times is large, there is a strong negative drift towards reducing this number, and the optimum requires all n distinct variables to appear an odd number of times in the solution. The negative drift stems primarily from the HVL-Prime insertion operator: if a large number of variables are represented an odd number of times, it is more likely to insert one of these variables when choosing a terminal uniformly at random.

Also when sampling solution fitness using a polynomial number of rows of the complete truth table, the outcome is underwhelming: if only a logarithmically small number of training set rows are sampled in each iteration, the algorithm will terminate in expected polynomial time with a nonoptimal solution that fits the sampled training set, while using training sets of super-logarithmic size will lead to superpolynomial optimization time. Thus, in any polynomial amount of time, the expected generalization ability of the GP systems considered on XOR is 1/2, i.e., they require in expectation a constant number of samples taken uniformly at random from C before a divergence from the target function is discovered.

There is also a straightforward extension of Theorem 11.5.14 to dynamic training sets of polynomial size, as such sampling provides no consistent indication of fitness.

Corollary 11.5.15. *The RLS-GP and* $(1+1)$ *GP algorithms sampling* $s \in \omega(\log n)$ *rows of the complete truth table in each iteration on* XOR_n *with* $F = \{XOR\}$ *and* $L = \{x_1, \ldots, x_n\}$ *with high probability do not reach the optimum in polynomial time.*

Proof. The RLS-GP and $(1+1)$ GP algorithms will accept *any* nonoptimal offspring of a nonoptimal parent with probability at least 1/2, as both the offspring and the parent are wrong on 2^{n-1} inputs, and there are exactly as many rows on which the offspring is correct while the parent is wrong as the converse, and the offspring is accepted in cases of tied fitness.

With $s \in \omega(\log n)$ rows sampled uniformly at random in each iteration, the probability that a nonoptimal solution is correct on all sampled rows is $2^{-\omega(\log n)} = n^{-\omega(1)}$, and, by a straightforward union bound, the GP algorithms do not terminate within polynomial time unless the optimal solution is found.

With the exception of any iterations in which the offspring individual is rejected, the algorithms behave identically to the RLS-GP and $(1+1)$ GP algorithms using the complete truth table to evaluate solution fitness (i.e., accepting offspring regardless of the effects of mutation), and thus cannot achieve better performance than these algorithms in terms of the number of iterations performed.

Theorem 11.5.14 provides a runtime bound for RLS-GP only. A similar result for the $(1+1)$ GP can be obtained by observing that the $(1+1)$ GP performs in expectation two HVL-Prime suboperations in each iteration, and hence, even if the algorithm terminated immediately upon constructing the optimal solution (even if this occurred in the middle of a mutation), it would in expectation be only a constant factor faster than RLS-GP in terms of the number of iterations required to find the optimum. □

11.5.3 Outlook

In this section, the available computational complexity results regarding the evolution of proper functions with input/output behavior have been overviewed. Simple GP systems equipped with the AND (or AND and OR) functions and positive literals (or possibly both positive and negative literals) can evolve conjunctions of arbitrary size with high probability if appropriate limits on maximum tree size are put in place. Important open problems are providing performance statements of the algorithms without tree size limits, and analyses of GP systems equipped with comprehensive function sets F, i.e., those that allow the expression of any Boolean function.

11.6 Other GP Algorithms

The previous sections have covered the available theoretical results for standard tree-based GP systems, which constitute the majority of theoretical complexity analysis results for GP. Several other GP paradigms have been proposed in the literature which use different representations for candidate solutions, e.g., Cartesian GP [31], Linear GP [1], PushGP [52], and Geometric Semantic GP (GSGP) [33]. Amongst these, the only class for which computational complexity analyses are available is GSGP. In this section, we present the available results concerning this different approach to GP system design which aims to evolve programs semantically rather than syntactically.

11.6.1 Geometric Semantic Genetic Programming

Standard tree-based GP evolves programs by applying mutation and crossover to their syntax. Programs that are considerably different syntactically may produce identical output, while introducing minimal syntactic mutations may completely change the output of a program. Moraglio et al. [33] introduced Geometric Semantic GP (GSGP) with the aim of focusing GP search on program behavior. In particular, GSGP mutation and crossover operators modify programs in a way that allows the GP system to search through the semantic neighborhood (which consists of programs with similar behavior) rather than their syntactic neighborhood (which consists of programs with similar syntax).

GSGP generally uses a natural program representation for the domain at hand (e.g., it represents programs using Boolean expressions when a Boolean expression is to be evolved), and uses specialized semantic mutation and crossover operators to produce offspring programs with *behavior* similar to that of their parents. These operators generally reproduce the parent programs in their entirety, adding to them to modify their behavior in a limited fashion. For example, the GSGP mutation operator could produce an offspring which contains an exact copy of its parent and a random element which overrides some portions of the parent's behavior, while the GSGP crossover operator could construct an offspring containing exact copies of both parents and a random element which switches between the two behaviors depending on the inputs. As both operators increase the size of the programs by adding additional syntax to the parent programs to encode the chosen random components (and the crossover includes exact copies of *both* parents), the programs produced by these operators need to be simplified in order for the algorithms to remain tractable. For some domains, such as Boolean functions, quick function-preserving simplifiers exist, while computer algebra systems and static analysis can be used to simplify more complex expressions and programs [33].

Semantic geometric crossover and mutation operators have been designed for many problem domains, including regression problems [34], learning classification trees [28], and Boolean functions [35]. Initial experimental results suggest that GSGP consistently finds solutions that fit the training sets used for a wide array of simple Boolean benchmark functions, regression problems for polynomials of degree up to 10, and various classification problems, outperforming standard tree-based GP with the same evaluation budget [33]. Theoretical guarantees have been derived regarding the number of generations it takes GSGP to construct a solution fitting the training set, or achieving an ε-small training set error in the case of regression problems [28, 34, 35]. In this section, we explore the available theoretical results focusing on applying geometric semantic search to evolving Boolean functions.

In the case of Boolean functions, the program semantics can be represented by a 2^n-row output vector, corresponding to the program output on all rows of the complete n-variable truth table. In this setting, the semantic crossover operator SGXB, acts on two parents T_1 and T_2, and produces an offspring solution $(T_1 \wedge T_\mathrm{R}) \vee (T_2 \wedge \overline{T_\mathrm{R}})$, where T_R is a randomly generated Boolean function. This offspring outputs the solution produced by T_1 if T_R evaluates to true, and the solution produced by T_2 if T_R evaluates to false, effectively performing crossover on the 2^n-row output vectors of the two parent solutions. The semantic mutation operator SGMB, acting on a single parent T_1, produces the offspring $T_1 \vee M$ with probability 0.5, and $T \wedge \overline{M}$ with probability 0.5, where M is a random minterm (a conjunction where each variable appears in either positive or negated form) of all input variables. This effectively copies the output vector of T_1, setting the rows on which M evaluates to true to either true or false.

These operators allow GSGP to always observe a cone fitness landscape on any Boolean function, i.e., the mutation operator is always able to improve the behavior of the parent program. This allows mutation-only GSGP to hill-climb its way up to the optimal program for any function in this domain. However, since the output vector contains 2^n rows, hill-climbing by applying SGMB, which only affects one row per iteration, would take $O(2^n \log(2^n)) = O(n2^n)$ iterations (by the coupon collector argument, or similarly to RLS on a 2^n-bit OneMax function).

For GSGP on any Boolean function, a polynomially sized training set can be viewed as a OneMax problem on a 2^n-bit string where only a polynomial number of bits are nonneutral (i.e., contribute to the solution's fitness). In that setting, the runtime can be improved by allowing mutations to flip more than one bit of the output vector per iteration (e.g., such that in expectation one nonneutral bit is affected per iteration). This setting was explored in [35], with various approaches to the design of mutation operators, establishing a hierarchy of operator expressiveness (based on how much of the search space they enable the GP system to explore), and considering the probability of fitting a training set of polynomial size. The following mutation operators, differing in how the random minterm M used to modify program behavior is constructed, were analyzed:

- Fixed Block Mutation (FBM), which picks the $v \leq n$ variables to use as the base for M *once* during the run,
- Fixed Alternative Block Mutation (FABM), which partitions the variables into v sets, and forms M by picking a variable from each set uniformly at random in each iteration,
- Varying Block Mutation (VBM), which in each iteration chooses $v \leq n$ variables uniformly at random to form the base for M.

For all three operators, v is a fixed parameter. The results show that while VBM is more expressive than FABM, which in turn is more expressive than FBM, there nevertheless exist training sets which GSGP using VBM cannot fit in any amount of time. Conversely, the less expressive FBM operator can with high probability fit a training set of polynomial size sampled uniformly at random from the complete truth table of any Boolean function [35].

Theorem 11.6.1 ([35]). *Let a training set T consist of n^c rows, with c a positive constant, the rows being sampled uniformly at random from the complete truth table of any Boolean function. Then GSGP using the FBM* *operator with $v = (2c+\varepsilon)\log_2(n)$ (for any $\varepsilon > 0$), is able to fit T with probability at least $1-\frac{1}{2}n^{-\varepsilon}$. Conditioning on this, a function that fits the training set is found in an expected $O(n^{2c}\log n)$ iterations.*

This result is proven by observing that FBM's initial choice of v variables (to use as the basis for the minterms) partitions the 2^n row output vector of P into 2^v blocks of equal size, each corresponding to a particular minterm of the v variables. Choosing $v > 2c\log_2 n$ partitions the output vector into more than $2^{2c\log_2 n} = n^{2c}$ blocks, ensuring that with high probability all n^c training set rows (chosen uniformly at random from the complete truth table) are in different blocks, and thus the training set can be satisfied by collecting the exact minterms corresponding to the blocks which contain the training set rows. When this condition holds, the expected runtime is obtained by a coupon collector argument.

Of course, if FBM chooses the v variables poorly with respect to the training set T (meaning that at least two training set rows demanding different output are contained in the same block), GSGP will not be able to fit it. More expressive operators such as FABM or VBM can minimize this probability at the cost of a mild runtime penalty by allowing the mutation operator to be more flexible when choosing which variables to use as the basis for the minterm (e.g., increasing the runtime by a factor of n/v, but improving the success probability from p to $1-(1-p)^{n/v}$, where v is the number of classes in the partition created by FABM).

There are also modifications of the GSGP mutation operators that are able to cover the entire search space of programs, eliminating the possibility of failure. There exist classes of Boolean functions for which such operators are effective, allowing GSGP to fit any training set in expected polynomial time (with no failure probability, unlike Theorem 11.6.1), as shown in the following theorem.

Theorem 11.6.2 ([35]). *Let ϕ be a formula in disjunctive normal form with $\alpha = \text{poly}(n)$ conjunctions, every conjunction containing at most $\beta = O(1)$ variables. Then GSGP with Multiple Size Block Mutation (MSBM) can fit any training set for ϕ in expected $O(\alpha n^{\beta+1}2^{\beta})$ iterations, i.e., polynomial time.*

The MSBM mutation operator is a modification of the VBM variant of the SGMB operator. It samples an integer v between 0 and n, selects v variables from the set of n input variables, and then generates uniformly at random an incomplete minterm M of these variables. This modified mutation operator essentially allows each clause of the target function to be "fixed" in the current solution in an expected polynomial number of iterations.

11.6.2 Outlook

GSGP systems have been proven to efficiently construct solutions which fit training sets of polynomial size for several function domains. In this section, we have covered the available results for the evolution of Boolean functions, although similar results are available for the other domains, such as learning classification trees and regression problems [28, 34].

At present though, there are no theoretical analyses of how the functions produced by GSGP generalize to unseen inputs. Experimental results concerning the generalization performance of GSGP systems yielded mixed conclusions [14, 45, 46].

11.7 Conclusion

We have presented an overview of the available results on the computational complexity analysis of GP algorithms. The results follow the blueprint suggested by Poli et al., starting with the analysis of simple GP systems based on mutation and stochastic-hill climbing on simple problems [48]. The complexity of the problems has gradually increased, from analyses focusing on the main characteristic difficulties of GP (i.e., variable solution length, and solution quality evaluations) to more recent results considering the evolution of functions with true input/output behavior and using realistically constrained fitness functions. The approach of gradually expanding the complexity of the systems analyzed was also endorsed by Goldberg and O'Reilly, who stated that "the methodology of using deliberately designed problems, isolating specific properties, and pursuing, in detail, their relationships in simple GP is more than sound; it is the only practical means of systematically extending GP understanding and design" [13].

The GP systems considered in theoretical analyses have remained relatively simple: the use of HVL-Prime mutation and limited, if any, populations with no crossover is a common setting. In many cases, an analysis that provides positive runtime results is only made tractable because "the fitness structure of the model problems is simple, and the algorithms use only a simple hierarchical variable length mutation operator" [9]. In particular, variable-length representations often complicate the analysis of GP systems, and require "rather deep insights into the optimization process and the growth of the GP-trees" [6].

The chapter has highlighted three different streams that have been followed for building the theoretical foundations of genetic programming. The first one is the design and analysis of benchmark functions with variable length representation for the analysis of tree structure growth. Three classes of such problems have been considered in the literature: ORDER, MAJORITY, and SORTING. While producing rigorous proofs is not easy, surprisingly simple hillclimbing GP systems optimize these problems efficiently without bloat seriously hindering their performance. Only recently has the 2/3-SUPERMAJORITY benchmark function been introduced as a benchmark problem where bloat provably is a major concern. Nevertheless, simple bloat control mechanisms address the issue effectively. As a result, there is a need for better benchmark functions to shed more light on how bloat affects evolution via GP.

The second line of research has addressed the evolution of toy programs of fixed size. The aim is to analyze GP behavior when tree structure is constrained (e.g., with tree size limits in place, as in the MAX problem) and solution quality estimation using training sets of limited size (i.e., how large training sets have to be for efficient evolution, e.g., the IDENTIFICATION problem). Only very preliminary results are available addressing these questions: tight bounds are unavailable for the MAX problem even for simple hillclimbing (1+1) GP algorithms and IDENTIFICATION problem results are available only for very simple linear functions.

The third line of research concerns the evolution of proper functions with inputs and outputs. Up to today, only conjunctions and parity Boolean functions have been considered for (1+1) GP systems using limited function sets (i.e., that do not have sufficient expressive power to express all Boolean functions). Nevertheless, such GP systems can provably evolve conjunctions of arbitrary sizes with proper tree size limits in place.

For GP systems utilizing geometric semantic mutation and crossover operators, analyses of the time required to produce a solution fitting the training set are available for wider classes of functions, and frequently do not require insight into the structure of the function considered. However, a rigorous understanding of how well the GSGP solutions generalize – how well they perform on inputs not included in the training set – remains a challenge.

While the results presented represent the first steps in the rigorous analysis of the behavior of GP systems, bridging the gap to the GP systems used in practice requires analyzing more complex GP algorithms on more realistic problems. Thus, extending the results presented to broader classes of problems (for instance, those allowing more flexibility in program behavior), to other problem classes on which GP experimentally performs well (such as symbolic regression), and to more realistic GP algorithms (introducing populations and crossover) constitute the main directions for further research.

Acknowledgements Financial support by the Engineering and Physical Sciences Research Council (EPSRC Grant No. EP/M004252/1) is gratefully acknowledged.

References

[1] Brameier, M., Banzhaf, W.: Linear Genetic Programming. Genetic and Evolutionary Computation. Springer (2007)

[2] Corus, D., Dang, D.C., Eremeev, A.V., Lehre, P.K.: Level-based analysis of genetic algorithms and other search processes. IEEE Transactions on Evolutionary Computation **22**(5), 707–719 (2017)

[3] Corus, D., Oliveto, P.S.: Standard steady state genetic algorithms can hillclimb faster than mutation-only evolutionary algorithms. IEEE Transactions on Evolutionary Computation **22**(5), 720–732 (2018)

[4] Corus, D., Oliveto, P.S., Yazdani, D.: On inversely proportional hypermutations with mutation potential. In: Proceedings of the Genetic and Evolutionary Computation Conference (GECCO 2019), pp. 215–223 (2019).

[5] Curry, R., Lichodzijewski, P., Heywood, M.I.: Scaling genetic programming to large datasets using hierarchical dynamic subset selection. IEEE Trans. Systems, Man, and Cybernetics, Part B **37**(4), 1065–1073 (2007)

[6] Doerr, B., Kötzing, T., Lagodzinski, J.A.G., Lengler, J.: Bounding bloat in genetic programming. In: Proceedings of the Genetic and Evolutionary Computation Conference (GECCO 2017), pp. 921–928 (2017)

[7] Doerr, B., Lissovoi, A., Oliveto, P.S.: Evolving boolean functions with conjunctions and disjunctions via genetic programming. In: Proceedings of the Genetic and Evolutionary Computation Conference (GECCO 2019), pp. 1003–1011 (2019).

[8] Droste, S., Jansen, T., Wegener, I.: On the analysis of the (1+1) evolutionary algorithm. Theoretical Computer Science **276**(1-2), 51–81 (2002)

[9] Durrett, G., Neumann, F., O'Reilly, U.: Computational complexity analysis of simple genetic programming on two problems modeling isolated program semantics. In: Proceedings of the 11th International Workshop on Foundations of Genetic Algorithms (FOGA 2011), pp. 69–80 (2011)

[10] Feldman, V.: A complete characterization of statistical query learning with applications to evolvability. Journal of Computer and System Sciences **78**(5), 1444–1459 (2012)

[11] Gathercole, C., Ross, P.: Dynamic training subset selection for supervised learning in genetic programming. In: Proceedings of the 3rd International Conference on Parallel Problem Solving from Nature (PPSN 1994), pp. 312–321 (1994)

[12] Gathercole, C., Ross, P.: An adverse interaction between crossover and restricted tree depth in genetic programming. In: Proceedings of the 1st Annual Conference on Genetic Programming, pp. 291–296. MIT Press, Cambridge, MA, USA (1996)

[13] Goldberg, D.E., O'Reilly, U.: Where does the good stuff go, and why? how contextual semantics influences program structure in simple genetic programming. In: Proceedings of Genetic Programming, First European Workshop (EuroGP 1998), pp. 16–36 (1998)

[14] Gonçalves, I., Silva, S., Fonseca, C.M.: On the generalization ability of geometric semantic genetic programming. In: Proceedings of Genetic Programming - 18th European Conference (EuroGP 2015), pp. 41–52 (2015)

[15] Jansen, T.: Analyzing Evolutionary Algorithms - The Computer Science Perspective. Natural Computing Series. Springer (2013)

[16] Jansen, T., Zarges, C.: Performance analysis of randomised search heuristics operating with a fixed budget. Theoretical Computer Science **545**, 39–58 (2014)

[17] Kötzing, T., Lagodzinski, J.A.G., Lengler, J., Melnichenko, A.: Destructiveness of lexicographic parsimony pressure and alleviation by a concatenation crossover in genetic programming. In: Proceedings of the 15th International Conference on Parallel Problem Solving from Nature (PPSN 2018), Part II, pp. 42–54 (2018)

[18] Kötzing, T., Neumann, F., Spöhel, R.: PAC learning and genetic programming. In: Proceedings of the Genetic and Evolutionary Computation Conference (GECCO 2011), pp. 2091–2096 (2011)

[19] Kötzing, T., Sutton, A.M., Neumann, F., O'Reilly, U.: The MAX problem revisited: The importance of mutation in genetic programming. Theoretical Computer Science **545**, 94–107 (2014)

[20] Koza, J.R.: Genetic programming - on the programming of computers by means of natural selection. Complex adaptive systems. MIT Press (1992)

[21] Koza, J.R.: Human-competitive results produced by genetic programming. Genetic Programming and Evolvable Machines **11**(3-4), 251–284 (2010)

[22] Koza, J.R., Al-Sakran, S.H., Jones, L.W.: Automated *ab initio* synthesis of complete designs of four patented optical lens systems by means of genetic programming. Artificial Intelligence for Engineering Design, Analysis and Manufacturing **22**(3), 249–273 (2008)

[23] Langdon, W.B., Poli, R.: Foundations of genetic programming. Springer (2002)

[24] Lipson, H.: Evolutionary synthesis of kinematic mechanisms. Artificial Intelligence for Engineering Design, Analysis and Manufacturing **22**(3), 195–205 (2008)

[25] Lissovoi, A., Oliveto, P.S.: On the time and space complexity of genetic programming for evolving boolean conjunctions. In: Proceedings of the Thirty-Second AAAI Conference on Artificial Intelligence, pp. 1363–1370. AAAI Press (2018)

[26] Lohn, J.D., Hornby, G., Linden, D.S.: Human-competitive evolved antennas. Artificial Intelligence for Engineering Design, Analysis and Manufacturing **22**(3), 235–247 (2008)

[27] Luke, S., Panait, L.: Lexicographic parsimony pressure. In: Proceedings of the Genetic and Evolutionary Computation Conference (GECCO 2002), pp. 829–836 (2002)

[28] Mambrini, A., Manzoni, L., Moraglio, A.: Theory-laden design of mutation-based geometric semantic genetic programming for learning classification trees. In: Proceedings of the Congress on Evolutionary Computation (CEC 2013), pp. 416–423 (2013)

[29] Mambrini, A., Oliveto, P.S.: On the analysis of simple genetic programming for evolving boolean functions. In: Proceedings of Genetic Programming - 19th European Conference (EuroGP 2016), pp. 99–114 (2016)
[30] McDermott, J., O'Reilly, U.M.: Genetic programming. In: Springer Handbook of Computational Intelligence, pp. 845–869. Springer (2015)
[31] Miller, J.F. (ed.): Cartesian Genetic Programming. Natural Computing Series. Springer (2011)
[32] Mitzenmacher, M., Upfal, E.: Probability and computing - randomized algorithms and probabilistic analysis. Cambridge University Press (2005)
[33] Moraglio, A., Krawiec, K., Johnson, C.G.: Geometric semantic genetic programming. In: Proceedings of the 12th International Conference on Parallel Problem Solving from Nature (PPSN 2012), pp. 21–31 (2012)
[34] Moraglio, A., Mambrini, A.: Runtime analysis of mutation-based geometric semantic genetic programming for basis functions regression. In: Proceedings of the Genetic and Evolutionary Computation Conference (GECCO 2013), pp. 989–996 (2013)
[35] Moraglio, A., Mambrini, A., Manzoni, L.: Runtime analysis of mutation-based geometric semantic genetic programming on boolean functions. In: Proceedings of the 12th International Workshop on Foundations of Genetic Algorithms (FOGA 2013), pp. 119–132 (2013)
[36] Myers, A.N., Wilf, H.S.: Some new aspects of the coupon collector's problem. SIAM Journal on Discrete Mathematics **17**(1), 1–17 (2003)
[37] Neumann, F.: Computational complexity analysis of multi-objective genetic programming. In: Proceedings of the Genetic and Evolutionary Computation Conference (GECCO 2012), pp. 799–806 (2012)
[38] Nguyen, A., Urli, T., Wagner, M.: Single- and multi-objective genetic programming: new bounds for weighted order and majority. In: Proceedings of the 12th International Workshop on Foundations of Genetic Algorithms (FOGA 2013), pp. 161–172 (2013)
[39] Oliveto, P.S., Witt, C.: On the runtime analysis of the simple genetic algorithm. Theoretical Computer Science **545**, 2–19 (2014)
[40] Oliveto, P.S., Witt, C.: Improved time complexity analysis of the simple genetic algorithm. Theoretical Computer Science **605**, 21–41 (2015)
[41] Oliveto, P.S., Yao, X.: Runtime analysis of evolutionary algorithms for discrete optimization. In: A. Auger, B. Doerr (eds.) Theory of Randomized Search Heuristics: Foundations and Recent Developments, chap. 2, pp. 21–52. World Scientific (2011)
[42] O'Reilly, U., Oppacher, F.: Program search with a hierarchical variable length representation: Genetic programming, simulated annealing and hill climbing. In: Proceedings of the 3rd International Conference on Parallel Problem Solving from Nature (PPSN 1994), pp. 397–406 (1994)
[43] O'Reilly, U.M.: An analysis of genetic programming. Ph.D. thesis, Carleton University, Ottawa-Carleton Institute for Computer Science, Ottawa, Ontario, Canada (1995)

[44] O'Reilly, U.M., Oppacher, F.: A comparative analysis of genetic programming. In: P.J. Angeline, K.E. Kinnear, Jr. (eds.) Advances in Genetic Programming 2, chap. 2, pp. 23–44. MIT Press, Cambridge, MA, USA (1996)

[45] Orzechowski, P., Cava, W.L., Moore, J.H.: Where are we now?: a large benchmark study of recent symbolic regression methods. In: Proceedings of the Genetic and Evolutionary Computation Conference (GECCO 2018), pp. 1183–1190 (2018)

[46] Pawlak, T.P., Krawiec, K.: Competent geometric semantic genetic programming for symbolic regression and boolean function synthesis. Evolutionary Computation **26**(2), 177–212 (2018)

[47] Poli, R., Langdon, W.B., McPhee, N.F.: A Field Guide to Genetic Programming. http://lulu.com (2008)

[48] Poli, R., Vanneschi, L., Langdon, W.B., McPhee, N.F.: Theoretical results in genetic programming: the next ten years? Genetic Programming and Evolvable Machines **11**(3-4), 285–320 (2010)

[49] Rowe, J.E., Sudholt, D.: The choice of the offspring population size in the $(1, \lambda)$ EA. In: Proceedings of the Genetic and Evolutionary Computation Conference (GECCO 2012), pp. 1349–1356 (2012)

[50] Scharnow, J., Tinnefeld, K., Wegener, I.: The analysis of evolutionary algorithms on sorting and shortest paths problems. Journal of Mathematical Modelling and Algorithms **3**(4), 349–366 (2004)

[51] Spector, L.: Automatic Quantum Computer Programming: A Genetic Programming Approach, *Genetic Programming*, vol. 7. Kluwer Academic Publishers, Boston/Dordrecht/New York/London (2004)

[52] Spector, L., Robinson, A.J.: Genetic programming and autoconstructive evolution with the push programming language. Genetic Programming and Evolvable Machines **3**(1), 7–40 (2002)

[53] Urli, T., Wagner, M., Neumann, F.: Experimental supplements to the computational complexity analysis of genetic programming for problems modelling isolated program semantics. In: Proceedings of the 12th International Conference on Parallel Problem Solving from Nature (PPSN 2012), pp. 102–112 (2012)

[54] Valiant, L.G.: A theory of the learnable. Communications of the ACM **27**(11), 1134–1142 (1984)

[55] Valiant, L.G.: Evolvability. Journal of the ACM **56**(1), 3:1–3:21 (2009)

[56] Wagner, M., Neumann, F.: Parsimony pressure versus multi-objective optimization for variable length representations. In: Proceedings of the 12th International Conference on Parallel Problem Solving from Nature (PPSN 2012), pp. 133–142 (2012)

[57] Wagner, M., Neumann, F., Urli, T.: On the performance of different genetic programming approaches for the SORTING problem. Evolutionary Computation **23**(4), 583–609 (2015)